T0192027

Lecture Notes in Computer Science 13137

More information about this subseries at https://link.springer.com/bookseries/7407

Sergei Artemov · Anil Nerode (Eds.)

Logical Foundations of Computer Science

International Symposium, LFCS 2022
Deerfield Beach, FL, USA, January 10–13, 2022
Proceedings

Editors
Sergei Artemov
CUNY Graduate Center
New York, NY, USA

Anil Nerode
Cornell University
Ithaca, NY, USA

ISSN 0302-9743 ISSN 1611-3349 (electronic)
Lecture Notes in Computer Science
ISBN 978-3-030-93099-8 ISBN 978-3-030-93100-1 (eBook)
https://doi.org/10.1007/978-3-030-93100-1

LNCS Sublibrary: SL1 – Theoretical Computer Science and General Issues

This Springer imprint is published by the registered company Springer Nature Switzerland AG
The registered company address is: Gewerbestrasse 11, 6330 Cham, Switzerland

Preface

The Symposium on Logical Foundations of Computer Science (LFCS) series provides a forum for the fast-growing body of work in the logical foundations of computer science, e.g., those areas of fundamental theoretical logic related to computer science. The LFCS series began with "Logic at Botik" held in Pereslavl-Zalessky, 1989, which was co-organized by Albert R. Meyer (MIT) and Michael Taitslin (Tver). After that, organization passed to Anil Nerode.

Currently, LFCS is governed by a Steering Committee consisting of Anil Nerode (General Chair), Samuel Buss, Stephen Cook, Dirk van Dalen, Yuri Matiyasevich, Andre Scedrov, and Dana Scott.

LFCS 2022 took place at the Wyndham Deerfield Beach Resort, Deerfield Beach, Florida, USA, during January 10–13, 2022. This volume contains the extended abstracts of talks selected by the Program Committee for presentation at LFCS 2022.

The scope of the symposium is broad and includes constructive mathematics and type theory; homotopy type theory; logic, automata, and automatic structures; computability and randomness; logical foundations of programming; logical aspects of computational complexity; parameterized complexity; logic programming and constraints; automated deduction and interactive theorem proving; logical methods in protocol and program verification; logical methods in program specification and extraction; domain theory logics; logical foundations of database theory; equational logic and term rewriting; lambda and combinatory calculi; categorical logic and topological semantics; linear logic; epistemic and temporal logics; intelligent and multiple-agent system logics; logics of proof and justification; non-monotonic reasoning; logic in game theory and social software; logic of hybrid systems; distributed system logics; mathematical fuzzy logic; system design logics; and other logics in computer science.

We thank the authors and reviewers for their contributions. We acknowledge the support of the U.S. National Science Foundation, The Association for Symbolic Logic, Cornell University, the Graduate Center of the City University of New York, and Florida Atlantic University.

November 2021

Anil Nerode
Sergei Artemov

Organization

Steering Committee

Samuel Buss	University of California, San Diego, USA
Stephen Cook	University of Toronto, Canada
Yuri Matiyasevich	Steklov Mathematical Institute, St. Petersburg, Russia
Anil Nerode (General Chair)	Cornell University, USA
Andre Scedrov	University of Pennsylvania, USA
Dana Scott	Carnegie Mellon University, USA
Dirk van Dalen	Utrecht University, The Netherlands

Program Committee

Antonis Achilleos	Reykjavik University, Iceland
Evangelia Antonakos	City University of New York, USA
Sergei Artemov (Chair)	City University of New York Graduate Center, USA
Steve Awodey	Carnegie Mellon University, USA
Matthias Baaz	Technical University of Vienna, Austria
Lev Beklemishev	Steklov Mathematical Institute, Moscow, Russia
Andreas Blass	University of Michigan, USA
Samuel Buss	University of California, San Diego, USA
Thierry Coquand	University of Gothenburg, Sweden
Melvin Fitting	City University of New York, USA
Sergey Goncharov	Sobolev Institute of Mathematics, Russia
Rosalie Iemhoff	Utrecht University, The Netherlands
Hajime Ishihara	Japan Advanced Institute of Science and Technology, Japan
Bakhadyr Khoussainov	University of Auckland, New Zealand
Roman Kuznets	Technical University of Vienna, Austria
Stepan Kuznetsov	Steklov Mathematical Institute, Moscow, Russia
Robert Lubarsky	Florida Atlantic University, USA
Lawrence Moss	Indiana University Bloomington, USA
Pavel Naumov	University of Southampton, UK
Anil Nerode	Cornell University, USA
Elena Nogina	City University of New York, USA
Hiroakira Ono	Japan Advanced Institute of Science and Technology, Japan
Valeria de Paiva	Topos Institute, USA
Alessandra Palmigiano	Vrije Universiteit Amsterdam, The Netherlands
Ramaswamy Ramanujam	Institute of Mathematical Sciences, India
Ruy de Queiroz	Federal University of Pernambuco, Brazil

Michael Rathjen	University of Leeds, UK
Sebastiaan Terwijn	Radboud University Nijmegen, The Netherlands
Ren-June Wang	National Chung Cheng University, Taiwan
Noson Yanofsky	City University of New York, USA
Junhua Yu	Tsinghua University, China

Additional Reviewers

Ellie Anastasiadi
Andrea De Domenico
Malvin Gattinger
Noah Kaufmann
Stanislav Kikot
Johannes Marti
Anantha Padmanabha
Mattia Panettiere
Mati Pentus
Michele Pra Baldi
Katsuhiko Sano
Sunil Simon
Stanislav Speranski
Thomas Studer
S. P. Suresh
Apostolos Tzimoulis

Contents

A Non-hyperarithmetical Gödel Logic

Juan Pablo Aguilera[1,2], Jan Bydzovsky[1(✉)], and David Fernández-Duque[2]

[1] Vienna University of Technology, Vienna, Austria
[2] Ghent University, Ghent, Belgium

Abstract. Let $\mathsf{G}_\downarrow$ be the Gödel logic whose set of truth values is $V_\downarrow = \{0\} \cup \{1/n : n \in \mathbb{N} \setminus \{0\}\}$. Baaz-Leitsch-Zach have shown that $\mathsf{G}_\downarrow$ is not recursively axiomatizable and Hájek showed that it is not arithmetical. We find the optimal strengthening of their theorems and prove that the set of validities of $\mathsf{G}_\downarrow$ is Π^1_1 complete and the set of satisfiable formulas in $\mathsf{G}_\downarrow$ is Σ^1_1 complete.

Keywords: Gödel logic · Fuzzy logic · Hyperarithmetical set

1 Introduction

A family of finite-valued logics was introduced by Gödel in [6] to show there are propositional logics weaker than classical but stronger than intuitionistic propositional logic. A natural extension of those logics to many-valued logic followed in the paper of Dummett [5] who also showed that they can be axiomatised by adding the axiom $(p \to q) \vee (q \to p)$ into intuitionistic logic. Today we call those logics Gödel logics. In particular, Gödel logics are intermediate logics where propositions take truth values in $[0, 1]$. Different Gödel logics arise by choosing a subset $V \subseteq [0, 1]$ as truth values. In the case of propositional Gödel logic, any infinite subset of $[0, 1]$ will yield the same set of valid formulas, but this is not the case for first order Gödel logic. In this case we require that V be a closed set, as suprema and infima are used to evaluate the quantifiers.

In particular, we are interested in $\mathsf{G}_\downarrow$, the Gödel logic whose set of truth values is

$$V_\downarrow = \{0\} \cup \left\{1/n : n \in \mathbb{N} \setminus \{0\}\right\}.$$

$\mathsf{G}_\downarrow$ is the same as the logic defined by *linearly ordered Kripke structures on constant domains* [1] - a fundamental concept in the definition of *Temporal logic of programs* [9], an origin of the study of program verification.

Baaz-Leitsch-Zach [1] have shown that $\mathsf{G}_\downarrow$ is not recursively axiomatizable and Hájek [8] showed that the sets of validities and satisfiable formulas are

J.P. Aguilera—Supported by FWF grant I4513N and FWO grant 3E017319.
J. Bydzovsky—Supported by FWF grant P31955 and I4427.
J.P. Aguilera and D. Fernández-Duque—Supported by FWO-FWF Lead Agency Grant G030620N.

S. Artemov and A. Nerode (Eds.): LFCS 2022, LNCS 13137, pp. 1–8, 2022.
https://doi.org/10.1007/978-3-030-93100-1_1

not arithmetical. We will show that they are Σ^1_1-complete and Π^1_1-complete, respectively. As satisfiability is Σ^1_1 and validity is Π^1_1, this yields the optimal strengthening of their theorems. We remark that each of the results for Σ^1_1 and Π^1_1 is not immediate from the other, because satisfiability and validity are not dual in Gödel logic as they are in classical logic.

2 Preliminaries

First order Gödel logic uses the syntax of intuitionistic predicate logic, where the set of formulas is defined according to the following clauses:

$$\bot \mid P(\vec{x}) \mid \phi \wedge \psi \mid \phi \vee \psi \mid \phi \rightarrow \psi \mid \exists x \phi \mid \forall x \phi.$$

Here, P is a predicate symbol of arity n in some predetermined alphabet, $\vec{x}$ is a tuple of n variables, ϕ, ψ are formulas and x is a variable.

Definition 1. *Fix a closed set $V \subseteq [0,1]$ with $0 \in V$; such a set is a* set of truth values. *A V-valued model is a pair $\mathfrak{M} = (D, \|\cdot\|)$, where D is a set of elements and $\|\cdot\|$ assigns to each n-ary predicate symbol P a function $\|P(\cdot)\|\colon D^n \rightarrow V$. We then extend $\|\cdot\|$ to complex formulas according to the following clauses:*

- $\|\bot(\vec{a})\| = 0$
- $\|\phi \wedge \psi(\vec{a})\| = \min\{\|\phi(\vec{a})\|, \|\psi(\vec{a})\|\}$
- $\|\phi \vee \psi(\vec{a})\| = \max\{\|\phi(\vec{a})\|, \|\psi(\vec{a})\|\}$
- $\|\phi \rightarrow \psi(\vec{a})\| = \begin{cases} 1 & \text{if } \|\phi(\vec{a})\| \leq \|\psi(\vec{a})\| \\ \|\psi(\vec{a})\| & \text{otherwise} \end{cases}$
- $\|\exists x \phi(x, \vec{a})\| = \sup_{b \in D} \|\phi(b, \vec{a})\|$
- $\|\forall x \phi(x, \vec{a})\| = \inf_{b \in D} \|\phi(b, \vec{a})\|$.

On occasion we may write $\|\cdot\|_{\mathfrak{M}}$ instead of $\|\cdot\|$ when we want to specify the model we are referring to. We write $\mathfrak{M} = (D, P_1, \ldots, P_n)$ instead of $\mathfrak{M} = (D, \|\cdot\|)$ to indicate that the alphabet of $\mathfrak{M}$ is $P_1, \ldots, P_n$. Given a closed set $V \subseteq [0,1]$ containing 0 and 1, we say that a sentence ϕ is *V-satisfiable* if there is a model $\mathfrak{M}$ such that $\|\phi\|_{\mathfrak{M}} = 1$ (in which case we write $\mathfrak{M} \models \phi$), and *weakly V-satisfiable* if there is a model $\mathfrak{M}$ such that $\|\phi\|_{\mathfrak{M}} > 0$. The formula ϕ is *V-valid* if for every model $\mathfrak{M}$, $\|\phi\|_{\mathfrak{M}} = 1$. A model $\mathfrak{M}$ is *crisp* if $V = \{0,1\}$; clearly, crisp models are equivalent to classical models. In the remainder of the text we fix $V = V_\downarrow = \{0\} \cup \{1/n : n \in \mathbb{N} \setminus \{0\}\}$, and *satisfiability,* etc. refer to this set of truth values. We will explicitly write e.g. *classical satisfiability* when referring to $V = \{0,1\}$.

A formula is $V_\downarrow$-satisfiable iff it is weakly $V_\downarrow$-satisfiable. In fact, a more general claim holds. Recall that a linear order is *Noetherian* if it contains no infinite strictly increasing sequences; note that $V_\downarrow$ is Noetherian.

Lemma 1. *Let V be a Noetherian set of truth values. Given a sentence φ and any set of truth values V, if there exists a model $\mathfrak{M}$ such that $\|\varphi\|_{\mathfrak{M}} > 0$, then there exists a model $\mathfrak{M}'$ such that for all formulas ψ and tuples $\vec{a}$,*

$$\|\psi(\vec{a})\|_{\mathfrak{M}'} = \begin{cases} \|\psi(\vec{a})\|_{\mathfrak{M}} & \text{if } \|\psi(\vec{a})\|_{\mathfrak{M}} < \|\varphi\|_{\mathfrak{M}} \\ 1 & \text{otherwise.} \end{cases} \tag{1}$$

Proof. Let $\mathfrak{M} = (D, \|\cdot\|)$ and define $\mathfrak{M}' = (D, \|\cdot\|')$ so that $\|\cdot\|$ is defined according to (1) for atomic formulas. Then proceed by induction on formula complexity. The assumption that V is Noetherian is used for the case of an existential quantifier, so we focus on this one. We have that $\|\exists x\psi(x,\vec{a})\|' = \sup_{b\in D}\|\psi(b,\vec{a})\|'$. By the induction hypothesis, $\|\psi(b,\vec{a})\|'$ satisfies (1) for all $b \in D$. Since V is Noetherian, there is $b_* \in D$ such that $\|\psi(b_*,\vec{a})\|' = \sup_{b\in D}\|\psi(b,\vec{a})\|'$. If $\|\psi(b_*,\vec{a})\| < \|\varphi\|$ then it is readily checked that for all $b \in D$, $\|\psi(b,\vec{a})\|' = \|\psi(b,\vec{a})\| < \|\varphi\|$, so $\|\exists x\psi(x,\vec{a})\| = \|\psi(b_*,\vec{a})\| = \|\psi(b_*,\vec{a})\|' = \|\exists x\psi(x,\vec{a})\|'$. Otherwise, $\|\psi(b_*,\vec{a})\| \geq \|\varphi\|$, so that $\|\exists x\psi(x,\vec{a})\| \geq \|\varphi\|$ and we must check that $\|\exists x\psi(x,\vec{a})\|' = 1$. But from $\|\psi(b_*,\vec{a})\| \geq \|\varphi\|$ we obtain $\|\psi(b_*,\vec{a})\|' = 1$ and thus $\|\exists x\psi(x,\vec{a})\|' = 1$, as needed.

We will make use of the abbreviation

$$\phi \prec \psi := (\psi \to \phi) \to \psi.$$

It can be checked by the semantics that $\phi \prec \psi$ evaluates to 1 if and only if either ϕ has a smaller value than ψ, or they both have value 1.

While first order logic in principle contains predicate symbols of all arities, in this paper it will suffice to work with a unary symbol N and binary symbols $=, \in$. The convention regarding equality is that it is interpreted as any fuzzy equivalence relation when $V \neq \{0,1\}$, but is true equality when $V = \{0,1\}$. The symbol $\in$ is meant to interpret Kripke-Platek set theory (KP), a weakening of ZFC which, in the version we consider, has the advantage of being finitely axiomatizable. In general, let $\mathcal{L}_{P_1\ldots P_n}$ be the language of first order Gödel logic whose predicate symbols are restricted to $P_1,\ldots,P_n$. To define KP, first say that a Δ_0 formula is a $\mathcal{L}_\in$ formula such that all quantifiers are of the form $\exists x(x \in y \wedge \phi)$ or $\forall x(x \in y \to \phi)$ (often abbreviated as e.g. $\exists x \in y\phi$ and $\forall x(x \in y \to \phi)$, respectively). A Σ_1 formula is one of the form $\exists x\phi$, where ϕ is Δ_0, and a Π_1 formula is one of the form $\forall x\phi$ with the same restriction.

We will use the version of KP with infinity axiomatized by all axioms of ZFC except for powerset, but with foundation restricted to Π_1 classes, separation restricted to Δ_0 formulas and replacement restricted to Δ_0-collection. We will also assume that KP contains equality axioms asserting that $=$ is an equivalence relation, as well as the axioms

$$x = y \to x \in z \leftrightarrow y \in z$$
$$x = y \to z \in x \leftrightarrow z \in y.$$

The precise definitions are not needed to follow the text, but we will use some properties of this version of KP, including that it is finitely axiomatizable.

The axiom of infinity asserts that the set of natural numbers exist as a set, and there is a formula of $\mathcal{L}_\in$ (which we denote by $x \in \mathbb{N}$) defining the set of natural numbers as von Neumann ordinals, which have the property that $n < m$ iff $n \in m$. We will usually write $<$ instead of $\in$ when working with natural numbers within KP. This definition allows us to quantify over the set of natural numbers and define quantifiers $\forall x \in \mathbb{N}$, $\exists X \subseteq \mathbb{N}$, etc. as abbreviations.

We will also use the fact that addition (along with other standard arithmetical operations) is readily interpretable in KP. An *arithmetical formula* is one where all quantifiers are of the form $\forall x \in \mathbb{N}$ or $\exists x \in \mathbb{N}$, a Π^1_1*-formula* is one of the form $\forall X \subseteq \mathbb{N}\phi$ where ϕ is arithmetical, and a Σ^1_1*-formula* is one of the form $\exists X \subseteq \mathbb{N}\phi$, also with ϕ arithmetical. A model $\mathfrak{M}$ is an ω-model if the natural numbers in $\mathfrak{M}$ are isomorphic to the standard natural numbers.

As for the symbol N, its intended meaning is that $\|N(x)\| > 0$ iff $x \in \mathbb{N}$, with larger natural numbers receiving smaller truth values. This will be made precise later in the text.

Some familiarity with the class of ordinals, as well as the constructible hierarchy $\{\mathbb{L}_\alpha \mid \alpha \text{ is an ordinal}\}$ is assumed. An ordinal α is *admissible* if $\mathbb{L}_\alpha$ is a (classical) model of KP; because every recursive ordinal is provably well-ordered in KP, the smallest admissible ordinal is the Church-Kleene ordinal ω_1^{CK}. Note that ω_1^{CK} is countable.

We will use the following two results involving admissible sets.

Theorem 1 (Ville [3]). *Let $\mathfrak{M}$ be any ω-model of* KP*. Then, the well-founded part of $\mathfrak{M}$ (with respect to $\in$) is admissible, and hence extends $\mathbb{L}_{\omega_1^{\mathrm{CK}}}$.*

Theorem 2 (Barwise-Gandy-Moschovakis [4]). *Given a Σ^1_1 formula ϕ, one can effectively and uniformly find a Π_1 $\mathcal{L}_\in$-formula $\psi(x)$ such that for every natural number n*

$$\mathbb{N} \models \phi(n) \leftrightarrow \mathbb{L}_{\omega_1^{\mathrm{CK}}} \models \psi(n).$$

3 Standard Models via Vagueness

Our proof of hardness follows by a variation of Hájek's proof. The high-level idea of Hájek's proof is to use the set of truth values to define an interpretation of the standard natural numbers. In our argument, we use the set of truth values to define an interpretation of the standard natural numbers in models of KP and then apply the theorem of Ville and of Barwise-Gandy-Moschovakis to them.

Recall that we are using a finitely axiomatizable presentation of KP, that $\mathbb{N}$ is definable in KP and that we write $<$ instead of $\in$ for natural numbers. Recall also that we are working with a monadic predicate N whose intended meaning is that $\|N(x)\| > 0$ iff $x \in \mathbb{N}$, and that we defined $\phi \prec \psi := (\psi \to \phi) \to \psi$. With this in mind, let Ψ be the sentence asserting the conjunction of the following statements:

(i) $=$ and $\in$ are crisp, i.e., they satisfy excluded middle;
(ii) KP holds of the predicate $\in$;
(iii) $\forall x, y\, (x = y \to (N(x) \leftrightarrow N(y)))$;
(iv) $\forall x, y \in \mathbb{N}\, (x < y \to (N(y) \prec N(x)))$;
(v) $\forall x\, \neg\neg N(x) \to x \in \mathbb{N}$;
(vi) $\neg \exists x \in \mathbb{N}\, \neg N(x)$;
(vii) $\neg \forall x \in \mathbb{N}\, N(x)$.

Let $\mathfrak{M} = (D, =, \in, N)$ be any model of Ψ. By (i), $\in$ is crisp, so for each $x \in D$, the formula $x \in \mathbb{N}$ has value 0 or 1. Formula (vi) asserts that whenever x is a natural number, $N(x)$ has a positive truth value. Conversely, formula (v) asserts that whenever $N(x)$ has positive truth value, then x is a natural number. Formula (vii) asserts that the infimum of the truth values of $N(x)$ is 0 as x ranges over the natural numbers. The intuition may be grasped easily through the following concrete construction.

Lemma 2. *Any model of Ψ satisfies the equality schema*

$$x = y \to \varphi(x) \leftrightarrow \varphi(y)$$

for all formulas φ in the vocabulary $\{=, \in, N\}$.

Proof. For atomic formulas involving the relations $\in$ and $=$ the result follows directly from the axioms of KP. For N, it follows from (iii). The general case follows by a straightforward induction.

Lemma 3. *Any ω-model of* KP *can be extended to a model of Ψ.*

Proof. Let $\mathfrak{M} = (D, =, \in)$ be an ω-model of KP. For $n \in \mathbb{N}$, define $\|N(n)\| = 1/n+1$. For $x \notin \mathbb{N}$, define $\|N(x)\| = 0$. The model $\mathfrak{M} = (D, =, \in, N)$ thus defined satisfies (i) and (ii) since the interpretation of $=, \in$ did not change, and (iii)–(vii) are readily checked to hold using the definition of N.

The key property of Ψ is that *only* ω-models of KP can be extended to models of Ψ.

Lemma 4. *If $\mathfrak{M} = (D, =, \in, N)$ is such that $\mathfrak{M} \models \Psi$ and $=$ is identity in D, then $(D, =, \in)$ is an ω-model of* KP.

Proof. Fix a model $\mathfrak{M} = (D, \|\cdot\|)$ over the signature $\{=, \in, N\}$. Noting that $=, \in$ (and hence $<$) are crisp by (i), we may reason classically about these relations. First note that by (vi), for every $a \in \mathbb{N}$ we have that $N(a) > 0$.

Claim. If $a < b$ are such that $\|N(a)\| < 1$, then $\|N(b)\| < \|N(a)\|$.

Proof of the Claim: By (iv) we have that $\|N(b) \prec N(a)\| = 1$. By the truth conditions of $\prec$ we have that either $\|N(b)\| = \|N(a)\| = 1$ or else $\|N(b)\| < \|N(a)\|$. As we do not have that $\|N(a)\| = 1$ by assumption, we conclude that $\|N(b)\| < \|N(a)\|$ as needed. This establishes the Claim.

By (vii), there is an $a_0 \in D$ with $\|N(a_0)\| < 1$. We claim that if $a \in D$ is such that $a_0 < a \wedge a \in \mathbb{N}$, then a is standard, in the sense that $\{b \in D : b < a\}$ is finite. This will conclude the proof, since $\mathfrak{M} \models \forall x \in \mathbb{N}(x \leq a_0 \vee a_0 < x)$, so then every natural number is standard.

So fix $a > a_0$. By the Claim, $\|N(a)\| < \|N(a_0)\|$. Now, let $c < d \in \mathbb{N}$. By the Claim once again, $\|N(a + d)\| < \|N(a + c)\|$. It follows that the sequence

$(\|N(a+n)\|)_{n\in\mathbb{N}}$ is strictly decreasing, hence its infimum is zero (as this is the case for any strictly decreasing sequence in $V_\downarrow$). Now, if a were non-standard, we would have that $\|N(a+a)\| < \|N(a+n)\|$ for all $n \in \mathbb{N}$, hence $\|N(a+a)\| = 0$, which contradicts (vi). We conclude that a is indeed standard, and hence $\mathfrak{M}$ is an ω-model of KP.

4 Satisfiability in $\mathbf{G}_\downarrow$

Lemma 4 suffices to establish our hardness results. We begin with satisfiability; recall that by this we mean $V_\downarrow$*-satisfiability.* In view of Lemma 1, satisfiability can be replaced by weak satisfiability in the theorem below.

Theorem 3. *The set of all (weakly)* $V_\downarrow$*-satisfiable formulas is* Σ^1_1*-complete.*

Proof. First, a formula ϕ is $V_\downarrow$-satisfiable if and only if it has a model. By downwards Löwenheim-Skolem (see e.g., Baaz et al. [2]), this is equivalent to it having a countable model. Hence, ϕ is satisfiable if and only if there is a subset of $\mathbb{N}$ coding a model of ϕ. This is clearly Σ^1_1.

Now, fix a Σ^1_1 formula $\phi(x)$. We find a many-one reduction of $\{n : \mathbb{N} \models \phi(n)\}$ to the set of satisfiable formulas of $\mathsf{G}_\downarrow$. By Lemma 2, one can effectively and uniformly find a Π_1 $\mathcal{L}_\in$-formula $\psi(x)$ such that for every natural number n

$$\mathbb{N} \models \phi(n) \leftrightarrow \mathbb{L}_{\omega_1^{\mathrm{CK}}} \models \psi(n).$$

We will show that for every standard natural number n, $\mathbb{N} \models \phi(n)$ if and only if $\Psi \wedge \psi(n)$ is $\mathsf{G}_\downarrow$-satisfiable. First, suppose that $\mathbb{N} \models \phi(n)$. Then, $\mathbb{L}_{\omega_1^{\mathrm{CK}}} \models \psi(n)$. By Lemma 3, $\mathbb{L}_{\omega_1^{\mathrm{CK}}}$ can be extended to a model $\mathfrak{M}$ of Ψ, and since $\psi(n)$ does not contain the symbol N, we have that $\mathfrak{M} \models \psi(n)$ as well.

Conversely, let $\mathfrak{M} = (D, \in, N)$ be a model of $\Psi \wedge \psi(n)$. By Lemma 3, $(D, =, \in)$ is an ω-model. By Theorem 1, the well-founded part of $\mathfrak{M}$ extends $\mathbb{L}_{\omega_1^{\mathrm{CK}}}$. Since $\mathfrak{M}$ is a model of $\Psi \wedge \psi(n)$, we have that $\mathfrak{M} \models \psi(n)$. Since ψ is Π_1, we have $\mathbb{L}_{\omega_1^{\mathrm{CK}}} \models \psi(n)$, as desired. This completes the proof.

5 Validity in $\mathbf{G}_\downarrow$

Finally, we show that validity is Π^1_1-complete.

Theorem 4. *The set of all* $V_\downarrow$*-valid formulas is* Π^1_1*-complete.*

Proof. Note that a formula is valid in $\mathsf{G}_\downarrow$ if and only if it holds in every model and this is equivalent to holding in every countable model, which is clearly Π^1_1. Thus the only problem will be to show the hardness. For this we use the complementary statement of Theorem 2, that is for any Π^1_1-formula $\phi(x)$ there is a Σ_1-formula $\psi(x)$ in the language of set theory such that for every natural number n

$$\mathbb{N} \models \phi(n) \leftrightarrow \mathbb{L}_{\omega_1^{\mathrm{CK}}} \models \psi(n).$$

We claim that $\mathbb{N} \models \phi(n)$ iff $\Psi \to \psi(n)$ is $V_\downarrow$-valid. For the easy direction assume $\Psi \to \psi(n)$ is $V_\downarrow$-valid. In that case extend $\mathbb{L}_{\omega_1^{\mathrm{CK}}}$ to a model of Ψ using Lemma 3 and call the resulting $V_\downarrow$-model $\mathfrak{M}$. By the assumption $\|\psi(n)\|_{\mathfrak{M}} = 1$, but since $\psi(n)$ does not contain the symbol N, it follows that $\mathbb{L}_{\omega_1^{\mathrm{CK}}} \models \psi(n)$, which gives $\mathbb{N} \models \phi(n)$.

For the other direction, assume that $\mathbb{N} \models \phi(n)$; we claim that $\Psi \to \psi(n)$ is $V_\downarrow$-valid. If it were not, by Lemma 1 there would be a model $\mathfrak{M} = (D, \|\cdot\|)$ with $\|\Psi\| = 1$ and $\|\psi(n)\| < \|\Psi\|$. We construct the model $\mathfrak{M}/E = (D/E, \|\cdot\|_E)$ by factorising D modulo the relation $E = \{(a, b) : \|a = b\| = 1\}$. In more details, let the universe of the new model be the set of all equivalence classes of E. We denote by $[a]$ the E-equivalence class of a and for any a, b we set $\|[a] \in [b]\|_E := \|a \in b\|$. To interpret N we fix for every $[a]$ a unique value from $\{\|N(b)\| : b \in [a]\}$ and let $\|N([a])\|_E$ be that value. Note that these values do not depend on the choice of representatives in view of (iii) and Lemma 2.

Claim. $\mathfrak{M}/E \models \Psi$

Proof of the Claim: By (i) E is a congruence for the interpretation of $\in$, so the definition of the model is correct and the model is a model of KP. For the other clauses, we use the fact that the value of $N([a])$ does not depend on the choice of representatives, as well as the fact that $\mathfrak{M}$ is a model of Ψ. This proves the Claim.

Since $\mathfrak{M}/E \models \Psi$, the natural numbers in $\mathfrak{M}/E$ are standard by Lemma 4. Then by Theorem 1, the well-founded part of $\mathfrak{M}/E$ is admissible and hence contains $\mathbb{L}_{\omega_1^{\mathrm{CK}}}$. Moreover if $\mathbb{N} \models \phi(n)$ then $\mathbb{L}_{\omega_1^{\mathrm{CK}}} \models \psi(n)$ and so $\mathfrak{M}/E \models \psi(n)$ as $\psi(n)$ is Σ_1. Since $\psi(n)$ does not contain the symbol N we get $\mathfrak{M} \models \psi(n)$, which was to be shown. This completes the proof of the Theorem.

6 Concluding Remarks

We have provided precise complexity bounds for $\mathsf{G}_\downarrow$, previously only known to be non-arithmetical. It is possible that similar results hold for other fuzzy logics. Hájek [7] showed that $\Pi\forall$SAT is non-arithmetical and Montagna [10] that $\Pi\forall$TAUT, BL$\forall$TAUT, and BL$\forall$SAT are non-arithmetical. We leave the question of whether these logics are Π^1_1/Σ^1_1 complete open.

References

1. Baaz, M., Leitsch, A., Zach, R.: Incompleteness of a first-order Gödel logic and some temporal logics of programs. In: Proceedings of CSL 1995, vol. 68 (1996)
2. Baaz, M., Preining, N., Zach, R.: Completeness of a hypersequent calculus for some first-order Godel logics with delta. In: ISMVL, p. 9. IEEE Computer Society (2006)
3. Barwise, J.: Admissible Sets and Structures. Perspectives in Mathematical Logic. Springer, Heidelberg (1975)

4. Barwise, K.J., Gandy, R., Moschovakis, Y.N.: The next admissible set. J. Symb. Log. **36**, 108–120 (1971)
5. Dummett, M.: A propositional calculus with denumerable matrix. J. Symb. Log. **24**, 97–106 (1959)
6. Gödel, K.: Zum intuitionistischen aussagenkalkül. Anzeiger der Akademie der Wissenschaften in Wien **69**, 65–66 (1932)
7. Hájek, P.: Fuzzy logic and the arithmetical hierarchy, III. Studia Log. **68**, 129–142 (2001). https://doi.org/10.1023/A:1011906423560
8. Hájek, P.: A non-arithmetical Gödel logic. Log. J. IGPL **13**, 435–441 (2005)
9. Kröger, F.: Temporal Logic of Programs. Monographs in Theoretical Computer Science (EATCS Series), vol. 8. Springer, Berlin (1987). https://doi.org/10.1007/978-3-642-71549-5
10. Montagna, F.: Three complexity problems in quantified fuzzy logic. Studia Log. **68**, 143–152 (2001). https://doi.org/10.1023/A:1011958407631

Andrews Skolemization May Shorten Resolution Proofs Non-elementarily

Matthias Baaz[1] and Anela Lolic[2(✉)]

[1] Institute of Discrete Mathematics and Geometry, TU Wien, Wien, Austria
baaz@logic.at
[2] Institute of Logic and Computation, TU Wien, Wien, Austria
anela@logic.at

Abstract. In this paper we construct a sequence of formulas $F_1, F_2, \ldots$ with resolution proofs $\pi_1, \pi_2, \ldots$ of these formulas after Andrews Skolemization, such that there is no elementary bound in the complexity of $\pi_1, \pi_2, \ldots$ of resolution proofs $\pi'_1, \pi'_2, \ldots$ after structural Skolemization. The proofs are based on the elementary relation of resolution derivations with Andrews Skolemization to cut-free $\mathbf{LK}^+$-derivations and of resolution derivations with structural Skolemization to cut-free $\mathbf{LK}$-derivations. Therefore, this paper develops an application of the concept of only globally sound calculi to automated theorem proving.

Keywords: Skolemization · Sequent calculus · Resolution

1 Introduction

The most prominent method of classical automated theorem proving in first-order logic is resolution. The resolution method is based on the negation of the target formula, elimination of strong quantifiers by the introduction of new function symbols (the Skolem functions), the deletion of weak quantifiers[1] and the transformation of the remaining formula in a conjunction of a disjunction of literals, the clause form.

The most prominent method of Skolemization is the so-called structural Skolemization, where the positive existential (negative universal) quantifiers are replaced by Skolem functions depending on all weak quantifiers in whose scope the replaced existential (universal) quantifier occurs[2].

In this paper structural Skolemization is compared to Andrews Skolemization [2,3], where the positive existential quantifier is replaced by a Skolem function

[1] In the negated formula, strong quantifiers are positive existential and negative universal quantifiers, and weak quantifiers are negative existential and positive universal quantifiers. (Otherwise, strong quantifiers are positive universal and negative existential quantifiers, and weak quantifiers are negative universal and positive existential quantifiers.).

[2] Prenexification before Skolemization is not recommendable [5].

S. Artemov and A. Nerode (Eds.): LFCS 2022, LNCS 13137, pp. 9–24, 2022.
https://doi.org/10.1007/978-3-030-93100-1_2

depending only on those weak quantifiers of the scope which bind in the subsequent formula. This sometimes reduces the dependencies of Skolem functions. (Note that Andrews Skolemization is a concept developed by the needs of automated theorem proving.)

Under assumption of any elementary clause form transformation it is shown that Andrews Skolemization might lead to a non-elementary speed-up compared to structural Skolemization.

2 The Sequent Calculi LK, LK$^+$ and LK^{++}

In this paper we will consider the sequent calculus **LK** based on sequents consisting of multisets (the exact formulation **LK** does not matter for this work).

This work is also based on the sequent calculi **LK**$^+$ and **LK**$^{++}$ introduced in [1]. They are obtained from **LK** by weakening the eigenvariable conditions. The resulting calculi are therefore globally but possibly not locally sound. This means that all derived statements are true but that not every sub-derivation is meaningful.

Definition 1 (side variable relation $<_{\varphi,\mathbf{LK}}$, cf. [1]). *Let φ be an* **LK**-*derivation. We say b is a side variable of a in φ (written $a <_{\varphi,\mathbf{LK}} b$) if φ contains a strong quantifier inference of the form*

$$\frac{\Gamma \vdash \Delta, A(a, b, \overline{c})}{\Gamma \vdash \Delta, \forall x A(x, b, \overline{c})} \forall_r$$

or of the form

$$\frac{A(a, b, \overline{c}), \Gamma \vdash \Delta}{\exists x A(x, b, \overline{c}), \Gamma \vdash \Delta} \exists_l$$

We may omit the subscript φ, **LK** *in $<_{\varphi,\mathbf{LK}}$ if it is clear from the context.*

In addition to strong and weak quantifier inferences we define **LK**$^+$-suitable quantifier inferences.

Definition 2 (LK$^+$-suitable quantifier inferences, cf. [1]). *We say a quantifier inference is suitable for a proof φ if either it is a weak quantifier inference, or the following three conditions are satisfied:*

- *(substitutability) the eigenvariable does not appear in the conclusion of φ.*
- *(side variable condition) the relation $<_{\varphi,\mathbf{LK}}$ is acyclic.*
- *(weak regularity) the eigenvariable of an inference is not the eigenvariable of another strong quantifier inference in φ.*

Definition 3 (LK$^+$, cf. [1]). **LK**$^+$ *is obtained from* **LK** *by replacing the usual eigenvariable conditions by* **LK**$^+$-*suitable ones.*

Similarly to **LK**$^+$, we define the calculus **LK**$^{++}$ by further weakening the eigenvariable conditions

Definition 4 ($\mathbf{LK}^{++}$-suitable quantifier inferences, cf. [1]). *We say a quantifier inference is suitable for a proof φ if either it is a weak quantifier inference, or it satisfies*

- *substitutability,*
- *the side variable condition, and*
- *(very weak regularity) the eigenvariable of an inference with main formula A is different to the eigenvariable of an inference with main formula A' whenever $A \neq A'$.*

Definition 5 ($\mathbf{LK}^{++}$, cf. [1]). *$\mathbf{LK}^{++}$ is obtained from $\mathbf{LK}$ by replacing the usual eigenvariable conditions by $\mathbf{LK}^{++}$-suitable ones.*

Theorem 1.

1. *If a sequent is $\mathbf{LK}^{+}$-derivable, then it is already $\mathbf{LK}$-derivable.*
2. *If a sequent is $\mathbf{LK}^{++}$-derivable, then it is already $\mathbf{LK}$-derivable.*

Proof (Proof Sketch). Consider an $\mathbf{LK}^{++}$-proof φ (an $\mathbf{LK}^{+}$ proof is also an $\mathbf{LK}^{++}$-proof). Replace every universal quantifier inference unsound w.r.t. $\mathbf{LK}$ by an $\rightarrow_l$ inference:

$$\frac{\Gamma \vdash \Delta, A(a) \qquad \forall x A(x) \vdash \forall x A(x)}{\Gamma, A(a) \rightarrow \forall x A(x) \vdash \Delta, \forall x A(x)}$$

Similarly, replace every existential quantifier inference unsound w.r.t. $\mathbf{LK}$ by an $\rightarrow_l$ inference:

$$\frac{\exists x A(x) \vdash \exists x A(x) \qquad A(a), \Gamma \vdash \Delta}{\Gamma, \exists x A(x), \exists x A(x) \rightarrow A(a) \vdash \Delta}$$

By doing this, we obtain a proof of the desired sequent, together with formulas of the form

$$A(a) \rightarrow \forall x A(x) \quad \text{or} \quad \exists x A(x) \rightarrow A(a)$$

on the left-hand side. Note that the resulting derivation does not contain any inference based on eigenvariable conditions. We can eliminate each of the additional formulas on the left-hand side by adding an existential quantifier inference and cutting with sequents of the form

$$\vdash \exists y (A(y) \rightarrow \forall x A(x)) \quad \text{or of the form} \quad \vdash \exists y (\exists x A(x) \rightarrow A(y)),$$

both of which are easily derivable. For more details see [1].

Example 1. Consider the following locally unsound but globally sound $\mathbf{LK}^{+}$-derivation φ:

$$\cfrac{\cfrac{\cfrac{A(a) \vdash A(a)}{A(a) \vdash \forall y A(y)}\ \forall_r}{\vdash A(a) \rightarrow \forall y A(y)}\ \rightarrow_r}{\vdash \exists x (A(x) \rightarrow \forall y A(y))}\ \exists_r$$

As a is the only eigenvariable the side variable relation $<_{\varphi,\mathbf{LK}}$ is empty.

For an $\mathbf{LK}^{++}$-derivation, which is not an $\mathbf{LK}^{+}$-derivation compare Example 2.

In the following the complexity is always understood as the number of symbols. Recall that a function on the natural numbers is *elementary* if it can be defined by a quantifier-free formula from $+, \cdot$, and the function $x \to 2^x$.

The focus in [1] has been on the strongly reduced complexity of cut-free $\mathbf{LK}^{+}$- and $\mathbf{LK}^{++}$-proofs (Theorem 2.6 and Corollary 2.7). From Theorem 2.6 it immediately follows that:

Theorem 2. *There is no elementary function bounding the complexity of the shortest cut-free* **LK***-proof of a formula in terms of its shortest cut-free* $\mathbf{LK}^{+}$*-proof.*

The proof is based on the worst case sequences of Orevkov [7] or Statman [8] as formalized in [5].

Note that for cut-free proofs in **LK**, $\mathbf{LK}^{+}$ and $\mathbf{LK}^{++}$ the complexity is elementarily bounded in the length of the proof, both defined as height and as numbers of inference nodes. (This does not hold in general for proofs with cuts.)

Theorem 3. *There is a transformation of a cut-free* $\mathbf{LK}^{++}$*-proof* φ *of the end-sequent* $\Pi \vdash \Gamma$ *into a cut-free* $\mathbf{LK}^{+}$*-proof of* $\Pi \vdash \Gamma$ *with an elementary bound in the complexity of* φ.

Proof. Let φ be a cut-free $\mathbf{LK}^{++}$-derivation with end-sequent $\Pi \vdash \Gamma$ and eigenvariables $a_1, \ldots, a_n$ of strong quantifier inferences. We replace the old inferences stepwise by inferences with the stronger regularity conditions of $\mathbf{LK}^{+}$. These new variables are of the form

$$x_{f(a_{i_1},\ldots,a_{i_k})},$$

where $a_{i_1}, \ldots, a_{i_k}$ are the old variables still present at the time of the transformation. Consider all inferences $\forall_r$ of the form

$$\frac{\Pi_j \vdash \Gamma_j, A(a)}{\Pi_j \vdash \Gamma_j, \forall x A(x)}$$

where this inference is an innermost inference not yet changed. Extract a proof $\varphi'(a)$ of

$$\Pi \vdash \Gamma, A(a)$$

by deleting all inferences working on $A(a)$ and resulting formulas. Locate all places in the proof where $\forall x A(x)$ is inferred from $A(a)$. Let i be such a position. We replace the sub-proof leading to the premise

$$\Pi \vdash \Gamma, A(a)$$

of this inference by $\varphi'(x_{f_i(a_{i_1},\ldots,a_{i_k})})$ of

$$\Pi(x_{f_i(a_{i_1},\ldots,a_{i_k})}) \vdash \Gamma(x_{f_i(a_{i_1},\ldots,a_{i_k})}), A(x_{f_i(a_{i_1},\ldots,a_{i_k})}).$$

Then we replace a in the indices of new variables everywhere by f_i. By weakening we obtain

$$\Pi, \Pi(x_{f_i(a_{i_1},\ldots,a_{i_k})}) \vdash \Gamma(x_{f_i(a_{i_1},\ldots,a_{i_k})}), \Gamma, A(x_{f_i(a_{i_1},\ldots,a_{i_k})})$$

and continue with the proof. ($\exists$ left inferences are dual.)

Note that by replacing a in the indices inferences already transformed to regular inferences remain regular. The side variable condition is extended correspondingly. Note that the final proof of the original end-sequent after contraction which obviously fulfils the strong regularity condition of $\mathbf{LK}^+$ also fulfils the non-occurrence of strong inference variables in the end-sequent and the side-variable conditions are acyclic, as $x_f < x_g$ only if g is a proper subterm of f.

Finally, we replace all occurrences of a (which are no eigenvariables of any inference anymore) by a fixed constant c.

The transformation is exponential in the number of variables w.r.t. the height if weakenings and contractions are not counted. It is therefore elementary in the complexity.

Example 2. Consider the following $\mathbf{LK}^{++}$-proof φ:

$$\dfrac{\dfrac{\dfrac{\dfrac{\dfrac{\dfrac{\dfrac{\dfrac{A(a,a,b) \vdash A(a,a,b)}{A(a,a,b) \vdash \forall z A(a,a,z)}}{A(a,a,b) \vdash \exists y \forall z A(a,y,z)}}{A(a,a,b) \vdash \forall x \exists y \forall z A(x,y,z)}}{C, A(a,a,b) \vdash \forall x \exists y \forall z A(x,y,z)} \qquad \dfrac{\dfrac{\dfrac{\dfrac{A(a,a,b) \vdash A(a,a,b)}{A(a,a,b) \vdash \forall z A(a,a,z)}}{A(a,a,b) \vdash \exists y \forall z A(a,y,z)}}{A(a,a,b) \vdash \forall x \exists y \forall z A(x,y,z)}}{D, A(a,a,b) \vdash \forall x \exists y \forall z A(x,y,z)}}{C \vee D, A(a,a,b) \vdash \forall x \exists y \forall z A(x,y,z)}}{C \vee D, \forall y A(a,a,y) \vdash \forall x \exists y \forall z A(x,y,z)}}{C \vee D, \forall x \forall y A(x,x,y) \vdash \forall x \exists y \forall z A(x,y,z)}$$

with $b <_{\varphi,\mathbf{LK}} a$.

In the following we will use the abbreviations Π for $C \vee D, \forall x \forall y A(x,x,y)$ and Γ for $\forall x \exists y \forall z A(x,y,z)$ (our end-sequent is $\Pi \vdash \Gamma$).

$\varphi(b)$ is:

$$\dfrac{\dfrac{\dfrac{\dfrac{\dfrac{A(a,a,b) \vdash A(a,a,b)}{C, A(a,a,b) \vdash A(a,a,b)} \qquad \dfrac{A(a,a,b) \vdash A(a,a,b)}{D, A(a,a,b) \vdash A(a,a,b)}}{C \vee D, A(a,a,b) \vdash A(a,a,b)}}{C \vee D, \forall y A(a,a,y) \vdash A(a,a,b)}}{C \vee D, \forall x \forall y A(x,x,y) \vdash A(a,a,b)}}{\Pi \vdash \Gamma, A(a,a,b)}$$

The new proof is obtain from $\varphi(x_{f(a)})$ and $\varphi(x_{g(a)})$:

$$
\begin{array}{c}
\begin{array}{c}
(\varphi(x_{f(a)})) \\ \hline
\Pi \vdash \Gamma, A(a,a,x_{f(a)}) \\ \hline
\Pi \vdash \Gamma, \forall z A(a,a,z) \\ \hline
A(a,a,b), \Pi \vdash \Gamma, \forall z A(a,a,z)
\end{array}
\quad
\begin{array}{c}
(\varphi(x_{g(a)})) \\ \hline
\Pi \vdash \Gamma, A(a,a,x_{g(a)}) \\ \hline
\Pi \vdash \Gamma, \forall z A(a,a,z) \\ \hline
A(a,a,b), \Pi \vdash \Gamma, \forall z A(a,a,z)
\end{array} \\ \hline\hline
\Pi, \Pi \vdash \Gamma, \Gamma \\ \hline
\Pi \vdash \Gamma
\end{array}
$$

where $x_{f(a)} < a$ and $x_{g(a)} < a$.

$\varphi'(a, x_{f(a)}, x_{g(a)})$ is:

$$
\begin{array}{c}
\begin{array}{c}
(\varphi(x_{f(a)})) \\ \hline
\Pi \vdash \Gamma, A(a,a,x_{f(a)}) \\ \hline
\Pi \vdash \Gamma, \forall z A(a,a,z) \\ \hline
A(a,a,b), \Pi \vdash \Gamma, \forall z A(a,a,z) \\ \hline
A(a,a,b), \Pi \vdash \Gamma, \exists y \forall z A(a,y,z) \\ \hline
C, A(a,a,b), \Pi \vdash \Gamma, \exists y \forall z A(a,y,z)
\end{array}
\quad
\begin{array}{c}
(\varphi(x_{g(a)})) \\ \hline
\Pi \vdash \Gamma, A(a,a,x_{g(a)}) \\ \hline
\Pi \vdash \Gamma, \forall z A(a,a,z) \\ \hline
A(a,a,b), \Pi \vdash \Gamma, \forall z A(a,a,z) \\ \hline
A(a,a,b), \Pi \vdash \Gamma, \exists y \forall z A(a,y,z) \\ \hline
D, A(a,a,b), \Pi \vdash \Gamma, \exists y \forall z A(a,y,z)
\end{array} \\ \hline
C \vee D, A(a,a,b), \Pi \vdash \Gamma, \exists y \forall z A(a,y,z) \\ \hline
C \vee D, \forall y A(a,a,y), \Pi \vdash \Gamma, \exists y \forall z A(a,y,z) \\ \hline
C \vee D, \forall x \forall y A(x,x,y), \Pi \vdash \Gamma, \exists y \forall z A(a,y,z)
\end{array}
$$

Finally,

$$
\begin{array}{c}
\begin{array}{c}
(\varphi'(x_d, x_{f(d)}, x_{g(d)})) \\ \hline
\Pi \vdash \Gamma, \exists y \exists z A(x_d,y,z) \\ \hline
\Pi \vdash \Gamma, \forall x \exists y \exists z A(x,y,z) \\ \hline
A(a,a,b), \Pi \vdash \Gamma, \forall x \exists y \exists z A(x,y,z) \\ \hline
C, A(a,a,b), \Pi \vdash \Gamma, \forall x \exists y \exists z A(x,y,z)
\end{array}
\quad
\begin{array}{c}
(\varphi'(x_e, x_{f(e)}, x_{g(e)})) \\ \hline
\Pi \vdash \Gamma, \exists y \exists z A(x_e,y,z) \\ \hline
\Pi \vdash \Gamma, \forall x \exists y \exists z A(x,y,z) \\ \hline
A(a,a,b), \Pi \vdash \Gamma, \forall x \exists y \exists z A(x,y,z) \\ \hline
D, A(a,a,b), \Pi \vdash \Gamma, \forall x \exists y \exists z A(x,y,z)
\end{array} \\ \hline
C \vee D, A(a,a,b), \Pi \vdash \Gamma, \forall x \exists y \exists z A(x,y,z) \\ \hline
C \vee D, \forall y A(a,a,y), \Pi \vdash \Gamma, \forall x \exists y \exists z A(x,y,z) \\ \hline
C \vee D, \forall x \forall y A(x,x,y), \Pi \vdash \Gamma, \forall x \exists y \exists z A(x,y,z) \\ \hline\hline
\Pi \vdash \Gamma
\end{array}
$$

where $x_{f(e)} < x_e, x_{g(e)} < x_e, x_{f(d)} < x_d, x_{g(d)} < x_d$.

3 Skolemization and Deskolemization

In this paper we are concentrating on the effects of Skolemization. Therefore, we will admit any form of clause transformation as long as this transformation is elementary.

We connect the Skolemization of formulas with the Skolemization of proofs. In proofs the Skolemization replaces strong quantifiers. This corresponds to the fact that as preprocessing of the resolution algorithm positive existential and negative universal quantifiers are replaced, because by assumption of its refutability the formula occurs on the left side of the sequent.

In this work we will focus on a different method of Skolemization. In Andrews' method [2,3] the introduced Skolem functions do not depend on the weak quantifiers $(Q_1x_1)\dots(Q_nx_n)$ dominating the strong quantifier (Qx), but on the subset of $\{x_, \dots, x_n\}$ appearing (free) in the subformula dominated by (Qx). This method leads to smaller Skolem terms in general.

Definition 6 (structural Skolem form). *Let A be a closed first-order formula. If A does not contain strong quantifiers, we define its structural Skolemization as* $\mathrm{sk}(A) = A$.

Suppose now that A contains strong quantifiers and (Qy) is the first strong quantifier occurring in A. If (Qy) is not in the scope of weak quantifiers, then its structural Skolemization is

$$\mathrm{sk}(A) = \mathrm{sk}(A_{-(Qy)}\{y \leftarrow c\}),$$

where $A_{-(Qy)}$ is the formula A after omission of (Qy) and c is a constant symbols not occurring in A. If (Qy) is in the scope of the weak quantifiers $(Q_1x_1)\dots(Q_nx_n)$, then its structural Skolemization is

$$\mathrm{sk}(A) = sk(A_{-(Qy)}\{y \leftarrow f(x_1, \dots, x_n)\}),$$

where f is a function symbol (Skolem function) not occurring in A.

If $F = \bigwedge \Gamma \rightarrow \bigvee \Delta$ and $\mathrm{sk}(F) = \bigwedge \Pi \rightarrow \bigvee \Lambda$ we define the structural Skolemization of the sequent $\Gamma \vdash \Delta$ as

$$\mathrm{sk}(\Gamma \rightarrow \Delta) = \Pi \vdash \Lambda.$$

Definition 7 (Andrews Skolem form). *Let A be a closed first-order formula. If A does not contain strong quantifiers, we define its Andrews Skolemization as* $\mathrm{sk}_A(A) = A$.

Suppose now that A contains strong quantifiers, $(Qy)B$ is a subformula of A and (Qy) is the first strong quantifier occurring in A (in a tree-like ordering). If $(Qy)B$ has no free variables which are weakly quantified, then its Andrews Skolemization is

$$\mathrm{sk}_A(A) = \mathrm{sk}_A(A_{-(Qy)}\{y \leftarrow c\}),$$

where $A_{-(Qy)}$ is the formula A after omission of (Qy) and c is a constant symbol not occurring in A. If $(Qy)B$ has n variables $x_1, \dots, x_n$ which are weakly quantified from outside, then its Andrews Skolemization is

$$\mathrm{sk}_A(A) = \mathrm{sk}_A(A_{-(Qy)}\{y \leftarrow f(x_1, \dots, x_n)\}),$$

where f is a function symbol not occurring in A.

If $F = \bigwedge \Gamma \rightarrow \bigvee \Delta$ and $\mathrm{sk}_A(F) = \bigwedge \Pi \rightarrow \bigvee \Lambda$ we define the Andrews Skolemization of the sequent $\Gamma \vdash \Delta$ as

$$\mathrm{sk}_A(\Gamma \rightarrow \Delta) = \Pi \vdash \Lambda.$$

Example 3. Consider the sequent

$$S\colon \forall x \forall y(\exists z P(x,z) \vee Q(y,x)) \vdash .$$

Then its structural Skolemization is

$$\forall x \forall y(P(x, f(x,y)) \vee Q(y,x)) \vdash$$

and its Andrews Skolemization is

$$\forall x \forall y(P(x, g(x)) \vee Q(y,x)) \vdash .$$

Theorem 4.

1. *An* **LK***-proof can be Skolemized by substitution w.r.t. the structural Skolemization conserving the length and without introducing additional cuts.*
2. *An* $\mathbf{LK}^+$*-proof can be Skolemized by substitution w.r.t. the Andrews Skolemization conserving the length and without introducing additional cuts.*

Proof.

1. Let φ be an **LK**-proof of an end-sequent S. Assume that S contains a positive occurrence of $\forall x A(x)$. Then there are three possibilities for the introduction of $\forall x A(x)$ in φ:
 case 1:

 $$\frac{\Gamma \vdash \Delta, A(\alpha)}{\Gamma \vdash \Delta, \forall x A(x)}\ \forall_r$$

 case 2:

 $$\frac{\Gamma \vdash \Delta, B}{\Gamma \vdash \Delta, B \vee C}\ \vee_r$$

 where $\forall x A(x)$ occurs in C.
 case 3:

 $$\frac{\Gamma \vdash \Delta}{\Gamma \vdash \Delta, B}\ w_r$$

 where $\forall x A(x)$ occurs in B.
 Case 1 is the interesting one, the other two cases are obvious. We will give a proof of case 2 here.
 Let $\rho[B \vee C]$ be the path connecting $\Gamma \vdash \Delta, B \vee C$ with the end-sequent S. Let $A(t)$ be the Skolemization of $\forall x A(x)$ in S. Then we define C' to be C where the occurrence of $\forall x A(x)$ is substituted with $A(t)$ and replace $\rho[B \vee C]$ by $\rho[B \vee C']$. Note that this transformation is not local but global.
 Now let us consider case 1. Let $\varphi(\alpha)$ denote the sub-proof ending in $\Gamma \vdash \Delta, A(\alpha)$. Let $\rho[\forall x A(x)]$ be the path connecting $\Gamma \vdash \Delta, \forall x A(x)$ with the end-sequent S. Locate all introductions of weak quantifiers $Q_i y_i$ in $\rho[\forall x A(x)]$

which dominate the occurrence of $\forall x A(x)$. All of them eliminate a term t_i, let $t_1, \ldots, t_n$ be all of these terms. Introduce a new function symbol f with arity n and replace the sub-proof $\varphi(\alpha)$ by $\varphi(f(t_1, \ldots, t_n))$, which is a proof of $\Gamma \vdash \Delta, A(f(t_1, \ldots, t_n))$. α is an eigenvariable and $\Gamma \vdash \Delta$ is left unchanged by this substitution. No eigenvariable conditions are violated in $\varphi(f(t_1, \ldots, t_n))$. Then we skip the $\forall_r$ inference rule and replace the path $\rho[\forall x A(x)]$ by $\rho[A(f(t_1, \ldots, t_n))]$. Note that the terms $t_1, \ldots, t_n$ in $\rho[A(f(t_1, \ldots, t_n))]$ are eliminated successively, therefore the occurrences of $A(\ldots)$ are of the form $A(f(y_1, \ldots, y_k, t_{k+1}, \ldots t_n))$.
Finally, the occurrence of $\forall x A(x)$ in the end-sequent is $A(f(y_1, \ldots, y_n))$, which is the Skolemized form of $\forall x A(x)$ in S.
Note that we have to be careful with contractions. If there are two predecessors $C[\forall x A(x)]$ on the same side of a sequent such that a contraction is applied on this formula we have to introduce the same Skolem symbols in both of them.
The case of a formula $\exists x A(x)$ occurring negatively in S is completely analogous.
The described transformation skips quantifier inference rules and performs term substitutions. Neither of these increases the proof complexity.
2. Analogous to 1., except that in the proof of case 1 we locate all introductions of weak quantifiers $Q_i y_i$ in $\rho[\forall x A(x)]$ such that $Q_i y_i$ dominates the occurrence of $\forall x A(x)$ and y_i appears in $\forall x A(x)$. a has to be replaced in the whole proof. Note that the side variable condition $<_\varphi$ excludes that the Skolem term $f(\ldots a \ldots)$ has to be substituted into a. As a is substituted everywhere by a term containing only free variables which are larger in the ordering $<_\varphi$ than a the side-variable condition does not loop by transitivity.

Remark 1. As $\mathbf{LK}^+$ is sound we obtain an additional soundness argument for Andrews Skolemization. The usual soundness argument in automated theorem proving is as follows:

Move all negations in the negated formula to be refuted to the atoms. Then search for an innermost existential subformula $\exists y A(\overline{x}, y)$ in whose scope only universal quantifiers occur, where $\overline{x}$ are the variables bound from outside. We obtain

$$\forall \overline{x}(\exists y A(\overline{x}, y) \rightarrow A(\overline{x}, f(\overline{x}))$$

by structural Skolemization from

$$\forall \overline{x}(\exists y A(\overline{x}, y) \rightarrow \exists y A(\overline{x}, y)).$$

Therefore,

$$\forall \overline{x}(\exists y A(\overline{x}, y)) \leftrightarrow A(\overline{x}, f(\overline{x}))$$

and we may replace $\exists y A(\overline{x}, y)$ by $A(\overline{x}, f(\overline{x}))$ in the context. We proceed to the next subformula dominated by $\exists$.

Remark 2. Herbrand disjunctions are generally shortened, cf the formula

$$\exists y(\exists x A(x) \rightarrow A(y)).$$

The structural Skolemization of this formula is $\exists y(A(f(y)) \to A(y))$, with shortest Herbrand disjunction

$$A(f(f(a)) \to A(f(a)) \vee A(f(a)) \to A(a).$$

The Andrews Skolemization of the above formula is $\exists y(A(c) \to A(y))$, with shortest Herbrand disjunction

$$A(c) \to A(c).$$

Example 4. Consider the following **LK**-proof (and consequently **LK**$^{+}$-proof):

$$\dfrac{\dfrac{\dfrac{\dfrac{P(a) \vdash P(a)}{P(a) \wedge Q(a) \vdash P(a)}}{\forall x(P(x) \wedge Q(x)) \vdash P(a)} \quad \dfrac{\dfrac{\dfrac{P(a) \vdash P(a)}{P(a) \wedge Q(a) \vdash Q(a)}}{\forall x(P(x) \wedge Q(x)) \vdash Q(a)}}{\forall x(P(x) \wedge Q(x)) \vdash \forall x Q(x)}}{\forall x(P(x) \wedge Q(x)) \vdash P(a) \wedge \forall x Q(x)}}{\forall x(P(x) \wedge Q(x)) \vdash \exists y(P(y) \wedge \forall x Q(x))}$$

The structural Skolemization of the proof is:

$$\dfrac{\dfrac{\dfrac{\dfrac{P(a) \vdash P(a)}{P(a) \wedge Q(a) \vdash P(a)}}{\forall x(P(x) \wedge Q(x)) \vdash P(a)} \quad \dfrac{\dfrac{Q(f(a)) \vdash Q(f(a))}{P(f(a)) \wedge Q(f(a)) \vdash Q(f(a))}}{\forall x(P(x) \wedge Q(x)) \vdash Q(f(a))}}{\forall x(P(x) \wedge Q(x)) \vdash P(a) \wedge Q(f(a))}}{\forall x(P(x) \wedge Q(x)) \vdash \exists y(P(y) \wedge Q(f(y)))}$$

The Andrews Skolemization of the proof is:

$$\dfrac{\dfrac{\dfrac{\dfrac{P(c) \vdash P(c)}{P(c) \wedge Q(c) \vdash P(c)}}{\forall x(P(x) \wedge Q(x)) \vdash P(c)} \quad \dfrac{\dfrac{Q(c) \vdash Q(c)}{P(c) \wedge Q(c) \vdash Q(c)}}{\forall x(P(x) \wedge Q(x)) \vdash Q(c)}}{\forall x(P(x) \wedge Q(x)) \vdash P(c) \wedge Q(c)}}{\forall x(P(x) \wedge Q(x)) \vdash \exists y(P(y) \wedge Q(c))}$$

Example 5. Consider the following **LK**$^{+}$-proof, which is not an **LK**-proof:

$$\dfrac{\dfrac{\dfrac{\dfrac{P(h(b), a) \vdash P(h(b), a)}{P(h(b), a) \vdash \forall y P(h(b), y)}}{P(h(b), a) \vdash \exists x \forall y P(x, y)}}{\forall y P(h(b), y) \vdash \exists x \forall y P(x, y)}}{\exists x \forall y P(h(x), y) \vdash \exists x \forall y P(x, y)}$$

with $a < b$. Its Andrews Skolemization is

$$\frac{\dfrac{P(h(c), f(h(c))) \vdash P(h(c), f(h(c)))}{P(h(c), f(h(c))) \vdash \exists x P(x, f(x))}}{\forall y P(h(c), y) \vdash \exists x P(x, f(x))}$$

Corollary 1.

1. *There is an elementary transformation of a cut-free* **LK**-*proof in its structural Skolemized version w.r.t the complexity.*
2. *There is an elementary transformation of a cut-free* **LK**$^+$-*proof in its Andrews Skolemized version w.r.t. the complexity.*

Note that the Skolemized versions are **LK**-proofs.

Theorem 5.

1. *There is an elementary transformation of a structural Skolemized cut-free* **LK**-*proof into a cut-free* **LK**-*proof with the original end-sequent depending on the complexity of the proof and the complexity of the original end-sequent.*
2. *There is an elementary transformation of an Andrews Skolemized cut-free* **LK**-*proof into a cut-free* **LK**$^{++}$-*proof with the original end-sequent depending on the complexity of the proof and the complexity of the original end-sequent.*

Proof.

1. See Theorem 2 in [4].
2. Assign to every Skolem term a free variable associated with the Skolem term. Introduce the strong quantifiers in any order which respects the structure of the formulas in the end-sequent, i.e. where the formula up to this quantifier is completed. The regularity conditions of **LK**$^{++}$ are fulfilled as always the same formula is generated from the same variable. The side variable conditions are fulfilled if we set $b > a$ if a is associated with $s(t)$ and b is associated with t. All these variables do not occur in the end-sequent.

Corollary 2. *There is no elementary transformation of a cut-free* **LK**$^+$-*proof in its structural Skolemized version.*

Proof. By 2. in the proof of Theorem 5 and Corollary 1.

Of course there is an elementary transformation of a cut-free **LK**$^+$-proof in its structured Skolemized version by adding cuts.

Example 6. Consider the following **LK**$^+$-derivation φ:

$$\frac{\dfrac{\dfrac{\dfrac{A(g(a)) \vdash A(g(a)) \qquad A(g(a)) \vdash A(g(a))}{A(g(a)) \vee A(g(a)) \vdash A(g(a)), A(g(a))}}{\forall x(A(x) \vee A(x)) \vdash A(g(a)), \forall x A(g(x))}}{\forall x(A(x) \vee A(x)) \vdash A(g(a)) \vee \forall x A(g(x))}}{\forall x(A(x) \vee A(x)) \vdash \exists y(A(y) \vee \forall x A(g(x)))}$$

φ has no structural Skolemization, as it would be necessary to substitute $f(g(a))$ for a. In contrast, φ can be Skolemized using Andrews Skolemization. The corresponding end-sequent is

$$\forall x(A(x) \vee A(x)) \vdash \exists y(A(y) \vee A(g(c)))$$

and the Andrews Skolemized proof is obtained by substituting the Skolem constant c for $g(a)$:

$$\dfrac{\dfrac{\dfrac{\dfrac{A(g(c)) \vdash A(g(c)) \qquad A(g(c)) \vdash A(g(c))}{A(g(c)) \vee A(g(c)) \vdash A(g(c)), A(g(c))}}{\forall x(A(x) \vee A(x)) \vdash A(g(c)), A(g(c))}}{\forall x(A(x) \vee A(x)) \vdash A(g(c)) \vee A(g(c))}}{\forall x(A(x) \vee A(x)) \vdash \exists y(A(y) \vee A(g(c)))}$$

Example 7. Consider the following derivation:

$$\dfrac{\dfrac{\dfrac{\dfrac{P(c,f(c)) \vdash P(c,f(c))}{P(c,f(c)) \vdash \exists x P(x,f(x))} \qquad \dfrac{P(c,f(c)) \vdash P(c,f(c))}{P(c,f(c)) \vdash \exists x P(x,f(x))}}{P(c,f(c)) \vee P(c,f(c)) \vdash \exists x P(x,f(x))}}{Q(f(c)) \wedge (P(c,f(c)) \vee P(c,f(c))) \vdash \exists x P(x,f(x))}}{\forall x(Q(x) \wedge (P(c,x) \vee P(c,x)) \vdash \exists x P(x,f(x))}$$

It's expansion is:

$$\dfrac{\dfrac{P(c,f(c)) \vdash P(c,f(c)) \qquad P(c,f(c)) \vdash P(c,f(c))}{P(c,f(c)) \vee P(c,f(c)) \vdash P(c,f(c))}}{Q(f(c)) \wedge (P(c,f(c)) \vee P(c,f(c))) \vdash P(c,f(c))}$$

The intended end-sequent is

$$\exists y \forall x(Q(x) \wedge (P(y,x) \vee P(y,x)) \vdash \exists x \forall y P(x,y).$$

We replace the Skolem term c by a and the Skolem term $f(c)$ by b:

$$\dfrac{\dfrac{P(a,b) \vdash P(a,b) \qquad P(a,b) \vdash P(a,b)}{P(a,b) \vee P(a,b) \vdash P(a,b)}}{Q(b) \wedge (P(a,b)) \vee P(a,b)) \vdash P(a,b)}$$

The quantifiers can be introduced in any order which respects the structure of the formulas in the end-sequent. (We assign indices to the quantifiers to express in which order and where they are introduced.)

$$\dfrac{\dfrac{\dfrac{\dfrac{\dfrac{\dfrac{P(a,b) \vdash P(a,b)}{P(a,b) \vdash \forall_1 x P(a,x)}}{P(a,b) \vdash \exists_3 y \forall x P(y,x)} \quad \dfrac{\dfrac{P(a,b) \vdash P(a,b)}{P(a,b) \vdash \forall_2 x P(a,x)}}{P(a,b) \vdash \exists_4 y \forall x P(y,x)}}{P(a,b) \vee P(a,b) \vdash \exists y \forall x P(y,x)}}{Q(b) \wedge (P(a,b)) \vee P(a,b)) \vdash \exists y \forall x P(y,x)}}{\forall_5 x(Q(x) \wedge (P(a,x)) \vee P(a,x))) \vdash \exists y \forall x P(y,x)}}{\exists_6 y \forall x(Q(x) \wedge (P(y,x)) \vee P(y,x))) \vdash \exists y \forall x P(y,x)}$$

with $b < a$ (the side variable condition is guaranteed by the inclusion relation of the Skolem terms).

Weak regularity and the condition that eigenvariables do not occur in the end-sequent are clearly fulfilled.

Note that this is an $\mathbf{LK}^{++}$ and not an $\mathbf{LK}^{+}$-derivation, by $\forall_1, \forall_2$.

Note that the immediate Deskolemization of cut-free **LK**-proofs into cut-free $\mathbf{LK}^{++}$-proofs is also a Deskolemization for the usual structural Skolemization. However, the transformation of $\mathbf{LK}^{++}$-proofs into **LK**-proofs is not elementarily bounded and therefore, the Deskolemization for structural Skolemization as in [4] is preferable, which is clearly elementary.

4 Cut-Free LK-Proofs with Weak Quantifiers and Resolution

As we are interested in this paper mainly in the impact of different forms of Skolemization we allow any elementary form of clause form constructions. (For the purpose of this paper it is not necessary to specify the exact formal of resolution proofs, as they simulate each other within elementary bounds in the complexity of the proofs.)

Definition 8. *Let A be a formula which contains only weak quantifiers when written left of the sequent and consequently only strong quantifiers when written on the right side of the sequent. An admissible clause form construction consists of sequents $A \vdash C$ and $C \vdash A$ elementary in the complexity of A, where*

1. *C (the clause form) is a conjunction of a universally quantified disjunctions of literals (negated or unnegated atomic formulas),*
2. *$A \vdash C$ and $C \vdash A$ are cut-free elementary derivable in the complexity A.*

Remark 3. Note that both, structural clause forms and standard clause forms, fall under this definition, together with clause forms which allow for atom evaluation ect. [6].

Theorem 6.

1. *Let φ be a cut-free* **LK***-proof of the sequent*

$$A_1, \ldots, A_n \vdash B_1, \ldots, B_m$$

with weak quantifiers only. Then there is a resolution refutation of an admissible clause form of

$$A_1 \wedge \ldots \wedge A_n \wedge \neg B_1 \wedge \ldots \wedge \neg B_m$$

elementary in the complexity of φ*.*

2. *Let* φ' *be a resolution refutation of an admissible clause form of*

$$A_1 \wedge \ldots \wedge A_n \wedge \neg B_1 \wedge \ldots \wedge \neg B_m.$$

Then there is a cut-free **LK***-proof of*

$$A_1, \ldots, A_n \vdash B_1, \ldots, B_m$$

with weak quantifiers only elementary in the complexity of φ'*.*

Proof.

1. We consider the **LK**-proof consisting of the proof of

$$A_1 \wedge \ldots \wedge A_n \wedge \neg B_1 \wedge \ldots \wedge \neg B_m \vdash$$

where we cut with $C \vdash A_1 \wedge \ldots \wedge A_n \wedge \neg B_1 \wedge \ldots \wedge \neg B_m$ for the admissible clause form C. Both proofs are elementary in the complexity of the proof of

$$A_1, \ldots, A_n \vdash B_1, \ldots, B_m,$$

as the admissible clause form leads to a cut-free derivation of

$$C \vdash A_1 \wedge \ldots \wedge A_n \wedge \neg B_1 \wedge \ldots \wedge \neg B_m$$

elementary in the proof of

$$A_1, \ldots, A_n \vdash B_1, \ldots, B_m.$$

(The complexity of $A_1 \wedge \ldots \wedge A_n \wedge \neg B_1 \wedge \ldots \wedge \neg B_m$ is majorized by the complexity of the proof of $A_1, \ldots, A_n \vdash B_1, \ldots, B_m$.) As

$$A_1 \wedge \ldots \wedge A_n \wedge \neg B_1 \wedge \ldots \wedge \neg B_m \vdash$$

contains only weak quantifiers it has an elementary expansion of the form $E \vdash$. We mimic this expansion with the strong quantifiers on the left side of the cut by duplicating them up to the limit of the expansion. This is clearly an elementary transformation.

We project the strong quantifiers on the left side with the uniquely corresponding terms of the expansion on the right side. We eliminate this propositional cut at most double exponential expense and obtain a proof of $C \vdash$ elementarily bounded in the complexity of the original proof of $A_1, \ldots, A_n \vdash B_1, \ldots, B_m$.

Again, we construct the elementary expansion of C', now the ground clauses. There is therefore an elementary bound in the number of different atoms in the grounded clause form which leads to an elementary resolution refutation by considering the Herbrand tree.

2. Assume a resolution refutation of the clauses in C originating from

$$C \rightarrow A_1 \wedge \ldots \wedge A_n \wedge \neg B_1 \wedge \ldots \wedge \neg B_m$$

and

$$A_1 \wedge \ldots \wedge A_n \wedge \neg B_1 \wedge \ldots \wedge \neg B_m \rightarrow C$$

being elementary derivable. We obtain therefore elementary cut-free proofs of

$$A_1 \wedge \ldots \wedge A_n \wedge \neg B_1 \wedge \ldots \wedge \neg B_m \rightarrow C_i$$

for the singular clauses C_i after deletion of universal quantifiers. We write unnegated atoms on the right side and negated atoms unnegated on the left side of the sequent. After instantiating the ground substitutions we obtain elementary derivations of

$$A_1 \wedge \ldots \wedge A_n \wedge \neg B_1 \wedge \ldots \wedge \neg B_m \vdash$$

with atomic cuts. These atomic cuts can be eliminated at exponential expense.

5 Andrews Skolemizations Allows for Non-elementarily Shorter Resolution Refutations

Independently of the concrete clause form transformation after the Skolemization we obtain the following result, which emphasizes the difference of structural and Andrews Skolemization.

Theorem 7. *There is a sequence of refutable formulas* $F_1, F_2, \ldots$ *with corresponding resolution refutations* $\pi_1, \pi_2, \ldots$ *with clause forms based on Andrews Skolemization such that no elementary bound in the complexity of* $\pi_1, \pi_2, \ldots$ *exists for the shortest sequence of corresponding resolution refutations based on structural Skolemization.*

Proof. Consider the sequence of **LK** and **LK**$^+$-proofs of Theorem 2. There is no elementary bound of the complexity of the smallest cut-free **LK**-proofs in terms of the corresponding cut-free **LK**$^+$-proofs. The resolution refutation generated from Andrews Skolemization corresponds elementarily to the complexity of **LK**$^+$-proofs by Theorem 4, 6, 3. The resolution refutation generated from structural Skolemization corresponds elementarily to the complexity of **LK**-proofs by Corollary 1 and [4].

Remark 4. The arguments of this paper are based on a comparison of usual Skolemization with **LK** and Andrews Skolemization with **LK**$^+$. It is work in progress to provide a direct construction which will be more involved but will provide sharper complexity bounds.

6 Conclusion

The worst case sequences constructed in this paper are highly artificial. It might be asked if they have an impact in the real world. It is however a known fact that worst case examples with extreme complexities correspond to practical examples which are not that bad, but bad enough.

References

1. Aguilera, J.P., Baaz, M.: Unsound inferences make proofs shorter. J. Symb. Log. **84**(1), 102–122 (2019)
2. Andrews, P.B.: Resolution in type theory. J. Symb. Log. **36**(3), 414–432 (1971)
3. Andrews, P.B.: Theorem proving via general matings. J. ACM **28**(2), 193–214 (1981)
4. Baaz, M., Hetzl, S., Weller, D.: On the complexity of proof deskolemization. J. Symb. Log. **77**(2), 669–686 (2012)
5. Baaz, M., Leitsch, A.: On skolemization and proof complexity. Fundam. Informaticae **20**(4), 353–379 (1994)
6. Eder, E.: Relative Complexities of First Order Calculi. Springer, Heidelberg (2013). https://doi.org/10.1007/978-3-322-84222-0
7. Orevkov, V.P.: Lower bounds for increasing complexity of derivations after cut elimination. J. Sov. Math. **20**(4), 2337–2350 (1982)
8. Statman, R.: Lower bounds on Herbrand's theorem. In: Proceedings of the American Mathematical Society, pp. 104–107 (1979)

The Isomorphism Problem for FST Injection Structures

Douglas Cenzer(✉) and Richard Krogman

University of Florida, Gainesville, FL 32611, USA
cenzer@ufl.edu
https://people.clas.ufl.edu/cenzer

Abstract. An injection structure $\mathcal{A} = (A, f)$ is a set A together with a one-place one-to-one function f. $\mathcal{A}$ is a Finite State Transducer (abbreviated FST) injection structure if A is a regular set, that is, the set of words accepted by some finite automaton, and f is realized by a deterministic finite-state transducer. Automatic relational structures have been well-studied along with the isomorphism problem for automatic structures. For an FST injection structure (A, f), the graph of f is not necessarily automatic. We continue the study of the complexity of FST injection structures by showing that the isomorphism problem for unary FST injection structures is decidable in quadratic time in the size (number of states) of the FST.

Keywords: Computability theory · Injection structures · Automatic structures · Finite state automata · Finite state transducers

1 Introduction and Preliminaries

The isomorphism problem for a class of structures is to determine from their presentations, whether two given structures are isomorphic. This is a fundamental problem in computable structure theory. It has been shown that for many classes of computable structures, the isomorphism problem is Σ^1_1 complete. This includes linear orders, trees, undirected graphs, Boolean algebras and Abelian p-groups as shown in the work of Calvert, Goncharov and Knight [4,8]. On the other hand, there are many classes of computable structures where the isomorphism problem has lower complexity. For example, it was shown by Calvert and Knight [4] that the isomorphism problems for vector spaces over a fixed infinite computable field and also for algebraically closed fields of a fixed characteristic, are Π^0_3 complete, and that the isomorphism problem for torsion-free abelian groups of rank 1 is Σ^0_3 complete. The isomorphism problem for computable equivalence structures was shown by Calvert, Cenzer, Harizanov and Morozov [3] to be Π^0_4 complete. The isomorphism problem for computable injection structures was shown by Cenzer, Harizanov and Remmel [6] to also be Π^0_4 complete.

S. Artemov and A. Nerode (Eds.): LFCS 2022, LNCS 13137, pp. 25–36, 2022.
https://doi.org/10.1007/978-3-030-93100-1_3

In some cases, the isomorphism problem for a class of automatic structures is just as complicated as for computable structures. It was shown by Khoussainov, Nies, Rubin and Stephan [12] that for many classes of automatic structures, the isomorphism problem is Σ^1_1 complete. This includes undirected graphs, successor trees, partial orders and lattices. However, there are many cases where the isomorphism problem for automatic structures has lower complexity. For example, the isomorphism problem for automatic Boolean algebras was shown to be decidable by Khoussainov, Nies, Rubin and Stephan [12]. The isomorphism problem for automatic ordinals was shown to be decidable by Khoussainov, Rubin and Stephan [14]. Liu and Minnes [16] showed that the isomorphism problem for unary automatic equivalence structures is decidable in linear time. The isomorphism problem for (binary) automatic equivalence structures was shown to be Π^0_1 complete by Kuske, Liu, and Lohrey [15]. They also showed that the isomorphism problem for automatic trees of height $n \geq 2$ is Π^0_{2n-3} complete.

The isomorphism problem for automatic injection structures is the focus of the present paper. It was shown by Blumensath and Gradel [1] and also by Khoussainov and Rubin [12] that the isomorphism problem for automatic injection structures is Π^0_1 complete. Here we will examine unary injection structures given by a Finite State Transducer, as introduced in the recent paper [2] by Buss, Cenzer, Minnes and Remmel. We show that the isomorphism problem for FST unary injection structures is decidable in quadratic time in the sizes of the two presentations.

An injection is a one-place one-to-one function and an injection structure $\mathcal{A} = (A, f)$ consists of a set A and an injection $f : A \to A$. Given $a \in A$, the orbit $\mathcal{O}_f(a)$ of a under f is

$$\mathcal{O}_f(a) = \{b \in A : (\exists n \in \mathbb{N})(f^n(a) = b \ \vee \ f^n(b) = a)\}.$$

We define the *character* $\chi(\mathcal{A})$ of an injection structure $\mathcal{A} = (A, f)$ by

$$\chi(\mathcal{A}) = \{(n, k) : \mathcal{A} \text{ has at least } n \text{ orbits of size } k\}$$

This is the natural definition since the character will be a c.e. set and furthermore any c.e. character may be realized by a computable injection structure. The set $\{(n, k) : \mathcal{A}$ has exactly n orbits of size $k\}$ will be the difference of two c.e. sets. An orbit of finite size $k \in \mathbb{N}$ will be a k-cycle of the form

$$\mathcal{O}_f(a) = \{f^i(a) : 0 \leq i \leq k - 1\},$$

where $f^k(a) = a$. Infinite orbits can have two forms. One is of type ζ, which are of the form

$$\mathcal{O}_f(a) = \{f^n(a) : n \in \mathbb{Z}\}$$

in which every element is in the range of f and $f^{-n}(a)$ for $n > 0$ refers to the unique element $b \in A$ with $f^n(b) = a$. The other is of type ω, which have the form

$$\mathcal{O}_f(a) = \{f^n(a) : n \in \mathbb{N}\}$$

for some $a \notin ran(f)$ which serves as the initial element. It is easy to see that the character of an injection structure plus the information about the number of ζ-orbits and ω-orbits completely characterizes its isomorphism type.

The algorithmic properties of injection structures were studied by Cenzer, Harizanov and Remmel [5,6]. They characterized computably categorical injection structures, and showed that they are all relatively computably categorical. Among other things, they proved that a computable injection structure $\mathcal{A}$ is computably categorical if and only if it has finitely many infinite orbits. They also showed that the character of any computable injection structure is a c.e. set and that any c.e. character may be realized by a computable injection structure. They found the complexity of a number of so-called index sets for computable injection structures, showing for example that the family of structures with at most m orbits of type ω is Π^0_3 complete and that the family of structures with finitely many orbits of type ω is Σ^0_4 complete. This culminated in the result that the isomorphism problem for computable injection structures is Π^0_4 complete. In the present paper, we will give an algorithm which computes the type of a given unary injection structure from its presentation in quadratic time, which implies that each of the above problems is decidable in quadratic time.

$\mathcal{A}$ is a Finite State Transducer (abbreviated FST) injection structure if A is a regular set, that is, the set of words accepted by some finite automaton, and f is realized by a finite-state transducer. It was shown in [2] that the model checking problem for FST injection structures is undecidable, contrasting with the fact that the model checking problem for automatic relational structures is decidable. They also explored which isomorphism types of injection structures can be realized by FST injections, in particular characterizing the types that can be realized by FST injection structures over a unary alphabet. They showed that any FST injection structure is isomorphic to an FST injection structure over a binary alphabet, and gave a number of results about the possible isomorphism types of FST injection structures over an arbitrary alphabet.

The authors continued this study in [7]. It was shown that the universal countable injection structure, which has infinitely many orbits of each possible finite and infinite type, has an FST presentation. There are in fact two such FST presentations such that the injection structures so defined are not computably isomorphic.

An injection structure $\mathcal{A}$ is said to be *graph relational* or *graph automatic* if the graph G_f is accepted by a two-tape DFA. It was shown in [7] that the notions of FST injection structures and graph automatic are distinct, that is, there is an FST computable injection structure which has no graph automatic presentation, and there is a graph automatic injection structure which has no FST computable presentation. An injection structure (A, f) is said to have *bounded growth* if there is a constant c such that, for all $w \in A$, $|w| - c \leq |f(w)| \leq |w| + c$. It was shown in [7] that an FST injection structure is graph relational if and only if it has bounded growth.

A structure $\mathcal{A} = (A, f)$ is said to be *unary* if the universe A is a subset of $\{1\}^* = \{1^n : n \in \mathbb{N}\}$. Note that in a unary alphabet, we may identify the string 1^n with the number n, and accordingly identify the isomorphic prefix relation $\sqsubseteq$ with the usual ordering on $\mathbb{N}$, $\leq$. The possible types of FST injection structures over a unary alphabet were characterized in [2] as one of three possible types:

1. A structure consisting of infinitely many cycles of length 1;
2. For each finite $m \geq 0$ and each $n \geq 1$, a structure with m orbits of length 1 and n orbits of type ω;
3. For each finite m, a structure with m orbits of length 1 and infinitely many orbits of type ω;

This was improved in [7] for graph relational unary FST structures, where only the first two types are possible. In either case, the character of an FST injection structure has trivial complexity. For binary structures, it was shown in [7] that the character of a graph relational FST injection structure is decidable in exponential time. In particular, $\{(a, k) : |\mathcal{O}_f(a)| = k\}$ is computable in exponential time and, for graph automatic FST structures, it is computable in quadratic time. This leads to some results about the complexity of isomorphisms between FST structures. It was shown that isomorphic unary FST injection structures are exponential time isomorphic, and furthermore, for graph relational structures, they are quadratic time isomorphic. It is shown that not all isomorphic pairs of FST injection structures are computably isomorphic. Better results are given for structures with full universe $\{0, 1\}^*$. Such structures are shown to have no orbits of type ζ. Any two such graph relational FST structures which are isomorphic are shown to be double exponential time isomorphic.

Here are some needed definitions and terminology. Let $\mathbb{N} = \{0, 1, 2, \ldots\}$ denote the natural numbers and $\mathbb{Z} = \{0, \pm 1, \pm 2, \ldots\}$ denote the integers. Let ω denote the order type of $\mathbb{N}$ under the usual ordering and ζ denote the order type of $\mathbb{Z}$ under the usual ordering. Let ϵ denote the empty word and for any word $w = w_1 \ldots w_n$, let $|w| = n$ denote the length of w. For any finite nonempty set Σ, let Σ^* denote the set of all words over the alphabet Σ, let $\Sigma^+ = \Sigma^* \setminus \{\epsilon\}$. For any $n \in \mathbb{N}$, let $\Sigma^n = \{w \in \Sigma^* : |w| = n\}$ and let $\Sigma^{\leq n} = \{w \in \Sigma^* : |w| \leq n\}$. For a string $w = a_0 a_1 \ldots a_k \neq \varepsilon$, we will let $w^- = a_1 a_2 \ldots a_k$.

A deterministic finite automaton (DFA) is specified by the tuple $M = (Q, \iota, \Sigma, \delta, F)$ where Q is the finite set of states, ι is the initial state, Σ is the input alphabet, $\delta : Q \times \Sigma \to Q$ is the (possibly partial) transition function, and $F \subseteq Q$ is the set of final, or accepting states. The transition function may be extended to $\delta : Q \times \Sigma^* \to \Sigma^*$ by recursion on the length of a word. For $q \in Q$, $\delta(q, \epsilon) = q$, and for $w \in \Sigma^*$ and a letter $a \in \Sigma$, $\delta(q, wa) = \delta(\delta(q, w), a)$. Then $\delta(\iota, w)$ represents the final state M reaches when scanning through w while transitioning states according to δ, starting at ι. A DFA M *accepts* a string w if $\delta(\iota, w) \in F$. The set $L(M) \subseteq \Sigma^*$ of strings accepted by M is the language *recognized* by M. A language $L \subseteq \Sigma^*$ is said to be *regular* or *automatic* if it is accepted by some

DFA. To recognize a relation $R \subseteq \Sigma^* \times \Sigma^*$, we use a two-tape synchronous DFA, where the transition function $\delta : Q \times \Sigma \cup \{\Box\} \times \Sigma \cup \{\Box\} \to Q$ and $\Box$ denotes a blank square. The blank square is needed in the case that one input is longer than the other. Then M halts after reaching the end of the longer word. Automatic relations and structures have been studied by Khoussainov, Liu, Minnes, Nies, Rubin, Stephan and others [9–13,16].

A finite-state transducer (FST) is specified by the tuple

$$M = (Q, \iota, \Sigma, \Gamma, \delta, \tau),$$

where Q is the finite set of states, ι is the initial state, Σ is the input alphabet, Γ is the output alphabet, $\delta : Q \times \Sigma \to Q$ is the (possibly partial) transition function, and $\tau : Q \times \Sigma \to \Gamma^*$ is the (possibly partial) output function. Here δ extends as with an automata, and τ is extended to $\tau : Q \times \Sigma^* \to \Gamma^*$ as follows: for $q \in Q$, $\tau(q, \epsilon) = \epsilon$, and $\tau(q, wa) = \tau(q, w)\tau(\delta(q, w), a)$ for word $w \in \Sigma^*$ and letter $a \in \Sigma$. A FST M defines a (possibly partial) function, $f_M : \Sigma^* \to \Gamma^*$ with $f_M(w) = \tau(\iota, w)$. We say that the FST M *realizes*, *computes*, or *generates* a function f on a set $U \subseteq \Sigma^*$ if $f_M \restriction U = f$.

A *state cycle* of the FST M is a pair $(q, y) \in Q \times \Sigma^*$ such that $|y| > 0$ and $\delta(q, y) = q$. A word $w \in \Sigma^*$ is said to contain a state cycle if $w = xyz$ with $(\delta(\iota, x), y)$ a state cycle.

It is possible to combine the underlying automaton that accepts the domain, and the transducer that computes the function as shown in [2]. Hence, we suppose from here on, without loss of generality, that (A, f) is computed by $T = (Q, \iota, \delta, \tau, F)$ where (Q, ι, δ, F) accepts A, and (Q, ι, δ, τ) computes f.

2 The Isomorphism Problem for FST Injection Structures

We now discuss the complexity of the isomorphism problem for unary FST injection structures. We will show that this problem of determining whether two given unary FST structures are isomorphic is decidable in quadratic time in the sizes of the presentations of the two structures.

Consider $T = (Q, \iota, \delta, \tau, F)$, a unary automatic FST with all states reachable. Suppose that $|Q| = s$ and let $q : \omega \to Q$ be $q_i := \delta(\iota, i)$. Then there must exist $t < s$ such that $q_s = q_t$. Distinguish now $l = s - t$. It may be verified that for $h \in [t, t + l - 1]$, and for any $m \in \omega$, $q_{h+ml} = q_h$ so q has period l for any $a \geq t$. Note that since, for a unary FST, there is only one possible input symbol and hence only one possible string of length l, the cycle from q_s to q_t must repeat every l steps once it occurs.

Consider now $H = \{h \in [t, t + l - 1] : q_h \in F\}$, at times enumerated as $\{h_i : i \in \eta\}$. For each $h \in H$, take the arithmetic progression of period l, $G_h = \{h + ml : m \in \omega\}$. Let $L = \{a \in A : a \geq t\}$ and observe that

$$L = \{a \in A : a \geq t\} = \bigcup_{h \in H} G_h$$

Then if we take $K = \{a \in A : a < t\}$, we have

$$A = K \cup (\bigcup_{h \in H} G_h).$$

This decomposition of the domain has the following consequences for the action of f on the loop L. If $h \in H$, then

$$f(h+l) = \tau(\iota, h+l) = \tau(\iota, h) + \tau(q_h, l) = f(h) + \tau(q_h, l).$$

Since $\tau(q_h, l) = \sum_{i \in l} \tau(q_{h+i}, 1)$, and q is l-periodic, we observe, that for any $h, k \in H$, $\tau(q_h, l) = \tau(q_k, l)$. That is, after looping around by l, the transducer T will produce the same total output no matter what state it started from. If we call this loop total output constant w, then it follows that for all $h \in H$, $f(h+l) = f(h) + w$. From $q_{h+l} = q_h$ it may be seen by induction that $f(h+ml) = f(h) + mw$. Hence, $f(h+ml) = f(h) + mw$, so that the action of f on the arithmetic progressions of the loop is simply a linear relation with new slope w.

This grants the following lemma.

Lemma 1. *Let (A, f) be a unary FST injection structure with infinite domain computed by an FST T with s states. Then (A, f) consists precisely of infinitely many 1-cycles (i.e. f is the identity on A) if and only if $f(a) = a$ for all $a \in A$ with $a \leq t + (2l - 1)$.*

Proof. The forward implication is obvious, so suppose that $a \leq t+(2l-1)$ implies $f(a) = a$. Then f is the identity on T, so we need check that f is the identity on L. Note that both h and $h+l$ are less than $t + (2l - 1)$ and so $f(h) = h$ and $h + w = f(h+l) = h + l$. Thus we see that $l = w$ so now $f(h+ml) = h + ml$ for all $m \in \omega$. Hence, $f(a) = a$ for any $a \in \bigcup_{h \in H} G_h = L$.

Corollary 1. *Let (A, f) and constants s and t be as above. If (A, f) has exactly n many 1-orbits for $n \in \mathbb{N}$, then $n < t + (2l - 1)$*

Proof. If (A, f) has finitely many 1-orbits, then there exists $a \in A$ with $|a| \leq 2s - (t+1)$ and $f(a) \neq a$.

In order to determine conditions for checking the number of ω-orbits, it is useful to consider the successor function $\sigma : A \to A$, with $\sigma(a) = \min(\{b \in A : b > a\})$. Observe that $h_{i+1} = \sigma(h_i)$ for $i + 1 \leq \eta - 1$ and $\sigma(h_{\eta-1}) = h_0 + l$. Thus through iteration we see $h_i = \sigma^i(h_0)$ and

$$\sigma^\eta(h_0) = \sigma(\sigma^{\eta-1}(h_0)) = \sigma(h_{\eta-1}) = h_0 + l.$$

The successor satisfies the following properties.

Lemma 2. *For $a \in L$, $m, j \in \omega$, and $i \in \eta$,*

1. $\sigma^j(a + ml) = \sigma^j(a) + ml$.
2. $\sigma^{(m\eta)}(a) = a + ml$.
3. $\sigma^{(m\eta+i)}(h_0) = h_i + ml$.

Proof. If $a+l < b = \sigma(a+l) < \sigma(a)+l$, then say $b = c+l$. Then $q_c = q_{c+l} \in F$, and we have $c \in A$ with $a < c < \sigma(a)$, which is absurd. Hence, $\sigma(a+l) = \sigma(a)+l$. Proceed by induction on m and j to acquire (1).

Take $a = h_i + ml \in L$. Now $\sigma^\eta(h_i + ml) = \sigma^\eta(h_i) + ml$ by (1). As $\sigma^\eta(h_i) = \sigma^\eta(\sigma^i(h_0)) = \sigma^i(\sigma^\eta(h_0)) = h_i + l$, we get $\sigma^\eta(a) = h_i + l + ml = a + l$. Once again, continue with induction to prove (2).

Part (3) is an obvious application of (1) and (2).

Definition 1. *Say that T has no loop gaps, if for all $h \in H$, $f(\sigma(h)) = \sigma(f(h))$. That is, f maps consecutive elements of H to consecutive elements of the range.*

That T has no loop gaps would guarantee that there will be no "gaps" in the range after a certain point, to avoid the possibility of having more least elements of a distinct ω-orbit. This is indeed the case, but note that a similar intuition could be held with the loop output constant w. If $w = l$, then the output cannot spread out and create more gaps. It turns out that this is equivalent.

Theorem 1. *For a transducer T decomposed as above, the following are equivalent:*

1. *T has no loop gaps.*
2. *$w = l$.*
3. *$A - f[A]$ is finite, and hence (A, f) has finitely many ω-orbits. In particular, the initial elements of these ω-orbits are $\leq f(h_0)$.*

Proof. We first establish the equivalence of 1 and 2, and then show that (1) is equivalent to (3).

(1) $\Rightarrow$ (2). Suppose that T has no loop gaps. For $i \in \eta$, let $k_i = f(h_i)$. By hypothesis, $\sigma(f(h_i)) = f(\sigma(h_i))$. Hence, $\sigma(k_i) = f(h_{i+1}) = k_{i+1}$ for $i+1 \leq \eta$. It follows via induction that $k_i = \sigma^i(k_0)$, so that $k_{\eta-1} = \sigma^{\eta-1}(k_0)$. Applying the hypothesis to $\eta - 1$ we see

$$\sigma^\eta(k_0) = \sigma(\sigma^{\eta-1}(k_0)) = \sigma(f(h_{\eta-1})) = f(\sigma(h_{\eta-1})) = f(h_0 + l) = k_0 + w.$$

But now $k_0 + l = \sigma^\eta(k_0) = k_0 + w$, so $l = w$.

(2) $\Rightarrow$ (1) Consider first an arbitrary sequence $\{a_i : i \in \eta\}$ of distinct elements of L with $a_i < a_{i+1}$ such that $a_{\eta-1} - a_0 < l$. If we take $p_i := q_{a_i}$, then from the facts that $a_{\eta-1} - a_0 < l$, $|H| = \eta$, and q has period l, we see that the p_i are distinct and comprise the entirety of the accepting loop states. If for some

$i+1 \leq \eta-1$, we have $b=\sigma(a_i)<a_{i+1}$, then $q_b \notin \{p_i : i \in \eta\}$, which is absurd. Hence, it must be that for such a sequence, $\sigma(a_i)=a_{i+1}$, and it can be similarly observed that $\sigma(a_{\eta-1})=a_0+l$.

Now suppose that $w=l$. Then $h_{\eta-1}<h_0+l$ implies that $k_{\eta-1}<f(h_0+l)=k_0+l$. If follows that $\{k_i\}_{i\in\eta}$ satisfies the above properties, and so $\sigma(k_i)=k_{i+1}$ and $\sigma(k_{\eta-1})=k_0+l$, which is equivalent to $\sigma(f(h_i))=f(\sigma(h_i))$ for $i\in\eta$.

(3) $\Rightarrow$ (1). We prove the contrapositive. Suppose that $i\in\eta$ is such that $f(h_i)<b=\sigma(f(h_i))<f(\sigma(h_i))$, so $b\in A-f[A]$. For any $m\in\omega$,

$$f(h_i+ml^2)=f(h_i)+\tau(q_{h_i},ml^2)=f(h_i)+mlw.$$

Likewise,

$$f(\sigma(h_i+ml^2))=f(\sigma(h_i)+ml^2)=f(\sigma(h_i))+mlw,$$

which implies

$$f(h_i+ml^2)<b+mlw<f(\sigma(h_i+ml^2)).$$

It follows that $\{b+mlw : m\in\omega\}\subseteq A-f[A]$ so $A-f[A]$ is infinite.

(1) $\Rightarrow$ (3). Suppose that T has no loop gaps, in which case we saw that $w=l$. Then from $\sigma(f(h_i))=f(\sigma(h_i))$ for $i\in\eta$, and induction, we have $k_i=\sigma^i(k_0)$, and hence for $m\in\omega$, $k_i+ml=\sigma^{(m\eta+i)}(k_0)$. As $w=l$,

$$f(\sigma^{(m\eta+i)}(h_0))=f(h_i+ml)=k_i+ml=\sigma^{(m\eta+i)}(k_0)=\sigma^{(m\eta+i)}(f(h_0)).$$

For any $j\in\omega$, we can divide by η to get $j=m\eta+i$ with $i\in\eta$, and so we have shown that $\sigma^i(f(h_0))=f(\sigma^i(h_0))\in f[A]$. As $a>f(h_0)$ if and only if $a=\sigma^j(f(h_0))$ for some j, we have shown that

$$\{a\in A : a\geq f(h_0)\}=\{\sigma^j(f(h_0)) : j\in\omega\}\subseteq f[A]$$

so $A-f[A]\subseteq\{a\in A : a<f(h_0)\}$ is finite.

Note that by the given types satisfiable by unary FST injection structures, we may encode the information about the type of a structure as a tuple (n,m) where n is the number of 1-orbits, and m is the number of ω-orbits. Then two structures are isomorphic if and only if their tuples are equal.

Theorem 2. *The isomorphism problem for unary FST injection structures is decidable in quadratic time in the sizes of the finite transducers.*

Proof. Recall that any FST injection structure has one of three forms:

1. A structure consisting of infinitely many cycles of length 1;
2. For each finite $m\geq 0$ and each $n\geq 1$, a structure with m orbits of length 1 and n orbits of type ω;

3. For each finite m, a structure with m orbits of length 1 and infinitely many orbits of type ω;

Thus it suffices to show that we can compute the type of a given FST injection structure from its presentation in quadratic time. Then to see whether two given FST injection structures are isomorphic we just compute the types and check that they have the same type.

Let the finite transducer T be given which computes an automatic injection structure (A, f). The presentation of T as a table lists the states and the values $\tau(q)$ and $\delta(q)$ for each state q when reading the input 1, and also indicates which states are accepting. Thus we can find the value of s and then compute the sequence

$$q_0 = \iota, q_1 = \tau(q_0), \ldots, q_s = \tau(q_{s-1})$$

At the same time we are computing the sequence

$$f(0) = 0, f(1) = \delta(q_0), f(2) = f(1) + \delta(q_1), \ldots, f(s) = f(s-1) + \delta(q_{s-1}).$$

Thus for each $i < s$, we can determine whether i is accepted by T and find the values $f(i)$. (Here we identify the string 1^i with the number i.) These computations require scanning the table s times and therefore can be done in quadratic time, since $s < |T|$. While doing this computation, we observe the value of $t < s$ such that $q_s = q_t$, so that the length ℓ of the loop is $s - t$ and the output w added during the loop is $f(s) - f(t)$.

To check whether (A, f) has finitely many or infinitely many infinite orbits, we simply observe whether $w = \ell$. If $w \neq \ell$, then (A, f) has infinitely many infinite orbits by Theorem 1. If $w = \ell$, then, again by Theorem 1, the number of infinite orbits is given by $n = |A \setminus \{f(a) : a \in A, a \leq h_0\}|$.

To determine the number of finite orbits, we continue the computation using T to find $f(s), f(s+1), \ldots, f(s+\ell-1)$ and check whether $f(a) = a$ for all $a \in A$ with $a < s+\ell$. If so, then (A, f) consists exactly of infinitely many orbits of size 1, by Lemma 1. If not, then the number of finite orbits is $m = |\{a \in A : a < s + \ell \;\&\; f(a) = a\}|$. It is clear that this additional computation can be done in quadratic time in the size of T.

The computation given in the proof above is formalized by the Algorithm 1 below.

Algorithm 1

```
Initiate p = (0,0), an empty array of tuples R, and m = 0;
if H = ∅ then
    let r = 0;
    for 0 ≤ i < t do
        if q_i ∈ F then
            add 1 to r;
        end
    end
    Output: p = (r, 0)
end
if H ≠ ∅ then
    set h = min(H);
    set m = 0;
    for 0 ≤ i ≤ s do
        add τ(q_i, 1) to m;
        if i = t − 1 then
            let a = m;
        end
        if i = h − 1 then
            let g = m;
        end
        if i = s − 1 then
            let b = m;
        end
    end
    if b − a > 0 then
        set p_2 = ∞;
    end
    set n = 0;
    for i < g do
        if n = i then
            increment p_1 by 1;
        end
        if n ≠ i & p_2 ≠ ∞ then
            append (i, n) to the end of R;
        end
        add τ(q_i, 1) to n;
    end
    if R = ∅ then
        Output: p = (∞, 0);
    end
    if p_2 ≠ ∞ then
        for (a_i, b_i) ∈ R do
            set c = 0;
            for i < j do
                if a_i = b_j then
                    increment c by 1;
                end
            end
            if c = 0 then
                increment p_2 by 1;
            end
        end
        Output: (p_1, p_2)
    end
end
```

Theorem 3. *Algorithm 1 computes the characteristic tuple of a structure.*

Proof. If $H = \emptyset$, then $L = \emptyset$ so we are working with a finite function on finite domain, and f is the identity. Hence, we must simply count the number of elements. If nonempty, the algorithm computes $f(h_0)$, and determines whether or not $\tau(q_t, l) = w$, so it can determine whether there are infinitely many ω orbits. It then counts the amount of 1-cycles before $f(h_0)$ and puts elements not mapped to themselves, along with their output, into the register R. If the register is empty, it must have been the case that f is the identity so there are infinitely many 1 cycles. If not, then we must count, using the register, the number of elements not mapped to. The above results show that this is sufficient.

3 Conclusions and Further Research

We have shown that the isomorphism problem for unary FST injection structures is decidable in quadratic time. Ongoing research continues on the isomorphism problem for unary graph automatic injection structures. We are also investigating special cases of binary FST injection structures, where the domain has a restricted form, such as the full binary set $\{0,1\}^*$ Other structures with functions are of interest, including trees given by a predecessor function, Boolean algebras, and various groups. Another topic of investigation is structures presented by other types of finite state machines, such as Wheeler automata, or pushdown automata.

An important question remains whether every FST injection structure (A, f) has a decidable theory.

References

1. Blumensath, A., Gradel, E.: Finite presentations of infinite structures: automata and interpretations. Theor. Comput. Syst. **6**, 641–674 (2004)
2. Buss, S., Cenzer, D., Minnes, M., Remmel, J.B.: Injection structures specified by finite state transducers. In: Day, A., Fellows, M., Greenberg, N., Khoussainov, B., Melnikov, A., Rosamond, F. (eds.) Computability and Complexity. LNCS, vol. 10010, pp. 394–417. Springer, Cham (2017). https://doi.org/10.1007/978-3-319-50062-1_24
3. Calvert, W., Cenzer, D., Harizanov, V., Morozov, A.: Effective categoricity of equivalence structures. Ann. Pure Appl. Logic **141**, 61–78 (2006)
4. Calvert, W., Knight, J.F.: Classification from a computable viewpoint. Ann. Pure Appl. Logic **141**, 191–218 (2006)
5. Cenzer, D., Harizanov, V., Remmel, J.B.: σ_1^0 and π_1^0 structures. Ann. Pure Appl. Logic **162**, 490–503 (2011)
6. Cenzer, D., Harizanov, V., Remmel, J.B.: Computability theoretic properties of injection structures. Algebra Logic **53**, 39–69 (2014)
7. Krogman, R., Cenzer, D.: Complexity and categoricity of injection structures induced by finite state transducers. In: De Mol, L., Weiermann, A., Manea, F., Fernández-Duque, D. (eds.) CiE 2021. LNCS, vol. 12813, pp. 106–119. Springer, Cham (2021). https://doi.org/10.1007/978-3-030-80049-9_10

8. Goncharov, S.S., Knight, J.F.: Computable structure and anti-structure theorems. Algebra Log. **6**, 351–373 (2002)
9. Khoussainov, B., Liu, J., Minnes, M.: Unary automatic graphs: an algorithmic perspective. Math. Struct. Comput. Sci. **19**(1), 133–152 (2009)
10. Khoussainov, B., Minnes, M.: Model-theoretic complexity of automatic structures. Ann. Pure Appl. Logic **161**(3), 416–426 (2009)
11. Khoussainov, B., Nerode, A.: Automatic presentations of structures. In: Leivant, D. (ed.) LCC 1994. LNCS, vol. 960, pp. 367–392. Springer, Heidelberg (1995). https://doi.org/10.1007/3-540-60178-3_93
12. Khoussainov, B., Nies, A., Rubin, S., Stephan, F.: Automatic structures: richness and limitations. Log. Meth. Comput. Sci. **2**(2), 18 (2007). Special issue: Conference "Logic in Computer Science 2004"
13. Khoussainov, B., Rubin, S., Stephan, F.: On automatic partial orders. In: Proceedings of the LICS 2003, pp. 168–177 (2003)
14. Khoussainov, B., Rubin, S., Stephan, F.: Automatic linear orders and trees. ACM Trans. Comput. Log. **6**, 675–700 (2005)
15. Kuske, D., Liu, J., Lohrey, M.: The isomorphism problem on classes of automatic structures with transitive relations. Trans. Amer. Math. Soc. **365**, 5103–5151 (2013)
16. Liu, J., Minnes, M.: Deciding the isomorphism problem in classes of unary automatic structures. Theoret. Comput. Sci. **412**(18), 1705–1717 (2011)

Justification Logic and Type Theory as Formalizations of Intuitionistic Propositional Logic

Neil J. DeBoer(✉)

The Ohio State University, Columbus, OH 43210, USA
deboer.15@osu.edu

Abstract. We explore two ways of formalizing Kreisel's addendum to the Brouwer-Heyting-Kolmogorov interpretation. To do this we compare Artemov's justification logic with simply typed λ calculus, by introducing a map from justification terms into λ terms, which can be viewed as a method of extracting the computational content of the justification terms. Then we examine the interpretation of Kreisel's addendum in justification logic along with the image of the resulting justification terms under our map.

Keywords: Justification logic · Type theory · Intuitionistic logic · λ calculus · BHK interpretation

1 Introduction

In [2] Artemov introduces justification logic, in particular the Logic of Proofs. In this paper he claims that the logic of proofs can be viewed as a formalization of the BHK interpretation of intuitionistic logic. In this paper we will examine this claim in detail. In particular we compare a fragment of the intutionistic propositional Logic of Proofs with the simply typed λ-calculus. To do this we will first present a general framework for examining formalizations of the BHK interpretation. Then we will introduce the implicational fragment of the Logic of Proofs, and provide a Kripke style semantics based on Fitting's semantics for classical justification logic. Then we examine how justification logic and the simply typed λ calculus are related as interpretation of the BHK interpretation for impicational propositional statements. These are results from my PhD thesis [7].

1.1 The BHK Interpretation and Its Formalizations

We are concerned with formalizations of the BHK interpretation, specifically for the implicational condition which is as follows:

$C_{\to}$: A proof of the proposition "P implies Q" is a construction that will take any proof of P and produce a proof of Q.

S. Artemov and A. Nerode (Eds.): LFCS 2022, LNCS 13137, pp. 37–51, 2022.
https://doi.org/10.1007/978-3-030-93100-1_4

This condition has been subject to criticism. One such criticism goes as follows: Suppose that we are given a construction, let us call it c, that is a purported proof of a proposition of the form "P implies Q" and we wish to check whether this is the case. We will have to look at every proof of P and check to see whether c applied to those proofs is always a proof of Q; the problem with this being that unless we have some knowledge about the collection of all proofs of P it is not immediately apparent that this is possible. This means that we won't always be able to determine whether a given proof of a proposition is valid, and could result in non-recursive systems that satisfy condition $C_{\rightarrow}$.

This criticism lead to a reformulation of the conditions for implication, attributed to Kreisel, which can be seen in [13], and we will formulate as follows:

$C'_{\rightarrow}$: A proof of the proposition "P implies Q" is a construction that will take any proof of P and produce a proof of Q, and a verification that the construction satisfies these conditions.

We will refer to the propositional BHK interpretation with conditions $C_{\rightarrow}$ replaced with $C'_{\rightarrow}$ as the Brouwer-Heyting-Kolmogorov-Kreisel interpretation or BHKK interpretation.

We will explore the terminology of conditions $C_{\rightarrow}$ and $C'_{\rightarrow}$ in more detail. The first word that occurs that we wish to discuss is "proof" and the conditions of the BHK interpretation can be seen as a description of what needs to be done in order to construct a "proof" of a "proposition". That is, we are giving a definition of a "proof".

The next word from the BHKK interpretation that we wish to discuss is "proposition" which has a long history dating back to Zeno of Citium and the Stoic philosophers. According to the Stanford Encyclopedia of Philosophy, [16], propositions are "the sharable objects of the attitudes and the primary bearers of truth and falsity" and in [12] Heyting characterizes mathematical propositions as "the expectation to find a certain condition fulfilled". These seem like two different definitions, but recall that according to Brouwer "there are no non-experienced truths" [6], so under the lens of intuitionism these definitions can be viewed as complementary. The precise nature of propositions with regards to the BHKK Interpretation has not been agreed upon, and later in this section we will explore different attitudes that people have taken with regards to the nature of propositions.

The next word we wish to discuss is "construction". This word has a dubious usage in mathematics; it is typically used as a synonym for the value of a function or procedure. We will be exploring different formalizations of "constructions" in the context of the BHKK interpretation.

In discussions of systems meant to formalize BHKK interpretation there are two important questions that help characterize our attempts at formalization.

Q1. Is the statement "the construction c proves the proposition P" a proposition?

Q2. From the perspective of the BHKK interpretation, is there a fundamental difference between verifications in $C'_{\rightarrow}$ and proofs? (That is, should they be treated as the same kind of object?)

We will explore these questions, while briefly mentioning some of the formal systems that represent various answers to these question, keeping in mind that we wish to further our understanding of the additional condition in $C'_{\rightarrow}$.

We will begin with **Q1**. For the sake of brevity let $c : P$ be an abbreviation of the statement "the construction c is a proof of the proposition P". In considering possible answers to this question, we could imagine the case where only some statements of the form $c : P$ are propositions. We are, however, unaware of any systems that have this property, so we will only consider the cases in which either all statements of the form $c : P$ are propositions or no statements of that form are propositions.

If we answer **Q1** in the affirmative, then statements of the forms "$d : c : P$" and "if $c : P$ then Q" are propositions as well. In terms of formal systems exploring the BHK interpretation that assume a 'Yes' answer to **Q1**, there is the Kreisel-Goodman formulation [11] which is related to a system given by Beeson in chapter XVII of [5]. More recently, in [2], Artemov created a formal system called Justification Logic, which has this property. This system will be the focus of our investigations. Justification Logic will be discussed in more detail in the following section.

If we answer question **Q1** in the negative, then we have started on the road to modern "propositions-as-types" type theory. To use the terminology of Martin-Löf, $c : P$ is a judgment (a statement about a proposition) and not a proposition (see the introduction of [15] for more on the distinction between propositions and judgments). This means that a statement of the form $d : c : P$ is no longer a proposition. There is an abundance of formal type theories that have this perspective on propositions, see [17] for some examples.

Now let us explore how the systems we have mentioned answer **Q2**. Both Artemov's and Beeson's systems treat proofs and verifications as the same kind of object (i.e. 'No' to **Q2**). Type theories, however, do not treat proofs and verifications as the same kind of object (i.e. 'Yes' to **Q2**). To illustrate this let us consider an example taken from simply typed λ-Calculus, which is defined in a later section:

$$\cfrac{\cfrac{x : P, y : Q \vdash x : P}{x : P \vdash \lambda y.x : Q \to P}}{\vdash \lambda xy.x : P \to Q \to P}$$

Here the proposition we wish to prove is $P \to Q \to P$, and our 'proof' of this proposition is the λ-term $\lambda xy.x$, whereas our 'verification' is the above derivation. We can organize this, along with these systems positions on **Q1**, into the following table:

Q1 \ **Q2**	Yes	No
Yes	**R1**	Justification Logic
No	Type Theory	**R4**

I am unaware of any systems in **R1** or **R4**, and in this paper we will not go into the reasons that people seem uninterested in systems with these properties.

2 Justification Logic

In this paper we will be concerned with Condition $C'_{\rightarrow}$ of the BHKK Interpretation, so our focus will be on implication. Since we are concerned with implication we will only use the implicational fragment of intuitionistic logic. We will abbreviate the implicational fragment of intuitionistic logic as $\text{IPL}_{\rightarrow}$ and we will write $\Gamma \vdash_i \varphi$ to indicate that the sequent $\Gamma \vdash \varphi$ of implicational formulas is derivable in the implicational fragment of intuitionistic logic.

Next we will introduce the implicational fragment of intuitionistic justification logic. In formalizing proofs for justification logic we have two notions of 'atomic' proofs, being proofs of axioms and arbitrary proofs of propositions we are assuming. We call the proofs of axioms justification constants and the proofs of assumptions justification variables. We then form the set of justification terms by including a binary operator representing the application of one proof to another proof $\bullet$, another binary operation representing the combination of two proofs $+$, and a unary operator representing checking a proof !. These definitions are taken from [2]. See [3] for a more recent exploration of justification logic.

Definition 1. *Let $c_0, c_1, \ldots$ be a set of symbols, which will be called the* ***justification constants****, abbreviated C_j. Let $x_0, x_1, \ldots$ be a list of symbols, which will be called the* ***justification variables****, abbreviated Var. The collection of* ***justification terms*** *are defined as follows:*

$$JT ::= C_j | Var | JT \bullet JT | JT + JT | !JT$$

Definition 2. *We define the set of free variables of a justification term r, written $FV(r)$, recursively on the structure of r as follows:*

- *If r is the variable x then $FV(r) = x$.*
- *If r is the constant c then $FV(r) = \varnothing$.*
- *If $r = u + t$ or $r = u \bullet t$ then $FV(r) = FV(u) \cup FV(t)$.*
- *If $r =!t$ then $FV(r) = FV(t)$.*

Definition 3. *Let $p_0, p_1, \ldots$ be a list of symbols, which will be called propositional variables, written as P_{Var}. The collection of* ***propositional justification formulas*** *are defined as follows:*

$$PJ ::= P_{Var} | PJ \rightarrow PJ | JT : PJ$$

Next we present the rules for the implicational fragment of intuitionistic justification logic. A presentation of the full propositional justification logic is given in [1].

Definition 4 (Axioms of $\text{ILP}_{\rightarrow}$).

Justification Axiom P1 $\varphi \rightarrow \psi \rightarrow \varphi$
Justification Axiom P2 $(\varphi \rightarrow \theta \rightarrow \psi) \rightarrow (\varphi \rightarrow \theta) \rightarrow \varphi \rightarrow \psi$
Justification Axiom J1 $t : \varphi \rightarrow \varphi$
Justification Axiom J2 $t : \varphi \rightarrow !t : t : \varphi$

Justification Axiom J3 $t : (\varphi \to \psi) \to r : \varphi \to t \bullet r : \psi$
Justification Axiom J4 $t : \varphi \to r + t : \varphi$
Justification Axiom J5 $t : \varphi \to t + r : \varphi$

Definition 5. *A* ***constant specification*** *is a set of pairs of the form (c, φ) where c is a justification constant and φ is an instance of an Axiom from Definition 4. A constant specification, CS, is said to be* ***axiomatically appropriate*** *if for each axiom φ that is an instance of a scheme from Definition 4 there is a constant c so that $(c, \varphi) \in CS$.*

Definition 6. *Given a constant specification CS we define a system of rules, called* ***intuitionistic implicational logic of proofs*** *with respect to CS, by the following rule schema:*

Identity: $\Gamma \vdash \varphi$ $(\varphi \in \Gamma)$.
Axiom: $\Gamma \vdash \varphi$ *where φ is an instance of an axiom scheme given in Definition 4.*
Necessitation: $\Gamma \vdash c : \varphi$ $((c, \varphi) \in CS)$.
Modus Ponens: $\dfrac{\Gamma \vdash \varphi \to \psi \qquad \Gamma \vdash \varphi}{\Gamma \vdash \psi}$

We use the notation $\Gamma \vdash^{CS}_{iLP} \varphi$ to indicate that the sequent $\Gamma \vdash \varphi$ is provable in the intuitionistic implicational logic of proofs with respect to CS.

We will abbreviate intuitionistic implicational logic of proofs with respect to the constant specification CS as $\mathrm{ILP}^{CS}_{\to}$.

The following two results were shown for full intitionistic justification logic in [1], and will be useful in our later analysis.

Theorem 1 (Deduction). *$\Gamma \vdash^{CS}_{iLP} \psi \to \varphi$ iff $\Gamma, \psi \vdash^{CS}_{iLP} \varphi$.*

Lemma 1. *If $\Gamma, \psi \vdash^{CS}_{iLP} \varphi$ then $\Gamma, x : \psi \vdash^{CS}_{iLP} \varphi$.*

Notice that $\mathrm{ILP}^{CS}_{\to}$ extends $\mathrm{IPL}_{\to}$, as we have the following result:

Lemma 2. *If $\Gamma \vdash_i \varphi$ then for any constant specification CS we have $\Gamma \vdash^{CS}_{iLP} \varphi$.*

Next we present a map from implicational justification formulas to implicational formulas, that can be used to show the conservativity of intuitionistic logic over justification logic.

Definition 7. *Let $\flat$ be the map from justification formulas into propositional formulas defined recursively by the following clauses:*

- $\flat(p_i) = p_i$
- $\flat(\varphi \to \psi) = \flat(\varphi) \to \flat(\psi)$
- $\flat(r : \varphi) = \flat(\varphi)$

We can extend this to a set of justification formulas Γ as follows: $\flat(\Gamma) = \{\flat(\gamma) | \gamma \in \Gamma\}$.

Lemma 3. *For any constant specification CS, if* $\Gamma \vdash^{CS}_{iLP} \varphi$ *then* $\flat(\Gamma) \vdash_i \flat(\varphi)$.

Corollary 1 (Conservativity of $\mathbf{IPL}_{\rightarrow}$ over $\mathbf{ILP}^{CS}_{\rightarrow}$). *Let CS be a constant specification. If* Γ *is a set of propositional formula and* φ *is a propositional formula so that* $\Gamma \vdash^{CS}_{iLP} \varphi$ *then* $\Gamma \vdash_i \varphi$.

Definition 8. *A justification formula* φ *is a* ***realization*** *of a propositional formula* ψ *if* $\flat(\varphi)$ *is* ψ.

Looking at the definition of $\flat$, realizations may seem like the simplest way to interpret an intuitionistic formula into our justification logic. However, after considering some of the history surrounding justification logic, it is possible to come to the conclusion that this map has a certain amount of robustness. Artemov was concerned with interpretations of intuitionistic logic that mapped through the modal logic **S4** into some formalized notion of provability. In [2] we are shown the following diagram:

$$\mathbf{Int} \hookrightarrow \mathbf{S4} \hookrightarrow \text{...?...} \hookrightarrow \mathit{REAL\ \ PROOFS}$$

Where **Int** means propositional intuitionistic logic, and the arrows represent maps that preserve provability. The map from **Int** to **S4** is Gödel's translation from intuitionistic logic to **S4**. The ...?... is revealed to be Artemov's Logic of Proofs, written here as **LP**, and REAL PROOFS is taken to be proofs in Peano Arithmetic. In the same paper Artemov also constructs the other two arrows giving us this diagram:

$$\mathbf{Int} \hookrightarrow \mathbf{S4} \hookrightarrow \mathbf{LP} \hookrightarrow \mathit{REAL\ \ PROOFS}$$

The map from **S4** to **LP** actually takes proofs in **S4** and gives provable formulas in **LP**, whereas the other maps take provable formulas in one system to provable statements in another system.

Artemov also states that the interpretation from intuitionistic logic to **LP**, given the composition of the first two maps, "provides an exact specification of **Int** by means of classical notion of proof consistent with BHK semantics". However the image of **Int** under Gödel's map can be shown to be contained in the intuitionistic fragment of **S4**, which we will denote **iS4**. Furthermore, we can show that the image of **iS4** under Artemov's map (from **S4** to **LP**) is contained in the intuitionistic fragment of **LP** (which we will denote **iLP**). This leads to the following diagram

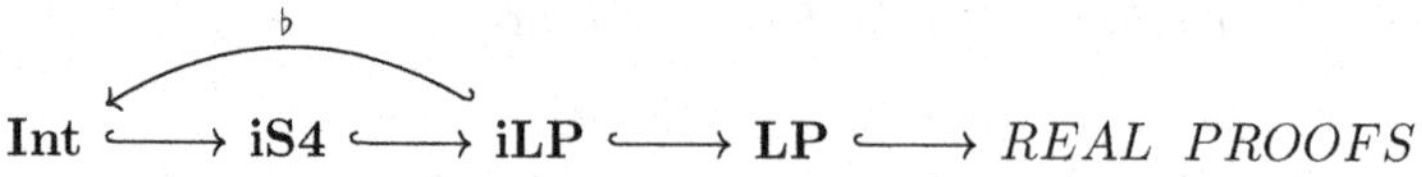

where the map from **iLP** to **LP** is inclusion, and if we compose these maps together with $\flat$, which we have depicted above, we get the identity map. Instead of the specific map from **Int** to **iLP** in the above diagram, we can concern ourselves with any map # from **Int** to **iLP**, so that $\flat \circ \#$ is the identity. In other words we are concerned with realizations that are provable in **iLP**.

2.1 Substitution

The final topic we wish to cover in this section is substitution for justification terms, as it will be useful in later discussions. These definitions were studied for classical justification logic in [14].

Definition 9 (Substitution for Justification terms). *If r and t are justification terms, and x is a variable then we define the justification term $r[x := t]$ by recursion on the structure of r, as follows:*

- *If r is a variable then* $r[x := t] = \begin{cases} t & \text{if } r = x \\ r & \text{otherwise} \end{cases}$
- *If $r = u + v$ then $r[x := t] = u[x := t] + v[x := t]$.*
- *If $r = u \bullet v$ then $r[x := t] = u[x := t] \bullet v[x := t]$.*
- *If $r = !u$ then $r[x := t] = !u[x := t]$*

Definition 10 (Substitution for Justification formulas). *If φ is a justification formula, t is a justification term, and x is a variable then we define the justification formula $\varphi[x := t]$, by recursion on the structure of φ as follows:*

- *If φ is a propositional variable then $\varphi[x := t] = \varphi$.*
- *If $\varphi = \theta \to \psi$ then $\varphi[x := t] = \theta[x := t] \to \psi[x := t]$.*
- *If $\varphi = r : \psi$ then $\varphi[x := t] = r[x := t] : \psi[x := t]$.*

We can also extend substitution to sets of justification formulas as follows. Let Γ be a set of justification formulas, then $\Gamma[x := t] = \{\gamma[x := t] | \gamma \in \Gamma\}$.

Lemma 4. *If CS is a constant specification, x is a variable, and t is a justification term then the set $CS[x := t] = \{(c, \varphi[x := t]) | (c, \varphi) \in CS\}$ is a constant specification. Moreover, if $\Gamma \vdash_{iLP}^{CS} \psi$ then $\Gamma[x := t] \vdash_{iLP}^{CS[x:=t]} \psi[x := t]$.*

Proof. This follows from a straightforward induction. □

3 Comparing Formalizations

The other way we will formalize proofs from the BHKK Interpretation is with λ calculus. In λ calculus we have variables that can represent an arbitrary proof, and we have an application operation, but we also have an abstraction operation that allows us to build function. These two operations are given computational meaning via β reduction.

The particular formal system we will be using is the Curry style simply typed λ calculus details of which can be found in [17]. We will not present the system here, and we will write $\Delta \vdash_{\lambda_\to} M : \varphi$ to indicate that for context Δ, λ term M, and implicational formula φ the sequent $\Gamma \vdash M : \varphi$ is provable in the simply typed λ calculus.

Now we can turn to the question of how the two systems we've introduced relate to the BHKK Interpretation. In order for the BHKK interpretation to interpreter intuitionistic logic it should satisfy the following condition: If $\vdash_i \varphi$ then there should be a BHKK 'proof' of φ.

For the Simply Typed λ Calculus our BHKK 'proof' is a λ term, so the above condition is satisfied by the following classic result:

Theorem 2 (Curry Howard Isomorphism). *$\Gamma \vdash_i \varphi$ iff there exists a λ term M so that $\Delta \vdash_{\lambda_\to} M : \varphi$, where Γ is the set of formulas in the context Δ.*

However at the moment we haven't show that intuitionistic justification logic has this property. In [2] Artemov showed the follow result for classical justification logic, and a similar method can be used for $\mathrm{ILP}_\to^{CS}$ for certain CS.

Lemma 5. *If CS is an axiomatically appropriate constant specification, and $x_1 : \psi_1 \ldots x_n : \psi_n, \theta_1 \ldots \theta_m \vdash_{iLP}^{CS} \varphi$ then there is a $+$-free justification term t so that $x_1 : \psi_1, \ldots, x_n : \psi_n, y_1 : \theta_1, \ldots, y_m : \theta_m \vdash_{iLP}^{CS} t : \varphi$ and $FV(t) \subseteq \{x_1, \ldots, x_n, y_1, \ldots, y_n\}$.*

By using Lemma 5 and Lemma 2 we get:

Corollary 2. *If CS is axiomatically appropriate and $\vdash_i \varphi$ then there is a $+$-free justification term r so that $\vdash_{iLP}^{CS} r : \varphi$.*

One potential objection to Theorem 2 and Corollary 2 meaning that the Simply Typed λ Calculus and $\mathrm{ILP}_\to^{CS}$ formalizing the BHKK interpretation is that our results do not yet guarantee that the BHKK 'proofs' have no free variables. The following well known result for the Simply Typed λ Calculus shows that this isn't an issue:

Lemma 6. *If $\Delta \vdash_{\lambda_\to} M : \varphi$ then $FV(M) \subseteq dom(\Delta)$, where $dom(\Delta)$ is the set of λ variable declared in Δ.*

And we can show the following result using induction over the sequent formulation of iLP_{CS} or by a trick involving Kripke style semantics for justification logic developed by Fitting in [8]:

Lemma 7. *If r is $+$-free and $\vdash_{iLP}^{CS} r : \varphi$ then $FV(r) = \varnothing$.*

Due to the expressiveness of justification logic we can attempt to formalize Condition $C'_\to$ of the BHKK interpretation in justification logic. In trying to translate this condition for the proposition $P \to Q$ in justification logic we get $t : (x : P \to r : Q)$, where x is an arbitrary proof of P, r is the construction applied to x, and t is the verification that r actually does what it is supposed to do. However this does not quite match up with the statement of $C'_\to$ since it says "...any proof of P..." which seems to indicate that we should quantify over all proofs of P. Unfortunately we have no way of quantifying over justification variables[1] in the system of justification logic we presented. We can, however, argue that our results surrounding substitution for justification variables allow us to bypass this objection.

[1] See [9] for a version of justification logic that allows this.

For the sake of simplicity let us take P and Q to be the formulas φ and ψ (respectively) both of which contain no justification terms. Furthermore let us assume that we have some constant specification CS and that we have shown $\vdash_{iLP}^{CS} t : (x : \varphi \to r : \psi)$. Let u be any justification term then by Lemma 4 we have $\vdash_{iLP}^{CS[x:=u]} t[x := u] : (u : \varphi \to r[x := u] : \psi)$. On the face of it this looks bad, since t is supposed to check that for any proof u of φ it is the case that $r[x := u]$ is a proof of ψ, so it should not depend on u. However, by Lemma 7 if we assume that t is +-free, then $FV(t) = \varnothing$ so it is straightforward to show that $t[x := u] = t$. Hence, for any justification term u we have $\vdash_{iLP}^{CS[x:=u]} t : (u : \varphi \to r[x := u] : \psi)$. Provided that $CS[x := u] \subseteq CS$ we will have $\vdash_{iLP}^{CS} t : (u : \varphi \to r[x := u] : \psi)$. This means that as long as our CS has the property that for all justification terms u we have $CS[x := u] \subseteq CS$ then we have $\vdash_{iLP}^{CS} t : (u : \varphi \to r[x := u] : \psi)$. This suggests the following definition:

Definition 11. *A* ***BHKK representation*** *of a propositional or justification formula $\varphi \to \psi$ is a justification formula of the form $t : (x : \varphi \to r : \psi)$ where r and t are +-free, $FV(t) = \varnothing$, and $FV(r) \subseteq \{x\}$.*

3.1 Comparing Proofs

Next we wish to create a framework for comparing the ways that justification terms formalize proofs with how λ-terms formalize proofs. First, we will consider a map from justification terms to λ-terms.

The naive idea is that our map, let us call it f, should respect the notion of provability. We have to decide how to formulate respecting provability for justification terms and λ-terms. Notice that for any constant specification CS, if we have $\vdash_{iLP}^{CS} r : \varphi$ then by Lemma 3 we have $\vdash_i \flat(r : \varphi)$, and since $\flat(r : \varphi) = \flat(\varphi)$ by applying Theorem 2 there is a λ-term M so that $\vdash_{\lambda_\to} M : \flat(\varphi)$. Therefore our condition on f should be that if $r : \varphi$ is provable in justification logic then the λ term $f(r)$ should inhabit $\flat(\varphi)$. Let CS be an arbitrary constant specification. This condition can be stated formally as:

Condition 1. If $\vdash_{iLP}^{CS} r : \varphi$ then $\vdash_{\lambda_\to} f(r) : \flat(\varphi)$.

There is an issue with Condition 1 in that it could be the case that r has the form $u + t$ where both $\vdash_{iLP}^{CS} u : \psi$ and $\vdash_{iLP}^{CS} t : \theta$. In such a case we have both that $\vdash_{\lambda_\to} f(u + t) : \flat(\psi)$ and that $\vdash_{\lambda_\to} f(u + t) : \flat(\theta)$. If we assume that CS is axiomatically appropriate and take ψ to be the formula $p \to p$ and take θ to be the formula $p \to q \to p$, where p and q are propositional variables, then by Lemma 5 we get the above u and t. We have both $\vdash_{\lambda_\to} f(u + t) : p \to p$ and $\vdash_{\lambda_\to} f(u+t) : p \to q \to p$, but by examining the possible normal forms of λ terms that inhabit $p \to p$ we get that $f(u+t)$ must reduce to $\lambda x.x$ (see Chap. 3 of [17] for more information about finding normal inhabitants of types). By Condition 1 we have $\vdash_{\lambda_\to} \lambda x.x : p \to q \to p$. However we cannot have $\vdash_{\lambda_\to} \lambda x.x : p \to q \to p$, since if we did then by examination of the rules for $\vdash_{\lambda_\to}$ our proof tree would have to have the following form:

$$\frac{x : p \vdash_{\lambda_\to} x : q \to p}{\vdash_{\lambda_\to} \lambda x.x : p \to q \to p}$$

But $x : p \vdash_{\lambda_\to} x : q \to p$ is not a valid instance of the identity rule, so $\nvdash_{\lambda_\to} \lambda x.x : p \to q \to p$.

There are a couple of ways to remedy this. One is to let f be a function from justification terms to sets of λ terms and rephrase Condition 1 as:

Condition 2. If $\vdash^{CS}_{iLP} r : \varphi$ then there exists a $M \in f(r)$ so that $\vdash_{\lambda_\to} M : \flat(\varphi)$.

One issue with this condition is that it is satisfied by the function f such that for each justification term r the set $f(r)$ is the set of all λ terms. This function satisfies Condition 2, but cannot be said to give us any information about the computational content of r. We could fix this by forcing $f(r)$ to be finite for each r, but this will eventually make stating theorems unnecessarily cumbersome. Instead we will only consider translations from +-free justification terms to λ terms. Thus Condition 1 becomes:

Condition 3. If $\vdash^{CS}_{iLP} r : \varphi$ and r is +-free then $\vdash_{\lambda_\to} f(r) : \flat(\varphi)$.

Functions that satisfy Condition 3 can still fail to exist for some constant specifications. For instance take $CS = \{(c, p \to p \to p), (c, c : p \to p)\}$ where c is a justification constant and p is a propositional variable. If f satisfies Condition 3 we would have $\vdash_{\lambda_\to} f(c) : p \to p \to p$ and $\vdash_{\lambda_\to} f(c) : p \to p$, which causes the same problem we had with Condition 1. Instead of attempting to refine this condition by adding a hypothesis regarding CS, we will consider this as a condition on both our constant specification and our map f.

We may wish for something stronger than Condition 3. For instance:

Condition 4. If $x_1 : \psi_1, \ldots, x_n : \psi_n \vdash^{CS}_{iLP} r : \varphi$ and r is +-free then $x_1 : \flat(\psi_1), \ldots, x_n : \flat(\psi_n) \vdash_{\lambda_\to} f(r) : \flat(\varphi)$.

This condition is not tenable. To explore this let r be any justification term, let φ be any justification formula. It is straightforward to show $x : r : \varphi \vdash^{CS}_{iLP} r : \varphi$, so Condition 4 implies $x : \flat(\varphi) \vdash_{\lambda_\to} f(r) : \flat(\varphi)$. This will cause problems for any inhabited constant specification, since if $(c, \alpha) \in CS$ then Condition 4 also implies $\vdash_{\lambda_\to} f(c) : \flat(\alpha)$. By Lemma 6 $FV(f(c)) = \varnothing$, but we also have that $x : \flat(\varphi) \vdash_{\lambda_\to} f(c) : \flat(\varphi)$, and since $x \notin FV(f(c))$ we have $\vdash_{\lambda_\to} f(c) : \flat(\varphi)$. This shows that every formula is inhabited, which in turn means that the implicational fragment of intuitionistic logic is inconsistent. This forces us to either have CS be empty or put some restrictions on $x_1 : \psi_1, \ldots, x_n : \psi_n$, and since Lemma 5 requires an inhabited constant specification the best option will be to restrict $x_1 : \psi_1, \ldots, x_n : \psi_n$.

Definition 12. *A map f from +-free propositional justification terms into λ-terms is said to be* ***justification preserving*** *with respect to a constant specification CS if it satisfies the following three properties:*

1. *If $(c, \varphi) \in CS$ then $\vdash_{\lambda_\to} f(c) : \flat(\varphi)$.*

2. *For any context Γ, if $\Gamma \vdash_{\lambda_{\rightarrow}} f(t) : \theta \rightarrow \varphi$ and $\Gamma \vdash_{\lambda_{\rightarrow}} f(r) : \theta$ then $\Gamma \vdash_{\lambda_{\rightarrow}} f(t \bullet r) : \varphi$.*
3. *For any context Γ, if $\Gamma \vdash_{\lambda_{\rightarrow}} f(r) : \varphi$ then $\Gamma \vdash_{\lambda_{\rightarrow}} f(!r) : \varphi$.*

Lemma 8. *If f is justification preserving with respect to CS, r is a $+$-free justification term, and $\vdash^{CS}_{iLP} r : \varphi$ then $\vdash_{\lambda_{\rightarrow}} f(r) : \flat(\varphi)$.*

Proof. This can be shown by induction on the lengths of proofs in the sequent calculus formulation of $\text{ILP}^{CS}_{\rightarrow}$ or by a trick involving Fitting's Kripke style semantics. □

Definition 13. *Let s, k, and i be specific justification constants (which are intended to emulate the S, K, and I from the SK combinators). We define the set IT as the collection of pairs below.*

- *(k, φ) where φ is an instance of Justification Axiom P1.*
- *(s, φ) where φ is an instance of Justification Axiom P2.*
- *(i, φ) where φ is an instance of Justification Axioms J1, J2, J3, J4, or J5.*

Lemma 9. *The constant specification IT is axiomatically appropriate.*

Lemma 10. *For all justification terms u we have $IT[x := u] \subseteq IT$.*

Definition 14. *The map $u \mapsto (u)_\lambda$ is defined recursively from $+$-free justification terms into λ-terms as follows:*

1. *Let $(i)_\lambda = \lambda x.x$ and $(k)_\lambda = \lambda xy.x$ and $(s)_\lambda = \lambda xyz.xz(yz)$*
2. *For any variable x let $(x)_\lambda = x$.*
3. *For any justification term r let $(!r)_\lambda = (r)_\lambda$.*
4. *For any justification terms r and t let $(r \bullet t)_\lambda = (r)_\lambda \cdot (t)_\lambda$.*

Lemma 11. *The map $r \mapsto (r)_\lambda$ is justification preserving with respect to IT.*

Proof. This follows from a straightforward induction. □

We want to explore the relationship between justification preserving maps and BHKK representations. In particular, we will restrict our discussion to IT and $()_\lambda$. Consider a propositional formula of the form $\varphi \rightarrow \psi$, by applying Lemma 5 and Theorem 1 we can get the following result:

Theorem 3. *Let $\varphi \rightarrow \psi$ be a propositional formula. $\varphi \rightarrow \psi$ has a iLP_{IT} provable BHKK representation iff $\vdash_i \varphi \rightarrow \psi$.*

So for our $\text{IPL}_{\rightarrow}$ provable $\varphi \rightarrow \psi$ we have r and t so that $\vdash^{IT}_{iLP} t : (x : \varphi \rightarrow r : \psi)$. This leads us to question how $()_\lambda$ would interpret the terms t and r. As it turns out, not only can we show that $(t)_\lambda$ and $(r)_\lambda$ inhabit the same type, but they are β equivalent. Due to Lemma 8 and Lemma 11 we know what type $(t)_\lambda$ would inhabit, but we do not have information about $(r)_\lambda$.

Proposition 1. *If $\vdash_i \varphi \rightarrow \psi$ and x is a justification variable then there exist $+$-free justification terms r and t so that:*

1. $\vdash^{IT}_{iLP} t : (x : \varphi \to r : \psi)$
2. $\vdash_{\lambda_\to} \lambda x(r)_\lambda : \varphi \to \psi$ *and* $\vdash_{\lambda_\to} (t)_\lambda : \varphi \to \psi$
3. $(t)_\lambda =_\beta \lambda x(r)_\lambda$

In fact we can prove something stronger than this. To see that the following Lemma is stronger than Proposition 1 recall that by Lemma 2 if $\vdash_i \varphi \to \psi$ then $\vdash^{IT}_{iLP} \varphi \to \psi$.

Lemma 12. *For all justification formulas* α *and* γ*, if* $\vdash^{IT}_{iLP} \alpha \to \gamma$ *then there exist* $+$*-free justification terms* r *and* t *so that:*

1. $\vdash^{IT}_{iLP} t : (x : \alpha \to r \bullet x : \gamma)$
2. $\vdash_{\lambda_\to} \lambda x(r \bullet x)_\lambda : \flat(\alpha \to \gamma)$ *and* $\vdash_{\lambda_\to} (t)_\lambda : \flat(\alpha \to \gamma)$
3. $(t)_\lambda =_\beta \lambda x(r \bullet x)_\lambda$

Proof. To begin with we will construct r and t. Since we have $\vdash^{IT}_{iLP} \alpha \to \gamma$ and IT is axiomatically appropriate, by Lemma 5 there is a $+$-free r so that $\vdash^{IT}_{iLP} r : (\alpha \to \gamma)$. Let t be $i \bullet !r$.

Next we have to show $\vdash^{IT}_{iLP} t : (x : \alpha \to r \bullet x : \gamma)$. By using Justification Axiom J2 we can show that $\vdash^{IT}_{iLP} !r : r : (\alpha \to \gamma)$. By the Necessitation Rule and the definition of IT we have $\vdash^{IT}_{iLP} i : (r : (\alpha \to \gamma) \to x : \alpha \to r \bullet x : \gamma)$, so by Justification Axiom J3 we have $\vdash^{IT}_{iLP} i \bullet !r : (x : \alpha \to r \bullet x : \gamma)$.

$\vdash_{\lambda_\to} \lambda x(r \bullet x)_\lambda : \flat(\alpha \to \gamma)$ can be shown using Lemma 8 and Lemma 11 and some basic reasoning about the Simply Typed λ Calculus.

Next we have to show $\vdash_{\lambda_\to} (t)_\lambda : \flat(\alpha \to \gamma)$, but this follows from $\vdash^{CS}_{iLP} t : (x : \alpha \to r : \gamma)$ and Lemma 8 and Lemma 11 (recall that $\flat(x : \alpha \to r : \gamma) = \flat(\alpha \to \gamma)$).

Finally, we have to show that $(t)_\lambda =_\beta \lambda x(r \bullet x)_\lambda$. First we reduce $(t)_\lambda$, notice that $(i \bullet !r)_\lambda =_\beta (r)_\lambda$. Next recall that a fundamental result about the Simply Typed λ Calculus implies $(r)_\lambda$ has a β normal form M. Since $\vdash_{\lambda_\to} M : \flat(\alpha \to \gamma)$ we have $FV(M) = \varnothing$, so M must[2] have the form $\lambda y N$. Therefore $\lambda x(r \bullet x)_\lambda = \lambda x.(r)_\lambda x =_\beta \lambda x.(\lambda y N)x =_\beta \lambda x.N[y := x] = \lambda y N =_\beta (r)_\lambda$. □

This applies to BHKK representations of propositional formulas, but we can extend this to BHKK representations if justification formulas. First notice the following:

Theorem 4. *Let* $\varphi \to \psi$ *be a justification formula, then* $\varphi \to \psi$ *has a* iLP_{IT} *provable BHKK representation iff* $\vdash^{IT}_{iLP} x : \varphi \to \psi$.

Proof. The reverse direction follows from Theorem 1 and Lemma 5. □

This indicates that we should replace the $\vdash^{IT}_{iLP} \alpha \to \gamma$ hypothesis from Lemma 12 with $\vdash^{IT}_{iLP} x : \alpha \to \gamma$.

Theorem 5. *For all justification formulas* α *and* γ*, and for all proof variables* x *so that* $\vdash^{IT}_{iLP} x : \alpha \to \gamma$ *there exist* $+$*-free proof terms* r *and* t *so that:*

1. $\vdash^{IT}_{iLP} t : (x : \alpha \to r : \gamma)$

[2] This requires a somewhat involved argument about normal forms.

2. $\vdash_{\lambda_{\rightarrow}} \lambda x(r)_{\lambda} : \flat(\alpha \rightarrow \gamma)$ *and* $\vdash_{\lambda_{\rightarrow}} (t)_{\lambda} : \flat(\alpha \rightarrow \gamma)$
3. $(t)_{\lambda} =_{\beta} \lambda x(r)_{\lambda}$

Proof. This can be proved by induction on the length of the proof of $\vdash^{IT}_{iLP} x : \alpha \rightarrow \gamma$. □

Consider the assumption $\vdash^{IT}_{iLP} x : \alpha \rightarrow \gamma$ from 5 and compare it to the assumption $\vdash^{IT}_{iLP} \alpha \rightarrow \gamma$ from 12. The first thing to notice is that $\vdash^{CS}_{iLP} x : \alpha \rightarrow \gamma$ is a weaker condition on α and γ than $\vdash^{CS}_{iLP} \alpha \rightarrow \gamma$, this can be shown by using Theorem 1 and Lemma 1. The second thing to notice is that $\vdash^{CS}_{iLP} x : \alpha \rightarrow \gamma$ is a *strictly* weaker condition on some α and γ, since if we take $\gamma = x : \alpha$ then $\vdash^{CS}_{iLP} x : \alpha \rightarrow x : \alpha$ but for all α we have $\nvdash^{CS}_{iLP} \alpha \rightarrow x : \alpha$.

Notice that if α and γ are propositional formulas then the conditions are equivalent, since if $\vdash^{IT}_{iLP} x : \alpha \rightarrow \gamma$ then by Lemma 3 we have $\vdash_i \flat(x : \alpha \rightarrow \gamma)$. Since α and γ are propositional we have $\flat(x : \alpha \rightarrow \gamma) = \alpha \rightarrow \gamma$. So $\vdash_i \alpha \rightarrow \gamma$, and by Lemma 2 we have $\vdash_{iLP} \alpha \rightarrow \gamma$.

The next thing we can examine is whether the conclusions of Theorem 5 are redundant. Consider the following conjecture.

Conjecture 1. For all justification formulas α and γ, for all proof variables x, and for all +-free proof terms r and t so that $\vdash^{IT}_{iLP} t : (x : \alpha \rightarrow r : \gamma)$[3] we have: $\vdash_{\lambda_{\rightarrow}} \lambda x(r)_{\lambda} : \flat(\alpha \rightarrow \gamma)$ and $\vdash_{\lambda_{\rightarrow}} (t)_{\lambda} : \flat(\alpha \rightarrow \gamma)$.

This conjecture is false. Consider $\alpha = y : p$ for some propositional variable p and justification variable y, $\gamma = p$, $r = y$, and $t = i$. Then $x : y : p \rightarrow y : p$ is an instance of Justification Axiom J1, so we have $\vdash^{IT}_{iLP} i : (x : y : p \rightarrow y : p)$. However $\lambda x(r)_{\lambda} = \lambda x.y$, which according to Lemma 6 there is no type φ so that $\vdash_{\lambda_{\rightarrow}} \lambda x.y : \varphi$, so the first condition fails badly.

We can still ask if we require the first 2 conclusions of Theorem 5 will we get the 3rd.

Conjecture 2. For all justification formulas α and γ, for all proof variables x, and for all +-free proof terms r and t so that $\vdash^{IT}_{iLP} t : (x : \alpha \rightarrow r : \gamma)$ and $\vdash_{\lambda_{\rightarrow}} \lambda x(r)_{\lambda} : \flat(\alpha \rightarrow \gamma)$ and $\vdash_{\lambda_{\rightarrow}} (t)_{\lambda} : \flat(\alpha \rightarrow \gamma)$ we have: $(t)_{\lambda} =_{\beta} \lambda x(r)_{\lambda}$.

This conjecture also fails. Consider $\alpha = i : (p \rightarrow p)$ for some propositional variable p, $\gamma = p \rightarrow p$, and $t = r = i$. Then $x : i : (p \rightarrow p) \rightarrow i : (p \rightarrow p)$ is an instance of Justification Axiom J1, so we have $\vdash^{IT}_{iLP} i : (x : i : (p \rightarrow p) \rightarrow i : (p \rightarrow p))$. However $\lambda x(r)_{\lambda} = \lambda xy.y$ and $(t)_{\lambda} = \lambda x.x$. Notice that $\flat(\alpha \rightarrow \gamma) = (p \rightarrow p) \rightarrow (p \rightarrow p)$. Next, notice that we have $\vdash_{\lambda_{\rightarrow}} \lambda x.x : (p \rightarrow p) \rightarrow (p \rightarrow p)$ and $\vdash_{\lambda_{\rightarrow}} \lambda xy.y : (p \rightarrow p) \rightarrow (p \rightarrow p)$, so both of our hypotheses are satisfied. Finally notice that $\lambda x.x$ and $\lambda xy.y$ are in β-normal form, so $\lambda x.x \neq_{\beta} \lambda xy.y$.

It is worth pointing out that all of the results in this section will still hold if we expand our language to include the connectives $\wedge$, $\vee$ and the symbol $\bot$, but it would involve expanding the λ terms and the type theory, as well as the justification constants.

[3] Notice that this is stronger than assuming $\vdash^{IT}_{iLP} x : \alpha \rightarrow \gamma$, so we do not need to include both assumptions.

4 Conclusion

When interpreting condition $C_{\rightarrow}$ of the BHK interpretation, we must decide on what a construction is. Artemov's justification logic used his justification terms, for which the output of a term r meant to represent a construction applied to a term a meant to represent an input is simply $r \bullet a$. Since justification terms have no associated reduction relation they are unable to serve as a model of computation.

We have constructed a way of interpreting justification terms into λ terms in a way that if we have a BHK 'proof' of an intuitionistic formula formalized in justification logic then we have a BHK 'proof' formalized in the Simply Typed λ Calculus. However when viewed through Kreisel's extra condition we get that some of these 'proofs' and 'verifications' formalized in justification logic get translated into the same λ term (with respect to β reduction). This obscures the nature of 'verifications' in these formalizations of the BHKK interpretation, as it is not clear whether a proper 'verification' should correspond to the same program as the 'proof', and leaves open the question as to what new information is provided by the 'verifications' since they do not normally provide any more computational information. Here are a few ways to interpret our result:

1. $C'_{\rightarrow}$ is not part of the original formulation of the BHK interpretation for a good reason. It has not even been solidly established that $C'_{\rightarrow}$ is immune to the criticisms that were levied against the original BHK interpretation, let alone that they do not admit any new criticisms. So we are not justified in adopting $C'_{\rightarrow}$ over $C_{\rightarrow}$.
2. The map from justification terms to λ-terms is erasing too much information, if you want a distinction between 'verifications' and 'constructions' you will need a more robust type system or a better map from justification terms to λ terms.
3. If we want $C'_{\rightarrow}$ to be distinct from $C_{\rightarrow}$ the 'verifications' cannot just be some rehash of the 'constructions', so $\mathrm{ILP}^{CS}_{\rightarrow}$ fails to fully capture conditions $C'_{\rightarrow}$.
4. The BHK interpretation started as an *informal* description of intuitionistic logic, and was never meant to be formalized. While certain formal systems may take certain principles from this interpretation, we are not required to take them seriously as 'formalizations of the BHK interpretation'.
   "Any premature attempt at a [formalization of the BHK/functional interpretation] could only weaken the idea: it is much more than that."-Girard in [10].
5. How do you check a calculation? You perform said calculation, so our result is not surprising.

Future Work

We would like to extend these result to first order justification logic (see [4]) and a type theory corresponding to first order intuitionistic logic (see Chapter 8 of [17]). Some progress has been made in [7] in particular an analog of Lemma 4 was shown, but the first order universal analog of Theorem 5 hasn't been proved.

It would also be interesting to explore possible systems that inhabit **R1** and **R4** mentioned in the introduction. This would give a more complete picture of the possible ways of formalizing the BHKK interpretation.

Acknowledgements. I would like to thank Tim Carlson, Neil Tennant, and Chris Miller for their advice. I would also like to thank the anonymous peer reviewers for their useful suggestions.

References

1. Artemov, S.: Proof polynomials: a unified semantics for modality and lambda-terms. Technical report, CFIS 98-06, Cornell University (1998)
2. Artemov, S.: Explicit provability and constructive semantics. Bull. Symbolic Logic **7**(1), 1–36 (2001). https://doi.org/10.2307/2687821
3. Artemov, S., Fitting, M.: Justification Logic: Reasoning with Reasons. Cambridge Tracts in Mathematics. Cambridge University Press (2019). https://doi.org/10.1017/9781108348034
4. Artemov, S., Yavorskaya (Sidon), T.: TR-2011005: First-Order Logic of Proofs. Technical report, CUNY Academic Works (2011)
5. Beeson, M.J.: Foundations of Constructive Mathematics: Metamathematical studies, vol. 6. Ergebnisse der Mathematik und ihrer Grenzgebiete. Springer, Berlin (1985). https://doi.org/10.1007/978-3-642-68952-9
6. Brouwer, L.E.J.: Consciousness, philosophy, and mathematics. In: Proceedings of the 10th International Congress of Philosophy, Amsterdam, 11–18 August 1948, pp. 1235–1249. North-Holland Publishing Company (1949)
7. DeBoer, N.: Justification logic, type theory and the BHK interpretation. Ph.D. thesis, Ohio State University (2020). http://rave.ohiolink.edu/etdc/view?acc_num=osu1598007830055549
8. Fitting, M.: The logic of proofs, semantically. Ann. Pure Appl. Logic **132**(1), 1–25 (2005). https://doi.org/10.1016/j.apal.2004.04.009
9. Fitting, M.: A quantified logic of evidence. Electron. Notes Theoret. Comput. Sci. **143**, 59–71 (2006). https://doi.org/10.1016/j.entcs.2005.04.038
10. Girard, J.Y.: The Blind Spot: Lectures on Logic. European Mathematical Society (2011)
11. Goodman, N.D.: A theory of constructions equivalent to arithmetic. In: Intuitionism and Proof Theory (Proc. Conf., Buffalo, N.Y., 1968), pp. 101–120. North-Holland, Amsterdam (1970). https://doi.org/10.1016/S0049-237X(08)70745-6
12. Heyting, A.: Die intuitionistische Grundlegung der Mathematik. Erkenntnis **2**(1), 106–115 (1931). https://doi.org/10.1007/BF02028143
13. Kreisel, G.: Mathematical Logic. In: Lectures on Modern Mathematics, vol. III, pp. 95–195. Wiley, New York (1965)
14. Kuznets, R., Studer, T.: Logics of Proofs and Justifications. College Publications (2019)
15. Martin-Löf, P.: Intuitionistic Type Theory. Volume 1 of Lecture notes, Studies in Proof Theory. Bibliopolis, Naples (1984). Sambin, G. (ed.)
16. McGrath, M., Frank, D.: Propositions. In: Zalta, E.N. (ed.) The Stanford Encyclopedia of Philosophy, spring 2018 edn. Metaphysics Research Lab, Stanford University (2018)
17. Sørensen, M., Urzyczyn, P.: Lectures on the Curry-Howard Isomorphism. Studies in Logic and the Foundations of Mathematics. Elsevier Science (2006)

Hyperarithmetical Worm Battles

David Fernández-Duque[1], Konstnatinos Papafilippou[1(✉)], and Joost J. Joosten[2]

[1] Department of Mathematics WE16, Ghent University, Ghent, Belgium
{David.FernandezDuque,Konstantinos.Papafilippou}@UGent.be
[2] Department of Philosophy, University of Barcelona, Catalonia, Spain
jjoosten@ub.edu

Abstract. Japaridze's provability logic GLP has one modality $[n]$ for each natural number and has been used by Beklemishev for a proof theoretic analysis of Peano arithmetic (PA) and related theories. Among other benefits, this analysis yields the so-called *Every Worm Dies* (EWD) principle, a natural combinatorial statement independent of PA. Recently, Beklemishev and Pakhomov have studied notions of provability corresponding to transfinite modalities in GLP. We show that indeed the natural transfinite extension of GLP is sound for this interpretation, and yields independent combinatorial principles for the second order theory ACA of arithmetical comprehension with full induction. We also provide restricted versions of EWD related to the fragments $\mathrm{I}\Sigma_n$ of Peano arithmetic.

Keywords: Provability logics · Independence results · Ordinal analysis

1 Introduction

It is an empirically observed phenomenon that 'natural' theories are linearly ordered by strength, suggesting that this strength could be quantified in some fashion. Much as real numbers are used to measure e.g. the distance between two points on the plane, proof theorists use ordinal numbers to measure the power of formal theories [18]. The precise relationship between these ordinals and their respective theories may be defined in various ways, each with advantages and disadvantages. One relatively recent and particularly compelling way to assign ordinals to a theory T lies in studying hierarchies of iterated consistency or reflection principles for a weaker base theory B that are provable in T. The work of Beklemishev [2] has shown how provability logic, particularly Japaridze's polymodal variant GLP [13], provides an elegant framework for analyzing theories in this fashion. GLP is a propositional logic which has one modality $[n]$ for each natural number. The expression $[n]\varphi$ is read *φ is n-provable,* where n-provability is defined by allowing any true Π_n sentence as an axiom. Dually, $\langle n\rangle\varphi$ denotes

Partially supported by the FWO-FWF Lead Agency Grant G030620N.

S. Artemov and A. Nerode (Eds.): LFCS 2022, LNCS 13137, pp. 52–69, 2022.
https://doi.org/10.1007/978-3-030-93100-1_5

the *n-consistency* of φ, which is equivalent to the schema stating that all Σ_n consequences of φ are true, also known as Σ_n*-reflection*.

This approach to ordinal analysis is based on special elements of the logic –the so-called *worms*. Formally, worms are formulas of the form $\langle n_1\rangle\ldots\langle n_m\rangle\top$, representing iterated reflection principles. However, worms can be interpreted in many ways: formulas of a logic, words over an infinite alphabet, special fragments of arithmetic [1,16], Turing progressions [14,19], worlds in a special model for the closed fragment of GLP, and also ordinals [10]. These interpretations of worms allowed Beklemishev [2] to give an ordinal analysis of Peano arithmetic (PA) and related systems, yielding as side-products a classification of provably total recursive functions, consistency proofs, and a combinatorial principle independent of PA, colloquially called *Every Worm Dies*.

Recently, Beklemishev and Pakhomov [5] extended the method of ordinal analysis via provability logics to predicative systems of second order arithmetic. It is important to investigate if said analysis also comes with the expected regular side-products for theories beyond the strength of PA. This paper is a first exploration in this direction, expanding on their analysis to provide combinatorial principles independent of standard extensions of PA.

Beklemishev and Pakhomov's analysis involves notions of provability naturally corresponding to modalities $[\lambda]$ for $\lambda \geq \omega$ in the natural transfinite extension of GLP. This extension is denoted GLP_Λ [4], where Λ is the supremum of all modalities allowed. The model theory of the logics GLP_Λ has been extensively studied [8,9], as have various proof-theoretic interpretations [6,11,15]. Our first result is that GLP_Λ is also sound for the notions of provability employed in [5].

The soundness of GLP_Λ allows us to develop combinatorial principles in the style of Beklemishev [2], an effort that was initiated in Papafilippou's master thesis [17]. We consider variants of the *Every Worm Dies* principle, denoted EWD^Λ for suitable Λ. Our main results are that over elementary arithmetic (EA), EWD^{ω^2} is equivalent to the 1-consistency of ACA, and that the principles EWD^{n+1} lie between the 1-consistency of $\mathrm{I\Sigma}_n$ and that of $\mathrm{I\Sigma}_{n+1}$.

2 Preliminaries

For first order arithmetic, we shall work with theories with identity in the language $\mathcal{L}_{\mathrm{PA}} := \{0; s, +, \cdot, exp\}$ with exp being the unary function for $x \mapsto 2^x$. We define $\Delta_0 = \Sigma_0 = \Pi_0$-formulas as those whose quantifiers occur in the form $\forall x{<}t\ \varphi$ or $\exists x{<}t\ \varphi$. Then inductively Σ_{n+1}/Π_{n+1}-formulas are those of the form $\exists x \varphi/\forall x \varphi$, where φ is a Π_n/Σ_n-formula, respectively. We may extend the above classes with a new predicate P by treating it as an atomic formula: the resulting classes are denoted $\Pi_n(P)$, $\Sigma_n(P)$, etc. More generally, for an extension $\mathcal{L} \supsetneq \mathcal{L}_{\mathrm{PA}}$ of the language of PA with new predicate symbols, we write $\Pi_n^{\mathcal{L}}$, $\Sigma_n^{\mathcal{L}}$, etc. to denote the corresponding classes of formulas with any new predicate symbols of $\mathcal{L}$ treated as atoms.

Elementary Arithmetic (EA) or *Kalmar Arithmetic* contains the basic axioms describing the non-logical symbols together with the induction axiom I_φ for every

Δ_0-formula φ, which as usual denotes $\mathrm{I}_\varphi := \varphi(0) \wedge \forall x\ (\varphi(x) \to \varphi(S(x))) \to \forall x\ \varphi(x)$. For a given complexity class Γ, we denote by $I\Gamma$ the theory extended EA with induction for all Γ-formulas. By EA^+ we denote the extension of EA by the axiom expressing the totality of the *super-exponentiation* function 2^x_y, which is defined inductively as: $2^x_0 := x$; $2^x_{n+1} := 2^{2^x_n}$. Finally PA can be seen as the union of all $\mathrm{I}\Sigma_n$ for every n. A theory S of a language $\mathcal{L} \supseteq \mathcal{L}_{\mathrm{PA}}$ is elementary axiomatizable if there is a Δ_0-formula $Ax_S(x)$ that is true iff x is the code of an axiom of S. By Craig's trick, all c.e. theories have an equivalent that is elementary axiomatizable.

The language of second order arithmetic is the extension of the language of first order arithmetic $\mathcal{L}_{\mathrm{PA}}$ by the addition of second order variables and parameters and the predicate symbol $\in$. The expression $t{\in}X$ is an atomic formula where t is a term and X a second order variable. We add no symbol for the second order identity and instead we express it in the language via extensionality.

Definition 1 (ACA). *The theory* ACA *is a theory in the language of second order arithmetic that extends* PA *by the induction schema for all second order formulas and the comprehension schema:* $\exists Y\, \forall x\ (x{\in}Y \leftrightarrow \varphi(x))$, *for every arithmetical formula with possibly both first and second order parameters (excluding* Y*).*

Definition 2. *For* Λ *an ordinal, the logic* GLP_Λ *is the propositional modal logic with a modality* $[\alpha]$ *for each* $\alpha < \Lambda$*. Each* $[\alpha]$ *modality satisfies the* GL *identities given by all tautologies, distribution axioms* $[\alpha](\varphi \to \psi) \to ([\alpha]\varphi \to [\alpha]\psi)$*, Löb's axiom scheme* $[\alpha]([\alpha]\varphi \to \varphi) \to [\alpha]\varphi$ *and the rules modus ponens and necessitation* $\varphi/[\alpha]\varphi$*. The interaction between modalities is governed by two schemes, monotonicity* $[\beta]\varphi \to [\alpha]\varphi$ *and, negative introspection* $\langle\beta\rangle\varphi \to [\alpha]\langle\beta\rangle\varphi$ *where in both schemes it is required that* $\beta < \alpha < \Lambda$*. As usual* $\langle\alpha\rangle\varphi$ *is a shorthand for* $\neg[\alpha]\neg\varphi$*.*

The closed fragment of GLP_Λ suffices for ordinal analyses and worms are its backbone.

Definition 3. *The class of* worms *of* GLP_Λ *is denoted* $\mathbb{W}^\Lambda$ *and defined by* $\top{\in}\mathbb{W}^\Lambda$*, and* $A{\in}\mathbb{W}^\Lambda \wedge \alpha{<}\Lambda \Rightarrow \langle\alpha\rangle A{\in}\mathbb{W}^\Lambda$*. By* $\mathbb{W}^\Lambda_\alpha$ *we denote the set of worms where all occurring modalities are at least* α*. We define an order* $<_\alpha$ *for each* $\alpha{<}\Lambda$ *by setting* $A{<_\alpha}B$ *if* $\mathrm{GLP}_\Lambda{\vdash}B{\to}\langle\alpha\rangle A$*.*

It will be convenient to introduce notation to compose and decompose worms. Let us write α instead of $\langle\alpha\rangle$ when this does not lead to confusion. For worms A and B we define the concatenation AB via $\top B := B$ and $(\alpha A)B := \alpha(AB)$. We define the α-head h_α of A inductively: $h_\alpha(\top) := \top$; $h_\alpha(\beta A) := \top$ if $\beta{<}\alpha$, and $h_\alpha(\beta A) := \beta h_\alpha(A)$ otherwise. Likewise, we define the α-remainder r_α of A as $r_\alpha(\top) := \top$ and, $r_\alpha(\beta A) := \beta A$ if $\beta{<}\alpha$ and $r_\alpha(\beta A) := r_\alpha(A)$ otherwise. We define the head h and remainder r of αA as $h(\alpha A) := h_\alpha(\alpha A)$ and $r(\alpha A) := r_\alpha(\alpha A)$. Further, $h(\top) := r(\top) := \top$.

Lemma 1. *The following formulas are derivable in* GLP_Λ*:*

(i) If $\alpha \leq \beta$ *and* $A \in \mathbb{W}^\Lambda$*, then* $\mathrm{GLP}_\Lambda \vdash \beta\alpha A \to \alpha A$*;*
(ii) If $\alpha < \beta$*, then* $\mathrm{GLP}_\Lambda \vdash \beta\varphi \wedge \alpha\psi \leftrightarrow \beta(\varphi \wedge \alpha\psi)$*;*
(iii) If $A \in \mathbb{W}^\Lambda_{\alpha+1}$*, then* $\mathrm{GLP}_\Lambda \vdash AC \wedge \alpha B \leftrightarrow A(C \wedge \alpha B)$*;*
(iv) If $A \in \mathbb{W}^\Lambda_{\alpha+1}$*, then* $\mathrm{GLP}_\Lambda \vdash A \wedge \alpha B \leftrightarrow A\alpha B$*.*

The proof of which follows successively from the axioms of GLP_Λ, details for which can be found in [3] and [4]. With this lemma in our tool-belt, we can prove the following proposition which will be of use to us later as we present worm battles.

Proposition 1. *Let* $n \in \mathbb{N}$*,* $\alpha < \Lambda$ *be ordinals,* $A \in \mathbb{W}^\alpha$*, and* $B \in \mathbb{W}^{\alpha+1}$ *be such that* $|B| \leq n$*.*[1] *Then,* $\mathrm{GLP}_\Lambda \vdash \langle\alpha\rangle^{n+1}\top \to AB$*.*

Proof. We will prove this fact through two external inductions, first we will show that for every n and B satisfying the above conditions, $\mathrm{GLP}_\Lambda \vdash \langle\alpha\rangle^n\top \to B$. If $n = 0$, then it is clear. Assume now that it holds for $n = k$. Let $B \in \mathbb{W}^{\alpha+1}$ with $|B| \leq k$ and let $\beta \leq \alpha$, then in GLP_Λ, $\langle\alpha\rangle^{k+1}\top \vdash \langle\alpha\rangle^{|B|+1}\top \vdash \langle\alpha\rangle\langle\alpha\rangle^{|B|}\top \vdash \langle\alpha\rangle B \vdash \langle\beta\rangle B$, where the first step uses at most k applications of the **4** axiom.

Now we will perform an external induction on $|A|$. If $A = \beta$ for some $\beta < \alpha$, then we fall in the case of the previous induction. If $A = \langle\beta\rangle C$, where $\beta < \alpha$ and $\mathrm{GLP}_\Lambda \vdash \langle\alpha\rangle^{n+1}\top \to CB$, then in GLP_Λ, $\langle\alpha\rangle^{n+1}\top \vdash (CB \wedge \langle\alpha\rangle\top) \vdash \langle\alpha\rangle CB$, using Lemma 1.

From [2] we know that $\langle\mathbb{W}^\omega_n/\equiv, <_n\rangle \cong \langle\varepsilon_0, \prec\rangle$ so that worms (modulo provable equivalence) can be used to denote ordinals. One can find analogs of fundamental sequences for ordinals by defining $Q^\alpha_0(\varphi) := \langle\alpha\rangle\varphi$; $Q^\alpha_{k+1}(\varphi) := \langle\alpha\rangle\big(\varphi \wedge Q^\alpha_k(\varphi)\big)$. By an easy induction on k one sees that $Q^\alpha_{k+1}(A) <_\beta \langle\alpha+1\rangle A$ for any $\beta \leq \alpha$ yielding a so-called step-down function.

This step-down function can be rewritten to get a more combinatorial flavour reminiscent of the Hydra battle. To this end we define the *chop-operator* c on worms that do not start with a limit ordinal by $c(\top) := \top; c(\langle 0\rangle A) := A$ and, $c(\langle\alpha+1\rangle A) := \langle\alpha\rangle A$. Now we define a stepping down function based on a combination of chopping a worm, the worm growing back and using a given and fixed fundamental sequence of the limit ordinals occurring in GLP_Λ for countable Λ.

Definition 4 (Step-down function). *For any number* k *let* $A[\![k]\!] := c(A)$ *for* $A = \top$ *or* $A = 0B$*,* $A[\![k]\!] := \big(c(h(A))\big)^{k+1} r(A)$ *for* $A = \langle\alpha+1\rangle B$ *and* $A[\![k]\!] := \langle\lambda[k]\rangle B$ *for* $A = \langle\lambda\rangle B$ *where* λ *is a limit ordinal and* $\lambda[k]$ *is the* k*-th element of the fundamental sequence of* λ*.*

The definition above relates to the functions Q^α_k and serves as a way to produce fundamental sequences of worms inside $\langle\mathbb{W}^\Lambda/\equiv, <_0\rangle$. With an easy induction on k, one can prove the following:

Lemma 2. *If* $A, B \in \mathbb{W}^\Lambda$ *and* $A = \langle\alpha+1\rangle B$ *then for every* $k \in \mathbb{N}$ *we have that,*

$$\mathrm{GLP}_\Lambda \vdash Q^\alpha_k(B) \leftrightarrow (\alpha h_{\alpha+1}(B))^{k+1} r_{\alpha+1}(B).$$

[1] The length of a worm $|B|$ is defined inductively as: $|\top| = 0$ and $|\alpha B| = 1 + |B|$.

Then, assuming that there is a Δ_1 coding of the ordinals present according to EA, the following is provable over EA:

Corollary 1. *For any $k \in \mathbb{N}$ and $A \in \mathbb{W}^\Lambda$ with $A \neq \top$, we have $A[\![k]\!] <_0 A$.*

Proof. Since for every natural number k: $\mathrm{GLP}_\Lambda \vdash \langle\alpha{+}1\rangle B \to Q_k^\alpha(B)$, for $A = \langle\alpha{+}1\rangle B$, in GLP_Λ, $A \vdash (\alpha h_{\alpha+1}(B))^{k+2} r_{\alpha+1}(B) \vdash \alpha A[\![k]\!] \vdash 0\, A[\![k]\!]$. The limit stage follows by the monotonicity axiom of GLP_Λ.

Given a worm $A \in W^\Lambda$, we now define a decreasing sequence (strictly as long as we have not reached $\top$) by $A_0 := A$ and $A_{k+1} := A_k[\![k+1]\!]$. We now define the principle EWD^Λ standing for *Every Worm Dies* as an arithmetisation of $\forall A \in W^\Lambda \exists k A_k = \top$. Note that here the worms are coded as sequences of ordinals which we achieve by assuming we have Δ_1 codings of the ordinals $< \Lambda$ and of the relation $\alpha < \beta$ in EA.

The modalities of GLP_ω can be linked to arithmetic by interpreting $\langle n\rangle\varphi$ for a given c.e. theory $S \supseteq \mathrm{EA}$ as the finitely axiomatisable scheme $\Sigma_n\text{-RFN}(S{+}\varphi^*) := \{\Box_{S+\varphi^*}\sigma \to \sigma \mid \sigma \in \Sigma_n\} \equiv \Pi_{n+1}\text{-RFN}(S{+}\varphi^*)$. The $\Box_S$ denotes the standard arithmetisation of formalised provability for the theory S and φ^* denotes an interpretation of φ in arithmetic, mapping propositional variables to sentences, commuting with the connectives and, translating the $\langle n\rangle$ as above. This interpretation is used to classify the aforementioned first order theories of arithmetic.

Theorem 1 (Leivant, Beklemishev [1,16]). *Provably in* EA^+, *for* $n{\geq}1$*:*

$$\mathrm{I}\Sigma_n \equiv \Sigma_{n+1}\text{-RFN}(\mathrm{EA}).$$

Further known results involving partial reflection are given by assessing the totality of certain functions. For a Σ_1-definable function f, by $f{\downarrow}$ we denote the arithmetical sentence $\forall x \exists y f(x) = y$ stating that f is defined everywhere and likewise, by $f(x){\downarrow}$ we denote $\exists y f(x) = y$.

Lemma 3 *([3]). Let f be a Σ_1-definable function that is non-decreasing and $f(x) \geq 2^x$. Then,*

$$\mathrm{EA} \vdash \lambda x.f^{(x)}(x){\downarrow} \leftrightarrow \langle 1\rangle_{\mathrm{EA}}\, f{\downarrow}.$$

If we substitute f with *exp*, we get:

Corollary 2. *Provably in* EA, *we have* $\mathrm{EA}^+ \equiv \mathrm{EA} + \Pi_2\text{-RFN}(\mathrm{EA})$.

3 Arithmetical Soundness of GLP_Λ

In our interest of expanding the worm principle, we have to first expand the interpretation of GLP in arithmetic for modalities $[\alpha]$ where $\alpha \geq \omega$.

Let $\mathcal{L}$ be a language of arithmetic with or without a unary predicate T and let S be a c.e. theory extending EA in a language extending $\mathcal{L}$. We will prove arithmetical soundness of GLP_Λ for a particular interpretation for which most of the work has already been done in [5] by proving arithmetical soundness for the weaker system of RC_Λ. As such, we will call onto many results from that paper, starting with some properties of partial reflection in potentially extended languages of arithmetic.

Lemma 4. *For all sentences $\varphi, \psi \in \mathcal{L}$ and for every $n \geq 0$, the following hold provably in* EA*:*

- *If $S \vdash \varphi \to \psi$ then $\Pi^{\mathcal{L}}_{n+1}$-RFN$(S+\varphi) \vdash \Pi^{\mathcal{L}}_{n+1}$-RFN$(S+\psi)$;*
- *$\Pi^{\mathcal{L}}_{n+1}$-RFN$(S+\varphi) \vdash \varphi$ if $\varphi \in \Pi^{\mathcal{L}}_{n+1}$;*
- *$\Pi^{\mathcal{L}}_{n+1}$-RFN$(S+\varphi) \vdash \Diamond_S \varphi$.*

It is a known result that $\Pi^{\mathcal{L}}_{n+1}$-RFN(S) is finitely axiomatizable over EA for $\mathcal{L} = \mathcal{L}_{\mathrm{PA}}$ which is achieved by using *truth-definitions* for $\Pi^{\mathcal{L}_{\mathrm{PA}}}_{n+1}$-formulas. For $\mathcal{L} \supsetneq \mathcal{L}_{\mathrm{PA}}$, we have the following properties for truth definitions (Theorems 12 & 13, [5]):

Theorem 2. *Let $\mathcal{L}$ be finite. There is a $\Pi^{\mathcal{L}}_1$-formula* Tr *such that for all $\Delta^{\mathcal{L}}_0$-formulas $\varphi(\vec{x})$,*

- $\mathrm{EA} \vdash \forall \vec{x}\ \big(\mathsf{Tr}(\varphi(\vec{x})) \to \varphi(\vec{x})\big)$;
- $\mathrm{EA}^{\mathcal{L}} \vdash \forall \vec{x}\ \big(\mathsf{Tr}(\varphi(\vec{x})) \leftrightarrow \varphi(\vec{x})\big)$.

Let Γ be either $\Pi^{\mathcal{L}}_n$ or $\Sigma^{\mathcal{L}}_n$ for $n > 0$, then there exists a Γ-formula Tr_Γ such that for each Γ-formula $\varphi(\vec{x})$,

$$\mathrm{EA}^{\mathcal{L}} \vdash \forall \vec{x}\ \big(\mathsf{Tr}_\Gamma(\varphi(\vec{x})) \leftrightarrow \varphi(\vec{x})\big).$$

For languages extending the language of arithmetic, we require a way to finitely axiomatize $\Delta^{\mathcal{L}}_0$-induction, which is given for finite $\mathcal{L}$ (Lemma 4.2 in [7]). Over EA we have the following theorem.

Theorem 3 (Theorem 3, [5]). *For finite $\mathcal{L}$ the schema $\Pi^{\mathcal{L}}_{n+1}$-RFN(S) is finitely axiomatizable*

$$i\delta^{\mathcal{L}} \wedge \forall \varphi \in \Pi^{\mathcal{L}}_{n+1}\ \big(\Box_S \varphi \to \mathsf{Tr}_{\Pi^{\mathcal{L}}_{n+1}}(\varphi)\big),$$

where $i\delta^{\mathcal{L}}$ is a $\Pi^{\mathcal{L}}_1$-axiomatization of $I\Delta^{\mathcal{L}}_0$ and $\mathsf{Tr}_{\Pi^{\mathcal{L}}_m}$ is the truth definition for $\Pi^{\mathcal{L}}_m$-formulas.

From here on, we will be using $\Pi^{\mathcal{L}}_{n+1}$-RFN(S) and the formula axiomatizing it interchangeably where applicable. By $i\delta^{\mathcal{L}}$ we will always denote the $\Pi^{\mathcal{L}}_1$-axiomatization of $I\Delta^{\mathcal{L}}_0$.

Notation 1. *Let $\mathcal{L}$ be finite, then given a formula $\varphi \in \mathcal{L}$, we write $[n]^{\mathcal{L}}_S \varphi$ as shorthand for $\exists \theta {\in} \Sigma^{\mathcal{L}}_{n+1} \big(\mathsf{Tr}_{\Sigma^{\mathcal{L}}_{n+1}}(\theta) \wedge \Box_S(\theta \to \varphi)\big)$.*

The lemma below corresponds to the distributivity axiom L1. and we will be using it to prove an arithmetical soundness of GLP_Λ.

Lemma 5. *If $\mathcal{L}$ is finite, then* $\mathrm{EA}^{\mathcal{L}} \vdash [n]^{\mathcal{L}}_S(\varphi \to \psi) \to ([n]^{\mathcal{L}}_S\varphi \to [n]^{\mathcal{L}}_S\psi)$.

Proof. Working within $\mathrm{EA}^{\mathcal{L}}$, assume that $\exists\,\theta_1 \in \Sigma^{\mathcal{L}}_{n+1}(\mathsf{Tr}_{\Sigma^{\mathcal{L}}_{n+1}}(\theta_1) \wedge \Box_S(\theta_1 \to (\varphi \to \psi)))$ and $\exists\,\theta_2{\in}\Sigma^{\mathcal{L}}_{n+1}\ (\mathsf{Tr}_{\Sigma^{\mathcal{L}}_{n+1}}(\theta_2) \wedge \Box_S(\theta_2 \to \varphi))$. Since $\mathrm{EA}^{\mathcal{L}} \vdash \mathsf{Tr}_{\Sigma^{\mathcal{L}}_{n+1}}(\varphi) \leftrightarrow \varphi$ for every $\Sigma^{\mathcal{L}}_{n+1}$-formula φ, it is then given that $\mathrm{EA}^{\mathcal{L}} \vdash \mathsf{Tr}_{\Sigma^{\mathcal{L}}_{n+1}}(\theta_1{\wedge}\theta_2) \leftrightarrow \mathsf{Tr}_{\Sigma^{\mathcal{L}}_{n+1}}(\theta_1){\wedge}\mathsf{Tr}_{\Sigma^{\mathcal{L}}_{n+1}}(\theta_2)$. Thus we get $\mathsf{Tr}_{\Sigma^{\mathcal{L}}_{n+1}}(\theta_1{\wedge}\theta_2){\wedge}\Box_S((\theta_1{\wedge}\theta_2) \to \psi)$.

We will be focusing on languages extending that of arithmetic via the addition of so-called truth predicates. These are unary predicates with the purpose of expressing the truth of formulas –a task achieved by expanding our base theories of arithmetic with the theory of the *Uniform Tarski Biconditionals.*

Definition 5. *Let* $\mathrm{UTB}_{\mathcal{L}}$ *be the* $\mathcal{L}{\cup}\mathsf{T}$ *theory –where* T *is a unary truth predicate not in* $\mathcal{L}$*– axiomatized by the schema* $\forall\vec{x}(\varphi(\vec{x}) \leftrightarrow \mathsf{T}(\ulcorner\varphi(\dot{\vec{x}})\urcorner))$*, for every* $\mathcal{L}$*-formula* φ.

This process of extending the base language via the addition of truth predicates can be iterated over ordinals. For that we assume that given an ordinal Λ there are Δ_1-formulas in the base language of arithmetic; $x <_\Lambda \Lambda$ and $x \leq_\Lambda y$, roughly expressing that "x is the code of an ordinal in Λ" and "x, y code ordinals α, β with $\alpha < \beta$" respectively. More formally, we want the following to hold:

- For every ordinal $\alpha < \Lambda$, it holds that $\mathbb{N} \models \ulcorner\alpha\urcorner <_\Lambda \Lambda$;
- for all ordinals $\alpha, \beta < \Lambda$, it holds that $\alpha \leq \beta$ iff $\mathbb{N} \models \ulcorner\alpha\urcorner \leq_\Lambda \ulcorner\beta\urcorner$;
- $\mathrm{EA} \vdash$ " $\leq_\Lambda$ is a partial order".

Notice that we make no demands on $\leq_\Lambda$ being a well order or even linear. Since both $x <_\Lambda \Lambda$ and $x \leq_\Lambda y$ are Σ_1-formulas, we can use Σ_1-completeness to have for every representable theory $S \supseteq \mathrm{EA}$ that $\mathbb{N} \models x <_\Lambda \Lambda$ implies $\Box_S \dot{x} <_\Lambda \Lambda$, and similarly $\mathbb{N} \models x <_\Lambda y$ implies $\Box_S \dot{x} <_\Lambda \dot{y}$. For the remainder of this paper we will write $\alpha < \beta$ instead of $\ulcorner\alpha\urcorner <_\Lambda \ulcorner\beta\urcorner$ and $\alpha < \Lambda$ instead of $\ulcorner\alpha\urcorner <_\Lambda \Lambda$.

With all that in mind, we can return to extending the base language with iterated truth predicates.

Definition 6. *Given an at most finite extension of the language of arithmetic* $\mathcal{L}$*, let* $\mathcal{L}_\alpha := \mathcal{L} \cup \{\mathsf{T}_\beta : \beta < \alpha\}$*. We then define* UTB_α *as the* $\mathcal{L}_{\alpha+1}$ *theory* $\mathrm{UTB}_{\mathcal{L}_\alpha}[\mathsf{T} \leftarrow \mathsf{T}_\alpha]$*. Additionally, we define:*

$$\mathrm{UTB}_{<\alpha} := \bigcup_{\beta<\alpha} \mathrm{UTB}_\beta, \qquad \mathrm{UTB}_{\leq\alpha} := \mathrm{UTB}_{<\alpha} \cup \mathrm{UTB}_\alpha.$$

Given an ordinal α*, we write*

$$\mathrm{UTB}_{\lfloor\alpha\rfloor} := \begin{cases} \mathrm{UTB}_\beta, & \text{if } \alpha = \omega(1+\beta)+n; \\ \emptyset, & \text{if } \alpha = n. \end{cases}$$

Observe that the β *above is unique for given* α.

Typically, the language $\mathcal{L}_\alpha$ is going to be infinite. So in order to make use of Theorem 3, we will use a translation of formulas to a finite fragment of the language as is done in [5]. Given an $\mathcal{L}$-formula φ and some ordinal α, let $\varphi^\bullet$ denote the result of the simultaneous substitution of $\mathsf{T}_\alpha(\mathsf{T}_\beta(\dot{t}))$ for $\mathsf{T}_\beta(t)$ in φ for every $\beta < \alpha$ (not substituting inside the terms t). Then we write $\mathrm{UTB}^\bullet_{\leq\alpha}$ to denote the $\mathcal{L}_{\alpha+1}$-theory axiomatized by $\{\varphi^\bullet : \varphi \in \mathrm{UTB}_{\leq\alpha}\}$.

Lemma 6. *For all* $\varphi \in \mathcal{L}_{\alpha+1}$,

- $\mathrm{EA} + \mathrm{UTB}_\alpha \vdash \varphi \leftrightarrow \varphi^\bullet$;
- $\mathrm{EA} + \mathrm{UTB}_{\leq\alpha} \vdash \varphi$ *iff* $\mathrm{EA} + \mathrm{UTB}^\bullet_{\leq\alpha} \vdash \varphi^\bullet$

It is formalizable in EA that for any c.e. $\mathcal{L}$-theory $S \supseteq \mathrm{EA}$, the theory $S + \mathrm{UTB}$ is a conservative extension over S for $\mathcal{L}$-formulas [12]. In particular, given $\alpha < \beta$ and $S \supseteq \mathrm{EA} + \mathrm{UTB}_{<\alpha}$ a c.e. $\mathcal{L}_\alpha$-theory, then $S + \mathrm{UTB}_{<\beta}$ is a conservative extension over S for $\mathcal{L}_\alpha$-formulas [5].

From here on, we will assume that $\mathcal{L} \supseteq \mathcal{L}_{\mathrm{PA}}$ is at most a finite extension of the language of arithmetic. For a given elementary well-ordering $(\Lambda, <)$, we expand it into an ordering of $(\omega(1+\Lambda), <)$ by encoding $\omega\alpha + n$ as pairs $\langle \alpha, n\rangle$ with the expected ordering on them.

Definition 7 (Hyperarithmetical hierarchy). *For ordinals up to* $\omega(1+\Lambda)$, *we define the hyperarithmetical hierarchy as (*Σ_α *is defined similarly):*

- $\Pi_n := \Pi_n^{\mathcal{L}}$, *for every* $n < \omega$;
- $\Pi_{\omega(1+\alpha)+n} := \Pi_{n+1}^{\mathcal{L}_\alpha}(\mathsf{T}_\alpha)$;
- *For* λ *a limit ordinal, we denote* $\Pi_{<\lambda} := \bigcup_{\alpha<\lambda} \Pi_\alpha$.

For any theory S and for every $\alpha, \lambda < \omega(1+\Lambda)$, where λ is a limit ordinal, we define $R_\alpha(S) := \Pi_{1+\alpha}\text{-RFN}(S)$ and $R_{<\lambda}(S) := \Pi_{<\lambda}\text{-RFN}(S)$. Using Lemma 6 and Theorem 3, we obtain:

Proposition 2 (Proposition 5.4 [5]).

(i) If $S \supseteq \mathrm{EA} + \mathrm{UTB}_\alpha$, *then over* $\mathrm{EA} + \mathrm{UTB}_\alpha$,

$$R_{\omega(1+\alpha)+n}(S) \equiv \Pi_{n+1}^{\mathcal{L}}(\mathsf{T}_\alpha)\text{-RFN}(S);$$

(ii) If $S \supseteq \mathrm{EA}+\mathrm{UTB}_\alpha$ *and* $\beta = \omega(1+\alpha)+n$, *then* $R_\beta(S)$ *is finitely axiomatizable over* $\mathrm{EA} + \mathrm{UTB}_\alpha$;

(iii) If $S \supseteq \mathrm{EA} + \mathrm{UTB}_{<\alpha}$, *then over* $\mathrm{EA} + \mathrm{UTB}_{<\alpha}$,

$$R_{<\omega(1+\alpha)}(S) \equiv \mathcal{L}_\alpha\text{-RFN}(S) \equiv \{R_\beta(S) : \beta < \omega(1+\alpha)\}.$$

Now we can define the interpretation of $[\alpha]\varphi$ that we will be using for the soundness proof.

Definition 8. *We will write $[\alpha]_S\varphi$ as a shorthand for the finite axiomatization of $\neg R_\alpha(S+\neg\varphi)$ given by Statement (ii) of Proposition 2. which for $\alpha=\omega(1+\beta)+n$, is the Σ_α-formula:*

$$i\delta^{\mathcal{L}(\mathsf{T}_\beta)} \to \exists\theta\in\Sigma_{n+1}^{\mathcal{L}(\mathsf{T}_\beta)}\big(\mathsf{Tr}_{\Sigma_{n+1}^{\mathcal{L}(\mathsf{T}_\beta)}}(\theta)\wedge\Box_S(\theta\to\varphi)\big),$$

where $i\delta^{\mathcal{L}(\mathsf{T}_\beta)}$ is a finite $\Pi_1^{\mathcal{L}(\mathsf{T}_\beta)}$-axiomatization of $I\Delta_0^{\mathcal{L}(\mathsf{T}_\beta)}$.
Similarly, by $\langle\alpha\rangle\varphi$ we will denote the finite axiomatization of $R_\alpha(S+\varphi)$.

An arithmetical realization is a function $(\cdot)^*_S$ from the language of GLP_Λ to $\mathcal{L}_\Lambda$, mapping propositional variables to formulas of $\mathcal{L}_\Lambda$ and preserving the logical operations: $(\varphi\wedge\psi)^*_S := \varphi^*_S\wedge\psi^*_S$, $(\neg\varphi)^*_S := \neg\varphi^*_S$, and $([\alpha]\varphi)^*_S := [\alpha]_S\varphi^*_S$. GLP_Λ is sound for this interpretation.

Theorem 4. *For every $S\supseteq \mathrm{EA}+\mathrm{UTB}_{<\Lambda}$ and every formula φ in the language of GLP_Λ,*

$$\mathrm{GLP}_\Lambda\vdash\varphi \Rightarrow \mathrm{EA}+\mathrm{UTB}_{<\Lambda}\vdash(\varphi)^*_S,\ \textit{for every realization } (\cdot)^*_S \textit{ of the variables of } \varphi.$$

The proof of soundness from here on is routine, starting with the corresponding provable completeness.

Lemma 7 (Provable Σ_α-completeness).

$$\mathrm{EA}+\mathrm{UTB}_{\lfloor\alpha\rfloor}\vdash\varphi\to[\alpha]_S\varphi,\ \textit{if } \varphi\in\Sigma_\alpha.$$

Proof. From Statement (ii) of Proposition 2, $[\alpha]_S\varphi$ is finitely axiomatizable in $\mathrm{EA}+\mathrm{UTB}_{\lfloor\alpha\rfloor}$. We will prove the contrapositive by reasoning within $\mathrm{EA}+\mathrm{UTB}_{\lfloor\alpha\rfloor}$. Assume the finite axiomatization of $R_\alpha(S+\neg\varphi)$, which implies $\Box_{S+\neg\varphi}\neg\varphi\to\neg\varphi$ because $\neg\varphi\in\Pi_\alpha$. Since $\Box_{S+\neg\varphi}\neg\varphi$ holds, $\neg\varphi$ follows.

Now we have all the tools to prove Löb's derivability conditions:

Lemma 8. *Let $\alpha<\beta$ and $\mathrm{EA}+\mathrm{UTB}_{\lfloor\alpha\rfloor}+\mathrm{UTB}_{\lfloor\beta\rfloor}\subseteq S$, then*

(i) If $S\vdash\varphi$ then $\mathrm{EA}+\mathrm{UTB}_{\leq\Lambda}\vdash[\alpha]_S\varphi$;
(ii) $\mathrm{EA}+\mathrm{UTB}_{\leq\Lambda}\vdash[\alpha]_S(\varphi\to\psi)\to([\alpha]_S\varphi\to[\alpha]_S\psi)$;
(iii) $\mathrm{EA}+\mathrm{UTB}_{\leq\Lambda}\vdash[\alpha]_S\varphi\to[\alpha]_S[\alpha]_S\varphi$;
(iv) $\mathrm{EA}+\mathrm{UTB}_{\leq\Lambda}\vdash[\alpha]_S\varphi\to[\beta]_S\varphi$;
(v) $\mathrm{EA}+\mathrm{UTB}_{\leq\Lambda}\vdash\langle\alpha\rangle_S\varphi\to[\beta]_S(\langle\alpha\rangle_S\varphi)$.

Proof. By statement (ii) of Proposition 2 the $[\alpha]_S\varphi$ and $[\beta]_S\varphi$ formulas are well defined as the finite axiomatizations of $\neg R_\alpha(S+\neg\varphi)$ and $\neg R_\beta(S+\neg\varphi)$ respectively.

(i) The assumption implies $\mathrm{EA}\vdash\Box_S\varphi$ and so statement (*i*) follows.
(ii) Immediate from the statement (i) of Proposition 2 and Lemma 5.
(iii) Follows from Lemma 7 as $[\alpha]_S\varphi$ is a Σ_α formula.

(iv) Assume that $\alpha = \omega\gamma + n$ and $\beta = \omega\delta + m$ with $\gamma < \delta$ and reasoning in $\mathrm{EA} + \mathrm{UTB}_{\lfloor\alpha\rfloor} + \mathrm{UTB}_{\lfloor\beta\rfloor}$ we remark that if a formula φ is Σ_α then it is equivalent to $\mathsf{T}_\delta(\varphi)$ which is a $\Sigma_{\omega\delta}$-formula.

(v) Since $\langle\alpha\rangle_S\varphi$ is a Π_α-formula then, reasoning as above, it is also a Σ_β-formula over
$\mathrm{EA} + \mathrm{UTB}_\gamma + \mathrm{UTB}_\delta$.

Lemma 9 **([11]).** *Let* $\mathrm{GL}_\blacksquare$ *be the extension of* GL *by a new modal operator* $\blacksquare$ *and the axioms* $\Box\varphi \to \blacksquare\varphi$, $\blacksquare\varphi \to \blacksquare\blacksquare\varphi$, *and* $\blacksquare(\varphi \to \psi) \to (\blacksquare\varphi \to \blacksquare\psi)$. *Then for all* φ, $\blacksquare(\blacksquare\varphi \to \varphi) \to \blacksquare\varphi$.

Since Löb's theorem holds for [0] as usual from the fixed point theorem, we conclude that it holds for all modalities, concluding our proof of Theorem 4.

Lemma 10. *Let* $\mathrm{EA}+\mathrm{UTB}_{\lfloor\alpha\rfloor} \subseteq S$. *Then,* $\mathrm{EA}+\mathrm{UTB}_{\leq\Lambda} \vdash [\alpha]_S([\alpha]_S\varphi \to \varphi) \to [\alpha]_S\varphi$.

4 Worm Battles Outside PA

Let $\equiv_\alpha$ and $\equiv_{<\lambda}$ denote equivalence for $\Pi_{1+\alpha}$ and $\Pi_{<\lambda}$-sentences respectively. In [5] two conservation results are proven to hold provably in EA^+: Theorems 5 and 6. We fix a particular Λ.

4.1 The Reduction Property

The first conservation result centers around the case for reflection on limit ordinals.

Theorem 5. *Let* $\lambda = \omega(1 + \alpha)$ *and* $S \supseteq \mathrm{EA} + \mathrm{UTB}_\alpha$. *Over* $\mathrm{EA} + \mathrm{UTB}_{<\Lambda}$, $R_\lambda(S) \equiv_{<\lambda} R_{<\lambda}(S)$.

The second conservation result centers around successors. It can be viewed as an extension of the so-called reduction property (cf. [3]) to cover all successor ordinals and not just the finite ones.

Theorem 6. *Let* V *be a* $\Pi_{1+\alpha+1}$*-axiomatized extension of* $\mathrm{EA} + \mathrm{UTB}_{<\Lambda}$ *and let provably* $S \supseteq V$. *Then, over* V, $R_{\alpha+1}(S) \equiv_\alpha \{R_\alpha(S), R_\alpha(S + R_\alpha(S)), \ldots\}$.

As in the case of GLP_ω we can recast these conservation results in terms of our interpreted modalities (using the same notation for the modality and its arithmetical denotation).

Corollary 3 (Reduction Property). *If* $\beta \leq \alpha, \lambda < \Lambda$ *with* λ *being a limit ordinal, then*

$$\mathrm{EA}^+ + \mathrm{UTB}_{<\Lambda} \vdash \langle\beta\rangle\langle\alpha{+}1\rangle\varphi \leftrightarrow \forall k\ \langle\beta\rangle Q_k^\alpha(\varphi);$$
$$\mathrm{EA} + \mathrm{UTB}_{<\Lambda} \vdash \langle\beta\rangle\langle\lambda\rangle\varphi \leftrightarrow \forall k\ \langle\beta\rangle\langle\lambda[k]\rangle\varphi.$$

Proof. By Theorem 6 for $V = \mathrm{EA} + \mathrm{UTB}_\Lambda$ and $S = V + \varphi$, we have

$$\{\langle\alpha+1\rangle\varphi\} \equiv_\beta \{Q^\alpha_k(\varphi) : k < \omega\}$$

holds over $\mathrm{EA}+\mathrm{UTB}_\Lambda$ and is an equivalence formalizable in $\mathrm{EA}^+ + \mathrm{UTB}_{<\Lambda}$. So over $\mathrm{EA}^+ + \mathrm{UTB}_{<\Lambda}$, a $\Pi_{1+\beta}$ sentence is provable from $\langle\alpha+1\rangle\varphi$ if and only if it is so from $Q^\alpha_k(\varphi)$, for some k which proves the first Reduction Property.

For the second, by Theorem 5 for $S = \mathrm{EA}+\mathrm{UTB}_{\lfloor\lambda\rfloor}+\varphi$, over $\mathrm{EA}+\mathrm{UTB}_{<\Lambda}$, $\{\langle\lambda\rangle\varphi\} \equiv_{<\lambda} \{\langle\gamma\rangle\varphi : \gamma < \lambda\}$, and the equivalence is also provable over $\mathrm{EA} + \mathrm{UTB}_{<\Lambda}$. So over $\mathrm{EA}+\mathrm{UTB}_{<\Lambda}$, a $\Pi_{1+\beta}$-sentence ψ is provable from $\langle\lambda\rangle\varphi$ if and only if it is so from $\langle\gamma\rangle\varphi$ for some $\gamma < \lambda$. Let k be such that $\gamma < \lambda[k]$, then ψ is also provable from $\langle\lambda[k]\rangle\varphi$.

In the first order language $\mathcal{L}(\mathsf{T})$, consider the following theory

$$\mathrm{PA}(\mathsf{T}) := \mathrm{EA}{+}\mathrm{UTB}_{\mathcal{L}(\mathsf{T})} \;+\; R_{<\omega 2}(\mathrm{EA}{+}\mathrm{UTB}_{\mathcal{L}(\mathsf{T})}),$$

equivalent (provably so in EA^+) to the corresponding $\mathrm{PA}(\mathsf{T})$ in [5] and to CT in [12]. We have the following well known result from [12]:

Theorem 7. $\mathrm{PA}(\mathsf{T})$ *and* ACA *are proof theoretically equivalent.*

As such, we are going to use $\mathrm{PA}(\mathsf{T})$ as a substitute for ACA in our theorem on the equivalence between it and the corresponding worm principle.

Theorem 8. $\mathrm{EWD}^{\omega 2}$ *is equivalent to* $1\text{-Con}(\mathrm{PA}(\mathsf{T}))$ *in* EA.

At the same time, we will prove a relationship between the worm principle and $\mathrm{I}\Sigma_n$. It is currently unknown whether the implication is in fact an equivalence.

Theorem 9. *Over* EA *the following hold:*

$$1\text{-Con}(\mathrm{I}\Sigma_n) \vdash \mathrm{EWD}^n;$$
$$\mathrm{EWD}^{n+1} \vdash 1\text{-Con}(\mathrm{I}\Sigma_n).$$

We will prove the theorems simultaneously since the proofs for both are similar. To this end, we will make an abuse of notation using the fact that provably over EA, $\mathrm{EA}+\mathrm{UTB}$ is a conservative extension of EA for $\mathcal{L}$-formulas. For the remainder of this paper we write $[\alpha]\varphi$ to mean $[\alpha]_{\mathrm{EA+UTB}}\varphi$. Note that if $\varphi \in \mathcal{L}$, then $[\alpha]\varphi$ is equivalent to $[\alpha]_{\mathrm{EA}}\varphi$ due to conservativity. We will make the same convention for the $\langle\alpha\rangle\varphi$ and the proof theoretic worms.

4.2 From 1-consistency to the Worm Principle

The proof for both directions will follow the structure of the corresponding proof in [3].

Proposition 3.

1. $\mathrm{EA} + 1\text{-Con}(\mathrm{PA}(\mathsf{T})) \vdash \mathrm{EWD}^{\omega 2}$;
2. $\mathrm{EA} + 1\text{-Con}(\mathrm{I}\Sigma_n) \vdash \mathrm{EWD}^{n}$.

There is a distinction in the first step of this proof, due to the fact that $\mathrm{I}\Sigma_n$ is an extension of EA with reflection for a successor ordinal, in comparison to PA or PA(T), which correspond to reflection for a limit.

Lemma 11. *For any* $A{\in}\mathbb{W}^{\omega 2}$, $\mathrm{PA}(\mathsf{T}) \vdash A$.

Proof. For every $A \in \mathbb{W}^{\omega 2}$, there is some $m > 0$ such that $A \in \mathbb{W}^{\omega+m}$ and so by Proposition 1,

$$\mathrm{GLP} \vdash \langle \omega + m \rangle \top \to A.$$

Therefore, by arithmetical soundness of GLP, it holds that $\mathrm{EA} + \mathrm{UTB} \vdash \langle \omega{+}m \rangle \top \to A$ and since $\mathrm{PA}(\mathsf{T}) \vdash \langle \omega{+}m \rangle \top$, the lemma follows and its proof is formalizable in EA (or EA^+ if we are to use the corresponding PA(T) in [5]).

Similarly for the $\mathrm{I}\Sigma_n$, we have the corresponding theorem giving us the proof theoretic worms we can make use of in its case.

Lemma 12. *For any* $A{\in}\mathbb{W}^{n+1}$, $\mathrm{I}\Sigma_n \vdash A$.

Proof. By Proposition 1 we have that for every $A \in \mathbb{W}^{n+1}$, $\mathrm{GLP} \vdash \langle n{+}1 \rangle \top \to A$. Therefore, by arithmetical soundness of GLP, it holds $\mathrm{EA} \vdash \langle n + 1 \rangle \top \to A$ and since $\mathrm{I}\Sigma_n \vdash \langle n + 1 \rangle \top$, the lemma follows and its proof is formalizable in EA^+.

Now we introduce a notation we will use for the remainder of the proof of this direction. Given a worm A, we define A^+ inductively by $\top^+ := \top$ and if $A = \langle \alpha \rangle B$ then $A^+ = \langle \alpha + 1 \rangle (B^+)$.

Lemma 13. $\mathrm{EA} \vdash \forall\, A{\in}\mathbb{W}^{\omega 2}\, \forall k\, \big(A_k{\neq}\top \to \Box(A_k^+ \to \langle 1 \rangle\, A_{k+1}^+)\big)$.

Proof. It is sufficient to prove in EA

$$\forall\, A{\neq}\top\ \forall k\ \mathrm{EA} + \mathrm{UTB} \vdash A^+ \to \langle 1 \rangle\, A[\![k]\!]^+.$$

For this, we will move over to $\mathrm{GLP}_{\omega 2}$ where we have that the following proof is bounded by a function elementary in A and k and hence it is formalizable in EA that $\mathrm{GLP}_{\omega 2} \vdash A \to \Diamond A[\![k]\!]$, and as theorems of $\mathrm{GLP}_{\omega 2}$ are stable under the $(\cdot)^+$ operator, $\mathrm{GLP}_{\omega 2} \vdash A^+ \to \langle 1 \rangle\, A[\![k]\!]^+$, which by the arithmetical soundness of $\mathrm{GLP}_{\omega 2}$, proves that for every $A \in \mathbb{W}^{\omega 2}$ with $A \neq \top$ and for every k,

$$\mathrm{EA} + \mathrm{UTB} \vdash A^+ \to \langle 1 \rangle\, A[\![k]\!]^+.$$

From here, we are of course unable to use Σ_1-induction to prove

$$\mathrm{EA} \vdash \forall k\, \big(A_k{\neq}\top \to \Box(A_k^+ \to \langle 1 \rangle\, A_{k+1}^+)\big),$$

which is how we would –in principle– expect to complete the proof. Instead we use the fact that for a given k, the proof of $A_k^+ \to \langle 1 \rangle\, A_{k+1}^+$ is bounded by an elementary function of A and k. The proof itself can be formalized within EA and therefore the formula $\Box(A_k^+ \to \langle 1\rangle A_{k+1}^+))$ can be written as a Δ_0-formula by placing the existential quantifier inside this bound. So we complete the proof with a Δ_0-induction.

Lemma 14. $\mathrm{EA} \vdash \forall A \in \mathbb{W}^{\omega 2}(\langle 1\rangle A_0^+ \to \exists m\ A_m = \top)$.

Proof. We prove the contrapositive. The first part of our reasoning will prepare for an application of Löb's theorem. Reasoning within EA,

$$[1]\forall m\ [1]\neg A_m^+ \vdash [1]\forall m\ [1]\neg A_{m+1}^+ \vdash \forall m\ [1][1]\neg A_{m+1}^+.$$

Therefore, using Lemma 13 in the form $\mathrm{EA} \vdash \forall k\ (A_k \neq \top \to [1]([1]\neg A_{k+1}^+ \to \neg A_k^+))$,

$$\forall m\ A_m \neq \top \wedge [1]\forall m\ [1]\neg A_m^+ \vdash \forall m\ A_m \neq \top \wedge \forall m\ [1][1]\neg A_{m+1}^+ \vdash \forall m\ [1]\neg A_m^+.$$

Thus $\mathrm{EA} \vdash \forall m\ A_m \neq \top \to ([1]\forall m\ [1]\neg A_m^+ \to \forall m\ [1]\neg A_m^+)$. Then, after necessitation on the [1]-modality and distribution we have $\mathrm{EA} \vdash [1]\forall m\ A_m \neq \top \to [1]([1]\forall m\ [1]\neg A_m^+ \to \forall m\ [1]\neg A_m^+)$, hence by Löb's theorem $\mathrm{EA} \vdash [1]\forall m\ A_m \neq \top \to [1]\forall m [1]\neg A_m^+$.

Now observe that $\forall m\ A_m \neq \top$ is Π_1, so certainly Σ_2 and hence by Σ_2-completeness

$$\mathrm{EA} \vdash \forall m\ A_m \neq \top \to [1]\forall m\ A_m \neq \top.$$

But then in EA,

$$\forall m\ A_m \neq \top \vdash \forall m\ A_m \neq \top \wedge [1]\forall m [1]\neg A_m^+ \vdash \forall m\ [1]\neg A_m^+ \vdash [1]\neg A_0^+.$$

By contraposition $\mathrm{EA} \vdash \langle 1\rangle A_0^+ \to \exists m\ A_m = \top$, as desired.

Note that the use of A^+ in the above lemma does not allow us to apply it to EWD^{n+1} in place of $\mathrm{EWD}^{\omega 2}$. Moreover, it cannot be avoided using the current proof. Now we prove Proposition 3: from Lemmata 11 and 14 we obtain that for each $A \in \mathbb{W}^{\omega 2}$, $\mathrm{PA}(\mathsf{T}) \vdash \langle 1\rangle A^+$ and over EA

$$\langle 1\rangle \mathrm{PA}(\mathsf{T}) \vdash \forall A \in \mathbb{W}^{\omega 2}\ \langle 1\rangle A^+ \vdash \forall A \in \mathbb{W}^{\omega 2}\ \exists m\ A_m = \top \vdash \mathrm{EWD}^{\omega 2}.$$

Similarly for the case of $\mathrm{I}\Sigma_n$, from Lemmata 12 and 14 we obtain that formalisably in EA, for each $A \in \mathbb{W}^n$, $\mathrm{I}\Sigma_n \vdash \langle 1\rangle A^+$ and $\mathrm{EA} \vdash \langle 1\rangle A^+ \to \exists m\ A_m = \top$. Hence, as before we obtain $\forall A \in \mathbb{W}^n\ \exists m\ A_m = \top$, which is EWD^n.

4.3 From the Worm Principle to 1-consistency

Now we prove the second direction of Theorem 8, proving independence of $\mathrm{EWD}^{\omega 2}$.

Proposition 4.

1. $\mathrm{EA} + \mathrm{EWD}^{\omega 2} \vdash$ 1-Con(PA(T));
2. $\mathrm{EA} + \mathrm{EWD}^{n+1} \vdash$ 1-Con($\mathrm{I}\Sigma_n$).

We use a Hardy functions' analogue $h_A(m)$ defined as the smallest k such that $A[\![{k \atop m}]\!] = \top$, where $A[\![{k \atop m}]\!] := A[\![m]\!] \dots [\![m{+}k]\!]$. Each function h_A is computable and hence there is a natural Σ_1 presentation of $h_A(m) = k$ in EA. We will use the following relation to prove monotonicity for the h_A function.

Definition 9. *For $A, B \in \mathbb{W}^{\omega 2}$, we define the partial ordering $B \trianglelefteq A$ iff $B = \top$ or $A = D\alpha C$ and $B = \beta C$ for some $\beta \leq \alpha$.*
For every natural number m, we define $B \trianglelefteq_m A$ iff $B \trianglelefteq A$ and additionally, if $B = nC$ with $n < \omega$ and, $A = D\alpha C$ with $\alpha \geq \omega$, then $n \leq m$.

Of course, by the definition, we immediately have that if $B \trianglelefteq_m A$ and $m \leq n$ then $B \trianglelefteq_n A$. Additionally, if $A = CB$ for some C then $B \trianglelefteq_m A$ for every $m \geq 0$. Over EA, and for worms in $\mathbb{W}^{\omega 2}$, we have the following:

Lemma 15. *If $h_A(m)$ is defined and $B \trianglelefteq_m A$, then $\exists\, k\ A[\![{k \atop m}]\!] = B$.*

Proof. The Definition of the step-down function $A[\![\cdot]\!]$ is such that an ordinal α_i of $A = \alpha_{|A|-1} \dots \alpha_0$ can only change if all elements to the left of it are deleted. So by the assumption of $A[\![{s \atop m}]\!] = \top$ there is some k_0 such that $A[\![{k_0 \atop m}]\!] = \alpha_{|B|-1} \dots \alpha_0$. We consider the case where $\alpha_{|B|-1} \geq \omega$ and the corresponding ordinal $\beta_{|B|-1}$ in $B = \beta_{|B|-1} \dots \beta_0$ is $<\omega$; the other cases are similar. Then by assumption of $B \trianglelefteq_m A$, the ordinal $\beta_{|B|-1}$ is also some $n \leq m$.

Let $\alpha_{|B|-1} = \omega + l$, then we can prove with Δ_0-induction on l bounded by s that there is some k_1 such that $A[\![{k_1 \atop m}]\!] = \langle\omega\rangle C$ where $B = \langle n\rangle C$ and $A = D\langle\omega + l\rangle C$. Then $A[\![{k_1+1 \atop m}]\!] = \langle m + k_1 + 1\rangle C$. With a second Δ_0-induction bounded by s, we can find as before some $k < s$ such that $A[\![{k \atop m}]\!] = B$.

A more detailed proof can be found in the proofs of lemmata 9.4.3 and 6.3.3 in [17].

The above can be easily expanded into the following:

Corollary 4. *If $h_A(n)$ is defined and $B \trianglelefteq_n A$, then $\forall\, m{\leq}n\ \exists\, k\ A[\![{k \atop n}]\!] = B[\![m]\!]$.*

Proof. By Lemma 15, there is k_1 such that $A[\![{k_1 \atop n}]\!] = B$. Then, as $k_1 + 1 > n \geq m$, there is some C such that $A[\![{k_1+1 \atop n}]\!] = CB[\![m]\!]$ and since $h_A(n)$ halts, we can use Lemma 15 once more to show that there is some k_2 such that $A[\![{k_1+k_2 \atop n}]\!] = B[\![m]\!]$.

Using this result, we have the following monotonicity statement:

Lemma 16. *If $h_A(y)$ is defined, $B \trianglelefteq_y A$ and $x \leq y$, then $h_B(x)$ is defined and $h_B(x) \leq h_A(y)$.*

Proof. By applying Corollary 4 several times, we obtain $s_0, s_1, \dots$ such that $A[\![{s_0 \atop y}]\!] = B[\![x]\!]$, where $y + s_0 \geq x$, $A[\![{s_0+s_1 \atop y}]\!] = B[\![x]\!][\![x+1]\!]$, where $y + s_0 + s_1 \geq x + 1$, etc. Hence all elements of the sequence starting with B occur in the sequence for A and since $h_A(y)$ is defined, so is $h_B(x)$.

Next we look into some results that bound the functions h_A from below and compare them with some fast growing functions.

Lemma 17. *For every* $A, B \in \mathbb{W}^{\omega 2}$, *if* $h_{B0A}(n)$ *is defined, then*

$$h_{B0A}(n) = h_A(n + h_B(n) + 2) + h_B(n) + 1 > h_A(h_B(n)).$$

Proof. Since $0A \trianglelefteq_0 B0A$, by Lemma 15 we have that $h_B(n)$ is defined. As $B0A$ first rewrites itself to $0A$ in $h_B(n)$ steps and then begins to rewrite A into $\top$ at step $n + h_B(n) + 2$, we have that $h_A(n + h_B(n) + 2)$ is then defined. Finally, by Lemma 16, $h_A(h_B(n)) \leq h_A(n + h_B(n) + 2)$ and it is also defined.

Seeing how easy it is to achieve a lower bound based on the composition of functions, we can proceed by trying to get in-series iterations of this. Since the h_A functions are in general strictly monotonous, we will be getting faster and faster growing functions by following this method.

Corollary 5. *If* $A \in \mathbb{W}_1^{\omega 2}$ *and* $h_{1A}(n)$ *is defined, then* $h_{1A}(n) > h_A^{(n)}(n)$.

Proof. Since $(1A)[\![n]\!] = (0A)^{n+1}$, we can perform induction on the number of in-series concatenations of $0A$ by applying Lemma 17.

As an application of this, we can see how quickly we reach superexponential growth.

Corollary 6. *If* $h_{1111}(n)$ *is defined then,* $h_{1111}(n) > 2_n^n$ *and* $h_{111}(n) > 2^n$.

Proof. We will make use of Corollary 5 multiple times. Clearly we first have that $h_{1111}(n) > h_{111}^{(n)}(n)$, then $h_{111}(n) > h_{11}^{(n)}(n)$ and $h_{11}(n) > h_1^{(n)}(n)$. We can easily prove by induction in EA that $h_1(n) = n + 1$. So by applying the compositions, $h_{11}(n) > 2n$ and so $h_{111}(n) > 2^n$ and finally $h_{1111} > 2_n^n$.

At this point we find ourselves equipped to tackle the main lemma on which the proof of this direction rests. Due to the complexity added by the limit ordinal ω, there is a technical addition in this proof when compared to the corresponding proof for PA in [3].

Lemma 18. $\mathrm{EA} \vdash \forall A{\in}\mathbb{W}_1^{\omega 2}\ (h_{A1111}{\downarrow}\ \rightarrow \langle 1\rangle A)$.

Proof. By Löb's Theorem, this is equivalent to proving

$$\mathrm{EA} \vdash \Box(\forall A{\in}\mathbb{W}_1^{\omega 2}\ (h_{A1111}{\downarrow}\ \rightarrow \langle 1\rangle A)) \rightarrow \forall A{\in}\mathbb{W}_1^{\omega 2}\ (h_{A1111}{\downarrow}\ \rightarrow \langle 1\rangle A). \quad (1)$$

We reason in EA. Let us take the antecedent of (1) as an additional assumption, which by the monotonicity axiom of GLP_ω interpreted in EA, implies $[1](\forall A{\in}\mathbb{W}_1^{\omega 2}\ (h_{A1111}{\downarrow}\ \rightarrow \langle 1\rangle A))$. This in turn implies:

$$\forall A{\in}\mathbb{W}_1^{\omega 2}\ [1](h_{A1111}{\downarrow}\ \rightarrow \langle 1\rangle A). \quad (2)$$

We make a case distinction on whether $A1111$ starts with a 1 or with an ordinal strictly larger than 1.

If $A1111 = 1B$ then by Corollary 5, we have $h_{1B}{\downarrow} \to \lambda x.h_B^{(x)}(x){\downarrow}$. The function h_B is increasing, has an elementary graph and grows at least exponentially as per Corollary 6, $h_{111} > 2^x$. So for $A = \top$ we have that $h_{1111}{\downarrow}$ implies the totality of 2^x_n and hence EA^+, which by Corollary 2, implies $\langle 1 \rangle \top$.[2] If A is nonempty, we reason as follows:

$$\begin{aligned} \lambda x.h_B^{(x)}{\downarrow} &\vdash \langle 1 \rangle\, h_B{\downarrow}, && \text{by Lemma 3} \\ &\vdash \langle 1 \rangle \langle 1 \rangle\, B, && \text{by Assumption (2)} \\ &\vdash \langle 1 \rangle\, A. \end{aligned}$$

If $A1111 = C$ starts with $\alpha > 1$, we have $h_C{\downarrow} \vdash \lambda x.h_{C[\![x]\!]}(x+1){\downarrow} \vdash \forall n\ h_{C[\![n]\!]}{\downarrow}$. The last implication is derived by application of Lemma 16 as for arbitrary n, if $x \le n$ then $h_{C[\![n]\!]}(x) \le h_{C[\![n]\!]}(n+1)$ and if $n \le x$ then $h_{C[\![n]\!]}(x) \le h_{C[\![x]\!]}(x+1)$. In both cases, the larger value is defined.

We can perform this line of argument a second time, something we will use for the case that $\alpha = \omega$, obtaining

$$\forall n\ h_{C[\![n]\!]}{\downarrow} \vdash \forall n\ \lambda x.h_{C[\![n]\!][\![x]\!]}(x+1){\downarrow} \vdash \forall n\ h_{C[\![n]\!][\![n+1]\!]}{\downarrow}.$$

Now notice that no matter what the α is, we will always have that either $1(C[\![n]\!]) \trianglelefteq_1 C[\![n+1]\!]$ or $1(C[\![n]\!]) \trianglelefteq_1 C[\![n+1]\!][\![n+2]\!]$. To prove this, let D be such that $C = \alpha D1111$.

If $\alpha = \omega$, then $1(C[\![n]\!]) = 1nD1111$ and $C[\![n{+}1]\!] = \langle n+1\rangle D1111$ therefore $C[\![n{+}1]\!][\![n{+}2]\!] = (nh_{n+1}(D1111))^{n+3} r_{n+1}(D1111) = (nh_{n+1}(D1111))^{n+2} nD1111$. So if $n = 0$ then since $D \in \mathbb{W}_1^{\omega 2}$, we have that $r_1(D1111) = \top$ and therefore,

$$C[\![n+1]\!][\![n+2]\!] = (0D1111)^{0+2}0D1111 = 0D11110D11110D1111 = 0D11110D1111C[\![n]\!].$$

If $n > 0$ then clearly $nh_{n+1}(D1111)$ has as its rightmost element something ≥ 1 and so $1(C[\![n]\!]) \trianglelefteq_1 C[\![n+1]\!][\![n+2]\!]$.

If $\alpha \ne \omega$ then, $1(C[\![n]\!]) = 1(\langle \alpha - 1\rangle h_\alpha(D1111))^{n+1} r_\alpha(D1111)$ and

$$C[\![n{+}1]\!] = (\langle \alpha - 1\rangle h_\alpha(D1111))^{n+2} r_\alpha(D1111).$$

So $1(C[\![n]\!]) \trianglelefteq_1 C[\![n+1]\!]$. Therefore we have:

$$\begin{aligned} \forall n\ h_C{\downarrow} &\vdash \forall n\ h_{1(C[\![n]\!])}{\downarrow} && \text{(by the above)} \\ &\vdash \forall n\ \lambda x.h^{(x)}_{C[\![n]\!]}(x){\downarrow} \\ &\vdash \forall n\ \langle 1 \rangle\, h_{C[\![n]\!]}{\downarrow} && \text{(by Lemma 3).} \end{aligned}$$

[2] Since over EA, it is provable that EA + UTB is conservative over EA, by their definition, $\langle 1 \rangle_{\mathrm{EA}} \top$ and $\langle 1 \rangle_{\mathrm{EA+UTB}} \top$ are equivalent.

Again observe that since A starts with something bigger than 1, we have $C[\![n]\!] = A[\![n]\!]1111$, hence we can apply our assumption. Hence the argument continues,

$$\vdash \forall n \langle 1 \rangle (\langle 1 \rangle A[\![n]\!]) \quad \text{by Assumption (2)}$$
$$\vdash \langle 1 \rangle A \quad \text{(by the reduction property).}$$

The last step is achieved because $h_C\!\downarrow$ implies $h_{1111}\!\downarrow$ which, as per our first step in this proof, implies EA^+, hence allowing the use of the reduction property.

Now to prove Proposition 4 assume that $\mathrm{EWD}^{\omega 2}$ holds. In EA + UTB we have that

$$\forall A \in \mathbb{W}^{\omega 2}\, \exists m\; A_m = \top \vdash \forall A \in \mathbb{W}_1^{\omega 2}\; h_A\!\downarrow\; \vdash \forall n\; \langle 1 \rangle \langle \omega + n \rangle \top \vdash \text{1-Con}(\mathrm{PA}(\mathrm{T})).$$

The first implication holds since for every worm A and every number x, there is a worm $A' = 0^x A$ where $A'[\![{}^{x-1}_{\;0}]\!] = A$ hence $\exists m\; A'_m = \top$ iff $h_A(x)$ is defined.

As for the case of EWD^{n+1}, assume that EWD^{n+1} holds. We have in EA that

$$\begin{aligned} \forall A \in \mathbb{W}^{n+1}\, \exists m\; A_m = \top &\vdash \forall A \in \mathbb{W}_1^{n+1}\; h_A\!\downarrow \\ &\vdash \forall k\; \langle 1 \rangle (\langle n+1 \rangle \top [\![k]\!]), \quad \text{by Lemma 18} \\ &\vdash \langle 1 \rangle \langle n+1 \rangle \top \quad \text{(by the reduction property)} \\ &\vdash \text{1-Con}(\mathrm{I}\Sigma_n). \end{aligned}$$

For the use of the reduction property, notice that here $\langle 1 \rangle \langle n+1 \rangle \top \rightarrow \langle 1 \rangle \top$ which in turn implies EA^+.

5 Concluding Remarks

We have shown that GLP_Λ is sound for the transfinite notions of provability studied by Beklemishev and Pakhomov [5], and with this we have shown that a natural extension of the *Every Worm Dies* principle is independent of ACA. Likewise, we have shown that restricted versions of this principle are related to the theories $\mathrm{I}\Sigma_n$, although in this case we do not obtain a precise equivalence. Whether EWD^{n+1} is indeed equivalent to one of 1-Con($\mathrm{I}\Sigma_n$) or 1-Con($\mathrm{I}\Sigma_{n+1}$) remains open.

Stronger theories of second order arithmetic should also be proof-theoretically equivalent to reflection up to a suitable ordinal Λ. These equivalences may then be used to provide new variants of EWD independent of stronger theories of second order arithmetic, including theories related to transfinite induction or iterated comprehension. We expect that this work will be an important step in this direction.

References

1. Beklemishev, L.D.: Induction rules, reflection principles, and provably recursive functions. Ann. Pure Appl. Log. **85**, 193–242 (1997)
2. Beklemishev, L.D.: Provability algebras and proof-theoretic ordinals, I. Ann. Pure Appl. Log. **128**, 103–124 (2004)
3. Beklemishev, L.D.: Reflection principles and provability algebras in formal arithmetic. Uspekhi Matematicheskikh Nauk **60**(2), 3–78 (2005). (in Russian). English translation. In: Russian Mathematical Surveys 60(2), 197–268 (2005)
4. Beklemishev, L.D., Fernández-Duque, D., Joosten, J.J.: On provability logics with linearly ordered modalities. Stud. Logica. **102**, 541–566 (2014)
5. Beklemishev, L.D., Pakhomov, F.N.: Reflection algebras and conservation results for theories of iterated truth. arXiv arXiv:1908.10302 [math.LO] (2019)
6. Cordón Franco, A., Fernández-Duque, D., Joosten, J.J., Lara Martín, F.: Predicativity through transfinite reflection. J. Symb. Log. **82**(3), 787–808 (2017)
7. Enayat, A., Pakhomov, F.: Truth, disjunction, and induction. Arch. Math. Logic **58**(5–6), 753–766 (2019)
8. Fernández-Duque, D.: The polytopologies of transfinite provability logic. Arch. Math. Logic **53**(3–4), 385–431 (2014)
9. Fernández-Duque, D., Joosten, J.J.: Models of transfinite provability logics. J. Symb. Log. **78**(2), 543–561 (2013)
10. Fernández-Duque, D., Joosten, J.J.: Well-orders in the transfinite Japaridze algebra. Logic J. IGPL **22**(6), 933–963 (2014)
11. Fernández-Duque, D., Joosten, J.J.: The omega-rule interpretation of transfinite provability logic. Ann. Pure Appl. Logic **169**(4), 333–371 (2018)
12. Halbach, V.: Axiomatic Theories of Truth. University of Oxford (2014)
13. Japaridze, G.: The polymodal provability logic. In: Intensional Logics and Logical Structure of Theories: Material from the 4th Soviet-Finnish Symposium on Logic. Metsniereba, Telaviv (1988). (in Russian)
14. Joosten, J.J.: Turing-Taylor expansions of arithmetic theories. Stud. Logica. **104**, 1225–1243 (2016)
15. Joosten, J.J.: Münchhausen provability. J. Symb. Log., 1–30 (2021). https://doi.org/10.1017/jsl.2021.44
16. Leivant, D.: The optimality of induction as an axiomatization of arithmetic. J. Symb. Log. **48**, 182–184 (1983)
17. Papafillipou, K.: Independent combinatoric worm principles for first order arithmetic and beyond. Master's thesis, Master of Pure and Applied Logic, University of Barcelona (2020). http://diposit.ub.edu/dspace/handle/2445/170755
18. Rathjen, M.: The art of ordinal analysis. In: Proceedings of the International Congress of Mathematicians, vol. 2, pp. 45–69. European Mathematical Society (2006)
19. Turing, A.: Systems of logics based on ordinals. Proc. Lond. Math. Soc. **45**, 161–228 (1939)

Parametric Church's Thesis: Synthetic Computability Without Choice

Yannick Forster(✉)

Saarland University, Saarland Informatics Campus, Saarbrücken, Germany
forster@cs.uni-saarland.de

Abstract. In synthetic computability, pioneered by Richman, Bridges, and Bauer, one develops computability theory without an explicit model of computation. This is enabled by assuming an axiom equivalent to postulating a function ϕ to be universal for the space $\mathbb{N} \rightarrow \mathbb{N}$ (CT_ϕ, a consequence of the constructivist axiom CT), Markov's principle, and at least the axiom of countable choice. Assuming CT and countable choice invalidates the law of excluded middle, thereby also invalidating classical intuitions prevalent in textbooks on computability. On the other hand, results like Rice's theorem are not provable without a form of choice.

In contrast to existing work, we base our investigations in constructive type theory with a separate, impredicative universe of propositions where countable choice does not hold and thus a priori CT_ϕ and the law of excluded middle seem to be consistent. We introduce various parametric strengthenings of CT_ϕ, which are equivalent to assuming CT_ϕ and an S^m_n operator for ϕ like in the S^m_n theorem. The strengthened axioms allow developing synthetic computability theory without choice, as demonstrated by elegant synthetic proofs of Rice's theorem. Moreover, they seem to be not in conflict with classical intuitions since they are consequences of the traditional analytic form of CT.

Besides explaining the novel axioms and proofs of Rice's theorem we contribute machine-checked proofs of all results in the Coq proof assistant.

The constructivist axiom CT ("Church's thesis") [24,41] states that every function $\mathbb{N} \rightarrow \mathbb{N}$ is computable in a fixed model of computation. In his 1992 book Odifreddi states that the consistency of CT has been established "for all current intuitionistic systems (not involving the concept of choice sequence)" [29, §1.8 pg. 122]. For constructive type theory the consistency question is not solved entirely, but recent breakthroughs include a consistency proof for univalent type theory [40] and Martin-Löf type theory [43].

Assuming CT enables developing computability theory in constructive logic in full formality without explicit encodings of programs in models of computation. However, philosophically CT is unpleasing since its definition still requires the definition of a model. In his seminal paper "Church's thesis without tears" [34], Richman introduces a purely synthetic form of CT not mentioning any model of computation, which is powerful enough to develop computability theory synthetically. The axiom is equivalent to assuming a function ϕ and an axiom CT_ϕ which

S. Artemov and A. Nerode (Eds.): LFCS 2022, LNCS 13137, pp. 70–89, 2022.
https://doi.org/10.1007/978-3-030-93100-1_6

postulates that ϕ is a step-indexed interpreter universal for the function space $\mathbb{N}\to\mathbb{N}$, i.e. that every $f:\mathbb{N}\to\mathbb{N}$ has a code $c:\mathbb{N}$ such that ϕ_c agrees with f [15]. Richman routinely assumes both Markov's principle MP and the axiom of countable choice $\mathsf{AC}_\mathbb{N}$ (or even stronger forms like dependent choice), as is the case in later work by Richman and Bridges [8] and Bauer [4]. As a consequence, the law of excluded middle LEM becomes disprovable, since LEM and $\mathsf{AC}_\mathbb{N}$ together entail that every predicate is decidable, which is clearly in contradiction to results of synthetic computability deducible from CT_ϕ. Text book presentations of computability however make crucial use of classical logic, rendering synthetic computability a constructivist niche.

However, Richman and Bridges state that $\mathsf{AC}_\mathbb{N}$ can "usually be avoided" [8, pg. 54] by postulating a composition function w.r.t. ϕ, which as consequence allows proving an S^m_n principle w.r.t. ϕ which we abbreviate as SMN_ϕ. In this paper, we show that their observation indeed holds true.

We work in the calculus of inductive constructions (CIC), the type theory underlying the Coq proof assistant,[1] which is a foundation for constructive mathematics where the axiom of countable choice is independent, i.e. can be consistently assumed but is *not* provable.

Contribution. We introduce several axioms equivalent to assuming ϕ such that $\mathsf{CT}_\phi \wedge \mathsf{SMN}_\phi$. Since working with S^m_n operators in applications explicitly is tedious, we define all axioms via a respective notion of parametric universality. As a consequence, the statements of the axioms become more uniform and compact. At the same time, the axioms become easier to use in applications. The resulting theory enables carrying out synthetic computability theory without any form of choice axiom, rendering the theory agnostic towards axioms like the law of excluded middle and thereby compatible with classical intuitions.

All axioms have in common that they allow defining a enumerable but undecidable predicate $\mathcal{K}$, where both enumerability and decidability are defined purely in terms of functions rather than computable functions. Since all axioms are a consequence of the constructivist axiom CT, they are consistent in CIC but not in contradiction to the law of excluded middle. Thus, our axioms allow developing synthetic computability, agnostic towards classical logic. As case studies we give two synthetic proofs of Rice's theorem [33]: One based on the axiom EPF, following the proof approach by Scott [36] relying on Rogers' fixed-point theorem [35], and one based on the axiom EA, establishing a many-one reduction from an undecidable problem.

Outline. We motivate and introduce CT_ϕ and SMN_ϕ in Sect. 2. SCT is introduced in Sect. 3, its variants EPF, $\mathsf{SCT}_\mathbb{B}$, and $\mathsf{EPF}_\mathbb{B}$ in Sect. 4. We introduce EA in Sect. 5 prove two synthetic versions of Rice's theorems in Sect. 6.

[1] The results in the pdf version of this paper are hyperlinked with the html version of the Coq source code, which can be found at https://github.com/yforster/coq-synthetic-computability/.

1 Preliminaries

We work in the Calculus of Inductive Constructions (CIC), the type theory underlying the Coq proof assistant.

1.1 Common Definitions in CIC

We rely on the inductive types of natural numbers $n{:}\mathbb{N} ::= 0 \mid \mathsf{S}\,n$, booleans $\mathbb{B} ::= \mathsf{true} \mid \mathsf{false}$, lists $l{:}\mathbb{L}X ::= [\,] \mid x :: l$ where $x{:}X$, options $\mathbb{O}X ::= \mathsf{None} \mid \mathsf{Some}\,x$ where $x{:}X$, pairs $X \times Y ::= (x, y)$ where $x{:}X$ and $y{:}Y$, sums $X + Y ::= \mathsf{inl}\,x \mid \mathsf{inr}\,y$ where $x{:}X$ and $y{:}Y$, and, for $p{:}X{\to}\mathbb{T}$ or $p{:}X{\to}\mathbb{P}$, $\Sigma x.px ::= (a, b)$ where $a : X$ and $b : px$. π_1 and π_2 denote the projections $\pi_1(a, b) := a$ and $\pi_2(a, b) := b$.

One can easily construct a pairing function $\langle\_, \_\rangle{:}\mathbb{N}{\to}\mathbb{N}{\to}\mathbb{N}$ and for all $f{:}\mathbb{N}{\to}\mathbb{N}{\to}X$ an inverse construction $\lambda\langle n, m\rangle.\ fnm$ of type $\mathbb{N} \to X$ such that $(\lambda\langle n, m\rangle.\ fnm)\langle n, m\rangle = fnm$.

For discrete X (e.g. $\mathbb{N}$, $\mathbb{O}\mathbb{N}$, $\mathbb{L}\mathbb{B}$, ...), $\equiv_X$ denotes equality, $\equiv_{\mathbb{P}}$ denotes logical equivalence, $\equiv_{A\to B}$ denotes an extensional lift of $\equiv_B$, $\equiv_{A\to\mathbb{P}}$ denotes extensional equivalence, and $\equiv_{\mathsf{ran}}$ denotes range equivalence. More formally, two functions $f, g{:}X \to \mathbb{O}Y$ are range equivalent if $f \equiv_{\mathsf{ran}} g := \forall y.\ (\exists x.\ fx = \mathsf{Some}\,y) \leftrightarrow (\exists x.\ gx = \mathsf{Some}\,y)$.

1.2 Partial Functions

We work with a type $\mathsf{part}\,A$ where $A : \mathbb{T}$ for partial values and a definedness relation $\triangleright : \mathsf{part}\,A \to A \to \mathbb{P}$ and write $A \rightharpoonup B$ for $A \to \mathsf{part}\,B$. We assume monadic structure for part (ret and $\gg\!=$), an undefined value (undef), a minimisation operation (μ), and a step-indexed evaluator (seval), see [15, fig. 2].

An equivalence relation on partial values can be defined as $x \equiv_{\mathsf{part}\,A} y := \forall a.\ x \triangleright a \leftrightarrow y \triangleright a$. Lifted to partial functions, we have $f \equiv_{A\rightharpoonup B} g := \forall ab.\ fa \triangleright b \leftrightarrow ga \triangleright b$.

One possible definition of $\mathsf{part}\,A$ is via stationary sequences. We call $f{:}\mathbb{N}{\to}\mathbb{O}A$ a *stationary sequence* (or just *stationary*) if $\forall na.\ fn = \mathsf{Some}\,a \to \forall m \geq n.\ fm = \mathsf{Some}\,a$. For instance, the always undefined function $\lambda n.\ \mathsf{None}$ is stationary.

One can then define $\mathsf{part}\,A := \Sigma f{:}\mathbb{N}{\to}\mathbb{O}A.\ f$ *is stationary* with

$$f \triangleright a := \exists n.\ \pi_1 fn = \mathsf{Some}\,a.$$

Note that one can turn any function $f{:}\mathbb{N}{\to}\mathbb{O}A$ into a stationary sequence via a transformer mkstat with the following property.

Fact 1. $\mathsf{mkstat} f \triangleright a \leftrightarrow \exists n.\ fn = \mathsf{Some} a \land \forall m < n.\ fn = \mathsf{None}$

1.3 The Universe of Propositions $\mathbb{P}$, Elimination, and Choice Principles

CIC has a separate, impredicative universe of propositions $\mathbb{P}$ and a hierarchy of type universes $\mathbb{T}_i$ (where the natural number index i is left out from now on).

The universe $\mathbb{P}$ is separated in the sense that the definition of functions of type $\forall x : P.\ A(x)$ for $P{:}\,\mathbb{P}$ and $A{:}\,P \to \mathbb{T}$ by case analysis on x are restricted.

In CIC, both dependent pairs (Σ) and existential quantification $\exists$ can be defined using inductive types. We verbalise $\exists x$ with "there exists x" and in contrast Σx as "one can construct x". Dependent pairs can be eliminated into arbitrary contexts, i.e. there is an elimination function of type

$$\forall p{:}\,(\Sigma x.Ax)\to\mathbb{T}.\ (\forall x{:}\,X.\forall y{:}\,Ax.\ p(x,y)) \to \forall s.\ ps.$$

In contrast, existential quantification can only be eliminated for $p{:}\,(\exists x.Ax)\to\mathbb{P}$. This is because CIC forbids so-called large eliminations [30] on the inductively defined $\exists$ predicate. To avoid dealing with Coq's `match`-construct for eliminations in detail, we instead talk about *large elimination principles.* A large elimination principle for $\exists$, which would have the following type, is *not* definable in CIC:

$$\forall p{:}\,(\exists x.Ax)\to\mathbb{T}.\ (\forall x{:}\,X.\forall y{:}\,Ax.\ p(x,y)) \to \forall s.\ ps.$$

In particular, this means that one *cannot* define a function of the following type in general

$$\forall p{:}\,Y\to\mathbb{P}.\ (\exists y.\ py) \to \Sigma y{:}\,Y.\ py.$$

However, such an elimination of $\exists$ into Σ is *admissible* in CIC. This means that any concretely given, fully constructive proof of $\exists y.\ py$ without assumptions can always be given as a proof of $\Sigma y.\ py$. Note that admissibility of a statement is strictly weaker than provability, and in general does not even entail its consistency.

Crucially, CIC allows defining large elimination principles for the falsity proposition $\bot$ and for equality. Additionally, for some restricted types Y and restricted predicates p, one *can* define a large elimination principle for existential quantification. In particular, this holds for $Y = \mathbb{N}$ and $p(n{:}\,\mathbb{N}) := fn = \mathsf{true}$ for a function $f{:}\,\mathbb{N}\to\mathbb{B}$.

Fact 2. *One can define functions of type*

1. $\forall A{:}\,\mathbb{T}.\ \bot \to A$
2. $\forall X{:}\,\mathbb{T}.\forall A{:}\,X \to \mathbb{T}.\forall x_1 x_2{:}\,X.\ x_1 = x_2 \to Ax_1 \to Ax_2.$
3. $\forall f{:}\,\mathbb{N}\to\mathbb{B}.\ (\exists n.\ fn = \mathsf{true}) \to \Sigma n.\ fn = \mathsf{true}$

We will not need any other large elimination principle in this paper. A restriction of large elimination in general is necessary for consistency of Coq [11]. As a by-product, the computational universe $\mathbb{T}$ is separated from the logical universe $\mathbb{P}$, allowing classical logic in $\mathbb{P}$ to be assumed while the computational intuitions for $\mathbb{T}$ remain intact.

The intricate interplay between Σ and $\exists$ is in direct correspondence to the status of the axiom of choice in CIC. The axiom of choice was first stated for set theory by Cantor. In the formulation by Cantor, it is equivalent to the statement that every total, binary relation contains the graph of a function, i.e.:

$$\forall R \subseteq X \times Y.\ (\forall x.\exists y.(x,y) \in R) \to \exists f{:}\,X\to Y.\forall x.\ (x, fx) \in R$$

Here $X \to Y$ is the set-theoretic function space. As usual, such a classical principle can also be stated in type theory. However, the concrete formalisation crucially depends on how the notion of a (set-theoretic) function is translated: While in set theory the term *function* is just short for *functional relation*, in CIC functions and (total) functional relations are different objects, we thus discuss both possible translations of the axiom of choice here.

The more common version, used e.g. by Bishop [7], is to use type-theoretic functions for set-theoretic functions, i.e. state the type-theoretic axiom of (functional) choice as

$$\forall R\colon X \to Y \to \mathbb{P}.\ (\forall x.\exists y.\ Rxy) \to \exists f\colon X \to Y.\forall x.\ Rx(fx)$$

Since in type theory proofs are first class object, one can equivalently state a principle postulating the inhabitedness of the following type:

$$\forall p\colon Y \to \mathbb{P}.\ (\exists y.\ py) \to \Sigma y\colon Y.\ py$$

Note how this is exactly the non-provable correspondence of $\exists$ and Σ discussed above. This formulation makes clears why in Martin-Löf type theory as implementation of Bishop's constructive mathematics, where one defines $\exists := \Sigma$, the axiom of choice is accepted since it can be proved. In the context of Church's simple type theory, this axiom is also known as axiom of indefinite description [1].

1.4 Notions of Synthetic Computability

We call a predicate $p\colon X \to \mathbb{P} \ldots$

- *decidable* if $\mathcal{D}p := \exists f\colon X \to \mathbb{B}.\forall x\colon X.\ px \leftrightarrow fx = \mathsf{true}$.
- *enumerable* if $\mathcal{E}p := \exists f\colon \mathbb{N} \to \mathbb{O}X.\ \forall x\colon X.\ px \leftrightarrow \exists n\colon \mathbb{N}.\ fn = \mathsf{Some}\,x$.
- *semi-decidable* if $\mathcal{S}p := \exists f\colon X \to \mathbb{N} \to \mathbb{B}.\ px \leftrightarrow \exists n.\ fxn = \mathsf{true}$.

A type X is discrete if $\lambda(x_1, x_2)\colon X \times X.\ x_1 = x_2$ is decidable and enumerable if $\lambda x : X.\ \top$ is enumerable. We repeat the following facts from [15]:

Fact 3. *The following hold:*

1. *Decidable predicates are semi-decidable and co-semi-decidable.*
2. *Semi-decidable predicates on enumerable types are enumerable.*
3. *Enumerable predicates on discrete types are semi-decidable.*
4. *The complement of semi-decidable predicates is co-semi-decidable.*

Fact 4. *Decidable predicates are closed under complements. Decidable, enumerable, and semi-decidable predicates are closed under conjunction and disjunction.*

One can also characterise the notions via partial functions:

Fact 5. *1.* $\mathcal{E}p \leftrightarrow \exists f\colon \mathbb{N} \rightharpoonup X.\forall x.\ px \leftrightarrow \exists n.\ fn \triangleright x$
2. $\mathcal{S}p \leftrightarrow \exists Y(f\colon X \rightharpoonup Y).\forall x.\ px \leftrightarrow \exists y.\ fx \triangleright y$

Lastly, we introduce many-one reducibility. A predicate $p\colon X \to \mathbb{P}$ is many-one reducible to a predicate $q\colon Y \to \mathbb{P}$ if

$$p \preceq_m q := \exists f\colon X \to Y.\ \forall x.\ px \leftrightarrow q(fx).$$

Fact 6. *1. If $p \preceq_m q$ and q is decidable, then p is decidable.*
2. Let X be enumerable, Y discrete, and $p\colon X \to \mathbb{P}$, $q\colon Y \to \mathbb{P}$. If $p \preceq_m q$ and q is enumerable then p is enumerable.

2 Church's Thesis

Textbooks on computability start by defining a model of computation, Rogers [35] uses μ-recursive functions. As center of the theory, Rogers defines a step-indexed interpreter ϕ of all μ-recursive functions. An application $\phi_c^n x$ denotes executing the μ-recursive function with code c on input x for n steps.

Once some evidence is gathered, Rogers (as well as other authors) introduce the Church-Turing thesis, stating that all intuitively calculable functions are μ-recursively computable. Using the Church-Turing thesis, ϕ has the following (informal) universal property:

$$\forall f{:}\,\mathbb{N} \to \mathbb{N}.\ \textit{intuitively computable}\ f \to \exists c{:}\,\mathbb{N}.\forall x{:}\,\mathbb{N}.\exists n.\ \phi_c^n x = \mathsf{Some}\,(f x)$$

Note that the property is really only pseudo-formal: The notion of intuitive calculability is not made precise, which is exactly what allows ϕ to stay abstract for most of the development. Every invocation of the universality could be replaced by an individual construction of a (μ-recursive) program, but relying solely on the notion of intuitive calculability allows Rogers to build a theory based on a function ϕ which could equivalently be implemented in any other model of computation. Since not every function $f{:}\,\mathbb{N}{\to}\mathbb{N}$ in the classical set theory Rogers works in[2] is intuitively computable, every invocation of the universality of ϕ has to be checked individually to ensure that it is indeed for an intuitively calculable function.

We however do not work in classical set theory, but in CIC, a constructive system. As in all constructive systems, every definable function is intuitively calculable. It is thus natural to assume that the universal function ϕ is universal for *all* functions $f{:}\,\mathbb{N}{\to}\mathbb{N}$. For historical reasons, this axiom is called CT ("Church's thesis") [24,41].

We define CT_ϕ parametric in a step-indexed interpreter $\phi{:}\,\mathbb{N}{\to}\mathbb{N}{\to}\mathbb{N}{\to}\mathbb{O}\mathbb{N}$. As before, we write an evaluation of code c on input x for n steps as $\phi_c^n x$ instead of $\phi\, c\, x\, n$. For step-indexed interpreters, the sequence $\lambda n.\phi_c^n x$ is always stationary:

$$\forall c x n_1 v.\ \phi_c^{n_1} x = \mathsf{Some}\, v \to \forall n_2.\ n_2 \geq n_1 \to \phi_c^{n_2} x = \mathsf{Some}\, v$$

Now CT_ϕ states that $\phi{:}\,\mathbb{N}{\to}\mathbb{N}{\to}\mathbb{N}{\to}\mathbb{O}\mathbb{N}$ is universal for *all* functions $f{:}\,\mathbb{N}{\to}\mathbb{N}$:

$$\mathsf{CT}_\phi := \forall f{:}\,\mathbb{N}{\to}\mathbb{N}.\exists c{:}\,\mathbb{N}.\forall x{:}\,\mathbb{N}.\exists n{:}\,\mathbb{N}.\ \phi_c^n x = \mathsf{Some}\,(f x)$$

One can also see ϕ as an enumeration of stationary functions from $\mathbb{N}$ to $\mathbb{N}$, which enumerates every total function f.

CT_ϕ is not provable in CIC, independent of the definition of ϕ. However, when ϕ is a step-indexed interpreter for a model of computation, CT_ϕ is the

[2] "We use the rules and conventions of classical two-valued logic (as is the common practice in other parts of mathematics), and we say that an object exists if its existence can be demonstrated within standard set theory. We include the axiom of choice as a principle of our set theory." [35, pg. 10, footnote †].

well-known constructivist axiom CT, see [15] for a treatment of its status in CIC. We give an overview over arguments why CT is consistent in CIC in Appendix A.

In contrast to textbook proofs, proofs of theorems based on CT_ϕ do not have to be individually checked for valid applications of the Church-Turing thesis.

As stated above, CT_ϕ applies to unary total functions, but is immediately extensible to n-ary functions $f: \mathbb{N}^n \to \mathbb{N}$ using pairing. Partial application for such n-ary functions is realised via the S^m_n theorem. We only state the case $m = n = 1$, which implies the general case:

$$\mathsf{SMN}_\phi := \Sigma\sigma: \mathbb{N}\to\mathbb{N}\to\mathbb{N}. \forall cxyv.\ (\exists n.\ \phi^n_{\sigma cx} y = \mathsf{Some}\, v) \leftrightarrow (\exists n.\ \phi^n_c \langle x, y\rangle = \mathsf{Some}\, v)$$

Note that we formulate SMN with a Σ rather than an $\exists$. For the results we consider in this paper, the different is largely cosmetic. The formulation with Σ allows the construction of functions accessing σ directly, rather than only being able to prove the existence of functions based on σ.

The key property of CT_ϕ is that it allows the definition of an enumerable but undecidable problem:

Lemma 7. *Let ϕ be stationary. Then* $\mathsf{CT}_\phi \to \Sigma p: \mathbb{N}\to\mathbb{P}.\ \mathcal{S}p \land \neg\mathcal{S}\overline{p} \land \neg\mathcal{D}p \land \neg\mathcal{D}\overline{p}$.

Proof. One can define $pc := \exists nm.\ \phi^m_c \langle c, n\rangle = \mathsf{Some}\, 0$ and clearly we have $\mathcal{S}p$. If $f: \mathbb{N}\to\mathbb{N}\to\mathbb{B}$ is a semi-decider for $\overline{p}$, let c be its code w.r.t. ϕ. Then $pc \leftrightarrow \neg pc$, contradiction. Thus p is also not decidable. □

3 Synthetic Church's Thesis

By keeping ϕ abstract and assuming CT_ϕ, one never has to deal with encodings in a model of computation. However, formal proofs involving the SMN_ϕ axiom are tedious. We identify the axiom *synthetic Church's thesis* SCT as a more convenient variant of $\mathsf{CT}_\phi \land \mathsf{SMN}_\phi$, which postulates a step-indexed interpreter ϕ *parametrically universal* for $\mathbb{N} \to \mathbb{N}$:

$$\begin{aligned} \mathsf{SCT} := \Sigma\phi: \mathbb{N}\to\mathbb{N}\to\mathbb{N}\to\mathbb{O}\mathbb{N}.\ & \\ & (\forall c\, x\, n_1\, n_2\, v.\ \phi^{n_1}_c x = \mathsf{Some}\, v \to n_2 \geq n_1 \to \phi^{n_2}_c x = \mathsf{Some}\, v) \land \\ & \forall f: \mathbb{N}\to\mathbb{N}\to\mathbb{N}.\ \exists\gamma: \mathbb{N} \to \mathbb{N}.\ \forall ix. \exists n.\ \phi^n_{\gamma i} x = \mathsf{Some}\, (f_i x) \end{aligned}$$

By parametrically universal we mean that for any family of functions $f_i: \mathbb{N}\to\mathbb{N}$ parameterised by $i: \mathbb{N}$, we obtain a coding function γ s.t. γi is the code of f_i, i.e. $\phi_{\gamma i}$ agrees with f_i.

The consistency of SCT follows from the consistency of CT formulated for a Turing-complete model of computation. For this purpose, we choose the weak call-by-value λ-calculus L, which we discuss in detail in Sect. 7. Conversely, one can recover non-parametric universality of ϕ from parametric universality:

Theorem 8. *Let ϕ_L be a step-indexed interpreter for* L. *For any ϕ such that $\lambda n.\phi^n_c x$ is stationary we have the following:*

1. $\mathsf{CT}_{\phi_\mathsf{L}} \to \Sigma\phi.\ \mathsf{CT}_\phi \wedge \mathsf{SMN}_\phi$
2. $\mathsf{CT}_\phi \to \mathsf{SMN}_\phi \to \mathsf{SCT}$
3. $\mathsf{SCT} \to \Sigma\phi.\ \mathsf{CT}_\phi$

Proof. (1) follows by proving $\mathsf{SMN}_{\phi_\mathsf{L}}$, which we do in Sect. 7, see Corollary 33.

For (2), let ϕ and σ be given. We prove that ϕ satisfies the condition in SCT. Let $f\colon \mathbb{N}\to\mathbb{N}\to\mathbb{N}$ be given. We obtain a code c for $\lambda\langle x, y\rangle.\ fxy$. Now define $\gamma x := \sigma c x$.

(3) is trivial by turning the unary function $f\colon\mathbb{N}\to\mathbb{N}$ into the (constant) family of functions $f'_x y := fy$. Now a coding function γ for f' allows to choose $\gamma 0$ as code for f. □

In Theorem 12 we prove that SCT also implies $\Sigma\phi.\ \mathsf{CT}_\phi \wedge \mathsf{SMN}_\phi$. Note that the consistency of $\mathsf{CT}_{\phi_\mathsf{L}}$ implies the consistency of $\Sigma\phi.\ \mathsf{CT}_\phi \wedge \mathsf{SMN}_\phi$ and SCT.

4 Variations of Synthetic Church's Thesis

We have defined SCT to postulate a step-indexed interpreter $\phi\colon\mathbb{N}\to(\mathbb{N}\to\mathbb{N}\to\mathbb{O}\mathbb{N})$, parametrically universal for $\mathbb{N}\to\mathbb{N}$. In this section, we develop equivalent variations of SCT. There are three obvious points where SCT can be modified.

1. The return type of ϕ can be stationary functions of type $\mathbb{N}\to(\mathbb{N}\to\mathbb{O}\mathbb{N})$ or $\mathbb{N}\to(\mathbb{N}\to\mathbb{O}\mathbb{B})$, or partial functions of type $\mathbb{N}\rightharpoonup\mathbb{N}$ or $\mathbb{N}\rightharpoonup\mathbb{B}$,
2. ϕ can be postulated to be parametrically universal for $\mathbb{N}\to\mathbb{N}$, $\mathbb{N}\to\mathbb{B}$, $\mathbb{N}\rightharpoonup\mathbb{N}$, $\mathbb{N}\rightharpoonup\mathbb{B}$, or stationary functions $\mathbb{N}\to(\mathbb{N}\to\mathbb{O}\mathbb{N})$ or $\mathbb{N}\to(\mathbb{N}\to\mathbb{O}\mathbb{B})$.
3. Coding functions γ can be existentially quantified ($\exists$), computably obtained (Σ), or classically existentially quantified $\neg\neg\exists$.

For SCT, the return type of ϕ is stationary functions $\mathbb{N}\to(\mathbb{N}\to\mathbb{O}\mathbb{N})$, ϕ is parametrically universal for $\mathbb{N}\to\mathbb{N}$, and γ is existentially quantified.

For (1), it is important to see that letting ϕ return total functions is no option, since such an enumeration is inconsistent,[3] even up to extensionality:

Fact 9. (Cantor) *There is no* $e\colon\mathbb{N}\to(\mathbb{N}\to A)$ *such that* $\forall f\colon\mathbb{N}\to A.\ \exists c. ec \equiv_{\mathbb{N}\to A} f$ *for* $A = \mathbb{N}$ *or* $A = \mathbb{B}$.

For (3), the variant with Σ is consistent, but negates functional extensionality [43]. Variants with $\neg\neg\exists$ are often called *Weak* CT [27], we refrain from discussing such variants in this paper.

In this section, we discuss how all other variations of SCT are equivalent, and single out three of them:

1. EPF, the enumerability of partial functions axiom, postulating $\theta\colon\mathbb{N}\to(\mathbb{N}\rightharpoonup\mathbb{N})$ parametrically universal for $\mathbb{N}\rightharpoonup\mathbb{N}$,
2. $\mathsf{SCT}_\mathbb{B}$, postulating $\phi\colon\mathbb{N}\to(\mathbb{N}\to(\mathbb{N}\to\mathbb{O}\mathbb{B}))$ universal for $\mathbb{N}\to\mathbb{B}$, and

[3] Note that conversely an injection of $(\mathbb{N}\to\mathbb{N})\to\mathbb{N}$ can likely be consistently assumed [5].

3. $\mathsf{EPF}_{\mathbb{B}}$, postulating $\theta: \mathbb{N} \to (\mathbb{N} \rightharpoonup \mathbb{B})$ universal for $\mathbb{N} \rightharpoonup \mathbb{B}$.

The *enumerability of partial functions axiom* EPF is defined as:

$$\mathsf{EPF} := \exists \theta: \mathbb{N} \to (\mathbb{N} \rightharpoonup \mathbb{N}). \forall f: \mathbb{N} \to \mathbb{N} \rightharpoonup \mathbb{N}. \exists \gamma: \mathbb{N} \to \mathbb{N}. \forall i.\ \theta_{\gamma i} \equiv_{\mathbb{N} \rightharpoonup \mathbb{N}} f_i$$

Instead of seeing θ as enumeration, we can also see θ as surjection from $\mathbb{N}$ to $\mathbb{N} \rightharpoonup \mathbb{N}$ up to $\equiv_{\mathbb{N} \rightharpoonup \mathbb{N}}$. Proving that SCT$\leftrightarrow$EPF amounts to showing that any implementation of partial functions is equivalent to the implementation based on stationary sequences, and that any stationary function can be encoded in a total function $\mathbb{N} \to \mathbb{N}$ via pairing.

Theorem 10. SCT $\leftrightarrow$ EPF

Proof. The direction from left to right is by observing that there is a function $\mathsf{mktotal}: (\mathbb{N} \to \mathbb{N} \rightharpoonup \mathbb{N}) \to \mathbb{N} \to \mathbb{N} \to \mathbb{N}$ such that $f_i x \triangleright v \leftrightarrow \exists n.\ \mathsf{mktotal}\, f\, i\, \langle x, n \rangle = \mathsf{S}\, v$ using seval. We then define

$$\begin{aligned} \theta c x := &(\mu(\lambda n.\ \textbf{if}\ \phi_c^n x\ \textbf{is}\ \mathsf{Some}\,(\mathsf{S}\, v)\ \textbf{then}\ \mathsf{ret\,true}\ \textbf{else}\ \mathsf{ret\,false})) \\ &\ggg \lambda n.\ \textbf{if}\ \phi_c^n x\ \textbf{is}\ \mathsf{Some}\,(\mathsf{S}\, v)\ \textbf{then}\ \mathsf{ret}\, v\ \textbf{else}\ \mathsf{undef} \end{aligned}$$

The direction from right to left constructs $\phi_c^n x := \mathsf{seval}\,(\theta_c y)\, n$. Let $f: \mathbb{N} \to \mathbb{N} \to \mathbb{N}$. Define the partial function $f_i' x := \mathsf{ret}\,(f_i x)$. Now a coding function γ for f' by EPF is a coding function for f to establish SCT. □

Instead of stating EPF as enumeration of partial functions, we can equivalently state it w.r.t. parameterised functional relations:

Fact 11. EPF *is equivalent to the following:*

$$\exists \theta: \mathbb{N} \to (\mathbb{N} \rightharpoonup \mathbb{N}).\ \forall R: \mathbb{N} \to (\mathbb{N} \leadsto \mathbb{N}).\ (\exists f. \forall i.\ f_i \textit{ computes } R_i) \to \exists \gamma. \forall i.\ \theta_{\gamma i} \textit{ computes } R_i$$

Theorem 12. $\mathsf{EPF} \to \Sigma \phi.\ \mathsf{CT}_\phi \wedge \mathsf{SMN}_\phi \wedge \forall c x.\ \lambda n. \phi_c^n x$ *is stationary*

Proof. Let θ be given as in EPF and define $\phi_c^n x := \mathsf{seval}\,(\theta_c y)\, n$, which allows proving CT_ϕ as in Theorem 10. Let furthermore $f_{\langle c, x \rangle} y := \theta c \langle x, y \rangle$ and γ be a coding function for f by EPF. Define $S c x := \gamma \langle c, x \rangle$. We have

$$\theta_{Scx} y \equiv \theta_{\gamma \langle c, x \rangle} y \equiv \theta c \langle x, y \rangle$$

□

We introduce $\mathsf{SCT}_{\mathbb{B}}$, postulating a step-indexed interpreter parametrically universal for $\mathbb{N} \to \mathbb{B}$:

$$\begin{aligned} \mathsf{SCT}_{\mathbb{B}} := \Sigma \phi: \mathbb{N} \to \mathbb{N} \to \mathbb{N} \to \mathbb{O}\mathbb{B}. & \\ &(\forall c x n_1 n_2 v.\ \phi_c^{n_1} x = \mathsf{Some}\, v \to n_2 \geq n_1 \to \phi_c^{n_2} x = \mathsf{Some}\, v) \wedge \\ &\forall f: \mathbb{N} \to \mathbb{N} \to \mathbb{B}.\ \exists \gamma. \forall i x. \exists n.\ \phi_{\gamma i}^n x = \mathsf{Some}\,(f_i x) \end{aligned}$$

$\mathsf{SCT}_{\mathbb{B}}$ is equivalent to SCT. One direction is immediate since $\mathbb{B}$ is a retract of $\mathbb{N}$ (i.e. can be injectively embedded). The other direction follows by mapping the infinite sequence $f0, f1, f2, \ldots$ to the sequence

$$\mathsf{false}^{f0}\, \mathsf{true}\, \mathsf{false}^{f1}\, \mathsf{true}\, \mathsf{false}^{f2}\, \mathsf{true} \ldots$$

Theorem 13. $\mathsf{SCT}_{\mathbb{B}} \leftrightarrow \mathsf{SCT}$

We define the *parametric enumerability of partial boolean functions* axiom

$$\mathsf{EPF}_{\mathbb{B}} := \Sigma\theta{:}\,\mathbb{N}{\to}(\mathbb{N} \rightharpoonup \mathbb{B}).\forall f{:}\,\mathbb{N}{\to}\mathbb{N} \rightharpoonup \mathbb{B}.\exists\gamma{:}\,\mathbb{N} \to \mathbb{N}.\forall i.\ \theta_{\gamma i} \equiv_{\mathbb{N}\rightharpoonup\mathbb{B}} f_i$$

Recall that $\theta_{\gamma i} \equiv_{\mathbb{N}\rightharpoonup\mathbb{B}} f_i$ if and only if $\forall xv.\ \theta_{\gamma i}x \triangleright v \leftrightarrow f_i x \triangleright v$. Proving $\mathsf{EPF}_{\mathbb{B}}$ equivalent to SCT is easiest done by proving the following:

Theorem 14. $\mathsf{EPF}_{\mathbb{B}} \leftrightarrow \mathsf{SCT}_{\mathbb{B}}$

Proof. Exactly as in Theorem 10. □

Using $\mathsf{EPF}_{\mathbb{B}}$ it is easy to establish an enumerable, undecidable problem:

Fact 15. $\mathsf{EPF}_{\mathbb{B}} \to \Sigma p{:}\,\mathbb{N}{\to}\mathbb{P}.\ \mathcal{E}p \wedge \neg\mathcal{E}\overline{p} \wedge \neg\mathcal{D}p$

Proof. Let θ be given as in EPF. Define $\mathcal{K}c := \exists v.\theta_c c \triangleright v$. $\mathcal{K}$ is semi-decided by $\lambda cn.\textbf{if}\ \mathsf{seval}(\theta_c c)n\ \textbf{is}\ \mathsf{Some}\,v\ \textbf{then}\ \mathsf{true}\ \textbf{else}\ \mathsf{false}$ and thus enumerable by Fact 3.

We prove that $\overline{\mathcal{K}}$ is not semi-decidable, yielding both $\neg\mathcal{E}\overline{\mathcal{K}}$ and $\neg\mathcal{D}\mathcal{K}$ by Fact 3. Let $\overline{\mathcal{K}}$ be semi-decidable, i.e. by Fact 5 (2) there is $f{:}\,\mathbb{N} \rightharpoonup Y$ such that $\neg\mathcal{K}x \leftrightarrow \exists y.\ fx \triangleright y$. Define $f'{:}\,\mathbb{N} \rightharpoonup \mathbb{B}$ as $f'x := fx \gg\!= \lambda\_.\ \mathsf{ret\,true}$. Now f' has a code c such that $\forall x.\ ecx \triangleright f'x$ by universality of θ.

We have a contradiction via

$$\neg\mathcal{K}c \leftrightarrow (\exists y.\ fc \triangleright y) \leftrightarrow (\exists y.\ f'x \triangleright y)(\exists y.ecc \triangleright y) \leftrightarrow \mathcal{K}c.$$

□

5 The Enumerability Axiom

Using EPF or SCT as basis for synthetic computability requires the manipulation of partial functions or stationary functions, which is tedious. Alternatively, synthetic computability can be presented even more elegantly by an equivalent axiom concerned with enumerable predicates rather than partial functions. A non-parametric enumerability axiom is used by Bauer [4] together with countable choice to develop synthetic computability results.

We introduce the *parametric enumerability axiom* postulating an enumerator $\varphi{:}\,\mathbb{N} \to (\mathbb{N} \to \mathbb{O}\mathbb{N})$ which is parametrically universal for all parametrically enumerable predicates $p{:}\,\mathbb{N}{\to}\mathbb{N}{\to}\mathbb{P}$:

$$\begin{aligned}\mathsf{EA} :=\ &\Sigma\varphi{:}\,\mathbb{N}{\to}(\mathbb{N}{\to}\mathbb{O}\mathbb{N}).\forall(p{:}\,\mathbb{N}{\to}\mathbb{N}{\to}\mathbb{P}).\\ &(\exists(f{:}\,\mathbb{N} \to \mathbb{N} \to \mathbb{O}\mathbb{N}).\forall i.\ f_i\ \textit{enumerates}\ p_i) \to \exists\gamma{:}\,\mathbb{N} \to \mathbb{N}.\forall i.\ \varphi_{\gamma i}\ \textit{enumerates}\ p_i\end{aligned}$$

That is, EA states that whenever p is parametrically enumerable, then $\lambda i.\ \varphi_{\gamma i}$ parametrically enumerates p for some γ.

Note the two different roles of natural numbers in the axiom: If we would consider predicates over a general type X we would have $\varphi{:}\,\mathbb{N}{\to}(\mathbb{N}{\to}\mathbb{O}X)$.

Equivalently, we could have required that p is enumerable:

Fact 16. EA *is equivalent to*

$$\Sigma\varphi.\ \forall p{:}\,\mathbb{N}{\to}\mathbb{N}{\to}\mathbb{P}.\ \mathcal{E}(\lambda(x,y).\ p(x,y)) \to \exists\gamma{:}\,\mathbb{N}{\to}\mathbb{N}.\forall i.\ \varphi_{\gamma i}\ \textit{enumerates}\ pi.$$

Again equivalently, EA can be stated to only mention enumerators instead of predicates, which is the formulation of EA used in [15].

Fact 17. $\mathsf{EA} \leftrightarrow \Sigma\varphi{:}\,\mathbb{N}{\to}(\mathbb{N}{\to}\mathbb{O}\mathbb{N}).\forall f{:}\,\mathbb{N}{\to}\mathbb{N}{\to}\mathbb{O}\mathbb{N}.\exists\gamma.\forall x.\ \varphi_{\gamma x} \equiv_{ran} fx$

In this formulation, φ is a surjection w.r.t. range equivalence $f \equiv_{\mathsf{ran}} g$, where $\varphi_c \equiv_{\mathsf{ran}} f \leftrightarrow \forall x.(\exists n.\varphi_c n = \mathsf{Some}\, x) \leftrightarrow (\exists n.fn = \mathsf{Some}\, x)$.

Given φ, we define $\mathcal{W}_c x := \exists n.\ \varphi_c n = \mathsf{Some}\, x$ and the problem $\mathcal{K}$ as the diagonal of $\mathcal{W}$, i.e. $\mathcal{K}c := \mathcal{W}_c c$. We call $\mathcal{W}$ a *universal table.* One can show that $\mathcal{W}$ and $\mathcal{K}$ are m-equivalent, and both are m-complete. For now we only use $\mathcal{K}$ to note the following result:

Lemma 18. $\mathsf{EA} \to \Sigma p{:}\,\mathbb{N} \to \mathbb{P}.\ \mathcal{E}p \wedge \neg\mathcal{E}\overline{p} \wedge \neg\mathcal{D}p \wedge \neg\mathcal{D}\overline{p}$

Proof. We pick p as $\mathcal{K}c := \mathcal{W}_c c$. $\mathcal{K}$ is enumerated by

$$\lambda\langle c,m\rangle.\ \textbf{if}\ \varphi_c m\ \textbf{is}\ \mathsf{Some}\, x\ \textbf{then}\ \textbf{if}\ x =_{\mathbb{B}} c\ \textbf{then}\ \mathsf{Some}\, c\ \textbf{else}\ \mathsf{None}\ \textbf{else}\ \mathsf{None}.$$

If $\overline{\mathcal{K}}$ would be enumerable, there would be a code c s.t. $\forall x.\ \mathcal{W}_c x \leftrightarrow \overline{\mathcal{K}}x$. In particular $\mathcal{W}_c c \leftrightarrow \overline{\mathcal{K}}c \leftrightarrow \neg\mathcal{W}_c c$. Contradiction. Then $\neg\mathcal{D}p \wedge \neg\mathcal{D}\overline{p}$ is trivial. □

Similarly to how SCT can be reformulated by letting ϕ be universal for unary functions and introducing an explicit S^1_1-operator, EA can also be stated in this fashion, with an S^1_1-operator w.r.t. $\mathcal{W}$.

Lemma 19. EA *is equivalent to*

$$\Sigma\varphi.(\forall p.\mathcal{E}p \to \exists c.\varphi_c\ \textit{enumerates}\ p) \wedge \Sigma\sigma{:}\,\mathbb{N}{\to}\mathbb{N}{\to}\mathbb{N}.\forall cxy.\mathcal{W}_{(\sigma cx)}y \leftrightarrow \mathcal{W}_c\langle x,y\rangle.$$

Proof. The direction from right to left is straightforward using Lemma 16.

For the direction from left to right, let φ be given.

For the first part of the conclusion let p be given and enumerable. Then $\lambda xy.\ py$ is parametrically enumerable, so let γ be given from EA. Then $\varphi_{\gamma 0}$ enumerates p. For the second part, let $p\langle c,x\rangle y := \exists n.\varphi_c n = \mathsf{Some}\,\langle x,y\rangle$. Since p is enumerable, by Lemma 16 and EA there is γ such that $\varphi_{\gamma\langle c,x\rangle}$ enumerates $p\langle c,x\rangle$. Now $Scx := \gamma\langle c,x\rangle$ is the wanted function. □

SCT and EA are equivalent. For the forwards direction, we show that enumerators $\mathbb{N} \to \mathbb{O}\mathbb{N}$ can be equivalently given as functions $\mathbb{N} \to \mathbb{N}$.

Theorem 20. $\mathsf{SCT} \to \mathsf{EA}$

Proof. Let a universal function ϕ be given. Define:

$$\varphi_c\langle n,m\rangle := \textbf{if}\ \phi^n_c m\ \textbf{is}\ \mathsf{Some}\,(\mathsf{S}\, x)\ \textbf{then}\ \mathsf{Some}\, x\ \textbf{else}\ \mathsf{None}$$

Let $f{:}\,\mathbb{N}{\to}\mathbb{N}{\to}\mathbb{O}\mathbb{N}$ be a parametric enumerator for p. We define $f'{:}\,\mathbb{N}{\to}\mathbb{N}{\to}\mathbb{N}$ as $f'xn :=$ **if** fxn **is** $\mathsf{Some}\, y$ **then** $\mathsf{S}\, y$ **else** 0. By SCT, we obtain a function γ for f', and we have $\forall xy.\ pxy \leftrightarrow \exists k.\ \varphi_{\gamma x}k = \mathsf{Some}\, y$ as wanted. □

For the converse direction, we use that the graph of functions is enumerable.

Theorem 21. EA $\to$ SCT

Proof. Let φ as in EA be given. Recall mkstat: $(\mathbb{N}\to\mathbb{O}X)\to\mathbb{N}\to\mathbb{O}X$ turning arbitrary $F:\mathbb{N}\to\mathbb{O}\mathbb{N}$ into stationary sequences. We define

$$\varphi_c^n x := \mathsf{mkstat}(\lambda n.\ \textbf{if}\ \varphi_c n\ \textbf{is}\ \mathsf{Some}\ \langle x', y\rangle\ \textbf{then}\ \textbf{if}\ x' =_{\mathbb{B}} x\ \textbf{then}\ \mathsf{Some}\, y\ \textbf{else}\ \mathsf{None}\ \textbf{else}\ \mathsf{None})n$$

Let $f:\mathbb{N}\to\mathbb{N}\to\mathbb{N}$ and let $\varphi_{\gamma x}$ enumerate $\lambda x\langle n, m\rangle.\ fxn = m$ via EA. Now γ serves as coding function for f by Fact 1. □

6 Rice's Theorem

One of the central results of every introduction to computability theory is Rice's theorem [33], stating that non-trivial semantic predicates on programs are undecidable. Two proof strategies can be found in the literature: By using a fixed-point theorem or by establishing a many-one reduction from $\overline{\mathcal{K}}$. We here give synthetic variants of both proofs.

We base the first proof on the axiom EPF, since the notion of a fixed-point is more natural there. We base the second proof on the axiom EA. Here the choice is less canonical, but using EA enables a comparison of EA and EPF as axioms for synthetic computability.

We start by assuming EPF and proving a fixed-point theorem due to Rogers [35].

Theorem 22. *Let θ be given as in* EPF *and $\gamma:\mathbb{N}\to\mathbb{N}$, then there exists c such that $\theta_{\gamma c} \equiv \theta_c$.*

Proof. Let $\gamma:\mathbb{N}\to\mathbb{N}$. Let $f_x z := \theta_x x \gg\!= \lambda y.\theta_y z$ and γ' via EPF be such that $\theta_{\gamma' x} \equiv f_x$ (1). Let c via EPF be such that $\forall x.\ \theta_c x \triangleright \gamma(\gamma' x)$ (2).

Now $f_c \equiv \theta_{\gamma' c}$ by (1).

Also $f_c z \equiv (\theta_c c \gg\!= \lambda y.\theta_y z) \equiv \theta_{\gamma(\gamma' x)} z$ by the definition of f and (2).

Now $\gamma' c$ is a fixed-point for $\lambda i.\theta_{\gamma i}$: $\theta_{\gamma(\gamma' c)} \equiv f_c \equiv \theta_{\gamma' c}$. □

Rice's theorem can then be stated and proved as follows:

Theorem 23. *Let θ be given as in* EPF *and $p:\mathbb{N}\to\mathbb{P}$. If p treats elements as codes w.r.t. θ and is non-trivial, then p is undecidable. Formally:*

$$(\forall cc'.\ \theta_c \equiv \theta_{c'} \to pc \leftrightarrow pc') \to \forall c_1 c_2.\ pc_1 \to \neg pc_2 \to \neg\mathcal{D}p$$

Proof. Let f decide p and let pc_1 and $\neg pc_2$. Define $h:\mathbb{N}\to\mathbb{N}\rightharpoonup\mathbb{N}$ as $h_x :=$ **if** fx **then** θ_{c_2} **else** θ_{c_1} and let γ via EPF be such that $\theta_{\gamma x} y \equiv h_x y$. Let c be a fixed-point for γ via Theorem 22, i.e. $\theta_{\gamma c} \equiv \gamma$.

Then either $fc = \mathsf{true}$ and thus pc, but $\theta_c \equiv \theta_{c_2}$ and thus pc_2. A contradiction. Or $fc = \mathsf{false}$ and thus $\neg pc$, but $\theta_c \equiv \theta_{c_1}$ and thus $\neg pc_1$. A contradiction. □

Rice's theorem is often also stated for predicates $p\colon (\mathbb{N} \rightharpoonup \mathbb{N}) \to \mathbb{N}$. This formulation has the advantage that the requirement on p does not have to mention θ.

Corollary 24. EPF *implies that if* $p\colon (\mathbb{N} \rightharpoonup \mathbb{N}) \to \mathbb{N}$ *is extensional and non-trivial, then* p *is undecidable. Formally:*

$$\mathsf{EPF} \to (\forall f f'\colon \mathbb{N} \rightharpoonup \mathbb{N}. f \equiv_{\mathbb{N} \rightharpoonup \mathbb{N}} f' \to pf \leftrightarrow pf') \to \forall f_1 f_2.\, pf_1 \to \neg pf_2 \to \neg \mathcal{D}p$$

Proof. Let p be decidable. We define the index predicate of p as $I_p := \lambda c\colon \mathbb{N}.\ p(\theta_c)$, and have $I_p \preceq_m p$. Thus since p is decidable, I_p is decidable. Since I_p treats elements as codes and is non-trivial using EPF, we have that I_p is undecidable by Theorem 23. Contradiction. □

A second proof strategy for Rice's theorem is by establishing a many-one reduction from a problem proved undecidable via diagonalisation. We could use $\mathcal{K}$ defined using EPF in Fact 15, but here use EA to compare the two axioms. Thus, we use the problem $\mathcal{K}$ as used in Fact 18. We follow Forster and Smolka [16], who mechanise a fully constructive proof of Rice's theorem based on the call-by-value λ-calculus by isolating a reduction lemma ("Rice's Lemma").

Lemma 25. *Let* φ *be as in* EA *and* $p\colon \mathbb{N} \to \mathbb{P}$. *If* p *treats elements as codes w.r.t.* φ, p *is non-trivial and the code for the empty predicates satisfies* p, *then* $\overline{\mathcal{K}} \preceq_m p$.
Formally let $\mathcal{W}_c x := \exists n.\ \varphi_c n = \mathsf{Some}\,x$. *We then have:*

$$(\forall cc'.\ (\forall x.\ \mathcal{W}_c x \leftrightarrow \mathcal{W}_{c'} x) \to pc \leftrightarrow pc') \to \forall c_\emptyset c_0.\ (\forall x. \neg \mathcal{W}_{c_\emptyset} x) \to pc_\emptyset \to \neg pc_0 \to \overline{\mathcal{K}} \preceq_m p$$

Proof. The predicate $q := \lambda xy.\ \mathcal{K}x \wedge \mathcal{W}_{c_0} y$ is enumerable, meaning we obtain γ from EA such that $\forall xy.\ \mathcal{W}_{\gamma x} y \leftrightarrow \mathcal{K}x \wedge \mathcal{W}_{c_0} y$.

Let $\neg \mathcal{K}x$. We have $\mathcal{W}_{\gamma x} y \leftrightarrow \bot \leftrightarrow \mathcal{W}_{c_\emptyset} y$. Since $pc_\emptyset$ and p is semantic also $p(\gamma x)$. Conversely, let $p(\gamma x)$ and $\mathcal{K}x$. We have $\mathcal{W}_{\gamma x} y \leftrightarrow \mathcal{W}_{c_0} y$. Since p is semantic, also pc_0. Contradiction. □

Theorem 26. *Let* φ *be given as in* EA *and* $p\colon \mathbb{N} \to \mathbb{P}$. *If* p *treats inputs as codes w.r.t.* φ *and* p *is non-trivial, then* p *is not bi-enumerable. Formally let* $\mathcal{W}_c x := \exists n.\ \varphi_c n = \mathsf{Some}\,x$ *be the universal table for* φ. *We then have:*

$$(\forall cc'.(\forall x.\ \mathcal{W}_c x \leftrightarrow \mathcal{W}_{c'} x) \to pc \leftrightarrow pc') \to \forall c_1 c_2.\ pc_1 \to \neg pc_2 \to \neg(\mathcal{E}p \wedge \mathcal{E}\overline{p})$$

Proof. Since $\lambda x\colon \mathbb{N}.\bot$ is enumerable, by EA there is $c_\emptyset$ such that $\forall x. \neg W_{c_\emptyset} x$. Now let pc_1, $\neg pc_2$, and let p be bi-enumerable.

If $pc_\emptyset$, we have $\overline{\mathcal{K}} \preceq_m p$, a contradiction since $\overline{\mathcal{K}}$ would be enumerable by Fact 6. If $\neg pc_\emptyset$ we have $\overline{\mathcal{K}} \preceq_m \overline{p}$, again a contradiction. □

Corollary 27. *Let* φ *be given as in* EA *and* $p\colon \mathbb{N} \to \mathbb{P}$. *If* p *treats inputs as codes w.r.t.* φ *and* p *is non-trivial, then* p *is undecidable.*

We can state this second version of Rice's theorem for $p\colon (\mathbb{N} \to \mathbb{P}) \to \mathbb{P}$.

Corollary 28. EA *implies that if p is extensional and non-trivial w.r.t. enumerable predicates, then p is undecidable. Formally under the assumption of* EA *we have*

$$(\forall qq'{:}\mathbb{N} \to \mathbb{P}.(\forall x.\ qx \leftrightarrow q'x) \to pq \leftrightarrow pq') \to \forall q_1 q_2.\ \mathcal{E}q_1 \to \mathcal{E}q_1 \to pq_1 \to \neg pq_2 \to \neg\mathcal{D}p$$

Proof. Let p be decidable. We define the index predicate of p as $I_p := \lambda c{:}\mathbb{N}.\ p(\mathcal{W}_c)$, and have $I_p \preceq_m p$. Thus since p is decidable, I_p is decidable. Since I_p treats elements as codes and is non-trivial using EA, we have that I_p is undecidable by Theorem 27. Contradiction. □

We have formulated both theorems to explicitly assume θ and φ and their respective specification, to contrast the axioms EPF and EA. One can however obtain Theorem 23 from Theorem 27 – and vice versa – constructing θ from φ and constructing a predicate q treating elements as indices w.r.t. θ from a predicate p treating elements as indices w.r.t. φ – and vice versa.

Proofs based on EPF require the manipulation of partial functions, which is formally tedious. We will thus use EA as basis for synthetic computability: In contrast to SCT, it does not force us to encode every computation as total function $\mathbb{N} \to \mathbb{N}$, and in contrast to EPF it does not force us to work with partial functions either.

Instead, we can simply consider enumerable predicates and their enumerators, which are total functions.

7 CT in the Weak Call-by-Value λ-Calculus

In this section we treat CT_L, the formulation of CT_ϕ where ϕ is a universal function for the weak call-by-value λ-calculus L [16,32]. We largely omit technical details in this section and refer the interested reader to the accompanying Coq code. We only need a stationary function $\mathsf{eval}{:}\,\mathsf{tm}_\mathsf{L}{\to}\mathbb{N}{\to}\mathbb{O}(\mathsf{tm}_\mathsf{L})$ such that $\exists n.l\mathsf{eval}\ s\ n = \mathsf{Some}\,t$ for a λ-term s if and only if t is the normal form of s w.r.t. weak call-by-value evaluation, and a function $\varepsilon_\mathbb{N}{:}\,\mathbb{N}{\to}\mathsf{tm}_\mathsf{L}$ encoding natural numbers as terms, e.g. using Scott encoding [36], and an inverse function unenc.

We define ϕ such that for an application $\phi^n_c x$ we translate the code c to a term t and then evaluate the application $t\ (\varepsilon_\mathbb{N} x)$ for n steps using the step-indexed interpreter eval. To interpret c as term, we need the following:

Fact 29. *There are functions* $R{:}\,\mathbb{N}{\to}\mathbb{O}(\mathsf{tm}_\mathsf{L})$ *and* $I{:}\,\mathsf{tm}_\mathsf{L}{\to}\mathbb{N}$ *such that we have* closed $s \leftrightarrow R(Is) = \mathsf{Some}\,s$.

We can then define:

$$\phi^n_c x := \textbf{if } Rc \textbf{ is } \mathsf{Some}\,t \textbf{ then } \textbf{if } \mathsf{eval}\ (t\ (\varepsilon_\mathbb{N} n))\ n \textbf{ is } \mathsf{Some}\,v \textbf{ then } \mathsf{unenc}\ v \textbf{ else } \mathsf{None} \textbf{ else } \mathsf{None}$$

Fact 30. *If* $f : \mathbb{N} \to \mathbb{N}$ *is computed by* t *then* $\forall x. \exists n.\ \phi^n_{It} x = \mathsf{Some}\,(f x)$.

We write CT_L instead of CT_ϕ. To define an S^m_n operator σ we need the following:

Fact 31. *There are* t_{embed} *and* t_{unembed} *computing* $\lambda\langle n, m\rangle.(n, m)$ *and* $\lambda nm.\langle n, m\rangle$.

We define $\sigma c x := \textbf{if } Rc \textbf{ is } \mathsf{Some}\,t \textbf{ then } I(\lambda y.\ t(t_{\mathsf{embed}}(\varepsilon_\mathbb{N} x) y)) \textbf{ else } c$. Note that here the function I is applied to an L-term.

Theorem 32. $\forall cxyv.\ (\exists n.\ \phi^n_{\sigma c x} y = \mathsf{Some} v) \leftrightarrow (\exists n.\ \phi^n_c \langle x, y\rangle = \mathsf{Some} v)$

Proof. We only prove the direction from right to left, the other direction is similar. Let $\phi^n_c\langle x, y\rangle = \mathsf{Some}\,v$. Then $Rc = \mathsf{Some}\,t$ and $t\ \varepsilon_\mathbb{N}(\langle x, y\rangle) \rhd v$ for some closed term t. Let $s := (\lambda y.\ t_{\mathsf{embed}}(\varepsilon_\mathbb{N} x) y)\ (\varepsilon_\mathbb{N} y)$. We have $t\ \varepsilon_\mathbb{N}(\langle x, y\rangle) \equiv s$ and thus $s \rhd v$. Thus there is m s.t. $\mathsf{eval}\ m\ s = \mathsf{Some}\,v$ and we have $\phi^m_{\sigma c x} = \mathsf{Some}\,v$. □

Corollary 33. $\mathsf{CT}_\mathsf{L} \to \Sigma\phi.\ \mathsf{CT}_\phi \wedge \mathsf{SMN}_\phi$

8 Related Work

Bauer [4] develops synthetic computability based on an axiom stating that the set of enumerable sets of natural numbers is enumerable. Translating to our type theoretic setting this yields the following axiom stating that there is an enumerator $\mathcal{W}$ of all enumerable predicates, up to extensionality.

$$\mathsf{EA}' := \exists \mathcal{W} : \mathbb{N} \to (\mathbb{N} \to \mathbb{P}). \forall p : \mathbb{N} \to \mathbb{P}.\ \mathcal{E} p \leftrightarrow \exists c.\ \mathcal{W}_c \equiv_{\mathbb{N} \to \mathbb{P}} p$$

Additionally to EA′, Bauer also assumes countable choice and Markov's principle. In general however, the assumption of countable choice makes the theory anti-classical, i.e. assuming LEM is inconsistent. The interplay of axioms like EA′, MP, LEM, and countable choice is discussed in [15]. Countable choice allows extracting the enumerator for every enumerable predicate in the range of $\mathcal{W}$ computationally, corresponding to a non-parametric version of our axiom EA. Countable choice also can be used to prove a synthetic S^m_n theorem w.r.t. $\mathcal{W}$.

Our parametric formulation of EA implies EA′, and conversely EA′ implies EA under the presence of countable choice.

Richman [34] introduces the axiom CPF ("Countability of Partial Functions"). It states that the set of partial functions is (extensionally) countable, i.e. there is a surjection $\mathbb{N} \to (\mathbb{N} \rightharpoonup \mathbb{N})$ w.r.t. equivalence on partial functions. Intensionally, Richman models the partial function space $\mathbb{N} \rightharpoonup \mathbb{N}$ as stationary functions. Thus, written out fully his axiom is a non-parametric version of EPF, just instantiated to the stationary functions model of partial functions.

Theory based on CPF is developed in the book by Bridges and Richman [8], where CPF is taken as basis for the constructivist system RUSS. In RUSS, the axiom of countable choice is also assumed. Bridges and Richman discuss that "in RUSS countable choice can usually be avoided" [8, p. 54] by postulating a composition operator for θ or, equivalently, an SMN operator.

Our two proofs of Rice's theorem are in strong support of this conjecture. Recall that the first proof is based on a parametrically universal partial function θ, while the second proof is based on a parametrically universal enumerator φ. The two proofs also use different proof strategies.

The second strategy establishes a reduction from $\mathcal{K}$. This strategy is used in the textbooks by Cutland [12], Odifreddi [29], Soare [39], and Cooper [10], whereas Rogers [35] and Sipser [37] pose Rice's theorem as an exercise.

The first strategy, based on Rogers' fixed-point theorem or equivalently based on Kleene's recursion theorem is less frequently found. It is however mentioned in the Wikipedia article on Rice's theorem [42]. The technique appears first in the lecture notes by Scott [36], who shows a variant of Rice's theorem for the λ-calculus. Scott's proof can also be found in [2,38].

We are aware of five machine-checked proofs of Rice's theorem: Norrish [28] proves Rice's theorem for the λ-calculus, formulated for predicates $p\colon (\mathbb{N} \to \mathbb{O}\mathbb{N}) \to \mathbb{P}$, using the proof strategy via reduction. Forster and Smolka [16] prove Rice's theorem for the weak call-by-value λ-calculus, formulated for predicates on terms of the considered calculus which have the same extensional behaviour, using the proof strategy via reduction. Forster [14] proves Scott's variant of Rice's theorem for the weak call-by-value λ-calculus, formulated for predicates on terms of the considered calculus which do not distinguish β-equivalent terms, using a fixed-point theorem. Carneiro [9] proves Rice's theorem for μ-recursive functions, formulated for predicates $p\colon (\mathbb{N} \rightharpoonup \mathbb{N}) \to \mathbb{P}$, using a fixed-point theorem. Ramos et al. [13] prove Rice's theorem for the functional language PVS0, formulated for predicates on PVS0, using an assumed fixed-point theorem.

Bauer [4] also presents a synthetic variant of Rice's theorem. His formulation reads "If A is a set such that all functions of type $A \to A$ have a fixed-point, every function $A \to \mathbb{B}$ is constant" and uses the enumerability axiom as discussed above, but does not rely on countable choice to the best of our knowledge. Note that our variants of Rice's theorem presented in this paper are trivialities in classical set theory, the foundation of textbook computability, since both EPF and EA are false in classical set theory where all problems have a characteristing decision function. In contrast, Bauer's theorem is a triviality in classical set theory in *two* ways: First, the enumerability axiom is contradictory in classical set theory, and second the statement of the theorem is a trivial even without axioms since if all functions $A \to A$ have a fixed-point, A is a sub-singleton: two distinct elements a_1, a_2 would allow constructing a fixed-point free function $\lambda x.\textsf{if}\ x = a_1\ \textsf{then}\ a_2\ \textsf{else}\ a_1$. In short, we sacrifice identifying the minimal essence of theorems to better preserve classical intuitions.

Acknowledgements. I would like to thank Gert Smolka, Dominik Kirst, and Dominique Larchey-Wendling for constructive feedback and productive discussions, as well as Andrej Bauer and Dominik Wehr for helpful advice on consistency proofs of CT in the literature.

A Consistency and Admissibility of **CT** in CIC

In 1943, Kleene conjectured that whenever $\forall x.\exists y.\ Rxy$ is constructively provable, there in fact exists a μ-recursive function f such that $\forall x.\ Rx(fx)$ [22]. This corresponds to a strong form of the admissibility of CT. In 1945, Kleene [23] proved his conjecture for Heyting arithmetic, using number realizability. An independent proof of this is due to Beth [6].

In this section, we use CT to denote the historical formulation of CT, e.g. using μ-recursive functions, which is however equivalent to CT_L.

Troelstra and van Dalen [41, §4.5.1 p. 204] state an even stronger result, using Gödel's Dialectia interpretation [17], namely that in Heyting arithmetic CT, MP and a restricted form of the independence of premise rule IP (with P logically decidable) are consistent as schemes.

Odifreddi states that "for all current intuitionistic systems (not involving the concept of choice sequence) the consistency with CT has actually been established" [29, §1.8 pg. 122]. We do not discuss other systems for constructive or intuitionistic mathematics in detail.

For CIC, the result is not explicitly stated in the literature. An admissibility proof of CT seems to be immediate as a consequence of Letouzey's semantics extraction theorem for Coq [25]. Regarding a consistency proof one cannot mirror the situation in Heyting arithmetic, since a Dialectia interpretation for Coq is not available [31].

However, several approaches seem to yield the result:

First, CT is consistent in intuitionistic set theory (e.g. IZF) [18], and IZF can be used to model CIC [3].

Secondly, realizability models based on the first Kleene algebra prove CT consistent. Luo constructs an ω-set model for the Extended Calculus of Constructions (ECC, a type theory with type universes and impredicative $\mathbb{P}$, but no inductive types), where "[t]he morphisms between ω-sets are'computable' in the sense that they are realised by partial recursive functions" [26, §6.1 pg. 118].

Thirdly, it is well known how to build topos models of the calculus of constructions [20]. The effective topos, due to Hyland [19], validates CT.

Fourthly, Swan and Uemura [40] give a sheaf model construction proving that CT is consistent in Martin Löf type theory, together with propositional truncation, Markov's principle, and univalence. It seems like the syntactic universe of propositions $\mathbb{P}$ does not hinder adapting their model construction to CIC.

Fivthly, Yamada [43] gives a game-semantic proof that a $\forall f.\Sigma c$ form of CT is consistent in intensional Martin Löf type theory, settling an open question of at least 15 years [21]. Note that this form is significantly stronger, since it allows defining a strictly intensional higher-order coding *function* of type $(\mathbb{N} \to \mathbb{N}) \to \mathbb{N}$, which is inconsistent under the assumption of functional extensionality [15]. It is not obvious how to extend Yamada's proof to our $\forall f.\exists c$ formulation of CT in CIC with the impredicative universe $\mathbb{P}$.

References

1. Andrews, P.B.: An Introduction to Mathematical Logic and Type Theory: To Truth Through Proof. APLS, vol. 27. Springer, Dordrecht (2002). https://doi.org/10.1007/978-94-015-9934-4
2. Barendregt, H., Dekkers, W., Statman, R.: Lambda Calculus with Types. Cambridge University Press, Cambridge (2013)
3. Barras, B.: Sets in Coq, Coq in sets. J. Formaliz. Reason. **3**(1), 29–48 (2010)
4. Bauer, A.: First steps in synthetic computability theory. Electron. Notes Theoret. Comput. Sci. **155**, 5–31 (2006)
5. Bauer, A.: An injection from the Baire space to natural numbers. Math. Struct. Comput. Sci. **25**(7), 1484–1489 (2015)
6. Beth, E.W.: Semantical considerations on intuitionistic mathematics. J. Symb. Log. **13**(3), 173–173 (1948)
7. Bishop, E., Bridges, D.: Constructive Analysis, vol. 279. Springer, Heidelberg (2012)
8. Bridges, D., Richman, F.: Varieties of Constructive Mathematics, vol. 97. Cambridge University Press, Cambridge (1987)
9. Carneiro, M.: Formalizing computability theory via partial recursive functions. In: Harrison, J., O'Leary, J., Tolmach, A. (eds.) 10th International Conference on Interactive Theorem Proving (ITP 2019), Dagstuhl, Germany, Leibniz International Proceedings in Informatics (LIPIcs), vol. 141, pp. 12:1–12:17. Schloss Dagstuhl-Leibniz-Zentrum fuer Informatik (2019)
10. Cooper, S.B.: Computability Theory. CRC Press, Boca Raton (2003)
11. Coquand, T.: Metamathematical investigations of a calculus of constructions. Technical report RR-1088, INRIA (1989)
12. Cutland, N.: Computability. Cambridge University Press, Cambridge (1980)
13. Ferreira Ramos, T.M., Almeida, A.A., Ayala-Rincón, M.: Formalization of rice's theorem over a functional language model. Technical report (2020)
14. Forster, Y.: A formal and constructive theory of computation. Bachelor's thesis, Saarland University (2014)
15. Forster, Y.: Church's thesis and related axioms in Coq's type theory. In: Baier, C., Goubault-Larrecq, J. (eds.) 29th EACSL Annual Conference on Computer Science Logic (CSL 2021), Dagstuhl, Germany, Leibniz International Proceedings in Informatics (LIPIcs), vol. 183, pp. 21:1–21:19. Schloss Dagstuhl-Leibniz-Zentrum für Informatik (2021)
16. Forster, Y., Smolka, G.: Weak call-by-value lambda calculus as a model of computation in Coq. In: Ayala-Rincón, M., Muñoz, C.A. (eds.) ITP 2017. LNCS, vol. 10499, pp. 189–206. Springer, Cham (2017). https://doi.org/10.1007/978-3-319-66107-0_13
17. Gödel, V.K.: Über eine bisher noch nicht benützte Erweiterung des finiten Standpunktes. Dialectica **12**(3–4), 280–287 (1958)
18. Hahanyan, V.: The consistency of some intuitionistic and constructive principles with a set theory. Stud. Logica. **40**(3), 237–248 (1981)
19. Hyland, J.M.E.: The effective topos. In: The L. E. J. Brouwer Centenary Symposium, Proceedings of the Conference held in Noordwijkerhout, pp. 165–216. Elsevier (1982)

20. Hyland, J.M.E., Pitts, A.M.: The theory of constructions: categorical semantics and topos-theoretic models. In: Gray, J.W., Scedrov, A. (eds.) Categories in Computer Science and Logic. Contemporary Mathematics, vol. 92, pp. 137–199. Amer. Math. Soc, Providence (1989)
21. Ishihara, H., Maietti, M.E., Maschio, S., Streicher, T.: Consistency of the intensional level of the minimalist foundation with church's thesis and axiom of choice. Arch. Math. Log. **57**(7–8), 873–888 (2018)
22. Kleene, S.C.: Recursive predicates and quantifiers. Trans. Am. Math. Soc. **53**(1), 41–73 (1943)
23. Kleene, S.C.: On the interpretation of intuitionistic number theory. J. Symb. Log. **10**(4), 109–124 (1945)
24. Kreisel, G.: Mathematical logic. Lect. Mod. Math. **3**, 95–195 (1965)
25. Letouzey, P.: Programmation fonctionnelle certifiée: l'extraction de programmes dans l'assistant Coq. Ph.D. thesis (2004)
26. Luo, Z.: Computation and Reasoning: A Type Theory for Computer Science. Oxford University Press Inc., Oxford (1994)
27. McCarty, D.C.: Incompleteness in intuitionistic metamathematics. Notre Dame J. Formal Log. **32**(3), 323–358 (1991)
28. Norrish, M.: Mechanised computability theory. In: van Eekelen, M., Geuvers, H., Schmaltz, J., Wiedijk, F. (eds.) ITP 2011. LNCS, vol. 6898, pp. 297–311. Springer, Heidelberg (2011). https://doi.org/10.1007/978-3-642-22863-6_22
29. Odifreddi, P.: Classical Recursion Theory: The Theory of Functions and Sets of Natural Numbers. Elsevier, Amsterdam (1992)
30. Paulin-Mohring, C.: Inductive definitions in the system Coq rules and properties. In: Bezem, M., Groote, J.F. (eds.) TLCA 1993. LNCS, vol. 664, pp. 328–345. Springer, Heidelberg (1993). https://doi.org/10.1007/BFb0037116
31. Pédrot, P.-M.: A Materialist Dialectica. Theses, Paris Diderot (2015)
32. Plotkin, G.D.: Call-by-name, call-by-value and the λ-calculus. Theoret. Comput. Sci. **1**(2), 125–159 (1975)
33. Rice, H.G.: Classes of recursively enumerable sets and their decision problems. Trans. Am. Math. Soc. **74**(2), 358–366 (1953)
34. Richman, F.: Church's thesis without tears. J. Symb. Log. **48**(3), 797–803 (1983)
35. Rogers, H.: Theory of Recursive Functions and Effective Computability (1987)
36. Scott, D.: A system of functional abstraction (1968). Lectures delivered at University of California, Berkeley, Cal., 1962/63. Photocopy of a preliminary version, issued by Stanford University, September 1963, furnished by author in 1968
37. Sipser, M.: Introduction to the Theory of Computation, vol. 2. Thomson Course Technology Boston (2006)
38. Smullyan, R.M.: Diagonalization and Self-reference. Oxford Science Publications, Clarendon Press, Oxford (1994)
39. Soare, R.I.: Recursively Enumerable Sets and Degrees: A Study of Computable Functions and Computably Generated Sets. Springer, Heidelberg (1999)
40. Swan, A., Uemura, T.: On Church's thesis in cubical assemblies. arXiv preprint arXiv:1905.03014 (2019)
41. Troelstra, A.S., van Dalen, D.: Constructivism in Mathematics. Studies in Logic and the Foundations of Mathematics, vol. I, 26 (1988)

42. Wikipedia contributors. Rice's theorem – Wikipedia, the free encyclopedia (2021). https://en.wikipedia.org/w/index.php?title=Rice's_theorem&oldid=1017713534. Accessed 31 May 2021
43. Yamada, N.: Game semantics of Martin-Löf type theory, part iii: its consistency with church's thesis (2020)

Constructive and Mechanised Meta-Theory of Intuitionistic Epistemic Logic

Christian Hagemeier and Dominik Kirst[(✉)]

Saarland Informatics Campus, Saarland University, Saarbrücken, Germany
christian@hagemeier.ch, kirst@cs.uni-saarland.de

Abstract. Artemov and Protopopescu proposed intuitionistic epistemic logic (IEL) to capture an intuitionistic conception of knowledge. By establishing completeness, they provided the base for a meta-theoretic investigation of IEL, which was continued by Krupski with a proof of cut-elimination, and Su and Sano establishing semantic cut-elimination and the finite model property. However, to the best of our knowledge, no analysis of these results in a constructive meta-logic has been conducted.

We aim to close this gap and investigate IEL in the constructive type theory of the Coq proof assistant. Concretely, we present a constructive and mechanised completeness proof for IEL, employing a syntactic decidability proof based on cut-elimination to constructivise the ideas from the literature. Following Su and Sano, we then also give constructive versions of semantic cut-elimination and the finite model property. Given our constructive and mechanised setting, all these results now bear executable algorithms. We expect that our methods used for mechanising cut-elimination and decidability also extend to other modal logics (and have verified this observation for the classical modal logic K).

Keywords: Epistemic logic · Completeness · Constructivisation

1 Introduction

Intuitionistic epistemic logic (IEL), introduced by Artemov and Protopopescu [1], is a relatively recent formalism modelling an intuitionistic conception of knowledge. While classical epistemic logics [14,23] typically include the *reflection principle* $\mathsf{K}\,A \supset A$, read as "known propositions must be true", IEL is based on the *co-reflection principle* $A \supset \mathsf{K}\,A$, read as "from the existence of proofs we can gain knowledge by verification". This striking disagreement is explained by the divergent notions of truth: while a proposition is determined classically true by its binary truth value, it is considered intuitionistically true if an (intuitionistic) proof in the prevailing Brouwer-Heyting-Kolmogorov (BHK) interpretation has been constructed. While the sole addition of co-reflection to intuitionistic propositional logic results in the logic of intuitionistic belief (IEL^-), Artemov and Protopopescu propose the further addition of *intuitionistic reflection* $\mathsf{K}\,A \supset \neg\neg A$ for IEL. This principle reestablishes, up to a double negation,

S. Artemov and A. Nerode (Eds.): LFCS 2022, LNCS 13137, pp. 90–111, 2022.
https://doi.org/10.1007/978-3-030-93100-1_7

the factivity of truth classically expressed by reflection, and therefore places intuitionistic knowledge as a modality between intuitionistic and classical truth.

Complementing the philosophical arguments for (and against) IEL, the original paper [1] already contains several technical results such as soundness and completeness with respect to a suitable Kripke semantics, as well as derived observations concerning the disjunction property and admissibility of reflection. This formal investigation has been carried on for instance by Su and Sano [27] with proofs of the finite model property and semantic cut-elimination, and by Krupski [18] with proofs of syntactic cut-elimination and decidability. However, especially the arguments for completeness relying on the Lindenbaum construction manifestly employ classical logic, leaving the current state of the meta-theory of IEL unsatisfactory: while the formalism itself successfully embraces intuitionistic principles to tackle classical knowability paradoxes, no visible attempts are made to describe its semantics in constructive terms.

With this paper, we hope to contribute to a more uniform picture by developing all mentioned results in a purely constructive setting. Concretely, we illustrate that by preparing an argument for the finite model property along the lines of Su and Sano by a syntactic decidability proof inspired by Smolka, Brown, and Dang [6,26], completeness of IEL with respect to finite contexts can be obtained without appeal to classical logic. Moreover, in the fashion of *constructive reverse mathematics* [15,16], we show that completeness with respect to possibly infinite contexts as entailed by the development in [1] is equivalent to the law of excluded middle (LEM), while even the restriction of completeness to enumerable contexts is still strong enough to imply Markov's principle (MP), both observations following similar arguments as applicable to first-order logic [11].

As a framework, we employ the constructive type theory CIC [4,20] implemented in the *Coq proof assistant* [30]. We deem this choice valuable for three reasons: First, CIC embodies a rather modest system free of debatable choice principles diluting the analysis [24]. Secondly, CIC is based on the same principles justifying IEL by internalising the BHK interpretation in a proof-relevant way and in fact modelling K by a truncation operation from computational types to the impredicative universe $\mathbb{P}$ of propositions, obeying co-reflection and intuitionistic reflection. Thirdly, we use its implementation in Coq as a tool to verify all proofs, track the usage of assumptions, and exhibit the algorithmic content of the constructive meta-theory for instance in the form of executable algorithms for completeness, cut-elimination, and decidability. The resulting Coq development is systematically hyperlinked with the PDF version of this paper.[1]

Contributions. To the best of our knowledge, we are the first to explicitly develop the meta-theory of IEL in a fully constructive setting. Moreover, all our results are mechanised using the Coq proof assistant and accompanied by similar proof-theoretic results for the classical modal logic K.

[1] See Appendix 2, also browsable at https://www.ps.uni-saarland.de/extras/iel.

Outline. In Sect. 2, we begin with some preliminary definitions concerning the constructive type theory we are working in. In Sect. 3, we introduce formulas and the natural deduction system of IEL and outline their encoding in constructive type theory. In Sect. 4, we introduce a sequent calculus suitable for mechanising cut-elimination which we use in Sect. 5 to prove decidability for IEL. Section 6 establishes constructive completeness and the finite model property. In Sect. 7, we report results about infinite theories and strong completeness. We close with a review of the literature and future work in Sect. 8.

2 Preliminaries

We work in the constructive type theory CIC [4,20] of the Coq proof assistant [30], with a predicative hierarchy of type universes $\mathbb{T}_i$ above a single impredicative universe $\mathbb{P}$. We will always omit the level and write $\mathbb{T}$ for any $\mathbb{T}_i$. On the type level, we have the unit type $\mathbb{1}$ with the single element $*$, the void type $\mathbb{0}$, function spaces $X \to Y$, products $X \times Y$, sums $X + Y$, dependent products $\forall x^X. F\,X$, and dependent sums $\Sigma x^X. F\,x$. On a propositional level, these types are denoted using by the usual logical notation $(\top, \bot, \to, \wedge, \vee, \exists, \forall)$. Elimination from $\mathbb{P}$ into $\mathbb{T}$ is restricted to hide the computational content of proofs.

Basic inductive types we use are natural numbers $\mathbb{N} ::= 0 \mid n+1\,(n \in \mathbb{N})$ and booleans $\mathbb{B} := \mathsf{tt} \mid \mathsf{ff}$. Furthermore given a type X, we define lists $\mathcal{L}(X) := \emptyset \mid x :: L$ for $x : X$ and $L : \mathcal{L}(X)$, and the option type $\mathcal{O}(X) := \emptyset \mid \ulcorner x \urcorner$. To ease notation we will oftentimes denote appending an element x to a list L by L, x.

Definition 1. Let X be a type and $p : X \to \mathbb{P}$ be a predicate. We call

- p enumerable, if there is $f : \mathbb{N} \to \mathcal{O}(X)$ with $\forall x^X. p\,x \leftrightarrow \exists n^{\mathbb{N}}. f\,n = \ulcorner x \urcorner$,
- p decidable, if there is some $f : X \to \mathbb{B}$ with $p\,x \leftrightarrow \forall x^X. f(x) = \mathsf{tt}$.

These notions generalise easily to predicates of higher arity. A type X is enumerable if the predicate $p : X \to \mathbb{P}$ defined by $p\,x := \top$ is enumerable. X is discrete if the predicate $\lambda xy. x = y$ is decidable.

One technique we will often use throughout this paper is reasoning classically locally whenever we prove a negative statement, captured by the following fact:

Lemma 2. *The statements* $\neg\neg(P \vee \neg P)$ *and* $((P \vee \neg P) \to \neg Q) \to \neg Q$ *hold for arbitrary propositions* $P, Q : \mathbb{P}$.

Non-classical-axioms. Especially important for our development is the law of excluded middle, $\mathsf{LEM} := \forall P : \mathbb{P}. P \vee \neg P$ and Markov's principle

$$\mathsf{MP} := \forall f : \mathbb{N} \to \mathbb{B}. \neg\neg(\exists n. f\,n = \mathsf{tt}) \to \exists n. f\,n = \mathsf{tt}.$$

It is well-known, that MP is weaker than LEM and has a computational justification based on linear search, which LEM completely lacks [5].

3 Basic Intuitionistic Epistemic Logic

This section introduces formulas of IEL, the natural deduction system, and its models closing with a statement of the classical completeness proof. We present nothing new, instead recapping material from Artemov and Protopopescu [1] adapted to the setting of constructive type theory.

Definition 3. The syntax of IEL is given by the following inductive datatype:

$$A, B : \mathcal{F} ::= A \vee B \mid A \wedge B \mid A \supset B \mid \mathsf{K}\, A \mid p_i \mid \bot \quad (i \in \mathbb{N})$$

Lemma 4. *The type $\mathcal{F}$ is discrete and enumerable.*

Proof. Both are established using standard techniques e.g. [10, Fact 3.19]. □

Since $\mathcal{F}$ is inhabited, we can even establish a stronger claim than enumerability, namely that a function $f : \mathbb{N} \to \mathcal{F}$ exists s.t. $\forall A^{\mathcal{F}}. \exists n^{\mathbb{N}}. f n = A$. In our formal setting, we model finite theories as lists of formulas. Throughout this paper, we refer to these as finite sets and use usual set-theoretic notation. Induction on a finite set, then, is just induction on the list representing the finite set. Infinite contexts, here called theories, are represented as predicates $\mathcal{T} : \mathcal{F} \to \mathbb{P}$, with the intended reading that $A \in \mathcal{T}$ iff $\mathcal{T}\, A$ holds.

The natural deduction calculus for IEL is encoded as an inductive predicate $\vdash: \mathcal{L}(\mathcal{F}) \to \mathcal{F} \to \mathbb{P}$. Natural deduction for IEL$^-$ was introduced by Rogozin [25], however our system is slightly different to ease the mechanisation. The idea is to extend a natural deduction calculus for intuitionistic propositional logic by rules for co-reflection *(KR)* and distribution *(KD)* to express IEL$^-$, and by a rule for intuitionistic reflection *(KF)* to express IEL. These rules are shown in Fig. 1; the full system can be found in Appendix 1. The main difference between our system and that of Rogozin is the distribution rule, as Rogozin's formulation equivalently allows for multiple applications of our *KD*-rule in one step.

In this paper, we will always state and prove results for IEL, the proofs for IEL$^-$ can be obtained from the proofs for IEL by omitting certain parts. In fact, the mechanisation contains formal proofs for both systems, avoiding code duplication with tagged deduction systems (see Appendix 2).

We naturally extend derivability to theories $\mathcal{T} : \mathcal{F} \to \mathbb{P}$ by writing $\mathcal{T} \vdash A$ if there is a finite set $\Gamma \subseteq \mathcal{T}$ with $\Gamma \vdash A$.

$$\frac{\Gamma \vdash A}{\Gamma \vdash \mathsf{K}\, A}\ (KR) \qquad \frac{\Gamma \vdash \mathsf{K}\,(A \supset B)}{\Gamma \vdash \mathsf{K}\, A \supset \mathsf{K}\, B}\ (KD) \qquad \frac{\Gamma \vdash \mathsf{K}\, A}{\Gamma \vdash \neg\neg A}\ (KF)$$

Fig. 1. Selected natural deduction rules for IEL

Models for IEL extend standard Kripke semantics by a verification relation. We refer to the reader to Wolter and Zakharyashchev [33], whose paper contains general results in the model theory of intuitionistic modal logics.

Definition 5. (Kripke Models) A *Kripke Model* for IEL, IEL$^-$ is a quadruple $(\mathcal{W}, \mathcal{V}, \leq, \leq_\mathsf{K})$ consisting of a type of worlds $\mathcal{W}$, and a valuation $\mathcal{V} : \mathcal{W} \to \mathbb{N} \to \mathbb{P}$, which must have the following properties:

1. $\leq$ is a preorder on $\mathcal{W}$,
2. If $w \leq v$ and $V(w, i)$ then $V(v, i)$ for any w, v, i,
3. $\leq \circ \leq_\mathsf{K} \subseteq \leq_\mathsf{K}$, i.e. if $w \leq u$ and $u \leq_\mathsf{K} v$ then $w \leq_\mathsf{K} v$ for any $w, u, v \in \mathcal{W}$,
4. $\leq_\mathsf{K} \subseteq \leq$, i.e. if $w \leq_\mathsf{K} v$ then $w \leq u$ for any $u, v \in \mathcal{W}$.

Property 2 in above definition is known as *persistence.* For IEL, additionally the models need to have a serial $\leq_\mathsf{K}$-relation, i.e. for all w there should be some v with $w \leq_\mathsf{K} v$.

Definition 6. (Forcing Relation) Let $\mathcal{M}$ be a Kripke model. We define the forcing relation by recursion on the formula:

$$
\begin{aligned}
w \Vdash p_i &:\Leftrightarrow \mathcal{V}(w, i) \\
w \Vdash A_0 \wedge A_1 &:\Leftrightarrow w \Vdash A_0 \wedge w \Vdash A_1 \\
w \Vdash A_0 \vee A_1 &:\Leftrightarrow w \Vdash A_0 \vee w \Vdash A_1 \\
w \Vdash A_0 \supset A_1 &:\Leftrightarrow \forall w'. w \leq w' \to w' \Vdash A_0 \to w' \Vdash A_1 \\
w \Vdash \mathsf{K}\, A_0 &:\Leftrightarrow \forall w'.\, w \leq_\mathsf{K} w' \to w' \Vdash A_0
\end{aligned}
$$

We can easily establish that the forcing relation is monotone.

Lemma 7. (Monotonicity) *Let $\mathcal{M}$ be an arbitrary model and A be any formula. If $w \leq v$ and $\mathcal{M}, w \Vdash A$ then $\mathcal{M}, v \Vdash A$.*

Proof. Induction on A utilising the persistence of $\mathcal{V}$. □

We use the standard notation $\Gamma \Vdash A$ to denote that any model forcing Γ forces A, too.[2] Note that this notation is monotone in the following sense: If $\Gamma \subseteq \Gamma'$ and $\Gamma \Vdash A$ then $\Gamma' \Vdash A$.

Soundness for Kripke models can be established by a simple induction.

Lemma 8. (Soundness) *If $\mathcal{T} \vdash A$ then $\mathcal{T} \Vdash A$.*

Proof. Assume $\mathcal{T} \vdash A$, thus there is a finite set $\Gamma \subseteq \mathcal{T}$ s.t. $\Gamma \vdash A$. Then by induction on $\Gamma \vdash A$ we show $\Gamma \Vdash A$, relying on Lemma 7. Thus also $\mathcal{T} \Vdash A$. □

With soundness, we can establish consistency of IEL.

Lemma 9. (Consistency) *IEL is consistent.*

Proof. For deriving a contradiction, assume $\vdash \bot$. Thus by soundness (Lemma 8) $\bot$ is entailed in any model at every world. But we can easily construct a model where $M, w \nVdash \bot$, contradicting the assumption. □

[2] Formally, define $\Gamma \Vdash A := \forall \mathcal{M} w.\, (\forall B \in \Gamma.\, \mathcal{M}, w \Vdash B) \to \mathcal{M}, w \Vdash A$.

Finally, we formulate a strong version of the classical completeness theorem, by composition of the Lindenbaum and Truth Lemma both established in [1]. Notably, the authors of [1] prove both lemmas using LEM to allow case distinctions whether a formula is contained in or provable from an infinite context.

Theorem 10. (Classical Completeness) *Let $\mathcal{T} : \mathcal{F} \to \mathbb{P}$ be an arbitrary predicate on formulas. Assuming* LEM*, if $\mathcal{T} \Vdash A$ then $\mathcal{T} \vdash A$.*

4 Cut-Free Sequent Calculus

Sequent calculus representations for IEL have been proposed by Krupski [18], Su and Sano [28], and more recently Fiorino [9].

A main challenge for us is to find an encoding suitable for proving termination of the proof search and structural properties in a proof assistant. We employ a sequent calculus similar to the GKI-calculus by Kleene [17] and extending it to cover IEL by using additional rules, similar to those used by Krupski [18]. Similar techniques have been used by Smolka, Brown, and Dang [6,26] to establish decidability of classical and intuitionistic propositional logic in Coq.

Let us highlight why this encoding is well-suited for mechanisation: In most textbooks [31] the GKI-calculus does not use membership but instead just keeps the principal formula in the premiss.

$$\frac{\Gamma, A \wedge B, A, B \Rightarrow C}{\Gamma, A \wedge B \Rightarrow C} \qquad \frac{A \wedge B \in \Gamma \qquad \Gamma, A, B \Rightarrow C}{\Gamma \Rightarrow C}$$

The left-hand side is the usual presentation, while the version on the right is the one we use. This change into using membership helps with automation.

The rules of the calculus are displayed in Fig. 2, where for a finite set Γ we denote the downward K-projection by $\Gamma_{\mathsf{K}} := \{A \mid \mathsf{K}\, A \in \Gamma\}$.

The cumulative character of the rules makes it possible to encode this calculus easily in a proof assistant, utilising list membership. This calculus is encoded as a predicate $\Rightarrow: \mathcal{L}(\mathcal{F}) \to \mathcal{F} \to \mathbb{P}$, we also define a height-bounded variant and use $\Gamma \overset{h}{\Rightarrow} A$ to denote that a derivation of $\Gamma \Rightarrow A$ of height less or equal to h exists. Our height encoding is inspired by Michaelis and Nipkow [19]. We assume that the heights of all derivations in the premisses are equal and we include an additional rule to increase the height of any derivation (see Appendix 2).

From a high-level view, our cut-admissibility proof follows the same structure employed by many textbooks (e.g. [31]), using a double induction on the sum of heights in the derivation and the formula size. However the lower level structure is different since we cannot perform case distinctions on principality and instead can only use case analyses on the last rule applied in a derivation. Following the traditional presentation, we first show depth-preserving weakening.

Lemma 11. (Weakening) *If $\Gamma \subseteq \Delta$ and $\Gamma \overset{n}{\Rightarrow} A$ then $\Delta \overset{n}{\Rightarrow} A$.*

Proof. The proof is by induction on the derivation $\Gamma \overset{n}{\Rightarrow} A$ with Δ quantified. □

$$\frac{p_i \in \Gamma}{\Gamma \Rightarrow p_i}\ (V) \qquad \frac{\bot \in \Gamma}{\Gamma \Rightarrow S}\ (F)$$

$$\frac{F \supset G \in \Gamma \qquad \Gamma \Rightarrow F}{\Gamma \Rightarrow G}\ (IL) \qquad \frac{\Gamma, F \Rightarrow G}{\Gamma \Rightarrow F \supset G}\ (IR)$$

$$\frac{F \wedge G \in \Gamma \qquad \Gamma, F, G \Rightarrow H}{\Gamma \Rightarrow H}\ (AL) \qquad \frac{\Gamma \Rightarrow F \qquad \Gamma \Rightarrow G}{\Gamma \Rightarrow F \wedge G}\ (AR)$$

$$\frac{F \vee G \in \Gamma \qquad \Gamma, F \Rightarrow H \qquad \Gamma, G \Rightarrow H}{\Gamma \Rightarrow H}\ (OL) \qquad \frac{\Gamma \Rightarrow F_i}{\Gamma \Rightarrow F_1 \vee F_2}\ (\mathrm{OR}_i)$$

$$\frac{\Gamma \cup \Gamma_{\mathsf{K}} \Rightarrow F}{\Gamma \Rightarrow \mathsf{K} F}\ (\mathrm{KI}) \qquad \frac{\Gamma \Rightarrow \mathsf{K}\bot}{\Gamma \Rightarrow A}\ (\mathrm{KF})$$

Fig. 2. Sequent system for IEL (GKIEL)

Note that this result is stronger than what is usually referred as weakening e.g. $\Gamma \Rightarrow A \rightarrow \Gamma, B \Rightarrow A$, since our version does allow to remove duplicate occurrences of formulas. Thus we do not prove what is usually referred to as the contraction rule.

Lemma 12. (Inversion) *The rules for conjunction, disjunction and implication are height-preserving invertible in the following sense:*

- *If $B \in \Gamma$ and $\Gamma, A \supset B \overset{n}{\Rightarrow} C$ then $\Gamma \overset{n}{\Rightarrow} C$.*
- *If $A \in \Gamma$ and $\Gamma, A \vee B \overset{n}{\Rightarrow} C$ then $\Gamma \overset{n}{\Rightarrow} C$.*
- *If $B \in \Gamma$ and $\Gamma, A \vee B \overset{n}{\Rightarrow} C$ then $\Gamma \overset{n}{\Rightarrow} C$.*
- *If $A \wedge B, \Gamma \Rightarrow C$ and $\Gamma, A, B \overset{n}{\Rightarrow} C$ then $\Gamma \overset{n}{\Rightarrow} C$.*

Proof. The proofs are by induction on the height with the formulas quantified. Most cases are solved by applying the rule used to obtain the derivation and using the inductive hypothesis afterwards. Only when the rule we are showing invertible is used on the same formulas (e.g. same A and B), it suffices to use the inductive hypothesis directly. □

Theorem 13. (Cut-Admissibility) *If $\Gamma \Rightarrow A$ and $A, \Gamma \Rightarrow B$ then $\Gamma \Rightarrow B$.*

Proof. The proof uses a strong induction on pairs of numbers (r, s), representing the cut-rank (sum of the depths of the derivation) and the size of the cut-formula s. Thus we have one inductive hypothesis allowing us to delete cuts on smaller formulas (e.g. formulas with smaller size; this includes subformulas) with arbitrary depths and a second hypothesis, allowing us to eliminate cuts with a smaller rank on the same formula.

We first do a case analysis on $\Gamma \Rightarrow A$, in some cases, we also need to do a second case analysis on $A, \Gamma \Rightarrow B$. Some illustrative cases can be found in Appendix 3. □

With cut-elimination, we can prove the agreement between natural deduction and the sequent calculus directly.

Theorem 14. (Agreement) *For any Γ and A, we have $\Gamma \vdash A$ iff $\Gamma \Rightarrow A$.*

Proof. Both directions are proven by induction on the derivation. The direction from natural deduction to the sequent calculus prominently uses the cut-admissibility result (Theorem 13), the converse direction is straightforward and does not need this result. □

Lemma 15. (Disjunction Property) *If $\Rightarrow A \vee B$ then $\Rightarrow A$ or $\Rightarrow B$.*

Proof. By induction on the derivation $\Rightarrow A \vee B$. □

Combining both Theorem 14 and Lemma 15 yields a proof of the disjunction property for natural deduction.

Corollary 16. (ND Disjunction Property) *If $\vdash A \vee B$ then $\vdash A$ or $\vdash B$.*

5 Decidability via Proof Search

We establish decidability of the natural deduction system for IEL by proving decidability for the cut-free sequent calculus and combining this with our equivalence proof (Theorem 14). The algorithm is an instance of Kleene-style fixed-point iteration. Crucial to this endeavor is the subformula property, which states that for a sequent $\Gamma \Rightarrow A$ there is a finite universe of sequents such that any backwards application of the rules stays within the universe.

Definition 17. (Subformula) The finite set $\mathrm{Subs}(A)$, containing all subformulas of a formula A, is defined by recursion on A:

$$\begin{aligned}\mathrm{Subs}(A_1 \circ A_2) &:= \mathrm{Subs}(A_1) \cup \mathrm{Subs}(A_2) \cup \{A_1 \circ A_2\}\\ \mathrm{Subs}(\mathsf{K}\, A_1) &:= \mathrm{Subs}(A_1) \cup \{\mathsf{K}\, A_1\}\\ \mathrm{Subs}(p_i) &:= \{p_i\}\end{aligned}$$

In the above definition, the circle $\circ$ is a placeholder for any binary connective. For a set of formulas Γ we define its *subformula universe* $\mathrm{Subs}(\Gamma) := \bigcup_{F \in \Gamma} \mathrm{Subs}(F)$. We call Γ subformula-closed if $\mathrm{Subs}(\Gamma) \subseteq \Gamma$.

Formally, we represent the sequents used during the proof search, also called goals, by members of the type $\mathcal{G} := \mathcal{L}(\mathcal{F}) \times \mathcal{F}$, so pairs (Γ, A) of a context Γ and a formula A. IEL enjoys the subformula property, since all derivations of $\Gamma \Rightarrow A$ only use formulas from $S := \mathrm{Subs}(\Gamma, A, \mathsf{K}\,\bot)$ and thus we can identify a universe $\mathcal{U} := \{(\Gamma, A) \mid \Gamma \subseteq S \wedge A \in S\}$ and restrict our proof search to $\mathcal{U}$-goals.

Having identified this set, we compute the set of derivable goals by a fixed-point iteration starting from the empty set. We can envision this process as iteratively expanding a candidate set of derivable goals until the set no longer changes. We always add a goal when it is possible to derive it using the previous goals, for example, assume that $\Gamma \Rightarrow A$ and $\Gamma \Rightarrow B$ are both derivable, then in the next step of the iteration, it would be possible to add $\Gamma \Rightarrow A \wedge B$. To formalise this extension process we define a decidable step relation $\mathsf{step} : \mathcal{L}(\mathcal{G}) \to \mathcal{G} \to \mathbb{P}$. This relation holds if, using the derivations in the list, the goal can be derived in a single step.

The algorithm now works by, in every iteration, checking if there is a goal in $\mathcal{U}$ which is in step relation with the set of currently known derivable sequents. If there is such a goal, it is added and the step is repeated, otherwise the algorithm terminates. Such a procedure will reach a fixed-point after at most $|\mathcal{U}|$ iterations. We denote the resulting list of goals by Λ.

Two crucial properties of Λ we need later are the closure property and induction principle.

Lemma 18. *The following hold for the list Λ obtained as fixed-point of* step*:*

- *Λ-Closure:* $\mathsf{step}\,\Lambda \subseteq \Lambda$
- *Λ-Induction: Let* $\mathsf{step}\,A \cap \mathcal{U} \subseteq p$ *for all* $A \subseteq p$ *and an arbitrary predicate* p*. Then* $\Lambda \subseteq p$.

Proof. See Lemma 12.4.2 in [26]. □

Lemma 19. *If* $(\Gamma, A) \in \Lambda$ *then* $\Gamma \Rightarrow A$.

Proof. By Λ-induction. Thus fix any set $\mathcal{U}'$ s.t. $(\Gamma, A) \in U' \to \Gamma \Rightarrow A$ and assume that the step relation holds for U' and $\Gamma' \Rightarrow A'$. We need to show $\Gamma' \Rightarrow A'$. We can analyse which rule caused the step relation to be fulfilled and the assumptions about $\mathcal{U}'$ to create the derivation. □

Lemma 20. *If* $\Gamma, A \in \mathcal{U}$ *and* $\Gamma \Rightarrow A$ *then* $(\Gamma, A) \in \Lambda$.

Proof. The proof is by induction on $\Gamma \Rightarrow A$. We use Λ-closure in every step and thus only need to prove that $\mathsf{step}\,\Lambda\,(\Gamma, A)$ holds.

Case AR: Assume $\Gamma \Rightarrow A_1$ and $\Gamma \Rightarrow A_2$, thus $(\Gamma, A_1) \in \Lambda$ and $(\Gamma, A_2) \in \Lambda$ by the inductive hypothesis. Thus the step relation holds between Λ and $(\Gamma, A_1 \wedge A_2)$. Since Λ is closed under the step relation, we are done.

Case AL: Assume there is $B \wedge C \in \Gamma$ and $B, C, \Gamma \Rightarrow A$. Thus by the inductive hypothesis $((B, C, \Gamma), A) \in \Lambda$. Thus the step relation holds between Λ and $((B \wedge C, \Gamma), A)$, since we can derive the goal in one step using AL.

□

Theorem 21. *The sequent calculus* $\Gamma \Rightarrow A$ *is decidable.*

Proof. Decide $(\Gamma, A) \in \Lambda$ and, depending on the outcome, apply Lemma 20 or Lemma 19 to obtain either $\Gamma \Rightarrow A$ or $\Gamma \not\Rightarrow A$. □

Corollary 22. *The natural deduction system* $\Gamma \vdash A$ *is decidable.*

Proof. A consequence of Theorem 21 and Theorem 14. □

6 Constructive Completeness

In this section, we detail the constructive proof of both the finite model property and completeness. Both properties are proven by constructing a finite canonical model, whose worlds consist of finite, prime, and consistent sets of formulas that are deductively closed with respect to a subformula-universe.

We begin by carrying out the Lindenbaum construction constructively. The key insight here is that due to decidability, we can actually represent the extension process as a computable function operating on finite contexts.

6.1 Lindenbaum Extension

We start by defining a function that extends a (finite) set of formulas Γ by a formula B, if non-derivability of $A_\bot$ is preserved.

$$\Gamma \oplus_{A_\bot} B := \begin{cases} \Gamma, B & \text{if } \Gamma, B \nvdash A_\bot \\ \Gamma & \text{otherwise} \end{cases}$$

Note that due to the decidability, we can actually compute this function for any finite set of formulas Γ. For a finite set $\mathcal{U}$ we use $\Gamma \oplus_{A_\bot} \mathcal{U}$ as notation for applying the extension procedure iteratively to every element from $\mathcal{U}$.

Definition 23. (Context Properties) Let $\mathcal{U}$ be a finite set of formulas. A set of formulas Γ is a $\mathcal{U}$-theory iff for any formula $A \in U$ derivability implies membership, i.e. $\Gamma \vdash A \rightarrow A \in \Gamma$.

Γ is $\mathcal{U}$-prime if for any $A \vee B \in \Gamma$ we have $A \in \Gamma \vee B \in \Gamma$ for any $A, B \in \mathcal{U}$.

We can now establish properties of the extension.

Lemma 24. *If $\Gamma \nvdash A_\bot$ then $\Gamma \oplus_{A_\bot} \mathcal{U} \nvdash A_\bot$ for any $\mathcal{U}$.*

Proof. The proof is by induction on $\mathcal{U}$. The case $\mathcal{U} = \emptyset$ is trivial. In the case where $\mathcal{U} = \mathcal{U}' \cup \{u\}$, we can decide $\Gamma \oplus_{A_\bot} \mathcal{U}', u \vdash A_\bot$. If $\Gamma \oplus_{A_\bot} \mathcal{U}', u \vdash A_\bot$, we know that $\Gamma \oplus_{A_\bot} \mathcal{U}'$ is extensionally equivalent to $\Gamma \oplus_{A_\bot} \mathcal{U}$ and thus can use the inductive hypothesis; in the other case we have $\Gamma \oplus_{A_\bot} \mathcal{U}' \nvdash A_\bot$ as a hypothesis. □

Lemma 25. *If $\Gamma \nvdash A_\bot$, $B \in \mathcal{U}$ and $\Gamma \oplus_{A_\bot} \mathcal{U} \nvdash B$ then $\Gamma \oplus_{A_\bot} \mathcal{U} \vdash B \supset A_\bot$.*

Lemma 26. *The extension is a $\mathcal{U}$-theory.*

Next, we can establish that the extension is $\mathcal{U}$-prime.

Lemma 27. (Primeness) *For any Γ,$\mathcal{U}$: If $\Gamma \nvdash A_\bot$ then $\Gamma \oplus_{A_\bot} \mathcal{U}$ is $\mathcal{U}$-prime.*

Proof. Let $A \vee B \in \Gamma \oplus_{A_\bot} \mathcal{U}$, furthermore assume $A \in \mathcal{U} \vee B \in \mathcal{U}$. Since we can compute the extension, we can decide wether A or B are contained in the extension. The cases where either are contained are easy. In the other case, we have both $A \notin \Gamma \oplus_{A_\bot} \mathcal{U}$ and $B \notin \Gamma \oplus_{A_\bot} \mathcal{U}$, thus by Lemma 25 we have both $\Gamma \oplus_{A_\bot} \mathcal{U} \vdash A \supset A_\bot$ and $\Gamma \oplus_{A_\bot} \mathcal{U} \vdash B \supset A_\bot$. Since $A \vee B \in \Gamma \oplus_{A_\bot} \mathcal{U}$ we can derive $A_\bot$ contradicting Lemma 24. □

So essentially, constructive primeness follows from decidable membership.

Lemma 28. (Lindenbaum) *Let Γ be a list of formulas s.t. $\Gamma \nvdash A$ and $\mathcal{U}$ arbitrary. We can compute a $\mathcal{U}$-prime $\mathcal{U}$-theory extending Γ not deriving A.*

Proof. Can be achieved by combining Lemmas 25, 24 and 27. □

6.2 Canonical Models

In this section, we construct a canonical model with respect to a finite formula universe $\mathcal{U}$. This universe will be instantiated to a concrete subformula universe in the proof of Theorem 32. The construction is inspired by both [1] and [27].

Definition 29. (Canonical Model) We define $\mathcal{M}_C = (\mathcal{W}_C, \mathcal{V}_C, \leq, \leq_{\mathsf{K}})$ by

- $\mathcal{W}_C := \{\Gamma \subseteq \mathcal{U} \mid \Gamma \text{ is a } \mathcal{U}\text{-prime, consistent } \mathcal{U}\text{-theory}\}$
- $\mathcal{V}_C(\Gamma, i) := p_i \in \Gamma$
- $\Gamma \leq \Delta := \Gamma \subseteq \Delta$
- $\Gamma \leq_{\mathsf{K}} \Delta := \Gamma \cup \Gamma_{\mathsf{K}} \subseteq \Delta$

We can easily establish that the defined model is actually a model for IEL, by showing that the $\leq_{\mathsf{K}}$-relation is serial (e.g. every world w has a $\leq_{\mathsf{K}}$-successor):

Lemma 30. *Every world has a $\leq_{\mathsf{K}}$-successor.*

Proof. The proof works by Lindenbaum-extending $\Gamma \cup \Gamma_{\mathsf{K}}$ to not derive $\bot$. This yields a world in the model, which is a $\leq_{\mathsf{K}}$-successor to Γ. □

Now we can show the following version of the Truth Lemma constructively.

Lemma 31. (Truth Lemma) *For any $\Gamma \in \mathcal{W}_C$ and $A \in \mathcal{U}$ we have*

$$A \in \Gamma \iff \mathcal{M}_C, \Gamma \Vdash A.$$

Proof. The proof is by induction on A. We only consider selected cases here.

$A = A_1 \vee A_2$: Assume $\mathcal{M}_C, \Gamma \Vdash A$ thus by definition, we have either $\mathcal{M}_C, \Gamma \Vdash A_1$ or $\mathcal{M}_C, \Gamma \Vdash A_2$, thus by the inductive hypothesis; we either have $A_1 \in \Gamma$ or $A_2 \in \Gamma$. Using that Γ is a $\mathcal{U}$-theory and $A_1 \vee A_2 \in \mathcal{U}$ we can arrive at the conclusion.
For the other direction, we assume $A_1 \vee A_2 \in \Gamma$. Since Γ is $\mathcal{U}$-prime, we have either $A_1 \in \Gamma$ or $A_2 \in \Gamma$. In both cases, we can establish $\Gamma \Vdash A$ using the inductive hypothesis and the definition of entailment.

$A = \mathsf{K}\, A_1$: Assume $\mathsf{K}\, A \in \Gamma$. Let Δ be an arbitrary $\leq_{\mathsf{K}}$-successor to Γ. We need to establish $\Delta \vdash A$, by the inductive hypothesis it suffices to establish $A \in \Delta$, which is simple using the definition of $\leq_{\mathsf{K}}$.
Assume $M_C, \Gamma \Vdash \mathsf{K}\, A$. Again using stability of membership, furthermore assume $\mathsf{K}\, A \notin \Gamma$. Now we can Lindenbaum-extend $\Gamma, \Gamma_{\mathsf{K}}$ to a world in the model that does not derive A. But this world is a $\leq_{\mathsf{K}}$-successor, contradicting $M_C, \Gamma \Vdash \mathsf{K}\, A$.

□

This allows us to prove completeness constructively, which can be interpreted as an algorithm reifying a proof term of the formal type-theoretic meta-logic into a derivation in natural deduction.

Theorem 32. (Constructive Completeness) *If $\Gamma \Vdash A$ then $\Gamma \vdash A$.*

Proof. Since $\vdash$ is stable under double negation (consequence of decidability, Corollary 22), we can assume both $\Gamma \Vdash A$ and $\Gamma \nvdash A$ and need to derive a contradiction. Using the Lindenbaum lemma (Lemma 28) Γ can be extended to a world $\Gamma'$with $A \notin \Gamma'$ of the canonical model for the subformula universe of (Γ, A) and therefore by Lemma 31, $\mathcal{M}_C, \Gamma' \nVdash A$. But this contradicts $\Gamma' \Vdash A$, which is easily obtained from $\Gamma \Vdash A$ using monotonicity. □

With the constructive completeness proof it is now possible to constructively derive admissibility results from [1], e.g. the admissibility of reflection (we don't repeat the proof from [1] here but refer to the Coq development).

6.3 Finite Model Property

Intuitively, the finite model property is a trivial consequence of the fact that the canonical model is finite, which is simple to observe since the worlds are subsets of a finite set and thus only finitely many of them exist. This has also been established by Su and Sano [27]. For IEL$^-$ the finite model property has already been established by Wolter and Zakharyaschev [32]. We first define entailment restricted to finite models:[3]

$$\Gamma \Vdash_{\mathsf{fin}} A \;:=\; \forall \mathcal{M}.\, \mathsf{fin}(\mathcal{M}) \to \mathcal{M} \Vdash \Gamma \to \mathcal{M} \Vdash A.$$

A logic now has the finite model property, if any formula entailed in all finite models is a theorem.[4]

Definition 33. A logic $\mathcal{L}$ has the finite model property if $\Gamma \Vdash_{\mathsf{fin}} A \to \Gamma \vdash A$.

To complete this definition, a suitable notion of *finite model* needs to be made. A straightforward choice would be to define that a model is finite if the type of worlds is finite. But since the world-type of the canonical model does not just contain the formulas, but also proofs about them (e.g. a proof that the finite set of formulas is consistent), an additional axiom, namely proof irrelevance, is needed. To avoid, the additional axiom, we introduce the property of being essentially finite.

Definition 34. A model $\mathcal{M} = (\mathcal{W}, \mathcal{V}, \leq, \leq_{\mathsf{K}})$ with world type $\mathcal{W}$ is essentially finite, if there is a list of worlds L s.t.

$$\forall w \exists v \in L.\, w \leq v \wedge v \leq w.$$

Theorem 35. *The canonical model is essentially finite.*

[3] The notation $\mathcal{M} \Vdash \Gamma$ is a short-hand for $\forall A \in \Gamma.\, \mathcal{M} \Vdash A$.

[4] Or, classically equivalent, if any formula valid in some model also has a finite model.

Proof. We can compute a list containing all finite, prime, consistent, $\mathcal{U}$-theories, which is possible since all aforementioned properties are decidable. From these we obtain a list of worlds which satisfies the essential finiteness property. □

Corollary 36. *IEL has the finite model property.*

Proof. Analogous to Theorem 32, utilising Theorem 35. □

There are different versions of the finite model property in the literature. One commonly used version is that any non-theorem must have a finite countermodel:

$$\mathsf{FMP} := \forall A.\ \nvdash A \rightarrow \exists \mathcal{M}.\ \mathcal{M} \nVdash A \wedge \mathrm{fin}(\mathcal{M})$$

We can actually establish this result, too, as the canonical model can be employed as the countermodel.

6.4 Semantic Cut-Elimination

Su and Sanno [27] use a slightly different construction to prove completeness of the cut-free sequent calculus. We can adapt their argument to be constructive by using the decidability of the cut-free sequent calculus.

Theorem 37. (Completeness for GKIEL) *If $\Gamma \Vdash A$ then $\Gamma \Rightarrow A$.*

Proof. We construct a canonical model with finite saturated theories as worlds as in [27] and then we proceed as in Theorem 32. □

Theorem 38. (Semantic Cut-Elimination) *If $\Gamma \vdash A$ then $\Gamma \Rightarrow A$.*

Proof. By composing Lemma 8 and Theorem 37. □

Corollary 39. *If $\Gamma \Rightarrow A$ and $A, \Gamma \Rightarrow B$ then $\Gamma \Rightarrow B$.*

Proof. Since the two premises can be replayed in the natural deduction system, they trivially entail $\Gamma \vdash B$ and thus $\Gamma \Rightarrow B$ by Theorem 38. □

Since completeness is constructive, this procedure bears an executable algorithm. Note that the presented semantic cut-elimination proof does not rely on the syntactic cut-elimination proof given in Sect. 4; thus in principle, we could as well obtain all main results without it. However, we view both proofs of cut-elimination as valuable, especially since the syntactic one does not rely on completeness and is overall shorter.

7 Completeness for Infinite Theories

In this section, we analyse the connections between *strong completeness*, i.e. completeness for infinite theories as stated in Theorem 10 based on [1], and non-constructive axioms. This is similar to the analysis by Forster et al. [11] for first-order logic and relies on the stability of semantic inconsistencies:

Lemma 40. *If* $\neg\neg(\mathcal{T} \Vdash \bot)$ *then* $\mathcal{T} \Vdash \bot$.

Proof. Assume $\neg\neg(\mathcal{T} \Vdash \bot)$ and let w be a world in an arbitrary model $\mathcal{M}$, that forces $\mathcal{T}$. We need to show $\mathcal{M}, w \vDash \bot$ which is by definition $\bot$. Using Lemma 2 we can reason classically and thus strip the double negation off $\neg\neg(\mathcal{T} \Vdash \bot)$. Now we can obtain the proof of $\mathcal{M}, w \Vdash \bot$ from $\mathcal{T} \vDash \bot$ since $\mathcal{M}$ forces $\mathcal{T}$. □

7.1 Arbitrary Theories

Our first result is the equivalence between strong completeness for arbitrary theories and the law of excluded middle, adding the converse of Theorem 10.

Lemma 41. *Assuming strong completeness, derivation of falsity is stable, i.e.* $\neg\neg(\mathcal{T} \vdash \bot)$ *implies* $\mathcal{T} \vdash \bot$ *for arbitrary* $\mathcal{T}$.

Proof. Assume $\neg\neg(\mathcal{T} \vdash \bot)$. By soundness we have $\neg\neg(\mathcal{T} \vDash \bot)$, by Lemma 40 we thus have $\mathcal{T} \vDash \bot$. Using strong completeness concludes the proof. □

Theorem 42. *Strong completeness implies* LEM.

Proof. Assume strong completeness and let P be arbitrary. We have to show $P \vee \neg P$. Consider the theory $\mathcal{T} := \{A | P \vee \neg P\}$. Let us first show $\mathcal{T} \vdash \bot$. For this, we can use stability of deriving falsity (Lemma 40) and are left with proving $\neg\neg(\mathcal{T} \vdash \bot)$. Since our goal is negated, we can assume $P \vee \neg P$ by Lemma 2. Now we can show $\mathcal{T} \vdash \bot$ using the assumption rule.

Having established that $\mathcal{T} \vdash \bot$, by definition a list $\Gamma \subseteq \mathcal{T}$ exists s.t. $\Gamma \vdash \bot$. We either have $\Gamma = \emptyset$ or $\Gamma = A, \Gamma'$. In the first case we thus have a derivation $\vdash \bot$, we can derive a contradiction using consistency (Lemma 9)). In the second case we know that $A \in \mathcal{T}$ is proven, but this yields a proof of $P \vee \neg P$. □

7.2 Enumerable Theories

Even if we restrict strong completeness to enumerable theories, we can still derive MP. Here we will need the fact that the type of formulas is enumerable (Lemma 4), we will denote the n-th formula in this enumeration by A_n. We can first adapt Lemma 41 to strong enumerable completeness.

Lemma 43. *Assuming strong enumerable completeness, derivation of falsity is stable for any enumerable theory* $\mathcal{T}$, *i.e.* $\neg\neg(\mathcal{T} \vdash \bot)$ *implies* $\mathcal{T} \vdash \bot$.

Proof. The proof is analogous to the proof of Lemma 41. □

Theorem 44. *Strong enumerable completeness implies* MP.

Proof. To show MP, let $f : \mathbb{N} \to \mathbb{B}$ be a boolean function s.t. $\neg\neg\exists n.\, f\, n = \mathsf{tt}$. We construct the enumerable theory $\{A_n \wedge \neg A_n | f\, n = \mathsf{tt}\}$. It is easy to verify that this theory is enumerable. In constructive type theory, we can encode this theory as $\lambda(F : \mathcal{F}).\, \exists n\, fn = \mathsf{tt} \wedge F = (A_n \wedge \neg A_n)$.

Assume that $\mathcal{T} \vdash \bot$, we will establish this fact shortly. Now we have that a finite subset $\Gamma \subseteq \mathcal{T}$ derives $\bot$. As before, this subset cannot be empty since otherwise consistency would be violated. But then we can derive that there is an n s.t. $f\,n = \mathsf{tt}$.

To conclude the proof, we need to establish $\mathcal{T} \vdash \bot$. Since $\mathcal{T}$ is enumerable, we can use Lemma 43 and use $\neg\mathcal{T} \vdash \bot$ as an additional hypothesis and need to derive a contradiction. Therefore we can strip the double-negation and know that there is a j s.t. $f\,j = \mathsf{tt}$, thus $F_j \wedge \neg F_j \in \mathcal{T}$. We can now show $\mathcal{T} \vdash \bot$ using the implication elimination rule with F_j and $\neg F_j$. □

Thus, having observed Theorems 42 and 44, we can conclude that the approach to completeness of IEL pursued e.g. by Artemov and Protopopescu [1] inherently relies on a classical meta-theory. Also, let us remark that for these observations we only used a modest propositional fragment of IEL, therefore they generalise to strong completeness of many other logical formalisms.

8 Conclusion

8.1 Related Work

Of course the main reference for IEL is the paper introducing the logic by Artemov and Protopoescu [1]. Protopopescu [21] furthermore proves soundness and completeness of embeddings from IEL to S4. His dissertation [22] consists of two more papers on IEL, one investigating the connection between IEL and modal logics of verifications and one about fallible knowledge.

Proof theory of IEL has been studied by Krupski [18], Su and Sano [27,28], and more recently Fiorino [9]. Su and Sano propose a cut-free sequent calculus for IEL (and an extension of IEL with quantifiers). Their calculus for IEL uses sets of formulas with at most 1 element in the succedent of some rules, which will probably makes it less convenient to mechanise in a proof assistant. Fiorino proposes a sequent calculus for IEL with linear depth.

Tarau [29] develops a theorem prover for IEL using Prolog and presents embeddings from IEL into IPC (intuitionistic propositional calculus), however soundness or completeness proofs about the embeddings are not given. We tried to investigate those, however in our setting, we were unable to come up with a completeness proof.

There is a lot of existing work on mechanising decidability and cut-elimination in Coq and other proof assistants. For instance, Bentzen [2] mechanises a completeness proof for S5 in the Lean proof assistant, which uses classical logic. Doczkal and Smolka present axiom-free Coq mechanisations of completeness with respect to Kripke semantics and decidability of the forcing relation for K*, an extension of the classical modal logic K [7], as well as for various temporal logics (e.g. CTL) [8]. There are many mechanisations of cut-elimination proofs, many use G3K-style calculi and embed these using permutations in a proof assistant. Michaelis and Nipkow [19] establish (among other results, such as completeness) cut-elimination of IPC using Isabelle/HOL formalising the rules using multisets. A somewhat similar approach (using permutations to express multisets) is also used by Chaudhuri and Lima [3]. Goré et al. [12] mechanise cut-elimination for the provability logic GL in multiset representation using Coq and discuss how their work benefits from using a proof assistant.

8.2 Future Work

There are several possible lines of future work. For one, it would be worthwhile to investigate whether, as the converse to Theorem 44, MP implies completeness for enumerable theories, or whether a stronger assumption is required. As indication for the latter, it seems not clear how MP would help with for instance establishing the Lindenbaum lemma constructively for infinite enumerable theories.

Secondly, it is certainly interesting to see if this general method for proving the finite model property and completeness in a constructive setting will also generalise to other modal logics. Here, we currently only have partial results: We verified that the cut-elimination and decidability proofs extend to the modal logic K (using an encoding based on a sequent calculus by Hakli and Negri [13]).

Lastly, our current decidability and cut-elimination proofs are not very efficient. Mechanising more efficient decidability procedures might be an interesting challenge (for example basing on Krupski's [18] decidability proof or Fiorino's [9] refutation calculus).

Acknowledgments. The authors would like to thank thank Yannick Forster, Marc Hermes, Jannik Kudla, Dominik Wehr and the anonymous reviewers for their helpful comments and suggestions on drafts of this paper.

Appendix 1 Natural Deduction System for IEL

$$\frac{A \in \Gamma}{\Gamma \vdash A}\ \mathrm{A} \qquad \frac{\Gamma \vdash \bot}{\Gamma \vdash A}\ \mathrm{E}$$

$$\frac{\Gamma, A \vdash B}{\Gamma \vdash A \supset B}\ \mathrm{II} \qquad \frac{\Gamma \vdash A \quad \Gamma \vdash A \supset B}{\Gamma \vdash B}\ \mathrm{IE}$$

$$\frac{\Gamma \vdash A}{\Gamma \vdash A \vee B}\ \mathrm{DIL} \qquad \frac{\Gamma \vdash B}{\Gamma \vdash A \vee B}\ \mathrm{DIR} \qquad \frac{\Gamma, A \vdash C \quad \Gamma, B \vdash C \quad \Gamma \vdash A \vee B}{\Gamma \vdash C}\ \mathrm{DE}$$

$$\frac{\Gamma \vdash A \quad \Gamma \vdash B}{\Gamma \vdash A \wedge B}\ \mathrm{CI} \qquad \frac{\Gamma \vdash A \wedge B}{\Gamma \vdash A}\ \mathrm{CEL} \qquad \frac{\Gamma \vdash A \wedge B}{\Gamma \vdash B}\ \mathrm{CER}$$

$$\frac{\Gamma \vdash A}{\Gamma \vdash \mathsf{K}\,A}\ \mathrm{KR} \qquad \frac{\Gamma \vdash \mathsf{K}\,(A \supset B)}{\Gamma \vdash \mathsf{K}\,A \supset \mathsf{K}\,B}\ \mathrm{KD} \qquad \frac{\Gamma \vdash \mathsf{K}\,A}{\Gamma \vdash \neg\neg A}\ \mathrm{KF}$$

Appendix 2 Coq Mechanisation

Component	Spec	Proof
preliminaries	121	93
natural deduction + lindenbaum	183	418
models	43	23
completeness	75	325
semantic cut-elimination	49	214
cut-elimination + decidability IEL	193	399
classical completeness / infinite theories	90	261
cut-elimination + decidability K	116	362
$\sum$	737	2194

Fig. 3. Overview of the mechanisation components

Our mechanisation compiles using Coq 8.13.2. It takes roughly 4 min to compile on a 2.6 GHz machine. An overview of the development with line counts can be found in Fig. 3.

We use a *parametrised deduction system* to represent natural deduction (and the sequent calculus) for both IEL and IEL^-. That is, formally our deduction system has type $\vdash : \mathbb{F} \to \mathcal{L}(\mathcal{F}) \to \mathcal{F} \to \mathbb{P}$, where $\mathbb{F}$ is a two-element type class, which is responsible for flagging whether IEL^- or IEL shall be used. This allows us to prove most results simultaneously for IEL and IEL^- as the lemmas are parametrised in the flag of the deduction system.

2.1 The Classical Modal Logic K

For the classical modal logic K, we prove cut-elimination and decidability by using a similar strategy as we used for IEL. Hakli and Negri [13] propose a G3C-style calculus for K; as in the case of IEL, we instead adopt a mechanisation-friendly variant of a G3I-calculus and introduce a single modal rule. A similar system for classical propositional logic was presented by Dang [6]. For the full system, we refer the reader to the Coq mechanisation. Here, we only present the modal rule and compare it with the one used by Hakli and Negri. Hakli and Negri use the following rule:[5]

$$\frac{\Gamma \Rightarrow A}{\Box\Gamma, \Theta \Rightarrow \Delta, \Box A}$$

In a similar spirit, as our modal rule, for IEL, we adopt the following rule for our system:

[5] In the following rules, we use a box instead of K as the modal operator, as this is standard for the classical modal logic K.

$$\frac{\Box A \in \Delta \qquad \Gamma_{\Box} \Rightarrow A}{\Gamma \Rightarrow \Delta}$$

This is easier to formalise as we use membership prominently and no longer split-up a context. The proofs for decidability and cut-elimination are similar as for IEL, we use the same induction on pairs of cut-rank and size of the cut-formula for the proof of cut-elimination. However the proofs are slightly more compact (due to more symmetric rules in the system).

2.2 Height-Encoding

To better illustrate the height-encoding, we consider how the right rule for conjunction is encoded.

$$\frac{\Gamma \stackrel{h}{\Rightarrow} A \qquad \Gamma \stackrel{h}{\Rightarrow} B}{\Gamma \stackrel{h+1}{\Rightarrow} A \wedge B} \qquad \frac{\Gamma \stackrel{h}{\Rightarrow} A}{\Gamma \stackrel{h+1}{\Rightarrow} A}$$

On the left side, we see how the conjunction rule is encoded, while on the right side, the step rule is given, which allows us to boost the height of any derivation.

One alternative would be to use a height-encoding using maximum on both sides, e.g.:

$$\frac{\Gamma \stackrel{h_1}{\Rightarrow} A \qquad \Gamma \stackrel{h_2}{\Rightarrow} B}{\Gamma \stackrel{\max(h_1,h_2)+1}{\Rightarrow} A \wedge B}$$

The encoding we used leads to easier proofs and inductions (since less arithmetical reasoning about maximum or minimum is needed).

Appendix 3 Cut-Elimination: Selected Cases

We shall showcase some cases of our cut-elimination proof.

Theorem 45. (Cut is admissible) *The cut rule is admissible.*

$$\frac{\begin{array}{c}[\delta_1] \\ \Gamma \stackrel{h_1}{\Rightarrow} B\end{array} \qquad \begin{array}{c}[\delta_2] \\ \Gamma, B \stackrel{h_2}{\Rightarrow} A\end{array}}{\Gamma \Rightarrow A} \; Cut$$

Proof. The proof is by induction on pairs (s, r) of formula-size s and cut-rank r. Here formula size is the size of the cut-formula B, and the cut-rank is the sum of the heights i.e. $r := h_1 + h_2$.

The induction principle gives us two inductive hypotheses, one which allows us to eliminate cuts of arbitrary height but with a cut formula of smaller size (s-cut) and another one, allowing us to eliminate cuts on formulas of the same size but with a smaller cut-rank (r-cut).

We now analyse which rule was used to derive δ_1. In two cases, namely the K introduction and right implication introduction rule we will need an additional case analysis (i.e. inversion) on δ_2.

AL-Rule: Assume δ_1 was derived using the left-rule for conjunction. Our derivation has the following form.

$$\dfrac{\dfrac{C_1 \wedge C_2 \in \Gamma \qquad \Gamma, C_1, C_2 \overset{h_1-1}{\Rightarrow} F}{\Gamma \overset{h_1}{\Rightarrow} B} \qquad \Gamma, B \overset{h_2}{\Rightarrow} A}{\Gamma \Rightarrow A}\ \text{r-cut}$$

We can permute the application of the left rule for conjunction downwards and use weakening on the derivation $C_1, C_2, \Gamma \overset{m}{\Rightarrow} \Delta$:

$$\dfrac{C_1 \wedge C_2 \in \Gamma \qquad \dfrac{\Gamma, C_1, C_2 \overset{n-1}{\Rightarrow} B \qquad B, \Gamma \overset{m}{\Rightarrow} A}{\Gamma, C_1, C_2 \Rightarrow A}}{\Gamma \Rightarrow A}$$

Note that the new cut is a cut on the same formula but of a smaller rank, thus we can eliminate it by the inductive hypothesis.

IR-Rule: Assume last rule used in the derivation of δ_1 was the right introduction rule for implication. Thus we know, that $B = B_1 \supset B_2$. We need to do a second case analysis on the derivation δ_2.

1. If δ_2 is an axiom, either $p_i = B$ or $p_i \in \Gamma$ and we know that $A = p_i$. The first case contradicts our assumption that $B = B_1 \supset B_2$ and in the second case we can directly use the variable rule.
2. Similarly, if the second premiss is derived using the falsity rule, either $F = \bot$ or $\bot \in \Gamma$.
3. An interesting case arises when the right premiss is proved using the left introduction rule for implication.

$$\dfrac{\dfrac{B_0, \Gamma \overset{h_1-1}{\Rightarrow} B_1}{\Gamma \overset{h_1}{\Rightarrow} B_0 \supset B_1} \qquad \dfrac{C_0 \supset C_1 \in \Gamma, B \qquad \Gamma, B \overset{h_2-1}{\Rightarrow} C_0 \qquad \Gamma, C_1, B \Rightarrow A}{\Gamma, B_0 \supset B_1 \overset{h_2}{\Rightarrow} A}}{\Gamma \Rightarrow A}$$

We have two cases: either $B = C_0 \supset C_1$ or $C_0 \supset C_1 \in \Gamma$.

(a) In the first case, we can build the following derivation:

$$\dfrac{\dfrac{\dfrac{\Gamma \overset{h_1}{\Rightarrow} B \qquad B, \Gamma \overset{h_2-1}{\Rightarrow} B_0}{\Gamma \Rightarrow B_0}\ \text{r-cut} \qquad \Gamma, B_0 \Rightarrow B_1}{\Gamma \Rightarrow B_1} \qquad \dfrac{B_1 \in B_1, \Gamma \qquad \Gamma, B_1, B \Rightarrow A}{\Gamma, B_1 \Rightarrow A}\ \text{IL-inv}}{\Gamma \Rightarrow A}$$

(b) In the second case, we can apply the left rule for implication first and do two cuts afterwards.

$$\dfrac{C_0 \supset C_1 \in \Gamma \qquad \dfrac{\Gamma \overset{h_1}{\Rightarrow} B \qquad \Gamma, B \overset{h_2-1}{\Rightarrow} C_0}{\Gamma \Rightarrow C_0}\ \text{r-cut} \qquad \dfrac{\dfrac{\Gamma \overset{h_1}{\Rightarrow} B}{\Gamma, C_1 \overset{h_1}{\Rightarrow} B}\ \text{weak.} \qquad \Gamma, B, C_1 \overset{h_2-1}{\Rightarrow} C_0}{\Gamma, C_1 \Rightarrow A}\ \text{r-cut}}{\Gamma \Rightarrow A}\ \text{IL}$$

KI-Rule: Assume the premiss was derived using the K-introduction rule. We need to make a second case distinction on the derivation of the right deduction. Most cases are similar to those obtained in the right rule for implication subcases, and we will not go into too much detail here.

$$\dfrac{\dfrac{\Gamma \cup \Gamma_{\mathsf{K}} \Rightarrow B_0}{\Gamma \Rightarrow \mathsf{K}\,B_0} \qquad \Gamma, \mathsf{K}\,B_0 \Rightarrow A}{\Gamma \Rightarrow A}$$

1. The right premise is an axiom. Either $p_i = \mathsf{K}\,B_0$ which is impossible (since the constructors of an inductive datatype are disjoint) or $A \in \Gamma$ in which case we can directly construct the derivation.
2. The most interesting case occurs when the KI-rule is used on both sides. We have the following derivation:

$$\dfrac{\dfrac{\Gamma \cup \Gamma_{\mathsf{K}} \overset{h_1-1}{\Rightarrow} B_0}{\Gamma \overset{h_1}{\Rightarrow} \mathsf{K}\,B_0} \qquad \dfrac{\Gamma \cup \Gamma_{\mathsf{K}}, \mathsf{K}\,B_0, B_0 \overset{h_2-1}{\Rightarrow} A_0}{\Gamma, \mathsf{K}\,B_0 \overset{h_2}{\Rightarrow} \mathsf{K}\,A_0}}{\Gamma \Rightarrow \mathsf{K}\,A_0}$$

We can build the following derivation:

$$\dfrac{\dfrac{\Gamma \cup \Gamma_{\mathsf{K}} \Rightarrow B_0 \qquad \dfrac{\dfrac{\Gamma \overset{h_1}{\Rightarrow} \mathsf{K}\,B_0}{\Gamma \cup \Gamma_{\mathsf{K}} \overset{h_1}{\Rightarrow} \mathsf{K}\,B_0}\text{weak.} \qquad \Gamma \cup \Gamma_{\mathsf{K}}, \mathsf{K}\,B_0, B_0 \overset{h_2-1}{\Rightarrow} A_0}{\Gamma \cup \Gamma_{\mathsf{K}}, B_0 \Rightarrow A_0}\text{r-cut}}{\Gamma \cup \Gamma_{\mathsf{K}} \Rightarrow A_0}\text{s-cut}}{\Gamma \Rightarrow \mathsf{K}\,A_0}\text{KI}$$

□

References

1. Artemov, S., Protopopescu, T.: Intuitionistic epistemic logic. Revi. Symbol. Logic **9**(2), 266–298 (2016). https://doi.org/10.1017/S1755020315000374
2. Bentzen, B.: A Henkin-style completeness proof for the modal logic s5. In: Baroni, P., Benzmüller, C., Wáng, Y.N. (eds.) Logic and Argumentation, pp. 459–467. Springer International Publishing, Cham (2021). https://doi.org/10.1007/978-3-030-89391-0_25
3. Chaudhuri, K., Lima, L., Reis, G.: Formalized Meta-Theory of Sequent Calculi for Substructural Logics. Electronic Notes in Theoretical Computer Science **332**, 57–73 (2017). https://doi.org/10.1016/j.entcs.2017.04.005
4. Coquand, T., Huet, G.: The calculus of constructions. Inf. Comput. **76**(2), 95–120 (1988). https://doi.org/10.1016/0890-5401(88)90005-3. https://www.sciencedirect.com/science/article/pii/0890540188900053
5. Coquand, T., Mannaa, B.: The independence of Markov's principle in type theory. arXiv preprint arXiv:1602.04530 (2016)
6. Dang, H.: Systems for Propositional Logics. Technical report, Saarland University (2015). https://www.ps.uni-saarland.de/~dang/ri-lab/propsystems/systems.pdf
7. Doczkal, C., Smolka, G.: Constructive completeness for modal logic with transitive closure. In: Hawblitzel, C., Miller, D. (eds.) Certified Programs and Proofs, pp. 224–239. Springer, Berlin, Heidelberg (2012)

8. Doczkal, C., Smolka, G.: Completeness and decidability results for CTL in constructive type theory. J. Autom. Reason. **56**(3), 343–365 (2016). https://doi.org/10.1007/s10817-016-9361-9
9. Fiorino, G.: Linear depth deduction with subformula property for intuitionistic epistemic logic (2021)
10. Forster, Y., Kirst, D., Smolka, G.: On synthetic undecidability in Coq, with an application to the Entscheidungs problem. In: CPP 2019 - Proceedings of the 8th ACM SIGPLAN International Conference on Certified Programs and Proofs, Co-located with POPL 2019 (2019). https://doi.org/10.1145/3293880.3294091
11. Forster, Y., Kirst, D., Wehr, D.: Completeness theorems for first-order logic analysed in constructive type theory. J. Logic Comput. (2021). https://doi.org/10.1093/logcom/exaa073
12. Goré, R., Ramanayake, R., Shillito, I.: Cut-elimination for provability logic by terminating proof-search: formalised and deconstructed using Coq. In: Das, A., Negri, S. (eds.) TABLEAUX 2021. LNCS (LNAI), vol. 12842, pp. 299–313. Springer, Cham (2021). https://doi.org/10.1007/978-3-030-86059-2_18
13. Hakli, R., Negri, S.: Does the deduction theorem fail for modal logic? Synthese (2012). https://doi.org/10.1007/s11229-011-9905-9
14. Hintikka, J.: Knowledge and belief: An introduction to the logic of the two notions. Studia Logica **16** (1962)
15. Ishihara, H.: Constructive reverse mathematics: compactness properties. From Sets and Types to Topology and Analysis. Oxford Logic Guides **48**, 245–267 (2005). https://doi.org/10.1093/acprof:oso/9780198566519.001.0001
16. Ishihara, H.: Reverse mathematics in bishop's constructive mathematics. Philosophia Scientiæ. Travaux d'histoire et de philosophie des sciences (CS 6), 43–59 (2006). https://doi.org/10.4000/philosophiascientiae.406
17. Kleene, S.C.: Introduction to Metamathematics, vol. 19. North Holland (1952)
18. Krupski, V.N.: Cut elimination and complexity bounds for intuitionistic epistemic logic. J. Logic Comput. **30**(1), 281–294 (2020). https://doi.org/10.1093/logcom/exaa012
19. Michaelis, J., Nipkow, T.: Formalized proof systems for propositional logic. In: Abel, A., Forsberg, F.N., Kaposi, A. (eds.) 23rd International Conference on Types for Proofs and Programs (TYPES 2017). Leibniz International Proceedings in Informatics (LIPIcs), vol. 104, pp. 5:1–5:16. Schloss Dagstuhl-Leibniz-Zentrum fuer Informatik, Dagstuhl, Germany (2018). https://doi.org/10.4230/LIPIcs.TYPES.2017.5. http://drops.dagstuhl.de/opus/volltexte/2018/10053
20. Paulin-Mohring, C.: Inductive definitions in the system Coq rules and properties. In: Bezem, M., Groote, J.F. (eds.) TLCA 1993. LNCS, vol. 664, pp. 328–345. Springer, Heidelberg (1993). https://doi.org/10.1007/BFb0037116
21. Protopopescu, T.: An arithmetical interpretation of verification and intuitionistic knowledge. Lecture Notes in Computer Science (including subseries Lecture Notes in Artificial Intelligence and Lecture Notes in Bioinformatics) **9537**, 317–330 (2016). https://doi.org/10.1007/978-3-319-27683-0_22
22. Protopopescu, T.: Three Essays in Intuitionistic Epistemology. Ph.D. thesis, CUNY (2016). https://academicworks.cuny.edu/gc_etds/1391
23. Rescher, N.: Epistemic Logic: A Survey of the Logic of Knowledge. University of Pittsburgh Press, Pittsburgh (2005)
24. Richman, F.: Constructive Mathematics without Choice, pp. 199–205. Springer, Netherlands, Dordrecht (2001). https://doi.org/10.1007/978-94-015-9757-9_17

25. Rogozin, D.: Categorical and algebraic aspects of the intuitionistic modal logic IEL− and its predicate extensions. J. Logic Comput. **31**(1), 347–374 (2021). https://doi.org/10.1093/logcom/exaa082. https://academic.oup.com/logcom/article/31/1/347/6049830
26. Smolka, G., Brown, C.E.: Introduction to Computational Logic (2012). http://www.ps.uni-saarland.de/courses/cl-ss12/script/icl.pdf
27. Su, Y., Sano, K.: Cut-free and analytic sequent calculus of intuitionistic epistemic logic. In: Sedlár, I., Blicha, M. (eds.) The Logica Yearbook 2019, pp. 179–193. College Publications (2019)
28. Su, Y., Sano, K.: First-order intuitionistic epistemic logic. In: Blackburn, P., Lorini, E., Guo, M. (eds.) LORI 2019. LNCS, vol. 11813, pp. 326–339. Springer, Heidelberg (2019). https://doi.org/10.1007/978-3-662-60292-8_24
29. Tarau, P.: Modality Definition Synthesis for Epistemic Intuitionistic Logic via a Theorem Prover (2019)
30. The Coq Development Team: The Coq proof assistant, January 2021. https://doi.org/10.5281/zenodo.4501022
31. Troelstra, A.S., Schwichtenberg, H.: Basic Proof Theory (2000). https://doi.org/10.1017/cbo9781139168717
32. Wolter, F., Zakharyaschev, M.: Intuitionistic modal logics as fragments of classical bimodal logics. Logic at Work (1999)
33. Wolter, F., Zakharyaschev, M.: Intuitionistic Modal Logic, pp. 227–238. Springer, Netherlands, Dordrecht (1999). https://doi.org/10.1007/978-94-017-2109-7_17

A Parametrized Family of Tversky Metrics Connecting the Jaccard Distance to an Analogue of the Normalized Information Distance

Bjørn Kjos-Hanssen(✉), Saroj Niraula, and Soowhan Yoon

University of Hawai'i at Mānoa, Honolulu, HI 96822, USA
{bjoernkh,sniraula,syoon2}@hawaii.edu
https://math.hawaii.edu/wordpress/bjoern/
http://math.hawaii.edu/~sniraula/

Abstract. Jiménez, Becerra, and Gelbukh (2013) defined a family of "symmetric Tversky ratio models" $S_{\alpha,\beta}$, $0 \leq \alpha \leq 1$, $\beta > 0$. Each function $D_{\alpha,\beta} = 1 - S_{\alpha,\beta}$ is a semimetric on the powerset of a given finite set.

We show that $D_{\alpha,\beta}$ is a metric if and only if $0 \leq \alpha \leq \frac{1}{2}$ and $\beta \geq 1/(1-\alpha)$. This result is formally verified in the Lean proof assistant.

The extreme points of this parametrized space of metrics are $J_1 = D_{1/2,2}$, the Jaccard distance, and $J_\infty = D_{0,1}$, an analogue of the normalized information distance of M. Li, Chen, X. Li, Ma, and Vitányi (2004).

Keywords: Jaccard distance · Normalized information distance · Metric space · Proof assistant

1 Introduction

Distance measures (metrics), are used in a wide variety of scientific contexts. In bioinformatics, M. Li, Badger, Chen, Kwong, and Kearney [12] introduced an information-based sequence distance. In an information-theoretical setting, M. Li, Chen, X. Li, Ma and Vitányi [13] rejected the distance of [12] in favor of a *normalized information distance* (NID). The Encyclopedia of Distances [3] describes the NID on page 205 out of 583, as

$$\frac{\max\{K(x \mid y^*), K(y \mid x^*)\}}{\max\{K(x), K(y)\}}$$

This work was partially supported by grants from the Simons Foundation (#704836 to Bjørn Kjos-Hanssen) and Decision Research Corporation (University of Hawai'i Foundation Account #129-4770-4).

S. Artemov and A. Nerode (Eds.): LFCS 2022, LNCS 13137, pp. 112–124, 2022.
https://doi.org/10.1007/978-3-030-93100-1_8

where $K(x \mid y^*)$ is the Kolmogorov complexity of x given a shortest program y^* to compute y. It is equivalent to be given y itself in hard-coded form:

$$\frac{\max\{K(x \mid y), K(y \mid x)\}}{\max\{K(x), K(y)\}}$$

Another formulation (see [13, p. 8]) is

$$\frac{K(x,y) - \min\{K(x), K(y)\}}{\max\{K(x), K(y)\}}.$$

The fact that the NID is in a sense a normalized metric is proved in [13]. Then in 2017, while studying malware detection, Raff and Nicholas [14] suggested Lempel–Ziv Jaccard distance (LZJD) as a practical alternative to NID. As we shall see, this is a metric. In a way this constitutes a full circle: the distance in [12] is itself essentially a Jaccard distance, and the LZJD is related to it as Lempel–Ziv complexity is to Kolmogorov complexity. In the present paper we aim to shed light on this back-and-forth by showing that the NID and Jaccard distances constitute the endpoints of a parametrized family of metrics.

For comparison, the Jaccard distance between two sets X and Y, and our analogue of the NID, are as follows:

$$J_1(X,Y) = \frac{|X \setminus Y| + |Y \setminus X|}{|X \cup Y|} = 1 - \frac{|X \cap Y|}{|X \cup Y|} \tag{1}$$

$$J_\infty(X,Y) = \frac{\max\{|X \setminus Y|, |Y \setminus X|\}}{\max\{|X|, |Y|\}} \tag{2}$$

Our main result Theorem 11 shows which interpolations between these two are metrics.

Incidentally, the names of J_1 and J_∞ come from the observation that they are special cases of J_p given by

$$J_p(A,B) = \left(2 \cdot \frac{|B \setminus A|^p + |A \setminus B|^p}{|A|^p + |B|^p + |B \setminus A|^p + |A \setminus B|^p}\right)^{1/p} = \begin{cases} J_1(A,B) & p = 1 \\ J_\infty & p \to \infty \end{cases}$$

We conjecture that J_p is a metric for each p, but shall not attempt to prove it here.

The way we arrived at Eq. 2 as an analogue of NID is via Lempel–Ziv complexity. While there are several variants [11,18,19], the LZ 1978 complexity [19] of a sequence is the cardinality of a certain set, the dictionary.

Definition 1. *Let* LZSet(A) *be the Lempel–Ziv dictionary for a sequence A. We define LZ–Jaccard distance* LZJD *by*

$$\text{LZJD}(A,B) = 1 - \frac{|\text{LZSet}(A) \cap \text{LZSet}(B)|}{|\text{LZSet}(A) \cup \text{LZSet}(B)|}.$$

It is shown in [12, Theorem 1] that the triangle inequality holds for a function which they call an information-based sequence distance. Later papers give it the notation d_s in [13, Definition V.1], and call their normalized information distance d. Raff and Nicholas [14] introduced the LZJD and did not discuss the appearance of d_s in [13, Definition V.1], even though they do cite [13] (but not [12]).

Kraskov et al. [9,10] use D and D' for continuous analogues of d_s and d in [13] (which they cite). The *Encyclopedia* calls it the normalized information metric,

$$\frac{H(X \mid Y) + H(X \mid Y)}{H(X,Y)} = 1 - \frac{I(X;Y)}{H(X,Y)}$$

or Rajski distance [15].

This d_s was called d by [12]—see Table 1. Conversely, [13, near Definition V.1] mentions mutual information.

Table 1. Overview of notation used in the literature. (It seems that authors use simple names for their favored notions.)

Reference	Jaccard notation	NID notation
[12]	d	
[13]	d_s	d
[9]	D	D'
[14]	LZJD	NCD

Remark 2. *Ridgway [4] observed that the entropy-based distance D is essentially a Jaccard distance. No explanation was given, but we attempt one as follows. Suppose X_1, X_2, X_3, X_4 are iid Bernoulli ($p = 1/2$) random variables, Y is the random vector (X_1, X_2, X_3) and Z is (X_2, X_3, X_4). Then Y and Z have two bits of mutual information $I(Y, Z) = 2$. They have an entropy $H(Y) = H(Z) = 3$ of three bits. Thus the relationship $H(Y, Z) = H(Y) + H(Z) - I(Y, Z)$ becomes a Venn diagram relationship $|\{X_1, X_2, X_3, X_4\}| = |\{X_1, X_2, X_3\}| + |\{X_2, X_3, X_4\}| - |\{X_2, X_3\}|$. The relationship to Jaccard distance may not have been well known, as it is not mentioned in [1, 2, 9, 12].*

A more general setting is that of STRM (Symmetric Tversky Ratio Models), Theorem 10. These are variants of the Tversky index (Theorem 4) proposed in [7].

Definition 3. *A* semimetric *on $\mathcal{X}$ is a function $d : \mathcal{X} \times \mathcal{X} \to \mathbb{R}$ that satisfies the first three axioms of a metric space, but not necessarily the triangle inequality: $d(x, y) \geq 0$, $d(x, y) = 0$ if and only if $x = y$, and $d(x, y) = d(y, x)$ for all $x, y \in \mathcal{X}$.*

Definition 4 (**[17]**). *For sets X and Y the Tversky index with parameters $\alpha, \beta \geq 0$ is a number between 0 and 1 given by*

$$S(X,Y) = \frac{|X \cap Y|}{|X \cap Y| + \alpha|X \setminus Y| + \beta|Y \setminus X|}.$$

We also define the corresponding Tversky dissimilarity $d^T_{\alpha,\beta}$ by

$$d^T_{\alpha,\beta}(X,Y) = \begin{cases} 1 - S(X,Y) & \text{if } X \cup Y \neq \emptyset; \\ 0 & \text{if } X = Y = \emptyset. \end{cases}$$

To motivate Theorem 3, we include the following lemma without proof.

Lemma 5. *Suppose d is a metric on a collection of nonempty sets $\mathcal{X}$, with $d(X,Y) \leq 2$ for all $X, Y \in \mathcal{X}$. Let $\hat{\mathcal{X}} = \mathcal{X} \cup \{\emptyset\}$ and define $\hat{d} : \hat{\mathcal{X}} \times \hat{\mathcal{X}} \to \mathbb{R}$ by stipulating that for $X, Y \in \mathcal{X}$,*

$$\hat{d}(X,Y) = d(X,Y); \quad d(X,\emptyset) = 1 = d(\emptyset, X); \quad d(\emptyset,\emptyset) = 0.$$

Then $\hat{d}$ is a metric on $\hat{\mathcal{X}}$.

Theorem 6 (**Gragera and Suppakitpaisarn** [5,6]). *The optimal constant ρ such that $d^T_{\alpha,\beta}(X,Y) \leq \rho(d^T_{\alpha,\beta}(X,Y) + d^T_{\alpha,\beta}(Y,Z))$ for all X, Y, Z is*

$$\frac{1}{2}\left(1 + \sqrt{\frac{1}{\alpha\beta}}\right).$$

Corollary 7. *$d^T_{\alpha,\beta}$ is a metric only if $\alpha = \beta \geq 1$.*

Proof. Clearly, $\alpha = \beta$ is necessary to ensure $d^T_{\alpha,\beta}(X,Y) = d^T_{\alpha,\beta}(Y,X)$. Moreover $\rho \leq 1$ is necessary, so Theorem 6 gives $\alpha\beta \geq 1$.

Definition 8. *The Szymkiewicz–Simpson coefficient is defined by*

$$\text{overlap}(X,Y) = \frac{|X \cap Y|}{\min(|X|,|Y|)}$$

We may note that $\text{overlap}(X,Y) = 1$ whenever $X \subseteq Y$ or $Y \subseteq X$, so that $1 - \text{overlap}$ is not a metric.

Definition 9. *The Sørensen–Dice coefficient is defined by*

$$\frac{2|X \cap Y|}{|X| + |Y|}.$$

Definition 10 (**[7, Section 2]**). *Let $\mathcal{X}$ be a collection of finite sets. We define $S : \mathcal{X} \times \mathcal{X} \to \mathbb{R}$ as follows. For sets $X, Y \in \mathcal{X}$ we define $m(X,Y) = \min\{|X \setminus Y|, |Y \setminus X|\}$ and $M(X,Y) = \max\{|X \setminus Y|, |Y \setminus X|\}$. The symmetric TRM is defined by*

$$S(X,Y) = \frac{|X \cap Y| + \text{bias}}{|X \cap Y| + \text{bias} + \beta\,(\alpha m + (1-\alpha)M)}$$

The unbiased symmetric TRM is the case where bias $= 0$, *which is the case we shall assume we are in for the rest of this paper. The Tversky semimetric $D_{\alpha,\beta}$ is defined by $D_{\alpha,\beta}(X,Y) = 1 - S(X,Y)$, or more precisely*

$$D_{\alpha,\beta} = \begin{cases} \beta\frac{\alpha m+(1-\alpha)M}{|X\cap Y|+\beta(\alpha m+(1-\alpha)M)}, & \text{if } X\cup Y \neq \emptyset;\\ 0 & \text{if } X = Y = \emptyset.\end{cases}$$

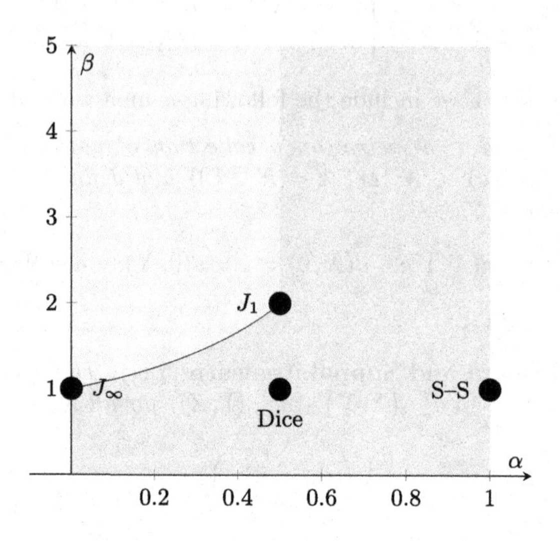

Fig. 1. A Tversky semimetric $D_{\alpha,\beta}$ is a metric if and only if (α, β) belongs to the green region. The parameter values corresponding to the Jaccard distance J_1, the analogue of normalized information distance analogue J_∞, the Sørensen–Dice semimetric, and the Szymkiewicz–Simpson semimetric are indicated. (Color figure online)

Note that for $\alpha = 1/2$, $\beta = 1$, the STRM is equivalent to the Sørensen–Dice coefficient. Similarly, for $\alpha = 1/2$, $\beta = 2$, it is equivalent to Jaccard's coefficient.

Our main result is (see Fig. 1):

Theorem 11. *Let $0 \leq \alpha \leq 1$ and $\beta > 0$. Then $D_{\alpha,\beta}$ is a metric if and only if $0 \leq \alpha \leq 1/2$ and $\beta \geq 1/(1-\alpha)$.*

Theorem 11 gives the converse to the Gragera and Suppakitpaisarn inspired Theorem 7:

Corollary 12. *The Tversky dissimilarity $d^T_{\alpha,\beta}$ is a metric iff $\alpha = \beta \geq 1$.*

Proof. Suppose the Tversky dissimilarity $d^T_{\alpha,\beta}$ is a semimetric. Let X, Y be sets with $|X \cap Y| = |X \setminus Y| = 1$ and $|Y \setminus X| = 0$. Then

$$1 - \frac{1}{1+\beta} = d^T_{\alpha,\beta}(Y,X) = d^T_{\alpha,\beta}(X,Y) = 1 - \frac{1}{1+\alpha},$$

hence $\alpha = \beta$. Let $\gamma = \alpha = \beta$.

Now, $d^T_{\gamma,\gamma} = D_{\alpha_0,\beta_0}$ where $\alpha_0 = 1/2$ and $\beta_0 = 2\gamma$. Indeed, let $m = \min\{|X \setminus Y|, |Y \setminus X|\}$ and $M = \max\{|X \setminus Y|, |Y \setminus X|\}$. Since

$$D_{\alpha_0,\beta_0} = \beta_0 \frac{\alpha_0 m + (1-\alpha_0)M}{|X \cap Y| + \beta_0 \left[\alpha_0 m + (1-\alpha_0)M\right]},$$

$$D_{\frac{1}{2},2\gamma} = 2\gamma \frac{\frac{1}{2}m + (1-\frac{1}{2})M}{|X \cap Y| + 2\gamma \left[\frac{1}{2}m + (1-\frac{1}{2})M\right]}$$

$$= \gamma \frac{|X \setminus Y| + |Y \setminus X|}{|X \cap Y| + \gamma[|X \setminus Y| + |Y \setminus X|]} = 1 - \frac{|X \cap Y|}{|X \cap Y| + \gamma|X \setminus Y| + \gamma|Y \setminus X|} = d^T_{\gamma,\gamma}.$$

By Theorem 11, $d^T_{\gamma,\gamma}$ is a metric if and only if $\beta_0 \geq 1/(1-\alpha_0)$. This is equivalent to $2\gamma \geq 2$, i.e., $\gamma \geq 1$.

The truth or falsity of Theorem 12 does not arise in Gragera and Suppakitpaisarn's work, as they require $\alpha, \beta \leq 1$ in their definition of Tversky index. We note that Tversky [17] only required $\alpha, \beta \geq 0$.

2 Results

Lemma 13. *Let $u, v, w, \epsilon > 0$. Then*

$$\frac{1}{u} \leq \frac{1}{v} + \frac{1}{w} \implies \frac{1}{u+\epsilon} \leq \frac{1}{v+\epsilon} + \frac{1}{w+\epsilon}.$$

Proof. It is of course equivalent to show

$$vw \leq uw + uv \implies (v+\epsilon)(w+\epsilon) \leq (u+\epsilon)(w+\epsilon) + (u+\epsilon)(v+\epsilon),$$

which reduces to

$$(v+w)\epsilon \leq (u+w)\epsilon + (u+v)\epsilon + \epsilon^2,$$

which is clearly the case.

Lemma 14. *Suppose $a(x,y) = a_{xy}$ and $b(x,y) = b_{xy}$ are functions. Suppose the function d given by $d(x,y) = a_{xy}/b_{xy}$ is a metric, and $\epsilon > 0$ is a real number. Let $\hat{d}(x,y) = \frac{a_{xy}}{b_{xy}+\epsilon a_{xy}}$. Then $\hat{d}$ is also a metric.*

Proof. The only nontrivial task is to verify the triangle inequality. Define further functions u, v, w by

$$u = b_{xy}/a_{xy}, \quad v = b_{xz}/a_{xz}, \quad w = b_{zy}/a_{zy}.$$

Since d is a metric we have

$$\frac{a_{xy}}{b_{xy}} \leq \frac{a_{xz}}{b_{xz}} + \frac{a_{zy}}{b_{zy}}$$

and hence $\frac{1}{u} \leq \frac{1}{v} + \frac{1}{w}$. We proceed by forward reasoning: we need the truth of the following equivalent conditions:

$$\frac{a_{xy}}{b_{xy} + \epsilon a_{xy}} \leq \frac{a_{xz}}{b_{xz} + \epsilon a_{xz}} + \frac{a_{zy}}{b_{zy} + \epsilon a_{zy}},$$
$$\frac{1}{u + \epsilon} \leq \frac{1}{v + \epsilon} + \frac{1}{w + \epsilon}.$$

By Theorem 13, we are done.

Theorem 15. *For each α, the set of β for which $D_{\alpha,\beta}$ is a metric is upward closed.*

Proof. Suppose D_{α,β_0} is a metric and $\epsilon = \beta - \beta_0 \geq 0$. Let $a_{XY} := \alpha m(X,Y) + (1-\alpha)M(X,Y)$. Since

$$D_{\alpha,\beta}(X,Y) = \beta \frac{a_{XY}}{|X \cap Y| + \beta a_{XY}}$$
$$= \beta \frac{a_{XY}}{|X \cap Y| + \beta_0 a_{XY} + \epsilon a_{XY}},$$

and since the upfront factor of β may be removed without loss of generality, the question reduces to Theorem 14.

Some convenient notation to be used below includes $\overline{\alpha} = 1 - \alpha$; $\gamma := \beta\alpha \leq 1$ with $\beta = 1/\overline{\alpha}$; $x_{\cap}y = |X \cap Y|$, $x = |X|$ etc.;

- $x_y = |X \setminus Y|$, $x_{zy} = |X \setminus (Z \cup Y)| = |(X \setminus Z) \setminus Y|$,
- $x_{000} = |\overline{X} \cap \overline{Y} \cap \overline{Z}|$, $x_{001} = |\overline{X} \cap \overline{Y} \cap Z|$, $x_{010} = |\overline{X} \cap Y \cap \overline{Z}|$, $x_{011} = |\overline{X} \cap Y \cap Z|$, $x_{100} = |X \cap \overline{Y} \cap \overline{Z}|$, $x_{101} = |X \cap \overline{Y} \cap Z|$, $x_{110} = |X \cap Y \cap \overline{Z}|$, $x_{111} = |X \cap Y \cap Z|$.

Theorem 16. *$\delta := \alpha m + \overline{\alpha} M$ satisfies the triangle inequality if and only if $\alpha \leq 1/2$.*

Proof. We first show the *only if* direction. Let $X = \{0\}$, $Y = \{1\}$, $Z = \{0,1\}$. Then

$$\alpha m(X,Y) + \overline{\alpha} M(X,Y) = 1,$$
$$\alpha m(X,Z) + \overline{\alpha} M(X,Z) = \alpha m(Y,Z) + \overline{\alpha} M(Y,Z) = 0 + \overline{\alpha}.$$

The triangle inequality then is equivalent to $1 \leq 2\overline{\alpha}$, i.e., $\alpha \leq 1/2$.

Now let us show the *if* direction. The triangle inequality says

$$\alpha \min\{x_y, y_x\} + \overline{\alpha} \max\{y_x, x_y\} \leq \alpha \min\{x_z, z_x\} + \overline{\alpha} \max\{z_x, x_z\} + \alpha \min\{z_y, y_z\} + \overline{\alpha} \max\{y_z, z_y\}$$

By symmetry between x and y, we may assume that $y \leq x$. Hence either $y \leq z \leq x$, $y \leq x \leq z$, or $z \leq y \leq x$. Thus our proof splits into three Cases, I, II, and III.

Case I: $y \leq z \leq x$: we must show that $\alpha y_x + \overline{\alpha} x_y \leq \alpha z_x + \overline{\alpha} x_z + \alpha y_z + \overline{\alpha} z_y$. Since $y_x \leq y_z + z_x$ and $x_y \leq x_z + z_y$, this holds for all α.
Case II: $y \leq x \leq z$: We want to show that $\alpha y_x + \overline{\alpha} x_y \leq \alpha x_z + \overline{\alpha} z_x + \alpha y_z + \overline{\alpha} z_y$. In terms of $\gamma = \alpha / \overline{\alpha}$ this says

$$0 \leq (y_z + x_z - y_x)\gamma + z_x + z_y - x_y = C\gamma + D.$$

The identity $x_y + y_z + z_x = x_z + z_y + y_x$ holds generally since both sides counts the elements that belong to exactly one of X, Y, Z once each, and counts the elements that belong to exactly two of X, Y, Z once each. Since $x \leq z$, it follows that

$$C = y_z + x_z - y_x \leq z_x + z_y - x_y = D.$$

Subcase II.1: $C \geq 0$. Then $C\gamma + D \geq 2C \geq 0$, as desired.
Subcase II.2: $C < 0$. In order to show $C\gamma + D \geq 0$ for all $0 \leq \gamma \leq 1$ it suffices that $C + D \geq 0$, since then $C\gamma + D = D - |C|\gamma \geq D - |C| \geq 0$.
We have $C + D = (x_z + z_y - x_y) + (y_z + z_x - y_x) \geq 0$.
Case III: $z \leq y \leq x$: We now need

$$\alpha y_x + \overline{\alpha} x_y \leq \alpha z_x + \overline{\alpha} x_z + \alpha z_y + \overline{\alpha} y_z,$$

$$0 \leq \gamma(z_x + z_y - y_x) + (x_z + y_z - x_y) = C\gamma + D.$$

The statement $C \leq D$ says $z_y + (z_x + x_y) \leq y_z + (y_x + x_z)$, which holds by the reasoning from Case II using now $z \leq y$. And now

$$C + D = (z_x + y_z - y_x) + (x_z + z_y - x_y) \geq 0.$$

Theorem 17. *The function $D_{\alpha,\beta}$ is a metric only if $\beta \geq 1/(1-\alpha)$.*

Proof. Consider the same example as in Theorem 16. Ignoring the upfront factor of β, we have

$$D = \frac{\delta}{|X \cap Y| + \beta\delta}.$$

In our example,

$$D(X,Y) = \frac{1}{0 + \beta \cdot 1} = \frac{1}{\beta},$$

$$D(X,Z) = D(Y,Z) = \frac{\overline{\alpha}}{1 + \beta \cdot \overline{\alpha}} = \frac{\overline{\alpha}}{1 + \beta\overline{\alpha}}.$$

The triangle inequality is then equivalent to:

$$\frac{1}{\beta} \leq 2\frac{\overline{\alpha}}{1+\beta\overline{\alpha}} \iff \beta \geq \frac{1+\beta\overline{\alpha}}{2\overline{\alpha}} \iff \beta \geq 1/(1-\alpha).$$

Theorem 18. *The function $D_{\alpha,\beta}$ is a metric on all finite power sets only if $\alpha \leq 1/2$.*

Proof. Suppose $\alpha > 1/2$. Let $Z_n = \{-(n-1), -(n-2), \dots, 0\}$, a set of cardinality n disjoint from $\{1,2\}$, and let $Y_n = Z_n \cup \{1\}$, $X_n = Z_n \cup \{2\}$. The triangle inequality says

$$\beta\frac{1}{n+\beta\cdot 1} = D(X_n, Y_n) \le D(X_n, Z_n) + D(Z_n, Y_n) = 2\beta\frac{\overline{\alpha}}{n+\beta\overline{\alpha}}$$
$$n + \beta\overline{\alpha} \le 2\overline{\alpha}(n+\beta)$$
$$n(1-2\overline{\alpha}) \le \beta\overline{\alpha}$$

Since $\alpha > 1/2$, we have $2\overline{\alpha} < 1$. Let $n > \frac{\beta\overline{\alpha}}{1-2\overline{\alpha}}$. Then the triangle inequality does not hold, so $D_{\alpha,\beta}$ is not a metric on the power set of $\{-(n-1), -(n-2), \dots, 0, 1, 2\}$.

Proof (Proof of Theorem 11). We saw in Theorem 16 that δ is a metric for $0 \le \gamma \le 1$. (Recall that $\beta = 1/(1-\alpha)$, so that $\gamma = \alpha/\overline{\alpha}$.) In general if d is a metric and a is a function, we may hope that $d/(a+d)$ is a metric. We shall use the observation, mentioned by [16], that in order to show

$$\frac{d_{xy}}{a_{xy}+d_{xy}} \le \frac{d_{xz}}{a_{xz}+d_{xz}} + \frac{d_{yz}}{a_{yz}+d_{yz}},$$

it suffices to show the following pair of inequalities:

$$\frac{d_{xy}}{a_{xy}+d_{xy}} \le \frac{d_{xz}+d_{yz}}{a_{xy}+d_{xz}+d_{yz}} \tag{3}$$
$$\frac{d_{xz}+d_{yz}}{a_{xy}+d_{xz}+d_{yz}} \le \frac{d_{xz}}{a_{xz}+d_{xz}} + \frac{d_{yz}}{a_{yz}+d_{yz}} \tag{4}$$

Here (3) follows from d being a metric, i.e., $d_{xy} \le d_{xz} + d_{yz}$, since $c \ge 0 < a \le b \implies \frac{a}{a+c} \le \frac{b}{b+c}$.

Next, (4) would follow from $a_{xy}+d_{yz} \ge a_{xz}$ and $a_{xy}+d_{xz} \ge a_{yz}$. By symmetry between x and y and since $a_{xy} = a_{yx}$ in our case, it suffices to prove the first of these, $a_{xy} + d_{yz} \ge a_{xz}$. This is equivalent to

$$x_{\cap}y + \gamma \min\{z_y, y_z) + \max\{z_y, y_z\} \ge x_{\cap}z,$$

which holds for all $0 \le \gamma \le 1$ if and only if $x_{\cap}y + \max\{z_y, y_z\} \ge x_{\cap}z$. There are now two cases.

Case $z \ge y$: We have

$$x_{\cap}y + z_y \ge x_{\cap}z$$

since any element of $X \cap Z$ is either in Y or not.

Case $y \ge z$:

$$x_{\cap}y + y_z \ge x_{\cap}z$$
$$x_{110} + x_{111} + x_{110} + x_{010} \ge x_{101} + x_{111}$$
$$x_{110} + x_{110} + x_{010} \ge x_{101}$$

This is true since $z_y \ge x_{\cap}z_y$.

3 Application to Phylogeny

The mutations of spike glycoproteins of coronaviruses are of great concern with the new SARS-CoV-2 virus causing the disease CoViD-19. We calculate several distance measures between peptide sequences for such proteins. The distance

$$Z_{2,\alpha}(x_0, x_1) = \alpha \min(|A_1|, |A_2|) + \overline{\alpha} \max(|A_1|, |A_2|)$$

where A_i is the set of subwords of length 2 in x_i but not in x_{1-i}, counts how many subwords of length 2 appear in one sequence and not the other.

We used the Ward linkage criterion for producing Newick trees using the `hclust` package for the Go programming language. The calculated phylogenetic trees were based on the metric $Z_{2,\alpha}$.

We found one tree isomorphism class each for $0 \leq \alpha \leq 0.21$, $0.22 \leq \alpha \leq 0.36$, and $0.37 \leq \alpha \leq 0.5$, respectively (Fig. 2, Fig. 3). In Fig. 3 we are also including the tree produced using the Levenshtein edit distance in place of $Z_{2,\alpha}$. We see that the various intervals for α can correspond to "better" or "worse" agreement with other distance measures. Thus, we propose that rather than focusing on $\alpha = 0$ and $\alpha = 1/2$ exclusively, future work may consider the whole interval $[0, 1/2]$.

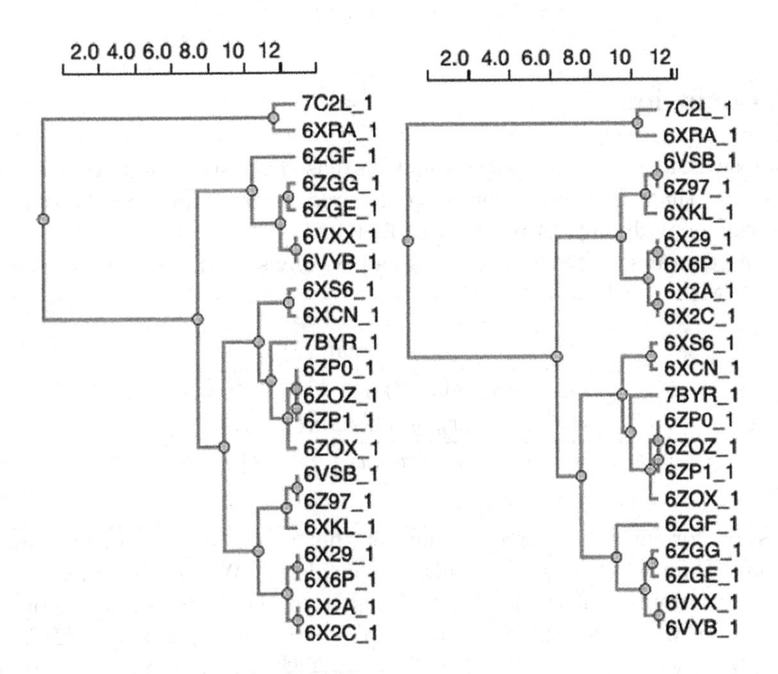

Fig. 2. $\alpha = 0.21$ and 0.36.

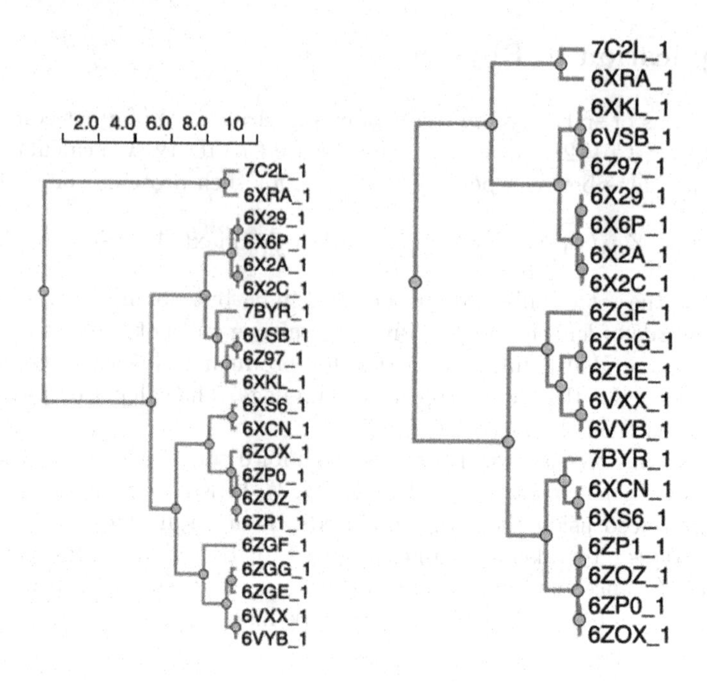

Fig. 3. $\alpha = 0.5$ and edit distance.

4 Conclusion

Many researchers have considered metrics based on sums or maxima, but we have shown that these need not be considered in "isolation" in the sense that they form the endpoints of a family of metrics.

More general set-theoretic metrics can be envisioned. The Steinhaus transform of δ with $\beta = 1/\overline{\alpha}$ is:

$$\delta'(X,Y) = \frac{2\delta(X,Y)}{\delta(X,\emptyset) + \delta(Y,\emptyset) + \delta(X,Y)}$$

$$= 2\frac{\gamma \min\{x_y, y_x) + \max\{x_y, y_x\}}{(x+y) + \gamma \min\{y_x, x_y\} + \max\{y_x, x_y\}}$$

A question for future research is whether this Steinhaus transform is more or less useful than what Jiménez et al. [7] considered. We can consider a general setting for potential metrics that contains both the Steinhaus transform of δ and the STRM metrics. In terms of $m(X,Y) = \min\{x_y, y_x\}$ and $M(X,Y) = \max\{x_y, y_x\}$, we can consider $\Delta_{\gamma,s} := \frac{\gamma m + M}{x \cap y + s(x \cup y) + (\gamma m + M)}$. When $s = 0$ this is our STRM metric. When $s = 1$ it is the Steinhaus transform, ignoring constant upfront factors.

Correctness of Results. We have formally proved Theorem 11 in the Lean theorem prover, with a more streamlined proof than that presented here. The Github repository can be found at [8].

References

1. Cilibrasi, R., Vitanyi, P.M.B.: Clustering by compression. IEEE Trans. Inf. Theor. **51**(4), 1523–1545 (2005)
2. Cilibrasi, R.L., Vitanyi, P.M.B.: The Google similarity distance. IEEE Trans. Knowl. Data Eng. **19**(3), 370–383 (2007). https://doi.org/10.1109/TKDE.2007.48
3. Deza, M.M., Deza, E.: Encyclopedia of Distances, 4th edn. Springer, Heidelberg (2016). https://doi.org/10.1007/978-3-662-52844-0
4. Ged Ridgway: Mutual information – Wikipedia, the Free Encyclopedia, Revision as of 14:55, 22 January 2010 (2010). https://en.wikipedia.org/w/index.php?title=Mutual_information&oldid=339351762. Accessed 14 May 2020
5. Gragera, A., Suppakitpaisarn, V.: Semimetric properties of Sørensen-dice and Tversky indexes. In: Kaykobad, M., Petreschi, R. (eds.) WALCOM 2016. LNCS, vol. 9627, pp. 339–350. Springer, Cham (2016). https://doi.org/10.1007/978-3-319-30139-6_27
6. Gragera, A., Suppakitpaisarn, V.: Relaxed triangle inequality ratio of the Sørensen-Dice and Tversky indexes. Theoret. Comput. Sci. **718**, 37–45 (2018). https://doi.org/10.1016/j.tcs.2017.01.004
7. Jiménez, S., Becerra, C.J., Gelbukh, A.F.: SOFTCARDINALITY-CORE: improving text overlap with distributional measures for semantic textual similarity. In: Diab, M.T., Baldwin, T., Baroni, M. (eds.) Proceedings of the 2nd Joint Conference on Lexical and Computational Semantics, *SEM 2013, Atlanta, Georgia, USA, 13–14 June 2013, pp. 194–201. Association for Computational Linguistics (2013). https://www.aclweb.org/anthology/S13-1028/
8. Kjos-Hanssen, B.: Lean project: a 1-parameter family of metrics connecting Jaccard distance to normalized information distance (2021). https://github.com/bjoernkjoshanssen/jaccard
9. Kraskov, A., Stögbauer, H., Andrzejak, R.G., Grassberger, P.: Hierarchical clustering using mutual information. Europhys. Lett. (EPL) **70**(2), 278–284 (2005). https://doi.org/10.1209/epl/i2004-10483-y
10. Kraskov, A., Stögbauer, H., Andrzejak, R.G., Grassberger, P.: Hierarchical clustering based on mutual information. arXiv arXiv:q-bio.QM/0311039 (2003)
11. Lempel, A., Ziv, J.: On the complexity of finite sequences. IEEE Trans. Inf. Theor. (IT) **22**(1), 75–81 (1976). https://doi.org/10.1109/tit.1976.1055501
12. Li, M., Badger, J.H., Chen, X., Kwong, S., Kearney, P.E., Zhang, H.: An information-based sequence distance and its application to whole mitochondrial genome phylogeny. Bioinformatics **17**(2), 149–54 (2001)
13. Li, M., Chen, X., Li, X., Ma, B., Vitányi, P.M.B.: The similarity metric. IEEE Trans. Inf. Theor. **50**(12), 3250–3264 (2004). https://doi.org/10.1109/TIT.2004.838101
14. Raff, E., Nicholas, C.K.: An alternative to NCD for large sequences, Lempel-Ziv Jaccard distance. In: Proceedings of the 23rd ACM SIGKDD International Conference on Knowledge Discovery and Data Mining (2017)

15. Rajski, C.: Entropy and metric spaces. In: Information Theory (Symposium, London, 1960), pp. 41–45. Butterworths, Washington, D.C. (1961)
16. Sra, S.: Is the Jaccard distance a distance? MathOverflow. https://mathoverflow.net/q/210750 (version: 2015-07-03)
17. Tversky, A.: Features of similarity. Psychol. Rev. **84**(4), 327–352 (1977). https://doi.org/10.1037/0033-295X.84.4.327
18. Ziv, J., Lempel, A.: A universal algorithm for sequential data compression. IEEE Trans. Inf. Theor. (IT) **23**(3), 337–343 (1977). https://doi.org/10.1109/tit.1977.1055714
19. Ziv, J., Lempel, A.: Compression of individual sequences via variable-rate coding. IEEE Trans. Inf. Theor. **24**(5), 530–536 (1978). https://doi.org/10.1109/TIT.1978.1055934

A Parameterized View on the Complexity of Dependence Logic

Juha Kontinen[1], Arne Meier[2], and Yasir Mahmood[2(✉)]

[1] Department of Mathematics and Statistics, University of Helsinki, Helsinki, Finland
juha.kontinen@helsinki.fi
[2] Institut für Theoretische Informatik, Leibniz Universität Hannover, Hannover, Germany
{meier,mahmood}@thi.uni-hannover.de

Abstract. In this paper, we investigate the parameterized complexity of model checking for Dependence Logic which is a well studied logic in the area of Team Semantics. We start with a list of nine immediate parameterizations for this problem, namely: the number of disjunctions (i.e., splits)/(free) variables/universal quantifiers, formula-size, the tree-width of the Gaifman graph of the input structure, the size of the universe/team, and the arity of dependence atoms. We present a comprehensive picture of the parameterized complexity of model checking and obtain a division of the problem into tractable and various intractable degrees. Furthermore, we also consider the complexity of the most important variants (data and expression complexity) of the model checking problem by fixing parts of the input.

Keywords: Team semantics · Dependence logic · Parameterized complexity · Model checking

1 Introduction

In this article, we explore the parameterized complexity of model checking for dependence logic ($\mathcal{D}$). We give a concise classification of this problem and its standard variants (expression and data complexity) with respect to several syntactic and structural parameters. Our results lay down a solid foundation for a systematic study of the parameterized complexity of team-based logics.

The introduction of Dependence Logic [27] in 2007 marks also the birth of the general semantic framework of team semantics that has enabled a systematic study of various notions of dependence and independence during the past decade. Team semantics differs from Tarski's semantics by interpreting formulas by sets of assignments instead of a single assignment as in first-order logic. Syntactically, dependence logic is an extension of first-order logic by new dependence atoms

First author funded by grants 308712 and 338259 of the Academy of Finland. Second and third authors funded by German Research Foundation (DFG), project ME 4279/1-2.

S. Artemov and A. Nerode (Eds.): LFCS 2022, LNCS 13137, pp. 125–142, 2022.
https://doi.org/10.1007/978-3-030-93100-1_9

$\mathsf{dep}(\mathbf{x};\mathbf{y})$ expressing that the values of variables $\mathbf{x}$ functionally determine values of the variables $\mathbf{y}$ (in the team under consideration). Soon after the introduction of dependence logic many other interesting team-based logics and atoms were introduced such as *inclusion*, *exclusion*, and *independence* atoms that are intimately connected to the corresponding inclusion, exclusion, and multivalued dependencies studied in database theory [9,13]. Furthermore, the area has expanded, e.g., to propositional, modal and probabilistic variants (see a selection of works from the literature [14,15,19] and the references therein).

For the applications, it is important to understand the complexity theoretic aspects of dependence logic and its variants. In fact, during the past few years, these aspects have been addressed in several studies. For example, on the level of sentences dependence logic and independence logic are equivalent to existential second-order logic while inclusion logic corresponds to positive greatest fixed point logic and thereby captures **P** over finite (ordered) structures [11]. Furthermore, there are (non-parameterized) studies that restrict the syntax and try to pin the intractability of a problem to a particular (set of) connective(s). For instance, Durand and Kontinen [5] characterize the data complexity of fragments of dependence logic with bounded arity of dependence atoms/number of universal quantifiers, and Grädel [12] characterizes the combined and the expression complexity of the model checking problem of dependence logic. These studies will be of great help in developing our parameterized approach.

A formalism to enhance the understanding of the inherent intractability of computational problems is brought by the framework of parameterized complexity [4]. Initiated by the founding fathers Downey and Fellows, in this area within computational complexity theory one strives for more structure within the darkness of intractability. Essentially, one tries to identify so-called parameters of a considered problem Π to find algorithms solving Π with runtimes of the form $f(k)\cdot|x|^{O(1)}$ for inputs x, corresponding parameter values k, and a computable function f. These kind of runtimes are called **FPT**-*runtimes* (from fixed-parameter tractable; short **FPT**) and tame the combinatoric explosion of the solution space to a function f in the parameter. As a very basic example in this vein, we can consider the propositional satisfiability problem SAT. An immediate parameter that pulls the problem into the class **FPT** is the number of variables, as one can solve SAT in time $2^k\cdot|\varphi|$ if k is the number of variables of a given propositional formula φ. Yet, this parameter is not very satisfactory as it neither is seen fixed nor slowly growing in its practical instances. However, there are several interesting other parameters under which SAT becomes fixed-parameter tractable, e.g., the so-called treewidth of the underlying graph representations of the considered formula [26]. This term was coined by Robertson and Seymour in 1984 [25] and established a profound position (currently DBLP lists 812 papers with treewidth in its title) also in the area of parameterized complexity in the last years [3,4].

Coming back to fpt-runtimes, a runtime of a very different quality (yet still polynomial for fixed parameters) than **FPT** is summarized by the complexity class **XP**: $|x|^{f(k)}$ for inputs x, corresponding parameter values k, and

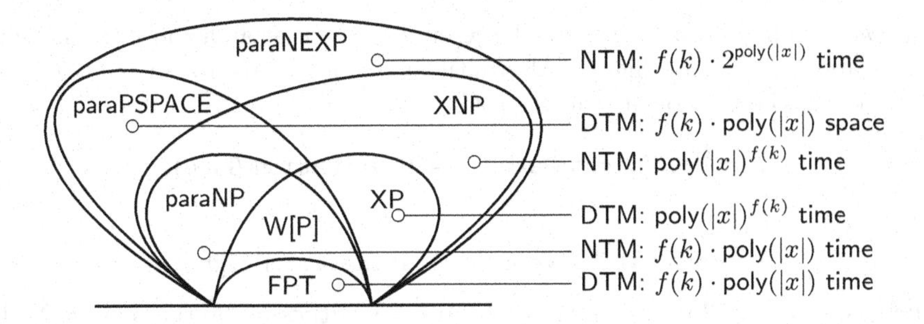

Fig. 1. Landscape showing relations of relevant parameterized complexity classes with machine definitions.

Table 1. An example flight departure screen at an airport

Flight	Destination	Gate	Date	Time
FIN-70	HEL – FI	C1	04.10.2021	09:55
SAS-475	OSL – NO	C3	04.10.2021	12:25
SAS-476	HAJ – DE	C2	04.10.2021	12:25
FIN-80	HEL – FI	C1	04.10.2021	19:55
KLM-615	ATL – USA	A5	05.10.2021	11:55
QR-70	DOH – QR	B6	05.10.2021	12:25
THY-159	IST – TR	A1	05.10.2021	15:55
FIN-80	HEL – FI	C1	05.10.2021	19:55

a computable function f. Furthermore, analogously as **XP** but on nondeterministic machines, the class **XNP** will be of interest in this paper. Further up in the hierarchy, classes of the form **para**$\mathcal{C}$ for a classical complexity class $\mathcal{C} \in \{\mathbf{NP}, \mathbf{PSPACE}, \mathbf{NEXP}\}$ play a role in this paper. Such classes intuitively capture all problems that are in the complexity class $\mathcal{C}$ after fpt-time preprocessing. In Fig. 1 an overview of these classes and their relations are depicted (for further details see, e.g., the work of Elberfeld et al. [7]).

Recently, the propositional variant of dependence logic ($\mathcal{PDL}$) has been investigated regarding its parameterized complexity [20,23]. Moreover, propositional independence and inclusion logic have also been studied from the perspective of parameterized complexity [21]. In this paper, we further pursue the parameterized journey through the world of team logics and will visit the problems of first-order dependence logic $\mathcal{D}$. As this paper is the first one that investigates $\mathcal{D}$ from the parameterized point of view, we need to gather the existing literature and revisit many results particularly from this perspective. As a result, this paper can be seen as a systematic study with some of the result following in a straightforward manner from the known non-parameterized results and some shedding light also on the non-parameterized view of model checking.

We give an example below to illustrate how the concept of dependence arises as a natural phenomenon in the physical world.

Example 1. The database in Table 1 presents a screen at an airport for showing details about departing flights. Alternatively, it can be seen as a team T over attributes in the top row as variables. Clearly

$$T \models \mathsf{dep}(\texttt{Flight}, \texttt{Date}, \texttt{Time}; \texttt{Destination}, \texttt{Gate}),$$

as well as

$$T \models \mathsf{dep}(\texttt{Gate}, \texttt{Date}, \texttt{Time}; \texttt{Destination}, \texttt{Flight}).$$

Whereas, $T \not\models \mathsf{dep}(\texttt{Destination}, \texttt{Gate}; \texttt{Time})$ as witnessed by the pair (FIN-70, HEL – FI, C1 , 04.10.2021, 09 : 55) and (FIN-80, HEL – FI, C1 , 04.10.2021, 19 : 55).

Contribution. Our classification is two-dimensional:

1. We consider the model checking problem of $\mathcal{D}$ under various parameterizations: number of split-junctions in a formula #splits, the length of the formula $|\Phi|$, number of free variables #free-variables, the treewidth of the structure $\mathsf{tw}(\mathcal{A})$, the size of the structure $|\mathcal{A}|$, the size of the team $|T|$, the number of universal quantifiers in the formula $\#\forall$, the arity of the dependence atoms dep-arity, as well as the total number of variables #variables.
2. We distinguish between expression complexity ec (the input structure is fixed), data complexity dc (the formula is fixed), and combined complexity cc.

The results are summarized in Table 2. For instance, parameters $\#\forall$, dep-arity, and #variables impact in lowering the complexity for ec (and not for cc or dc), while the parameter $|\mathcal{A}|$ impacts for dc but not for cc or ec.

Related Work. The parameterized complexity analyses in the propositional setting [20,21,23] have considered the combined complexity of model checking and satisfiability as problems of interest. On the cc-level, the picture there is somewhat different, e.g., team size as a parameter for propositional dependence logic enabled a **FPT** algorithm while in our setting it has no effect on the complexity (**paraNEXP**). Grädel [12] studied the expression and the combined complexity for $\mathcal{D}$ in the classical setting, whereas the data complexity was considered by Kontinen [16].

Organization of the Paper. In Sect. 2, we introduce the foundational concepts of dependence logic as well as parameterized complexity. In Sect. 3 our results are presented while Sect. 4 concludes the article.

2 Preliminaries

We require standard notions from classical complexity theory [24]. We encounter the classical complexity classes $\mathbf{P}, \mathbf{NP}, \mathbf{PSPACE}, \mathbf{NEXP}$ and their respective completeness notions, employing polynomial time many-one reductions ($\leq_m^{\mathbf{P}}$).

Parameterized Complexity Theory. A *parameterized problem* (PP) $P \subseteq \Sigma^* \times \mathbb{N}$ is a subset of the crossproduct of an alphabet and the natural numbers. For an *instance* $(x, k) \in \Sigma^* \times \mathbb{N}$, k is called the (value of the) *parameter*. A *parameterization* is a polynomial-time computable function that maps a value from $x \in \Sigma^*$ to its corresponding $k \in \mathbb{N}$. The problem P is said to be *fixed-parameter tractable* (or in the class **FPT**) if there exists a deterministic algorithm $\mathcal{A}$ and a computable function f such that for all $(x, k) \in \Sigma^* \times \mathbb{N}$, algorithm $\mathcal{A}$ correctly decides the membership of $(x, k) \in P$ and runs in time $f(k) \cdot |x|^{O(1)}$. The problem P belongs to the class **XP** if $\mathcal{A}$ runs in time $|x|^{f(k)}$ on a deterministic machine, whereas **XNP** is the non-deterministic counterpart of **XP**. Abusing a little bit of notation, we write $\mathcal{C}$-machine for the type of machines that decide languages in the class $\mathcal{C}$, and we will say a function f is "$\mathcal{C}$-computable" if it can be computed by a machine on which the resource bounds of the class $\mathcal{C}$ are imposed.

Also, we work with classes that can be defined via a precomputation on the parameter.

Definition 2. *Let $\mathcal{C}$ be any complexity class. Then* **para**$\mathcal{C}$ *is the class of all PPs $P \subseteq \Sigma^* \times \mathbb{N}$ such that there exists a computable function $\pi\colon \mathbb{N} \to \Delta^*$ and a language $L \in \mathcal{C}$ with $L \subseteq \Sigma^* \times \Delta^*$ such that for all $(x, k) \in \Sigma^* \times \mathbb{N}$ we have that $(x, k) \in P \Leftrightarrow (x, \pi(k)) \in L$.*

Notice that **paraP** = **FPT**. The complexity class $\mathcal{C} \in \{\textbf{NP}, \textbf{PSPACE}, \textbf{NEXP}\}$ is used in the **para**$\mathcal{C}$ context by us.

A problem P is in the complexity class **W[P]**, if it can be decided by a NTM running in time $f(k) \cdot |x|^{O(1)}$ steps, with at most $g(k)$-many non-deterministic steps, where f, g are computable functions. Moreover, **W[P]** is contained in the intersection of **paraNP** and **XP** (for details see the textbook of Flum and Grohe [8]).

Let $c \in \mathbb{N}$ and $P \subseteq \Sigma^* \times \mathbb{N}$ be a PP, then the *c-slice of P*, written as P_c is defined as $P_c := \{(x, k) \in \Sigma^* \times \mathbb{N} \mid k = c\}$. Notice that P_c is a classical problem then. Observe that, regarding our studied complexity classes, showing membership of a PP P in the complexity class **para**$\mathcal{C}$, it suffices to show that for each slice $P_c \in \mathcal{C}$ is true.

Definition 3. *Let $P \subseteq \Sigma^* \times \mathbb{N}, Q \subseteq \Gamma^*$ be two PPs. One says that P is* fpt-reducible *to Q, $P \leq^{\textbf{FPT}} Q$, if there exists an* **FPT**-*computable function $f\colon \Sigma^* \times \mathbb{N} \to \Gamma^* \times \mathbb{N}$ such that*

- *for all $(x, k) \in \Sigma^* \times \mathbb{N}$ we have that $(x, k) \in P \Leftrightarrow f(x, k) \in Q$,*
- *there exists a computable function $g\colon \mathbb{N} \to \mathbb{N}$ such that for all $(x, k) \in \Sigma^* \times \mathbb{N}$ and $f(x, k) = (x', k')$ we have that $k' \leq g(k)$.*

Finally, in order to show that a problem P is **para**$\mathcal{C}$-hard (for some complexity class $\mathcal{C}$) it is enough to prove that for some $c \in \mathbb{N}$, the slice P_c is $\mathcal{C}$-hard in the classical setting.

Dependence Logic. We assume basic familiarity with predicate logic [6]. We consider first-order vocabularies τ that are sets of *function* symbols and *relation* symbols with an equality symbol $=$. Let VAR be a countably infinite set of *first-order variables.* Terms over τ are defined in the usual way, and the set of well-formed formulas of first order logic ($\mathcal{FO}$) is defined by the following BNF:

$$\psi ::= t_1 = t_2 \mid R(t_1, \dots, t_k) \mid \neg R(t_1, \dots, t_k) \mid \psi \wedge \psi \mid \psi \vee \psi \mid \exists x \psi \mid \forall x \psi,$$

where t_i are terms $1 \leq i \leq k$, R is a k-ary relation symbol from σ, $k \in \mathbb{N}$, and $x \in \mathrm{VAR}$. If ψ is a formula, then we use $\mathrm{VAR}(\psi)$ for its set of variables, and $\mathrm{Fr}(\psi)$ for its set of free variables. We evaluate $\mathcal{FO}$-formulas in τ-structures, which are pairs of the form $\mathcal{A} = (A, \tau^{\mathcal{A}})$, where A is the *domain* of $\mathcal{A}$ (when clear from the context, we write A instead of $\mathrm{dom}(\mathcal{A})$), and $\tau^{\mathcal{A}}$ interprets the function and relational symbols in the usual way (e.g., $t^{\mathcal{A}}\langle s\rangle = s(x)$ if $t = x \in \mathrm{VAR}$). If $\mathbf{t} = (t_1, \dots, t_n)$ is a tuple of terms for $n \in \mathbb{N}$, then we write $\mathbf{t}^{\mathcal{A}}\langle s\rangle$ for $(t_1^{\mathcal{A}}\langle s\rangle, \dots, t_n^{\mathcal{A}}\langle s\rangle)$.

Dependence logic ($\mathcal{D}$) extends $\mathcal{FO}$ by dependence atoms of the form $\mathsf{dep}(\mathbf{t}; \mathbf{u})$ where $\mathbf{t}$ and $\mathbf{u}$ are tuples of terms. The semantics is defined through the concept of a team. Let $\mathcal{A}$ be a structure and $X \subseteq \mathrm{VAR}$, then an *assignment* s is a mapping $s\colon X \to A$.

Definition 4. *Let* $X \subseteq \mathrm{VAR}$. *A* team T *in* $\mathcal{A}$ *with domain* X *is a set of assignments* $s\colon X \to A$.

For a team T with domain $X \supseteq Y$ define its *restriction* to Y as $T \upharpoonright Y := \{\, s \upharpoonright Y \mid s \in T \,\}$. If $s\colon X \to A$ is an assignment and $x \in \mathrm{VAR}$ is a variable, then $s_a^x\colon X \cup \{x\} \to A$ is the assignment that maps x to a and $y \in X \setminus \{x\}$ to $s(y)$. Let T be a team in $\mathcal{A}$ with domain X. Then we define $f\colon T \to \mathcal{P}(A) \setminus \{\emptyset\}$ as the *supplementing function* of T. This is used to extend or modify T to the *supplementing team* $T_f^x := \{\, s_a^x \mid s \in T, a \in f(s) \,\}$. For the case $f(s) = A$ is the constant function we simply write T_A^x for T_f^x. The semantics of $\mathcal{D}$-formulas is defined as follows.

Definition 5. *Let* τ *be a vocabulary,* $\mathcal{A}$ *be a* τ*-structure and* T *be a team over* $\mathcal{A}$ *with domain* $X \subseteq \mathrm{VAR}$. *Then,*

$$
\begin{array}{ll}
(\mathcal{A}, T) \models t_1 = t_2 & \textit{iff } \forall s \in T : t_1^{\mathcal{A}}\langle s\rangle = t_2^{\mathcal{A}}\langle s\rangle \\
(\mathcal{A}, T) \models R(t_1, \dots, t_n) & \textit{iff } \forall s \in T : (t_1^{\mathcal{A}}\langle s\rangle, \dots, t_n^{\mathcal{A}}\langle s\rangle) \in R^{\mathcal{A}} \\
(\mathcal{A}, T) \models \neg R(t_1, \dots, t_n) & \textit{iff } \forall s \in T : (t_1^{\mathcal{A}}\langle s\rangle, \dots, t_n^{\mathcal{A}}\langle s\rangle) \notin R^{\mathcal{A}} \\
(\mathcal{A}, T) \models \mathsf{dep}(\mathbf{t}; \mathbf{u}) & \textit{iff } \forall s_1, s_2 \in T : \mathbf{t}^{\mathcal{A}}\langle s_1\rangle = \mathbf{t}^{\mathcal{A}}\langle s_2\rangle \Rightarrow \mathbf{u}^{\mathcal{A}}\langle s_1\rangle = \mathbf{u}^{\mathcal{A}}\langle s_2\rangle \\
(\mathcal{A}, T) \models \phi_0 \wedge \phi_1 & \textit{iff } (\mathcal{A}, T) \models \phi_0 \quad \textit{and} \quad (\mathcal{A}, T) \models \phi_1 \\
(\mathcal{A}, T) \models \phi_0 \vee \phi_1 & \textit{iff } \exists T_0 \exists T_1 : T_0 \cup T_1 = T \textit{ and} (\mathcal{A}, T_i) \models \phi_i \textit{ for } i = 0, 1 \\
(\mathcal{A}, T) \models \exists x \phi & \textit{iff } (\mathcal{A}, T_f^x) \models \phi \textit{ for some } f\colon T \to \mathcal{P}(A) \setminus \{\emptyset\} \\
(\mathcal{A}, T) \models \forall x \phi & \textit{iff } (\mathcal{A}, T_A^x) \models \phi
\end{array}
$$

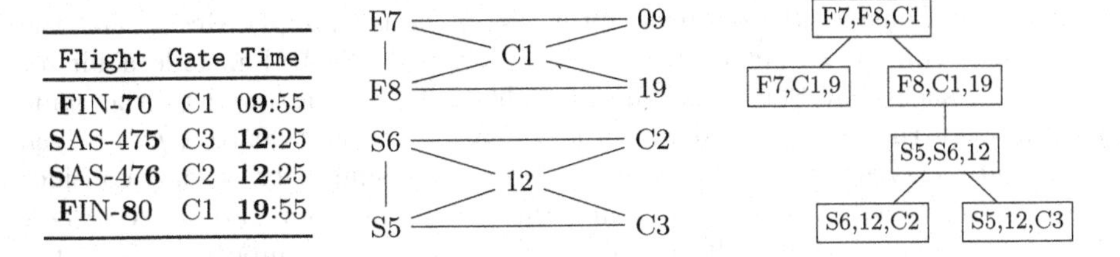

Flight	Gate	Time
FIN-70	C1	09:55
SAS-475	C3	12:25
SAS-476	C2	12:25
FIN-80	C1	19:55

Fig. 2. An $\mathcal{FO}$-structure $\mathcal{A} = (A, S^{\mathcal{A}}, R^{\mathcal{A}})$ (Left) with the Gaifman graph $G_{\mathcal{A}}$ (Middle) and a possible treedecomposition of $G_{\mathcal{A}}$ (Right) of Example 8. For brevity, universe elements are written in short forms.

Notice that we only consider formulas in negation normal form (NNF) as any formula of dependence logic can be transformed into logically equivalent NNF-form. Further note that $(\mathcal{A}, T) \models \phi$ for all ϕ when $T = \emptyset$ (this is also called the *empty team property*). Furthermore, $\mathcal{D}$-formulas are *local*, that is, for a team T in $\mathcal{A}$ over domain X and a $\mathcal{D}$-formula ϕ, we have that $(\mathcal{A}, T) \models \phi$ if and only if $(\mathcal{A}, T \upharpoonright \mathrm{Fr}(\phi)) \models \phi$. Finally, every $\mathcal{D}$-formula ϕ, if $(\mathcal{A}, T) \models \phi$ then $(\mathcal{A}, P) \models \phi$ for every $P \subseteq T$. This property is known as the downwards closure.

Definition 6. (Gaifman graph). *Given a vocabulary τ and a τ-structure $\mathcal{A}$, the Gaifman graph $G_{\mathcal{A}} = (A, E)$ of $\mathcal{A}$ is defined as*

$$E := \big\{ \{u, v\} \bigm| \text{ if there is an } R^n \in \tau \text{ and } \mathbf{a} \in A^n \text{ with } R^{\mathcal{A}}(\mathbf{a}) \text{ and } u, v \in \mathbf{a} \big\}.$$

That is, there is a relation $R \in \tau$ of arity n such that u and v appear together in $R^{\mathcal{A}}$.

Intuitively, the Gaifman graph of a structure $\mathcal{A}$ is an undirected graph with the universe of $\mathcal{A}$ as vertices and connects two vertices when they share a tuple in a relation (see also Fig. 2).

Definition 7. (Treewidth). *The* tree decomposition *of a given graph $G = (V, E)$ is a tree $T = (B, E_T)$, where the vertex set $B \subseteq \mathcal{P}(V)$ is the collection of* bags *and E_T is the edge relation such that the following is true.*

- $\bigcup_{b \in B} = V$,
- *for every $\{u, v\} \in E$ there is a bag $b \in B$ with $u, v \in b$, and*
- *for all $v \in V$ the restriction of T to v (the subset with all bags containing v) is connected.*

The width *of a given tree decomposition $T = (B, E_T)$ is the size of the largest bag minus one:* $\max_{b \in B} |b| - 1$. *The* treewidth *of a given graph G is the minimum over all widths of tree decompositions of G.*

Observe that if G is a tree then the treewidth of G is one. Intuitively, one can say that treewidth accordingly is a measure of tree-likeness of a given graph.

Example 8. Consider the database form our previous example. Recall that the universe A consists of entries in each row. Let $\tau = \{\mathrm{S}^2, \mathrm{R}^3\}$ include a binary relation S such that $\mathrm{S}(x, y)$ is true iff flights x and y are owed by the same company. Furthermore, consider a ternary relation R such that $\mathrm{R}(x, y, z)$ is true iff the gate x is reserved by the flight y at time z. For simplicity, we only consider first four rows with the corresponding three columns from Table 3, see Fig. 2 for an explanation. Since the largest bag size in our decomposition is 3, the treewidth of this decomposition is 2. Furthermore, the presence of cycles of length 3 suggests that there is no better decomposition. As a consequence the given structure has treewidth 2.

The decision problem to determine whether the treewidth of a given graph $\mathcal{G} = (V, E)$ is at most k, is **NP**-complete [1]. See Bodlaender's Guide [2] for an overview of algorithms that compute tree decompositions. When considering the parameter treewidth, one usually assumes it as a given value and does not need to compute it. We consider only the model checking problem (MC) and two variants in this paper. First, let us define the most general version.

Problem: cc (combined complexity of model checking)

Input: a structure $\mathcal{A}$, team T and a $\mathcal{D}$-formula Φ.
Question: $(\mathcal{A}, T) \models \Phi$?

We further consider the following two variants of the model checking problem.

Problem: dc (data complexity of model checking, Φ is fixed)

Input: a structure $\mathcal{A}$, team T.
Question: $(\mathcal{A}, T) \models \Phi$?

Problem: ec (expression complexity of model checking, $\mathcal{A}, T$ are fixed)

Input: a $\mathcal{D}$-formula Φ.
Question: $(\mathcal{A}, T) \models \Phi$?

List of Parameterizations. Now let us turn to the parameters that are under investigation in this paper. We study the model checking problem of $\mathcal{D}$ under nine various parameters that naturally occur in an MC-instance. Let $\langle \mathcal{A}, T, \Phi \rangle$ be an instance of MC, where Φ is a $\mathcal{D}$-formula, $\mathcal{A}$ is a structure and T is a team over $\mathcal{A}$. The parameter #splits denotes the number of occurrences of the split operator ($\vee$), $\#\forall$ is the number of universal quantifiers in Φ. Moreover, #variables (resp., #free-variables) denotes the total number of (free) variables in Φ. The parameter $|\Phi|$ is the size of the input formula Φ, and similarly the two other size parameters are $|\mathcal{A}|$ and $|T|$. The treewidth of the structure $\mathcal{A}$ (see Definition 7) is defined as the treewidth of $G_{\mathcal{A}}$ and denoted by $\mathsf{tw}(\mathcal{A})$. Note that for formulas using the dependence atom $\mathsf{dep}(\mathbf{x}; \mathbf{y})$, one can translate to a formula using only dependence atoms where $|\mathbf{y}| = 1$ (via conjunctions). That is why the arity of a dependence atom $\mathsf{dep}(\mathbf{x};\mathbf{y})$ is defined as $|\mathbf{x}|$ and dep-arity is the maximum arity of any dependence atom in Φ.

Let k be any parameterization and $P \in \{\mathsf{dc}, \mathsf{ec}, \mathsf{cc}\}$, then by k-P we denote the problem P when parameterized by k. If more than one parameterization is considered, then we use '+' as a separator and write these parameters in brackets, e.g., ($|\Phi|$ + #free-variables)-dc as the problem dc with parameterization $|\Phi|$ + #free-variables. Finally, notice that since the formula Φ is fixed for dc this implies that $|\Phi|$-dc is nothing but dc. That is, bounding the parameter does not make sense for dc as the problem dc remains **NP**-complete.

3 Complexity Results

Table 2. Complexity classification overview. A suffix -h represents the hardness result, whereas other results are completeness. The numbers in the exponent point to the corresponding result (Lx means Lemma x, Px means Proposition x, Rx means Remark x). Figure 3 on page 16 is a graphical presentation of this table with a different angle.

Parameter	cc	dc	ec
#splits	**paraPSPACE**-h^{L}17	**paraNP**^{L}12	**paraPSPACE**-h^{L}17
$\|\Phi\|$	**paraNP**^{L}18	**paraNP**^{R}13	**FPT**19
#free-variables	**paraNEXP**^{L}16	**paraNP**^{L}12	**paraNEXP**^{L}16
tw($\mathcal{A}$)	**paraNEXP**^{L}16	**paraNP**^{P}11	**paraNEXP**^{L}16
$\|\mathcal{A}\|$	**paraNEXP**^{L}16	**FPT**^{L}14	**paraNEXP**^{L}16
$\|T\|$	**paraNEXP**^{L}16	**paraNP**^{L}15	**paraNEXP**^{L}16
#∀	**paraNP**-h^{L}22	**paraNP**^{L}12	**paraNP**^{L}20
dep-arity	**paraPSPACE**-h^{L}25	**paraNP**^{L}12	**paraPSPACE**^{L}23
#variables	**paraNP**^{L}27	**paraNP**^{L}12	**FPT**^{L}28

We begin by proving relationships between various parameterizations.

Lemma 9. *The following relations among parameters hold.*

1. $|\Phi| \geq k$ *for any* $k \in \{$#splits, #∀, dep-arity, #free-variables, #variables$\}$,
2. $|\mathcal{A}| \geq \mathsf{tw}(\mathcal{A})$. *Moreover, for* dc, $|\mathcal{A}|^{O(1)} \geq |T|$,
3. *For* ec, #free-variables *is constant.*

Proof. 1. Clearly, the size of the formula limits all parts of it including the parameters mentioned in the list.

2. Notice that for data complexity, the formula Φ and consequently the number of free variables in Φ is fixed. Moreover, due to locality of $\mathcal{D}$ it holds that $T \subseteq A^r$, where r is the number of free variables in Φ. That is, the team T can be considered only over the free variables of Φ. This implies that teamsize is polynomially bounded by the universe size, as $|T| \leq |\mathcal{A}|^r$. Finally, the result for tw($\mathcal{A}$) follows due to Definition 7. This is due to the reason that in the worst case all universe elements belong to one bag in the decomposition and $\mathsf{tw}(\mathcal{A}) = |\mathcal{A}| - 1$.

Table 3. An example team for $(p_1 \vee \neg p_2 \vee \neg p_3)$

$x =$ 'variable'	$y =$ 'parity'	$u =$ 'clause'	$v =$ 'position'
p_1	1	1	0
p_2	0	1	1
p_3	0	1	2

3. Notice that the team T is fixed in ec. Together with the locality of $\mathcal{D}$-formulas (see Definition 5), this implies that the domain of T (which is same as the set of free variables in the formula Φ) is also fixed and as a result, of constant size.

□

Remark 10. If the number of free variables (#free-variables) in a formula Φ is bounded then the total number of variables (#variables) in Φ is not necessarily bounded, on the other hand, bounding #variables also bounds #free-variables.

3.1 Data Complexity (dc)

Classically, the data complexity of model checking for a fixed $\mathcal{D}$-formula Φ is **NP**-complete [27].

Proposition 11. *For a fixed formula, the problem whether an input structure $\mathcal{A}$ and a team T satisfies the formula is* **NP***-complete. That is, the data complexity of dependence logic is* **NP***-complete.*

In this section we prove that none of the considered parameter lowers this complexity, except $|\mathcal{A}|$. The proof relies on the fact that the complexity of model checking for already a very simple formula (see below) is **NP**-complete.

Lemma 12. *Let $k \in \{$#splits, #free-variables, #variables, #$\forall$, dep-arity, tw$(\mathcal{A})\}$. Then the problem k-dc is* **paraNP***-complete.*

Proof. The upper bound follows from Proposition 11. Kontinen [16, Theorem 4.9] proves that the data complexity for a fixed $\mathcal{D}$-formula of the form $\mathsf{dep}(x; y) \vee \mathsf{dep}(u; v) \vee \mathsf{dep}(u; v)$ is already **NP**-complete. For clarity, we briefly sketch the reduction presented by Kontinen [16]. Let

$$\phi := \bigwedge_{i \leq m} (\ell_{i,1} \vee \ell_{i,2} \vee \ell_{i,3})$$

be an instance of 3-SAT. Consider the structure $\mathcal{A}$ over the empty vocabulary, that is, $\tau = \emptyset$. Let $A = \mathrm{Var}(\phi) \cup \{0, 1, \dots, m\}$. The team T is constructed over variables $\{x, y, u, v\}$ that take values from A. As an example, the clause $(p_1 \vee \neg p_2 \vee \neg p_3)$ gives rise to assignments in Table 3. Notice that, a truth assignment θ for ϕ is constructed using the division of T according to each split. That is, $T \models \mathsf{dep}(x;y) \vee \mathsf{dep}(u;v) \vee \mathsf{dep}(u;v)$ if and only if $\exists P_0, P_1, P_2$ such that $\cup_i P_i = T$ for $i \leq 2$ and each P_i satisfies ith dependence atom. Let P_0 be such that $P_0 \models \mathsf{dep}(x;y)$, then we let $\theta(p_j) = 1 \iff \exists s \in P$, s.t. $s(x) = p_j$ and $s(y) = 1$. That is, one literal in each clause must be chosen in such a way that satisfies this clause, whereas, the remaining two literals per each clause are allowed to take values that does not satisfy it. As a consequence, each clause is satisfied by the variables chosen in this way, which proves correctness.

This implies that the 2-slice (for #splits-dc), 4-slice (for #free-variables-dc as well as #variables-dc), 0-slice (for #∀-dc), and 1-slice (for dep-arity-dc) are **NP**-complete. Consequently, the **paraNP**-hardness for these cases follow. Finally, the case for $\mathsf{tw}(\mathcal{A})$ also follows due to the reason that the vocabulary of the reduced structure is empty. As a consequence, our Definition 7 yields a tree decomposition of width 1 trivially as no elements of the universe are related.

This completes the proof to our lemma. □

Remark 13. Recall that $|\Phi|$ as a parameter for dc does not make sense as the input consists of $\langle \mathcal{A}, T \rangle$. That is, the formula Φ is already fixed which is stronger than fixing the size of Φ.

We now prove the only tractable case for the data complexity.

Lemma 14. $|\mathcal{A}|$-dc $\in$ **FPT**.

Proof. Notice first that restricting the universe size $|\mathcal{A}|$ polynomially bounds the teamsize $|T|$, due to Lemma 9. This implies that the size of whole input is (polynomially) bounded by the parameter $|\mathcal{A}|$. The result follows because any PP P is **FPT** when the input size is bounded by the parameter [8, Proposition 1.7]. □

Lemma 15. $|T|$-dc *is* **paraNP**-*complete.*

Proof. For a fixed sentence $\Phi \in \mathcal{D}$ (that is, with no free variables) and for all models $\mathcal{A}$ and team T we have that $(\mathcal{A}, T) \models \Phi \iff (\mathcal{A}, \{\emptyset\}) \models \Phi$. As a result, the problem $\leq^{\mathbf{FPT}}$-reduces to the model checking problem with $|T| = 1$. Consequently, 1-slice of $|T|$-dc is **NP**-complete because model checking for a fixed $\mathcal{D}$-sentence is also **NP**-complete [27, Corollary 6.3]. This results in **paraNP**-hardness.

For the membership, note that given a structure $\mathcal{A}$ and a team T then for a fixed formula Φ the question whether $(\mathcal{A}, T) \models \Phi$ is in **NP**. Consequently, giving **paraNP**-membership. □

A comparison with the propositional dependence logic ($\mathcal{PDL}$) at this point might be interesting. If the formula size is a parameter then the model checking for $\mathcal{PDL}$ can be solved in **FPT**-time [20]. However, this is not the case for $\mathcal{D}$ even if the formula is fixed in advance.

3.2 Expression and Combined Complexity (ec, cc)

Now we turn towards the expression and combined complexity of model checking for $\mathcal{D}$. Here again, in most cases the problem is still intractable for the combined complexity. However, expression complexity when parameterized by the formula size ($|\Phi|$) and the total number of variables (#variables) yields membership in **FPT**. Similar to the previous section, we first present results that directly translate from the known reductions for proving the **NEXP**-completeness for $\mathcal{D}$.

Lemma 16. *Let* $k \in \{\,|\mathcal{A}|, \mathsf{tw}(\mathcal{A}), |T|, \#\mathsf{free\text{-}variables}\,\}$. *Then both* k-cc *and* k-ec *are* **paraNEXP**-*complete.*

Proof. In the classical setting, **NEXP**-completeness of the expression and the combined complexity for $\mathcal{D}$ was shown by Grädel [12, Theorem 5.1]. This immediately gives membership in **paraNEXP**. Interestingly, the universe in the reduction consists of $\{0, 1\}$ with empty vocabulary and the formula obtained is a $\mathcal{D}$-sentence. This implies that 2-slice (for $|\mathcal{A}|$), 1-slice (for $\mathsf{tw}(\mathcal{A})$), 1-slice (for $|T|$), and 0-slice (for the number of free variables) are **NEXP**-complete. As a consequence, **paraNEXP**-hardness for the mentioned cases follows and this completes the proof. □

For the number of splits as a parameterization, we only know that this is also highly intractable, with the precise complexity open for now.

Lemma 17. #splits-ec *and* #splits-cc *are both* **paraPSPACE**-*hard.*

Proof. Consider the equivalence of $\{\exists, \forall, \wedge\}$-$\mathcal{FO}$-MC to quantified constraint satisfaction problem (QCSP) [22, p. 418]. That is, the fragment of $\mathcal{FO}$ with only operations in $\{\exists, \forall, \wedge\}$ allowed. Then QCSP asks, whether the conjunction of quantified constraints ($\mathcal{FO}$-relations) is true in a fixed $\mathcal{FO}$-structure $\mathcal{A}$. This implies that already in the absence of a split operator (even when there are no dependence atoms), the model checking problem is **PSPACE**-hard. Consequently, the mentioned results follow. □

The formula size as a parameter presents varying behaviour depending upon if we consider the expression or the combined complexity.

Lemma 18. $|\Phi|$-cc *is* **paraNP**-*complete.*

Proof. Notice that, due to Lemma 9, the size k of a formula Φ also bounds the maximum number of free variables in any subformula of Φ. This gives the membership in conjunction with [12, Theorem 5.1]. That is, the combined complexity of $\mathcal{D}$ is **NP**-complete if maximum number of free variables in any subformuala of Φ is fixed. The lower bound follows because of the construction by Kontinen [16] (see also Lemma 12) since for a fixed formula (of fixed size), the problem is already **NP**-complete. □

Lemma 19. $|\Phi|$-ec *is in* **FPT**.

Proof. Recall that in expression complexity, the team T and the structure $\mathcal{A}$ are fixed. Whereas, the size of the input formula Φ is a parameter. The result follows trivially because any PP P is **FPT** when the input size is bounded by the parameter. □

The expression complexity regarding the number of universal quantifiers as a parameter drops down to **paraNP**-completeness, which is still intractable but much lower than **paraNEXP**-completeness. However, regarding the combined complexity we can only prove the membership in **XNP**, with **paraNP**-lower bound.

Lemma 20. $\#\forall$-ec *is* **paraNP**-*complete.*

Proof. We first prove the lower bound through a reduction form the satisfiability problem for propositional dependence logic ($\mathcal{PDL}$). That is, given a $\mathcal{PDL}$-formula ϕ, whether there is a team T such that $T \models \phi$? Let ϕ be a $\mathcal{PDL}$-formula over propositional variables $p_1, \ldots, p_n$. For $i \leq n$, let x_i denote a variable corresponding to the proposition p_i. Let $\mathcal{A} = \{0, 1\}$ be the structure over empty vocabulary. Clearly ϕ is satisfiable iff $\exists p_1 \ldots \exists p_n \phi$ is satisfiable iff $(\mathcal{A}, \{\emptyset\}) \models \exists x_1 \ldots \exists x_n \phi'$, where ϕ' is a $\mathcal{D}$-formula obtained from ϕ by simply replacing each proposition p_i by the variable x_i. Notice that the reduced formula does not have any universal quantifier, that is $\#\forall(\phi') = 0$. This gives **paraNP**-hardness since the satisfiability for $\mathcal{PDL}$ is **NP**-complete [18].

For membership, notice that a $\mathcal{D}$-sentence Φ with k universal quantifiers can be reduced in **P**-time to an $\mathcal{ESO}$-sentence Ψ of the form $\exists f_1 \ldots \exists f_r \forall x_1 \ldots \forall x_k \psi$ by Durand and Kontinen [5, Cor. 3.9], where ψ is a quantifier free $\mathcal{FO}$-formula, $r \in \mathbb{N}$, and each function symbol f_i is at most k-ary for $1 \leq i \leq r$. Finally, we have that

$$(\mathcal{A}, \{\emptyset\}) \models \Phi \iff \mathcal{A} \models \bigvee_{f_1} \cdots \bigvee_{f_r} \forall x_1 \ldots \forall x_k \psi'.$$

Where the latter question can be solved by guessing an interpretation for each function symbol f_i and $i \leq r$. This requires $r \cdot |\mathcal{A}|^k$ guessing steps, and can be achieved in **paraNP**-time for a fixed structure $\mathcal{A}$ (as we consider expression complexity). Consequently, the membership in **paraNP** follows. Notice that the arity of function symbols in the **paraNP**-membership above is bounded by k if Φ is a $\mathcal{D}$-sentence. However, if Φ is a $\mathcal{D}$-formulas with m free variables then the arity of function symbols as well as the number of universal quantifiers in the reduction, both are bounded by $k + m$ where $k = \#\forall(\Phi)$ and $m = \#\text{free-variables}(\Phi)$. Nevertheless, recall that for ec, the team is also fixed. Moreover, due to Lemma 9 the collection of free variables in Φ has constant size. This implies that the reduction above provides an $\mathcal{ESO}$-sentence with $k + m$ universal quantifiers as well as function symbols of arity $k + m$ at most. Finally, guessing the interpretation for functions still takes **paraNP**-steps (because m is constant) and consequently, we get **paraNP**-membership for open formulas as well. □

The following corollary immediately follows from the proof above.

Corollary 21. $(\#\forall + \#\textsf{free-variables})$-ec *is* **paraNP**-*complete.*

Lemma 22. $\#\forall$-cc *is* **paraNP**-*hard. Moreover, for sentences of* $\mathcal{D}$, $\#\forall$-cc *is in* **XNP**.

Proof. The **paraNP**-lower bound follows due to the fact that the expression complexity of $\mathcal{D}$ is already **paraNP**-complete when parameterized by $\#\forall$ (see Lemma 20).

For sentences, similar to the proof in Lemma 20, a $\mathcal{D}$-sentence Φ can be translated to an equivalent $\mathcal{ESO}$-sentence Ψ in polynomial time. However, if the structure is not fixed as for expression complexity, then the computation of interpretations for functions can no longer be done in **paraNP**-time, but requires non-deterministic $|\mathcal{A}|^k$-time for each guessed function, where $k = \#\forall$. Consequently, we reach only membership in **XNP** for sentences. □

For open formulas, we do not know if $\#\forall$-cc is also in **XNP**. Our proof technique does not immediately settle this case as the team is not fixed for cc.

Similar to the case of universal quantifiers, the arity as a parameter also reduces the complexity but not as much as the universal quantifiers. Moreover, the precise combined complexity when parameterized by the arity is also open.

Lemma 23. dep-arity-ec *is* **paraPSPACE**-*complete.*

Proof. Notice that a $\mathcal{D}$-sentence Φ with k-ary dependence atoms can be reduced in **P**-time to an $\mathcal{ESO}$-sentence Ψ of the form $\exists f_1 \ldots \exists f_r \psi$ [5, Thm. 3.3], where ψ is an $\mathcal{FO}$-formula and each function symbol f_i is at most k-ary for $1 \leq i \leq r$.

Finally,

$$\mathcal{A} \models \Phi \iff \mathcal{A} \models \bigvee_{f_1} \cdots \bigvee_{f_r} \psi'.$$

That is, one needs to guess the interpretation for each function symbol f_i, which can be done in **paraNP**-time. Finally, evaluating an $\mathcal{FO}$-formula ψ' for a fixed structure $\mathcal{A}$ can be done in **PSPACE**-time. This yields membership in **paraPSPACE**. Moreover, if Φ is an open $\mathcal{D}$-formula then the result follows due to a similar discussion as in the prof of Lemma 20.

For hardness, notice that the expression complexity of $\mathcal{FO}$ is **PSPACE**-complete. This implies that already in the absence of any dependence atoms, the complexity remains **PSPACE**-hard, as a consequence, the 0-slice of dep-arity-ec is **PSPACE**-hard. This proves the desired result. □

The combination (dep-arity + #free-variables) also does not lower the expression complexity as discussed before in the case of $\#\forall$.

Corollary 24. (dep-arity + #free-variables)-ec *is* **paraPSPACE**-*complete.*

Lemma 25. dep-arity-cc *is* **paraPSPACE**-*hard.*

Proof. Consider the fragment of $\mathcal{D}$ with only dependence atoms of the form $\mathsf{dep}(;x)$, the so-called constancy logic. The combined complexity of constancy logic is **PSPACE**-complete [12, Theorem 5.3]. This implies that the 0-slice of dep-arity-cc is **PSPACE**-hard, proving the result. □

The combined complexity of model checking for constancy logic is **PSPACE**-complete [12, Thm. 5.3]. Aiming for an **paraPSPACE**-upper bound via squeezing the fixed arity of dependence atoms (in some way) into constancy atoms is unlikely to happen as $\mathcal{D}$ captures $\mathcal{ESO}$ whereas constancy logic for sentences (and also open formulas) collapses to $\mathcal{FO}$ [10].

Notice that a similar reduction as in the proof of Lemma 20 holds from $\mathcal{PL}$, in which both parameters ($\#\forall$ and dep-arity) are bounded. This implies that there is no hope for tractability even when both parameters are considered together. That is, the complexity of expression complexity remains **paraNP**-complete when parameterized by the combination of parameters ($\#\forall$, dep-arity).

Corollary 26. ($\#\forall$ + dep-arity)-ec *is also* **paraNP**-*complete.*

Finally, for the parameter total number of variables, the expression complexity drops to **FPT** whereas, the combined complexity drops to **paraNP**-complete. The case of expression complexity is particularly interesting. This is due to the reason that it was posed as an open question in [28, Page 88] whether the expression complexity of the fixed variable fragment of dependence logic ($\mathcal{D}^k$) is **NP**-complete similar to the case of the combined complexity therein. We answer this negatively by stating **FPT**-membership for #variables-ec, which as a corollary proves that the expression complexity of $\mathcal{D}^k$ is in **P** for each $k \geq 1$.

Lemma 27. #variables-cc *is* **paraNP**-*complete.*

Proof. Notice that if the total number of variables in Φ is fixed, then the number of free variables in any subformula ψ of Φ is also fixed. This implies the membership in **paraNP** due to [12, Theorem 5.1]. On the other hand, by [28, Theorem 3.9.6] we know that the combined complexity of $\mathcal{D}^k$ is **NP**-complete. This implies that for each k, the k-slice of the problem is **NP**-hard. This gives the desired lower bound. □

Lemma 28. #variables-ec *is* **FPT**.

Proof. Given a formula Φ of dependence logic with k variables, we can construct an equivalent formula Ψ of $\mathcal{ESO}^{k+1}$ in polynomial time [28, Theorem 3.3.17]. Moreover, since the structure $\mathcal{A}$ is fixed, there exists a reduction of Ψ to an $\mathcal{FO}$-formula ψ with $k+1$ variables (big disjunction on the universe elements for each second order existential quantifier). Finally, the model checking for $\mathcal{FO}$-formulas with k variables is solvable in time $O(|\psi| \cdot |A|^k)$ [17, Prop. 6.6]. This implies the membership in **FPT**. □

Corollary 29. *The expression complexity of* $\mathcal{D}^k$ *is in* **P** *for every* $k \geq 1$.

Proof. Since both, the number of variables and the universe size is fixed. The runtime of the form $O(|\psi| \cdot |A|^k)$ in Lemma 28 implies membership in **P**. □

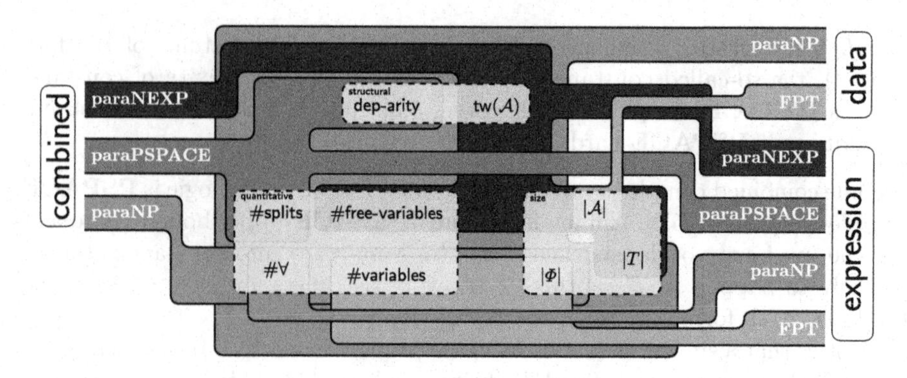

Fig. 3. Complexity classification overview for model checking problem of dependence logic, that takes grouping of parameters (quantitative, size, structural) and complexity classes into account.

4 Conclusion

In this paper, we started the parameterized complexity classification of model checking for dependence logic $\mathcal{D}$ with respect to nine different parameters (see Table 2 for an overview of the results). In Fig. 3 we depict a different kind of presentation of our results that also takes the grouping of parameters into quantitative, size related, and structural into account. The data complexity of $\mathcal{D}$ shows a dichotomy (**FPT** vs./**paraNP**-complete), where surprisingly there is only one case ($|\mathcal{A}|$) where one can reach **FPT**. This is even more surprising in the light of the fact that the expression (ec and the combined (cc) complexities under the same parameter are still highly intractable. Furthermore, there are parameters when cc and ec vary in the complexity (#variables). The combined complexity of $\mathcal{D}$ stays intractable under any of the investigated parameterizations. It might be interesting to study combination of parameters and see their joint effect on the complexity (yet, Corollaries 21, 24, 26 tackle already some cases).

We want to close this presentation with some further questions and topics that emerged of undertaking this study and should be tackled in the future:

- What other parameters could be meaningful (e.g., number of conjunctions, number of existential quantifiers, treewidth of the formula)?
- What is the exact complexity of #∀-cc, #splits-ec/-cc, dep-arity-cc?
- The parameterized complexity analysis for other team-based logics, such as independence logic and inclusion logic.

References

1. Arnborg, S., Corneil, D.G., Proskurowski, A.: Complexity of finding embeddings in a k-tree. SIAM J. Algebraic Discrete Meth. (1987). https://doi.org/10.1137/0608024

2. Bodlaender, H.L.: A tourist guide through treewidth. Acta Cybern., **11**(1–2), 1–21 (1993). https://cyber.bibl.u-szeged.hu/index.php/actcybern/article/view/3417
3. Bodlaender, H.L.: Discovering treewidth. In: SOFSEM, vol. 3381 of Lecture Notes in Computer Science, pp. 1–16. Springer (2005). https://doi.org/10.1007/978-3-540-30577-4_1
4. Fundamentals of Parameterized Complexity. TCS. Springer, London (2013). https://doi.org/10.1007/978-1-4471-5559-1
5. Durand, A., Kontinen, J.: Hierarchies in dependence logic. ACM Trans. Comput. Logic (TOCL) **13**(4), 31 (2012). https://doi.org/10.1145/2362355.2362359
6. Ebbinghaus, H.D., Flum, J.: Finite model theory. In: Perspectives in Mathematical Logic. Springer (1995). 978-3-540-60149-4, https://doi.org/10.1007/978-3-662-03182-7
7. Elberfeld, M., Stockhusen, C., Tantau, T.: On the space and circuit complexity of parameterized problems: classes and completeness. Algorithmica **71**(3), 661–701 (2014). https://doi.org/10.1007/s00453-014-9944-y
8. Parameterized Complexity Theory. TTCSAES. Springer, Heidelberg (2006). https://doi.org/10.1007/3-540-29953-X
9. Galliani, P.: Inclusion and exclusion dependencies in team semantics: on some logics of imperfect information. Ann. Pure Appl. Logic **163**(1), 68–84 (2012). https://doi.org/10.1016/j.apal.2011.08.005
10. Galliani, P.: On strongly first-order dependencies. In: Dependence Logic, pp. 53–71. Springer (2016). https://doi.org/10.1007/978-3-319-31803-5_4
11. Galliani, P., Hella, L.: Inclusion logic and fixed point logic. In: Rocca, S.R.D. (ed.) Computer science logic 2013 (CSL 2013), vol. 23 of Leibniz International Proceedings in Informatics (LIPIcs), pp. 281–295, Dagstuhl, Germany, 2013. Schloss Dagstuhl-Leibniz-Zentrum fuer Informatik. https://doi.org/10.4230/LIPIcs.CSL.2013.281
12. Grädel, E.: Model-checking games for logics of imperfect information. Theor. Comput. Sci. **493**, 2–14 (2013). https://doi.org/10.1016/j.tcs.2012.10.033
13. Grädel, E., Väänänen, J.: Dependence and independence. Studia Logica **101**(2), 399–410 (2013). https://doi.org/10.1007/s11225-013-9479-2
14. Hannula, M., Kontinen, J., Bussche, J.V., Virtema, J.: Descriptive complexity of real computation and probabilistic independence logic. In: LICS, pp. 550–563. ACM (2020). https://doi.org/10.1145/3373718.3394773
15. Hannula, M., Kontinen, J., Virtema, J., Vollmer, H.: Complexity of propositional logics in team semantic. ACM Trans. Comput. Log. **19**(1), 2:1–2:14 (2018). https://doi.org/10.1145/3157054
16. Kontinen, J.: Coherence and computational complexity of quantifier-free dependence logic formulas. Studia Logica **101**(2), 267–291 (2013). https://doi.org/10.1007/s11225-013-9481-8
17. Libkin, L.: Elements of finite model theory. In: Texts in Theoretical Computer Science. An EATCS Series. Springer (2004). https://doi.org/10.1007/978-3-662-07003-1
18. Lohmann, P., Vollmer, H.: Complexity results for modal dependence logic. Stud. Logica **101**(2), 343–366 (2013). https://doi.org/10.1007/s11225-013-9483-6
19. Lück, M.: Canonical models and the complexity of modal team logic. Log. Methods Comput. Sci. **15**(2) (2019). https://doi.org/10.23638/LMCS-15(2:2)2019

20. Mahmood, Y., Meier, A.: Parameterised complexity of model checking and satisfiability in propositional dependence logic. In: Foundations of Information and Knowledge Systems - 11th International Symposium, FoIKS 2020, 17–21 February 2020, Dortmund, Germany, Proceedings, pp. 157–174 (2020). https://doi.org/10.1007/978-3-030-39951-1_10
21. Mahmood, Y., Virtema, J.: Parameterised complexity of propositional logic in team semantics. CoRR. arXiv: 2105.14887
22. Martin, B.: First-order model checking problems parameterized by the model. In: CiE, volume 5028 of Lecture Notes in Computer Science, pp. 417–427. Springer (2008). https://doi.org/10.1007/978-3-540-69407-6_45
23. Meier, A., Reinbold, C.: Enumeration complexity of poor man's propositional dependence logic. In: FoIKS, volume 10833 of Lecture Notes in Computer Science, pp. 303–321. Springer (2018). https://doi.org/10.1007/978-3-319-90050-6_17
24. Papadimitriou, C.H.: Computational Complexity (1994). 978-0-201-53082-7 25
25. Robertson, N., Seymour, P.D.: Graph minors. III. planar tree-width. J. Comb. Theory Ser. B **36**(1), 49–64 (1984). https://doi.org/10.1016/0095-8956(84)90013-3
26. Samer, M., Szeider, S.: Fixed-parameter tractability. In: Handbook of Satisfiability, volume 185 of Frontiers in Artificial Intelligence and Applications, pp. 425–454. IOS Press (2009). https://doi.org/10.3233/978-1-58603-929-5-425
27. Väänänen, J.A.: Dependence Logic - A New Approach to Independence Friendly Logic, volume 70 of London Mathematical Society student texts. Cambridge University Press (2007). 978-0-521-70015-3
28. Virtema, J.: Approaches to Finite Variable Dependence: Expressiveness and Computational Complexity. PhD thesis, School of Information Sciences of the University of Tampere (2014). 978-951-44-9472-7

A Logic of Interactive Proofs

David Lehnherr[1], Zoran Ognjanović[2], and Thomas Studer[1(✉)]

[1] Institute of Computer Science, University of Bern, Bern, Switzerland
{david.lehnherr,thomas.studer}@inf.unibe.ch
[2] Mathematical Institute of Serbian Academy of Sciences and Arts, Belgrade, Serbia
zorano@mi.sanu.ac.rs

Abstract. We introduce the probabilistic two-agent justification logic IPJ, a logic in which we can reason about agents that perform interactive proofs. In order to study the growth rate of the probabilities in IPJ, we present a new method of parametrizing IPJ over certain negligible functions. Further, our approach leads to a new notion of zero-knowledge proofs.

Keywords: Interactive proof system · Zero-knowledge proof · Epistemic logic · Justification logic · Probabilistic logic

1 Introduction

An interactive proof system [6,8] is a protocol between two agents, the prover and the verifier. The aim of the protocol is that the prover can prove its knowledge of a secret to the verifier. To achieve this, the prover must answer a challenge provided by the verifier. Usually, the protocols are such that the verifier only knows with high probability that the prover knows the secret, that is the probability is a negligible function in the length of the challenge. The aim of the present paper is to introduce an epistemic logic $\mathsf{IPJ_I}$ to model interactive proof systems.

Our logic of interactive proofs and justifications $\mathsf{IPJ_I}$ will be a combination of modal logic, justification logic, and probabilistic logic. The logic includes two agents, P (the prover) and V (the verifier). The modal part of $\mathsf{IPJ_I}$ consists of two S4 modalities $\Box_P$ and $\Box_V$. As usual, $\Box_a$ means *agent a knows that.* Justification logic adds explicit reasons for the agents' knowledge [4,14]. We have formulas of the form $t{:}_a\alpha$, which stand for *agent a knows α for reason t.* The reason represented by the term t, can be a formal proof as in the first justification logic, the Logic of Proofs [2,13], the execution of an interactive proof protocol, the result of an agent's reasoning, or any other justification of knowledge like, e.g., direct observation. For $\mathsf{IPJ_I}$, we will use a two-agent version of the logic of proofs together with the justification yields belief principle $t{:}_a\alpha \rightarrow \Box_a\alpha$. The third ingredient of $\mathsf{IPJ_I}$ are probability operators of the form $\mathcal{P}_{\geq r}$ and $\mathcal{P}_{\approx r}$ meaning *with probability greater than or equal to r* and *with probability approximately r*,

Supported by the Swiss National Science Foundation grant 200020_184625 and by the Science Fund of the Republic of Serbia project AI4TrustBC.

S. Artemov and A. Nerode (Eds.): LFCS 2022, LNCS 13137, pp. 143–155, 2022.
https://doi.org/10.1007/978-3-030-93100-1_10

respectively. For the probabilistic part, we use the approach of [15,16], which has been adapted to justification logic in [10,11]. In order to deal with approximate probabilities, we need probability measures that can take non-standard values. Logics of this kind have been investigated in [17,18].

Goldwasser et al. [8] introduced interactive proof systems as follows. Let $\mathcal{L}$ be a language and P and V a pair of interacting (probabilistic) Turing machines, where P has unrestricted computational power and V is polynomial time. $\langle P, V\rangle$ is an interactive proof system for $\mathcal{L}$ if the following conditions hold:

1. Completeness: For all $k \in \mathbb{N}$, there exists $m \in \mathbb{N}$ such that for all inputs $x \in \mathcal{L}$ with $|x| > m$, the probability of $\langle P, V\rangle$ accepting x is at least $1 - |x|^{-k}$.
2. Soundness: For all $k \in \mathbb{N}$, there exists $m \in \mathbb{N}$ such that for all inputs $x \notin \mathcal{L}$ with $|x| > m$ and any interactive Turing machine P', the probability of $\langle P', V\rangle$ accepting x is at most $|x|^{-k}$.

Less formally, the agent P tries to prove its knowledge about a proposition α to the agent V. They may do that by following a challenge-response scheme. That is, V sends a challenge to P who then tries to answer it using his knowledge about α. On success, V's confidence in P knowing α is increased. Moreover, the harder the challenge, the stronger is V's belief. However, P may be dishonest and hence V may be convinced (with a low probability) that a wrong statement is true.

In order to model this in $\mathsf{IPJ_I}$, we introduce terms of the form f^n_t that represents V's view of the run of the protocol where P has evidence t and n is a measure for the complexity of the run (this may refer to the complexity of the challenge in a challenge response scheme). The outcome of a run will be formalized as $\mathcal{P}_{\geq r}(f^n_t{:}_V\Box_P\alpha)$ meaning that with probability greater than or equal to r, the run of the protocol with complexity n provides a justification for V that P knows α. Note that we are abstracting away the concrete protocol. Moreover, the subscript t in f^n_t does not imply that V has access to t; it only states that P's role in the protocol depends on t. We say that a formula α is interactively provable if the following two conditions hold:

1. Completeness: Assume $t{:}_P\alpha$. For all $k \in \mathbb{N}$, there exists a degree of complexity $m \in \mathbb{N}$ such that, for $n > m$ the probability of f^n_t justifying $\Box_P\alpha$ from V's view is at least $1 - n^{-k}$.
2. Soundness: Assume $\neg t{:}_P\alpha$. For all $k \in \mathbb{N}$, there exists a degree of complexity $m \in \mathbb{N}$ such that, for $n > m$ the probability of f^n_t justifying $\Box_P\alpha$ from V's view is at most n^{-k}.

Since $\mathsf{IPJ_I}$ is a propositional logic, we need a way to express the soundness and completeness condition without quantifiers. For integers m, k, we start with sets of formulas $\mathsf{I}_{m,k}$ and define the set of interactively provable formulas

$$\mathsf{I} := \bigcap_k \bigcup_m \mathsf{I}_{m,k}.$$

If a formula α belongs to $\mathsf{I}_{m,k}$, then the following two conditions must hold for $n > m$:

1. $t{:}_P\alpha \to \mathcal{P}_{\geq 1-\frac{1}{n^k}}(f_t^n{:}_V\Box_P\alpha)$
2. $\neg(t{:}_P\alpha) \to \mathcal{P}_{\leq \frac{1}{n^k}}(f_t^n{:}_V\Box_P\alpha)$.

Therefore, if $\alpha \in \mathsf{I}$ and $t{:}_P\alpha$ then, for every k, there exists an m such that $\alpha \in \mathsf{I}_{m,k}$ and thus $\mathcal{P}_{\geq 1-\frac{1}{n^k}}(f_t^n{:}_V\Box_P\alpha)$. Observe that this closely resembles the previously stated completeness property of interactive proof systems. The soundness property is obtained analogously.

Furthermore, we allow the probability operators to take non-standard values and consider protocols with transfinite complexity ω to capture the notion of a limit. Hence we can express statements of the form

if $t{:}_P\alpha$, then the probability of $f_t^\omega{:}_V\Box_P\alpha$ is almost 1.

Using the operator $\mathcal{P}_{\approx r}$, we add two more conditions for interactively provable formulas:

3. $t{:}_P\alpha \to \mathcal{P}_{\approx 1}(f_t^\omega{:}_V\Box_P\alpha)$ if $\alpha \in \mathsf{I}$;
4. $\neg(t{:}_P\alpha) \to \mathcal{P}_{\approx 0}(f_t^\omega{:}_V\Box_P\alpha)$ if $\alpha \in \mathsf{I}$.

We also include a principle saying that the justifications f_t^n are monotone in the complexity n:

5. $f_t^m{:}_a\alpha \to f_t^n{:}_a\alpha$ if $m < n$.

Justification logics with interacting agents are not new. Yavorskaya [20] introduced the evidence verification operator $!_P^V$ that can be used by V to verify P's evidence, i.e. her system includes the axiom $t{:}_P\alpha \to\ !_P^V t{:}_V t{:}_P\alpha$. This resembles the definition of the complexity class NP as interactive proof system, see, e.g., [1]. There, the verifier is a deterministic Turing machine. The prover generates a proof certificate t for α (where the complexity of t is polynomial in α), i.e. we have $t{:}_P\alpha$. Now P sends this certificate t to V and V checks it (which can be done in polynomial time). A successful check results in $!_P^V t$ being a justification for V that P knows the proof certificate t for α, i.e. $!_P^V t{:}_V t{:}_P\alpha$.

2 Syntax

Let $\mathbb{N}$ be the set of natural numbers and $\mathbb{N}^+ := \mathbb{N} \setminus \{0\}$. We define

$$\mathsf{Comp} := \mathbb{N} \cup \{\omega\}$$

where $\omega > n$ for each $n \in \mathbb{N}$.

We start with a countable set of justification variables and justification constants. Further we have a symbol f^n for each $n \in \mathsf{Comp}$. The set of *terms* Tm is given by the following grammar

$$t ::= c \mid x \mid t \cdot t \mid t + t \mid\ !t \mid f^n t$$

where c is a justification constant and x is a justification variable. In the following, we usually write f_t^n for $f^n t$.

Our language is based on two agents, the prover P and the verifier V. We write a for an arbitrary agent, i.e. either P or V. Further, we use a countable set of atomic propositions Prop. The set of *epistemic formulas* eFml is given by the following grammar:

$$\alpha ::= p \mid \neg\alpha \mid \alpha \wedge \alpha \mid \Box_a\alpha \mid t{:}_a\alpha$$

where p is an atomic proposition, t is a term and a is an agent.

For our formal approach, we consider probabilities that range over the unit interval of a non-archimedean recursive field that contains all rational numbers. We proceed as in [18] by choosing the unit interval of the Hardy field $\mathbb{Q}[\epsilon]$. The set $\mathbb{Q}[\epsilon]$ consists of all rational functions of a fixed non-zero infinitesimal $\epsilon \in \mathbb{R}^*$, where $\mathbb{R}^*$ is a non-standard extension of $\mathbb{R}$ (see [19]) for further details). Its positive elements have the form:

$$\epsilon^k \frac{\sum_{i=0}^{n} a_i \epsilon^i}{\sum_{i=0}^{m} b_i \epsilon^i},$$

where $a_i, b_i \in \mathbb{Q}$ for all $i \geq 0$ and $a_0 \cdot b_0 \neq 0$. We use S to denote the unit interval of $\mathbb{Q}[\epsilon]$.

The set of *formulas* Fml is given by the following grammar:

$$A ::= \alpha \mid \mathcal{P}_{\geq s}\alpha \mid \mathcal{P}_{\approx r}\alpha \mid \neg A \mid A \wedge A$$

where α is an epistemic formula, $s \in S$, and $r \in \mathbb{Q} \cap [0, 1]$.

Since any epistemic formula is a formula, we sometimes use latin letters to denote epistemic formulas, e.g. in $t{:}A \rightarrow \mathcal{P}_{\approx 1}B$, the letters A and B stand for epistemic formulas.

The remaining propositional connectives are defined as usual. Further we use the following syntactical abbreviations:

$$\mathcal{P}_{<s}\alpha \text{ denotes } \neg\mathcal{P}_{\geq s}\alpha \qquad \mathcal{P}_{\leq s}\alpha \text{ denotes } \mathcal{P}_{\geq 1-s}\neg\alpha$$
$$\mathcal{P}_{>s}\alpha \text{ denotes } \neg\mathcal{P}_{\leq s}\alpha \qquad \mathcal{P}_{=s}\alpha \text{ denotes } \mathcal{P}_{\leq s}\alpha \wedge \mathcal{P}_{\geq s}\alpha$$

Our Logic of Interactive Proofs $\mathsf{IPJ_l}$ depends on a parameter l. We will introduce that parameter later when it will be relevant. We start with presenting the axioms of $\mathsf{IPJ_l}$, which are divided into three groups: epistemic axioms, probabilistic axioms, interaction axioms.

Epistemic Axioms

For both modal operators $\Box_P$ and $\Box_V$ we have the axioms for the modal logic S4.

(p) all propositional tautologies
(k) $\Box_a(A \rightarrow B) \rightarrow (\Box_a A \rightarrow \Box_a B)$
(t) $\Box_a A \rightarrow A$
(4) $\Box_a A \rightarrow \Box_a\Box_a A$

For both agents, we have the axioms for the Logic of Proofs [2] and the connection axiom (jyb). This yields the system S4LP from [5].

(j) $s{:}_a(A \to B) \to (t{:}_a A \to_a s \cdot t{:}_a B)$
(j+) $(s{:}_a A \vee t{:}_a A) \to (s+t){:}_a A$
(jt) $t{:}_a A \to A$
(j4) $t{:}_a A \to\, !\, t{:}_a t{:}_a A$
(jyb) $t{:}_a A \to \Box_a A$

Probabilistic Axioms

The probabilistic axioms correspond to the axiomatization of approximate conditional probabilities used in [17,18] adapted to the unconditional case.

(p1) $P_{\geq 0} A$
(p2) $P_{\leq s} A \to P_{<t} A$, where $s < t$
(p3) $P_{<s} A \to P_{\leq s} A$
(p4) $P_{\geq 1}(A \leftrightarrow B) \to (P_{=s} A \to P_{=s} B)$
(p5) $P_{\leq s} A \leftrightarrow P_{\geq 1-s} \neg A$
(p6) $(P_{=s} A \wedge P_{=t} B \wedge P_{\geq 1} \neg(A \wedge B)) \to P_{=\min(1,s+t)}(A \vee B)$
(pa1) $P_{\approx r} A \to P_{\geq r_1} A$, for every rational $r_1 \in [0, r)$
(pa2) $P_{\approx r} A \to P_{\leq r_1} A$, for every rational $r_1 \in (r, 1]$

Interaction Axioms

So far, we have axioms for an epistemic justification logic with approximate probabilities. Let us now add axioms for terms of the form f_t^n that model interactive proof protocols. These axioms depend on the parameter I in $\mathsf{IPJ_I}$, which we introduce next.

An *interaction specification* I is a function $\mathsf{I} : \mathbb{N} \times \mathbb{N} \to \mathcal{P}(\mathsf{eFml})$, i.e. to each $m, k \in \mathbb{N}$ we assign a set of epistemic formulas $\mathsf{I}(m, k)$. In the following, we write $\mathsf{I}_{m,k}$ for $\mathsf{I}(m, k)$. Further, we overload the notation and use I also to denote the set

$$\mathsf{I} := \bigcap_k \bigcup_m \mathsf{I}_{m,k}.$$

The interaction axioms are:

(m) $f_t^m{:}_a \alpha \to f_t^n{:}_a \alpha$ for all $m, n \in \mathsf{Comp}$ such that $m < n$
(c) $t{:}_P \alpha \to P_{\geq 1 - \frac{1}{n^k}}(f_t^n{:}_V \Box_P \alpha)$ if $n > m$ and $\alpha \in \mathsf{I}_{m,k}$
(s) $\neg(t{:}_P \alpha) \to P_{\leq \frac{1}{n^k}}(f_t^n{:}_V \Box_P \alpha)$ if $n > m$ and $\alpha \in \mathsf{I}_{m,k}$
(cω) $t{:}_P \alpha \to P_{\approx 1}(f_t^\omega{:}_V \Box_P \alpha)$ if $\alpha \in \mathsf{I}$
(sω) $\neg(t{:}_P \alpha) \to P_{\approx 0}(f_t^\omega{:}_V \Box_P \alpha)$ if $\alpha \in \mathsf{I}$

Inference Rules

The rules of $\mathsf{IPJ_I}$ are the following. We have modus ponens:

$$\frac{A \quad A \to B}{B}$$

$\mathsf{IPJ_I}$ also includes the modal necessitation rule as well as the axiom necessitation rule from justification logic:

$$\frac{A}{\Box A} \qquad\qquad \frac{A \text{ is an axiom of } \mathsf{IPJ_I}}{c_1{:}_{a_1} c_2{:}_{a_2} \cdots c_n{:}_{a_n} A}$$

for arbitrary constants c_i and agents a_i. Of course, it would be possible to parameterize $\mathsf{IPJ_I}$ additionally by a constant specification as it is often done in justification logic. This would not affect our treatment of interactive proofs.

We have the following rules for the probabilistic part:

1. From A infer $P_{\geq 1} A$
2. From $B \to P_{\neq s} A$ for all $s \in S$ infer $B \to \bot$
3. From $B \to P_{\geq r - \frac{1}{n}} A$ and $B \to P_{\leq r + \frac{1}{n}} A$ for all integer $n \geq \frac{1}{1-r}$, infer

$$B \to P_{\approx r} A$$

3 Semantics

For this section, we assume that we are given an arbitrary interaction specification I. Many notions in this chapter will depend on that parameter. For any set X we use $\mathcal{P}(X)$ to denote the power set of X. We will use a Fitting-style semantics [7] for justification logic, but modular models [3,12] would work as well.

Definition 1 (Evidence relation). *An* evidence relation *is a mapping*

$$\mathcal{E} : \mathsf{Tm} \to \mathcal{P}(\mathsf{eFml})$$

from terms to sets of epistemic formulas such that for all $s, t \in \mathsf{Tm}$, $\alpha \in \mathsf{eFml}$, *constants* c_i, *and agents* a_i:

1. $\mathcal{E}(s) \cup \mathcal{E}(t) \subseteq \mathcal{E}(s+t)$;
2. $\mathcal{E}(s) \cdot \mathcal{E}(t) \subseteq \mathcal{E}(s \cdot t)$;
3. $t{:}\mathcal{E}(t) \subseteq \mathcal{E}(!t)$;
4. $c_2{:}_{a_2} \cdots c_n{:}_{a_n} A \in \mathcal{E}(c_1)$ *if* α *is an axiom;*
5. $\alpha \in \mathcal{E}(f_t^n)$, *if* $\alpha \in \mathcal{E}(f_t^m)$ *for* $n > m$.

Definition 2 (Epistemic model). *An* epistemic model *for* $\mathsf{IPJ_I}$ *is a tuple* $M = \langle W, R, \mathcal{E}, V \rangle$ *where:*

1. W *is a non-empty set of objects called worlds.*

2. *R maps each agent a to a reflexive and transitive accessibility relation R_a on W.*
3. *$\mathcal{E}$ maps each world w and each agent a to an evidence relation $\mathcal{E}^a_w$.*
4. *V is a valuation mapping each world to a set of atomic propositions.*

Definition 3 (Truth within a world). *Let $M = \langle W, R, \mathcal{E}, V\rangle$ be an epistemic model for $\mathsf{IPJ_I}$ and let w be a world in W. For an epistemic formula $\alpha \in \mathsf{eFml}$, we define $M, w \Vdash \alpha$ inductively by:*

1. *$M, w \Vdash \beta$ iff $\beta \in V(w)$ for $\beta \in \mathsf{Prop}$*
2. *$M, w \Vdash \neg\beta$ iff $M, w \nVdash \beta$*
3. *$M, w \Vdash \beta \wedge \gamma$ iff $M, w \Vdash \beta$ and $M, w \Vdash \gamma$*
4. *$M, w \Vdash \Box_a\beta$ iff $M, u \Vdash \beta$ for all $u \in W$ with $R_a wu$*
5. *$M, w \Vdash t{:}_a\beta$ iff $\beta \in \mathcal{E}^a_w(t)$ and $M, u \Vdash \beta$ for all $u \in W$ with $R_a wu$.*

Definition 4 (Algebra). *Let U be a non-empty set and let H be a non-empty subset of $\mathcal{P}(U)$. H will be called an algebra over U if the following hold:*

- $U \in H$
- $X, Y \in H \rightarrow X \cup Y \in H$
- $X \in H \rightarrow U \setminus X \in H$

Definition 5 (Finitely additive measure). *Let H be an algebra over U and $\mu : H \rightarrow S$, where S is the unit interval of the hardy field $\mathbb{Q}[\epsilon]$. We call μ a finitely additive measure if the following hold:*

1. *$\mu(U) = 1$*
2. *$X \cap Y = \emptyset \implies \mu(X \cup Y) = \mu(X) + \mu(Y)$ for all $X, Y \in H$.*

Definition 6 (Probability space). *A* probability space *is a triple $\langle U, H, \mu\rangle$ where:*

1. *U is a non-empty set*
2. *H is an algebra over U*
3. *$\mu : H \rightarrow S$ is a finitely additive measure.*

Definition 7 (Quasimodel). *A quasimodel for $\mathsf{IPJ_I}$ is a tuple*

$$M = \langle W, R, \mathcal{E}, V, U, H, \mu, w_0\rangle$$

such that

1. *$\langle W, R, \mathcal{E}, V\rangle$ is an epistemic model for $\mathsf{IPJ_I}$*
2. *$U \subseteq W$*
3. *$\langle U, H, \mu\rangle$ is a probability space*
4. *$w_0 \in U$.*

Let $M = \langle W, R, \mathcal{E}, V, U, H, \mu, w_0\rangle$ be a quasimodel, $w \in W$, and $\alpha \in \mathsf{eFml}$. Since M contains an epistemic model, we write $M, w \Vdash \alpha$ for $\langle W, R, \mathcal{E}, V\rangle, w \Vdash \alpha$.

Definition 8 (Events). *Let $M = \langle W, R, \mathcal{E}, V, U, H, \mu, w_0 \rangle$ be a quasimodel. For an epistemic formula $\alpha \in \mathsf{eFml}$, we define the event that α occurs as*

$$[\alpha]_M := \{u \in U \mid M, u \Vdash \alpha\}$$

We use $[\alpha]_M^C$ for the complement event $U \setminus [\alpha]_M$.

When the quasimodel M is clear from the context, we often drop the subscript M in $[\alpha]_M$.

Definition 9 (Independent events). *Let M be a quasimodel. We say that two events $S, T \in H$ are independent in M if*

$$\mu(S \cap T) = \mu(S) \cdot \mu(T).$$

Definition 10 (Probability almost r). *Let $\langle U, H, \mu \rangle$ be a probability space. For $r \in \mathbb{Q} \cap [0,1]$, we say that $X \in H$ has probability almost r ($\mu(X) \approx r$) if for all $n \in \mathbb{N}^+$ $\mu(X) \in \left[r - \frac{1}{n}, r + \frac{1}{n}\right]$.*

Definition 11 (Truth in a quasimodel). *Let*

$$M = \langle W, R, \mathcal{E}, V, U, H, \mu, w_0 \rangle$$

be quasimodel for $\mathsf{IPJ_I}$. We define $M \models A$ inductively by:

1. *$M \models A$ iff $M, w_0 \Vdash A$ for $A \in \mathsf{eFml}$; otherwise*
2. *$M \models \neg B$ iff $M \not\models B$*
3. *$M \models B \wedge C$ iff $M \models B$ and $M \models C$*
4. *$M \models \mathcal{P}_{\geq s}\alpha$ iff $\mu([\alpha]) \geq s$*
5. *$M \models \mathcal{P}_{\approx r}\alpha$ iff $\mu([\alpha]) \approx r$.*

Definition 12 Measurable model). *A quasimodel*

$$M = \langle W, R, \mathcal{E}, V, U, H, \mu, w_0 \rangle$$

is called measurable *if $[\alpha] \in H$ for all $\alpha \in \mathsf{eFml}$.*

Definition 13 (Model). *A* model *for $\mathsf{IPJ_I}$ is a measurable quasimodel M for $\mathsf{IPJ_I}$ that satisfies:*

1. *$M \models t :_P \alpha \rightarrow \mathcal{P}_{\geq 1 - \frac{1}{n^k}}(f_t^n :_V \Box^P \alpha)$ if $n > m$ and $\alpha \in \mathsf{I}_{m,k}$;*
2. *$M \models \neg(t :_P \alpha) \rightarrow \mathcal{P}_{\leq \frac{1}{n^k}}(f_t^n :_V \Box^P \alpha)$ if $n > m$ and $\alpha \in \mathsf{I}_{m,k}$.*

We say that a formula A is $\mathsf{IPJ_I}$-valid if $M \models A$ for all models M for $\mathsf{IPJ_I}$.

4 Properties and Results

We can read the operator $\mathcal{P}_{\approx 1}$ as *it is almost certain that.* As a first result, we observe that this operator provably behaves like a normal modality.

Lemma 1. *Let α, β be epistemic formulas.*

1. $\mathsf{IPJ_I}$ *proves* $\mathcal{P}_{\approx 1}(\alpha \to \beta) \to (\mathcal{P}_{\approx 1}\alpha \to \mathcal{P}_{\approx 1}\beta)$.
2. *The rule* $\dfrac{\alpha}{\mathcal{P}_{\approx 1}\alpha}$ *is derivable in* $\mathsf{IPJ_I}$.

Corollary 1. *For $\alpha \in \mathsf{I}$, $\mathsf{IPJ_I}$ proves $t{:}_P\alpha \to P_{\approx 1}(c \cdot f_t^\omega{:}_V\alpha)$ for any constant c.*

The deductive system $\mathsf{IPJ_I}$ is sound with respect to $\mathsf{IPJ_I}$-models.

Theorem 1 (Soundness). *Let I be an arbitrary interaction specification. For any formula F we have that*

$$\vdash F \quad \textit{implies} \quad F \textit{ is } \mathsf{IPJ_I}\textit{-valid.}$$

Proof As usual by induction on the length of the derivation. The interesting case is when F is an instance of $(\mathsf{c}\omega)$. But first note that axioms (m) and (c) are $\mathsf{IPJ_I}$-valid because of Definition 1 and Definition 13, respectively.

Now let F be an instance of $(\mathsf{c}\omega)$. Then F is of the form

$$t{:}_P\alpha \to P_{\approx 1}(f_t^\omega{:}_V\Box_P\alpha)$$

for some $\alpha \in \mathsf{I}$. Let $M = \langle W, R, \mathcal{E}, V, U, H, \mu, w_0\rangle$ be an arbitrary model for $\mathsf{IPJ_I}$ and assume $M \models t{:}_P\alpha$. We need to show

$$\mu([f_t^\omega{:}_V\Box_P\alpha]) \in \left[1 - \frac{1}{n}, 1\right] \quad \text{for all } n \in \mathbb{N}^+. \tag{1}$$

We fix an arbitrary $n \in \mathbb{N}^+$. Because of $\alpha \in \mathsf{I}$, we know that there exists an m such that $\alpha \in \mathsf{I}_{m,1}$. By soundness of axiom (c) we find that for each $n' > m$

$$\mu([f_t^{n'}{:}_V\Box_P\alpha]) \geq 1 - \frac{1}{n'}.$$

Let $n'' \in \mathbb{N}$ be such that $n'' > m$ and $n'' \geq n$. We find

$$\mu([f_t^{n''}{:}_V\Box_P\alpha]) \geq 1 - \frac{1}{n''} \geq 1 - \frac{1}{n}. \tag{2}$$

By soundness of axiom (m) we get that for each $w \in W$

$$M, w \Vdash f_t^{n''}{:}_V\Box_P\alpha \text{ implies } M, w \Vdash f_t^\omega{:}_V\Box_P\alpha.$$

Therefore, and by finite additivity of μ, we obtain

$$\mu([f_t^\omega{:}_V\Box_P\alpha]) \geq \mu([f_t^{n''}{:}_V\Box_P\alpha]). \tag{3}$$

Taking (2) and (3) together yields (1). □

In practice, one often considers interactive proofs systems that are round-based, see [1].

Definition 14 (Round-based interactive proof system). *An interactive protocol $\langle P, V\rangle$ is called* round-based *if the following two conditions hold:*

1. *Completeness: Let $x \in \mathcal{L}$. There exists a polynomial $p(x)$ such that the probability that $\langle P, V\rangle$ halts in an accepting state after $p(x)$ many messages is at least $\frac{2}{3}$.*
2. *Soundness: Let $x \notin \mathcal{L}$ and let $p(x)$ be any polynomial. For any interactive Turing machine P', the probability that $\langle P', V\rangle$ halts in an accepting state after $p(x)$ many messages is at most $\frac{1}{3}$.*

This definition achieves negligible (resp. overwhelming) probabilities by repeating the protocol several times and deciding based on a majority vote. Although this definition is simple to model in IPJ_I, it is not suitable for a limit analysis because our measure is not σ-additive. Note that to properly formalize σ-additivity one needs countable conjunctions and disjunctions [9], which we do not want to include here. However, for finitely many rounds, we can describe how the probability increases throughout the rounds (given that they are pairwise independent).

Lemma 2. *Let M be an IPJ_I-model for an arbitrary interaction specification I. Consider justification terms $s_1, \ldots, s_n$ and an epistemic formula α such that*

1. *$M \models s_i{:}_V\alpha$ for each s_i;*
2. *$[s_i{:}_V\alpha]$ and $[s_j{:}_V\alpha]$ are independent events for all $i \neq j$.*

We find that $M \models \bigwedge_{i=1,\ldots,n} \mathcal{P}_{\geq 1-r}(s_i{:}_V\alpha) \rightarrow \mathcal{P}_{\geq 1-r^n}\alpha$.

Proof. Whenever $s_i{:}_V\alpha$ is true at a world w, α is true at w by soundness of axiom (jt). Hence, by monotonicity of μ we find

$$\mu([\alpha]) \geq \mu\left(\bigcup_{i=1}^{n}[s_i{:}_V\alpha]\right) = 1 - \mu\left(\bigcap_{i=1}^{n}[s_i{:}_V\alpha]^C\right) \overset{indep.}{\geq} 1 - \prod_{i=1}^{n} r = 1 - r^n$$

□

An interactive proof protocol for a language $\mathcal{L}$ has the zero-knowledge property if, from a successful execution, the verifier only learns that x belongs to $\mathcal{L}$ but nothing else. Formally, a protocol is perfectly zero-knowledge if there exists a probabilistic Turing machine T that generates proof transcripts[1] that are indistinguishable from original ones. If the verifier can obtain additional information with negligible probability, then the protocol is said to be statistically zero-knowledge.

[1] In the setting of interactive Turing machines, a proof transcript is everything that V sees on the public tapes during the protocol.

However, we cannot directly implement this definition because it would require to model the Turing machine T as an agent and we would need to reason about something like indistinguishable terms. Simplified, a protocol is zero-knowledge if the verifier cannot compute the prover's secret. In our setting the prover's secret is represented by the term t. Hence, $f_t^n{:}_V t{:}_P \alpha$ means that the prover's secret has been revealed to the verifier. In fact, $f_t^n{:}_V t{:}_P \alpha$ being unlikely is a direct consequence of the protocol being statistically zero-knowledge because the probability of the verifier knowing the prover's secret is bound by its ability to distinguish between proof transcripts. This gives rise to the following definition of zero-knowledge in $\mathsf{IPJ_I}$.

Definition 15 (Evidentially zero-knowledge). *A protocol is* evidentially zero-knowledge *if for all inputs x belonging to $\mathcal{L}$, the probability of the verifier knowing the prover's evidence for x belonging to $\mathcal{L}$ is negligible.*

To address evidentially zero-knowledge protocols, we add the following two axioms to $\mathsf{IPJ_I}$:

1. $t{:}_P \alpha \to \mathcal{P}_{\leq \frac{1}{n^k}}(f_t^n{:}_V t{:}_P \alpha)$ if $n > m$ and $\alpha \in \mathsf{I}_{m,k}$;
2. $t{:}_P \alpha \to \mathcal{P}_{\approx 0}(f_t^\omega{:}_V t{:}_P A)$ if $\alpha \in \mathsf{I}$.

Models for $\mathsf{IPJ_I}$ are adjusted by requiring the condition:

$$M \models t{:}_P \alpha \to \mathcal{P}_{\leq \frac{1}{n^k}}(f_t^n{:}_V t{:}_P \alpha) \text{ if } n > m \text{ and } \alpha \in \mathsf{I}_{m,k}.$$

It is easy to show that this extension is sound with respect to its models. The proof of soundness for the second axiom is similar to the soundness proof of $(\mathsf{c}\omega)$.

5 Conclusion

We presented the probabilistic two-agent justification logic $\mathsf{IPJ_I}$, in which we can reason about agents that perform interactive proofs. The foundation of this work is based on probabilistic justification logic combined with interacting evidence systems. We further proposed a new technique that asserts a countable axiomatization and makes it possible to reason about the growth rate of a probability measure. Intuitively, the set $\mathsf{I} = \bigcap_k \bigcup_m \mathsf{I}_{m,k}$ can be thought of as the set of all formulas that are known to be interactively provable. For a formula $\alpha \in \mathsf{I}_{m,k}$ and a term t with $t{:}_P \alpha$,

$$\mathcal{P}_{\geq 1 - \frac{1}{n^k}}(f_t^n{:}_V \Box_P \alpha)$$

holds for all $n > m$. Hence, if $\alpha \in \mathsf{I}$, then the following first order sentence is true

$$\forall k \exists m \forall (n > m) \mu([f_t^n{:}_V \Box_P \alpha]) \geq 1 - \frac{1}{n^k},$$

which is the definition of an overwhelming function.

Our approach of modelling limits with the help of specification sets is quite versatile as the following example shows.

Example 1. Consider a sequence of the form:

$$\mathcal{P}_{=L+0.5}(f_t^1{:}_V\alpha) \quad \mathcal{P}_{=L+0.25}(f_t^2{:}_V\alpha) \quad \mathcal{P}_{=L+0.125}(f_t^3{:}_V\alpha) \quad \cdots$$

The sentence we want to model is:

$$(\forall \epsilon > 0)(\exists m \geq 0)(\forall n > m)(\mathcal{P}_{\leq L+\epsilon}(f_t^n{:}_V\alpha) \wedge \mathcal{P}_{\geq L-\epsilon}(f_t^n{:}_V\alpha))$$

Again, for $\epsilon, L \in \mathbb{Q}$ and $m \in \mathbb{N}$, we define sets $\mathsf{Conv}_{\epsilon,m}^L$ and let

$$\mathsf{Conv}^L := \bigcap_{\epsilon\in\mathbb{Q}} \bigcup_{m\in\mathbb{N}} \mathsf{Conv}_{\epsilon,m}^L.$$

With the following formulas, we can express that a sequence of probabilities converges:

1. $\mathcal{P}_{\leq L+\epsilon}(f_t^n{:}_V\alpha) \wedge \mathcal{P}_{\geq L-\epsilon}(f_t^n{:}_V\alpha)$ if $n > m$ and $\alpha \in \mathsf{Conv}_{\epsilon,m}^L$;
2. $\mathcal{P}_{\approx L}(f_t^\omega{:}_V\alpha)$ if $\alpha \in \mathsf{Conv}^L$.

Additionally, we showed that a round-based definition of interactive proofs can be addressed in our model, but only for finitely many rounds since our measure is not σ-additive. Further, we also investigated zero-knowledge proofs. As it turns out, IPJ_I cannot model the original definition because we cannot compare justification terms in IPJ_I. However, we introduced the notion of evidentially zero knowledge, which fits nicely in our framework.

Moreover, we established soundness of IPJ_I. Our axiomatization is a combination of systems that are known to be complete and we conjecture that IPJ_I is complete, too.

References

1. Arora, S., Barak, B.: Computational Complexity: A Modern Approach. Cambridge University Press (2009)
2. Artemov, S.: Explicit provability and constructive semantics. Bull. Symb. Log. **7**(1), 1–36 (2001)
3. Artemov, S.: The ontology of justifications in the logical setting. Stud. Logica. **100**(1–2), 17–30 (2012). https://doi.org/10.1007/s11225-012-9387-x
4. Artemov, S., Fitting, M.: Justification Logic: Reasoning with Reasons. Cambridge University Press (2019)
5. Artemov, S., Nogina, E.: Introducing justification into epistemic logic. J. Log. Comput. **15**(6), 1059–1073 (2005). https://doi.org/10.1093/logcom/exi053
6. Babai, L.: Trading group theory for randomness. In: Proceedings of the 17th Annual ACM Symposium on Theory of Computing, STOC 1985, pp. 421–429. Association for Computing Machinery (1985). https://doi.org/10.1145/22145.22192
7. Fitting, M.: The logic of proofs, semantically. Ann. Pure Appl. Logic **132**(1), 1–25 (2005). https://doi.org/10.1016/j.apal.2004.04.009

8. Goldwasser, S., Micali, S., Rackoff, C.: The knowledge complexity of interactive proof-systems. In: Proceedings of the 17th Annual ACM Symposium on Theory of Computing, STOC 1985, pp. 291–304. Association for Computing Machinery (1985). https://doi.org/10.1145/22145.22178
9. Ikodinović, N., Ognjanović, Z., Perović, A., Rašković, M.: Completeness theorems for σ-additive probabilistic semantics. Ann. Pure Appl. Log. **171**(4), 102755 (2020)
10. Kokkinis, I., Maksimović, P., Ognjanović, Z., Studer, T.: First steps towards probabilistic justification logic. Log. J. IGPL **23**(4), 662–687 (2015)
11. Kokkinis, I., Ognjanović, Z., Studer, T.: Probabilistic justification logic. J. Log. Comput. **30**(1), 257–280 (2020)
12. Kuznets, R., Studer, T.: Justifications, ontology, and conservativity. In: Bolander, T., Braüner, T., Ghilardi, S., Moss, L. (eds.) Advances in Modal Logic, vol. 9, pp. 437–458. College Publications (2012)
13. Kuznets, R., Studer, T.: Weak arithmetical interpretations for the logic of proofs. Log. J. IGPL **24**(3), 424–440 (2016). https://doi.org/10.1093/jigpal/jzw002
14. Kuznets, R., Studer, T.: Logics of Proofs and Justifications. College Publications (2019)
15. Ognjanović, Z., Rašković, M.: Some first order probability logics. Theoret. Comput. Sci. **247**, 191–212 (2000)
16. Ognjanović, Z., Rašković, M., Marković, Z.: Probability Logics: Probability-Based Formalization of Uncertain Reasoning. Springer, Cham (2016). https://doi.org/10.1007/978-3-319-47012-2
17. Ognjanović, Z., Savić, N., Studer, T.: Justification logic with approximate conditional probabilities. In: Baltag, A., Seligman, J., Yamada, T. (eds.) LORI 2017. LNCS, vol. 10455, pp. 681–686. Springer, Heidelberg (2017). https://doi.org/10.1007/978-3-662-55665-8_52
18. Rašković, M., Marković, Z., Ognjanović, Z.: A logic with approximate conditional probabilities that can model default reasoning. Int. J. Approx. Reason. **49**(1), 52–66 (2008). https://doi.org/10.1016/j.ijar.2007.08.006
19. Robinson, A.: Non-standard Analysis. Princeton University Press (1996)
20. Yavorskaya (Sidon), T.: Interacting explicit evidence systems. Theor. Comput. Syst. **43**(2), 272–293 (2008). https://doi.org/10.1007/s00224-007-9057-y

Recursive Rules with Aggregation: A Simple Unified Semantics

Yanhong A. Liu(✉) and Scott D. Stoller

Computer Science Department, Stony Brook University, Stony Brook, NY, USA
{liu,stoller}@cs.stonybrook.edu

Abstract. Complex reasoning problems are most clearly and easily specified using logical rules, but require recursive rules with aggregation such as counts and sums for practical applications. Unfortunately, the meaning of such rules has been a significant challenge, leading to many disagreeing semantics.

This paper describes a unified semantics for recursive rules with aggregation, extending the unified founded semantics and constraint semantics for recursive rules with negation. The key idea is to support simple expression of the different assumptions underlying different semantics, and orthogonally interpret aggregation operations using their simple usual meaning. We present formal definition of the semantics, prove important properties of the semantics, and compare with prior semantics. In particular, we present an efficient inference over aggregation that gives precise answers to all examples we have studied from the literature. We also applied our semantics to a wide range of challenging examples, and performed experiments on the most challenging ones, all confirming our analyzed results.

1 Introduction

Many computation problems, including complex reasoning problems in particular, such as program analysis, networking, and decision support, are most clearly and easily specified using logical rules [39]. However, such reasoning problems in practical applications, especially for large applications and when faced with uncertain situations, require the use of recursive rules with aggregation such as counts and sums. Unfortunately, the meaning of such rules has been challenging and remains a subject with significant complication and disagreement.

As a simple example, consider a single rule for Tom to attend the logic seminar: "Tom will attend the logic seminar if the number of people who will attend it is at least 20." What does the rule mean? If 20 or more other people will attend, then surely Tom will attend. If only 10 others will attend, then Tom will not attend. What if only 19 other people will attend? Will Tom attend, or not? Although simple, this example already shows that, when aggregation is

This work was supported in part by NSF under grants CCF-1954837, CCF-1414078, and IIS-1447549 and by ONR under grants N00014-20-1-2751 and N00014-21-1-2719.

S. Artemov and A. Nerode (Eds.): LFCS 2022, LNCS 13137, pp. 156–179, 2022.
https://doi.org/10.1007/978-3-030-93100-1_11

used in recursive rules—here count is used in a rule that defines "will attend" using "will attend"—the semantics can be tricky.

Some might say that this statement about Tom is ambiguous or ill-specified. However, it is a statement allowed by logic rule languages with predicates, sets, and counts. For example, let predicate `will_attend(p)` denote that `p` will attend the logic seminar; then the statement can be written as `will_attend(Tom) if count` $\{$`p: will_attend(p)`$\} \geq$ `20`. So the statement must be given a meaning. Indeed, "ambiguous" is a possible meaning, indicating there are two or more answers, and "ill-specified" is another possible meaning, indicating there is no answer. So which one should it be? Are there other possible meanings?

In deductive databases, to avoid challenging cases of aggregation as well as negation, processing of recursive rules with aggregation is largely limited to monotonic programs, i.e., adding a new fact used in a hypothesis cannot make a conclusion change from true to false. However, note that the rule about Tom attending the logic seminar is actually monotonic: adding `attend(p)` for a new `p` can not make the conclusion change from true to false. So, even restricted deductive databases must give a meaning to this rule. What should it be?

In fact, the many different semantics of recursive rules with aggregation are more complex and trickier than even semantics of recursive rules with negation. The latter was already challenging for over 120 years, going back at least to Russell's paradox, for which self-reference with negation is believed to form vicious circles [35]. Many different semantics, which disagree with each other, have been studied for recursive rules with negation, as summarized in Sect. 6. Two of them, well-founded semantics (WFS) [64,66] and stable model semantics (SMS) [26], became dominant since about 30 years ago.

Semantics of recursive rules with aggregation has been studied continuously since about 30 years ago, and more intensively in more recent years, as discussed in Sect. 6, especially as they are needed in graph analysis and machine learning applications. However, the many different semantics proposed, e.g., [30,63], are even more intricate than WFS and SMS for recursive rules with negation, and include even different extensions for WFS, e.g., [36,59,63], and for SMS, e.g., [30, 36,47]. Some authors also changed their own minds about the desired semantics, e.g., [25,30]. Such intricate and disagreeing semantics would be too challenging to use correctly.

This paper describes a simple unified semantics for recursive rules with aggregation as well as negation and quantification. The semantics is built on and extends the founded semantics and constraint semantics of logical rules with negation and quantification developed recently by Liu and Stoller [41,42]. The key idea is to capture, and to express in a simple way, the different assumptions underlying different semantics, and orthogonally interpret aggregation operations using their simple usual meaning. We present formal definition of the semantics and prove important properties of the semantics. In particular, we present an efficient derivability relation for comparisons containing aggregations; it can be computed in linear time and gives precise answers on all examples we have studied from the literature.

We also compared with main prior semantics for rules with aggregation, and showed how our semantics is direct and follows precisely from usual meanings of aggregations. We further applied our semantics to a wide range of challenging examples, and showed that our semantics is simple and matches the desired semantics in all cases. Additionally, we performed experiments on two most challenging examples, confirming the correctness of our computed results, while also discovering worse performance and some wrong results from well-known systems. Additional results from these comparisons, examples, and experiments are described in [44].

2 Problem and Solution Overview

The semantics of recursion with negation and aggregation is challenging for several reasons. First, recursion involves self-referencing and cyclic reasoning, for which it is already non-trivial to properly start and finish. Then, negation in recursion incurs self-denying and conflict in cyclic reasoning, which can lead to contradiction. Finally, aggregation generalizes negation to give rise to even greater challenges in recursion, because a negation essentially corresponds to only the simple case of a count being zero.

The first reason alone already called for a least fixed point semantics, which is beyond first-order logic. The second reason led to major different semantics that are sophisticated and disagreeing when trying to solve conflicts differently. The third reason exacerbated the sophistication and variety to tackle the even greater challenges.

A Smallest Example. Consider the following recursive rule with aggregation. It says that `p` is true for value `a` if the number of `x`'s for which `p` is true equals `1`:

```
p(a) ← count {x: p(x)} = 1
```

This rule is recursive because inferring a conclusion about `p` requires using `p` in a hypothesis. It uses an aggregation of `count` over a set. While each of recursion and aggregation by itself has a simple meaning, allowing recursion with aggregation is tricky, because recursion is used to define a predicate, which is equivalent to a set, but aggregation using a set requires the set to be already defined.

We use this example in addition to our Tom example in Sect. 1, for two reasons. First, this and similar small examples are used for comparisons in previous papers, e.g., [19,27,29]. Second, this example differs from the Tom example in that the comparison with count in this example is non-monotonic, i.e., adding more `x`'s for which `p(x)` is true can change the value of the comparison, and thus the conclusion, from true to false; using only one example is insufficient to show the main different cases.

- **Two models: Kemp-Stuckey 1991, Gelfond 2002.** According to Kemp and Stuckey [36] and Gelfond [25], the above rule has two models: one empty model, i.e., a model in which nothing is true and thus `p(a)` is false, and one containing only `p(a)` being true.
- **One model: Faber et al. 2011, Gelfond-Zhang 2014–2019.** According to Faber, Pfeifer, and Leone [19] and Gelfond and Zhang [27, Examples 2 and 7], [29, Examples 4 and 6], and [30, Example 9], the rule above has only one model: the empty model.

As one of the several main efforts investigating aggregation, Gelfond and Zhang [25,27–30,73] have studied the challenges and solutions extensively, presenting dozens of definitions and propositions and discussing dozens of examples [30]. Their examples where count is used in inequalities, greater than, etc., with additional variables, with more hypotheses in a rule, or with more rules and facts, are even more complicated.

Extending Founded Semantics and Constraint Semantics for Aggregation. Aggregation, such as count, is a simple concept that even kids understand. So it is stunning to see so many sophisticated treatments for figuring out its meaning when it is used in rules, and to see the many disagreeing semantics resulting from those.

We develop a simple and unified semantics for rules with aggregation as well as negation and quantification by building on founded semantics and constraint semantics [41,42] for rules with negation and quantification. The key insight is that disagreeing complex semantics for rules with aggregation are because of different underlying assumptions, and these assumptions can be captured using the same simple binary declarations about predicates as in founded semantics and constraint semantics but generalized to include the meaning of aggregation.

Certain. First, if there is no potential non-monotonicity, including no aggregation in recursion, then the predicate in the conclusion can be declared "certain".

Being certain means that assertions of the predicate are given true or inferred true by simply following rules whose hypotheses are given or inferred true, and the remaining assertions of the predicate are false. This is both the founded semantics and constraint semantics.

For the Tom example, there is no potential non-monotonicity; with this declaration, when given that only 19 others will attend, the hypothesis of the rule is not true, so the conclusion cannot be inferred. Thus Tom will not attend.

Uncertain. Regardless of monotonicity, a predicate can be declared "uncertain".

It means that assertions of the predicate can be given or inferred true or false using what is given, and any remaining assertions of the predicate are undefined. This is the founded semantics.

If there are undefined assertions from founded semantics, all combinations of true and false values are checked against the rules and declarations as constraints, yielding a set of possible satisfying combinations. This is the constraint semantics.

Complete, or not complete. An uncertain predicate can be further declared "complete" or not.

Being complete means that all rules that can conclude assertions of the predicate are given. Thus a new rule, called completion rule, can be created to infer negative assertions of the predicate when none of the given rules apply.

Being not complete means that negative assertions cannot be inferred using completion rules, and thus all assertions of the predicate that were not inferred to be true are undefined.

For the Tom example, the completion rule implies: Tom will not attend the logic seminar if the number of people who will attend it is less than 20.

When given that only 19 others will attend, due to the uncertainty of whether Tom will attend, neither the given rule nor the completion rule will fire. So whether one uses the declaration of complete or not, there is no way to infer that Tom will attend, or Tom will not attend. So, founded semantics says it is undefined.

Then constraint semantics tries both for it to be true, and for it to be false; both satisfy the rule, so there are two models: one where Tom will attend, and one where Tom will not attend.

Closed, or not closed. Finally, an uncertain complete predicate can be further declared "closed" or not.

Being closed means that an assertion of the predicate is made false if inferring it to be true requires itself to be true.

Being not closed means that such assertions are left undefined.

For the Tom example, with this declaration, if there are only 19 others attending, then Tom will not attend in both founded semantics and constraint semantics. This is because inferring that Tom will attend requires Tom himself to attend to make the count to be 20, so it should be made false, meaning that Tom will not attend.

Note that this is the same result as using "certain". Because the rule for deciding whether Tom will attend has no potential non-monotonicity, using "certain" is much simpler and has the same meaning as using "closed", as stated in general in Sect. 4.5.

For the smallest example about `p` near the beginning of this section, the equality comparison is not monotonic. Thus `p` must be declared uncertain. This example also shows different semantics when using the declarations of not complete and complete, unlike the Tom example.

- **Not complete.** Suppose `p` is declared not complete. Founded semantics does not infer `p(a)` to be true using the given rule because `count {x: p(x)} = 1` cannot be determined to be true, and nothing infers `p(a)` to be false. Thus `p(a)` is undefined. So is `p(b)` for any constant `b` other than `a` because nothing infers `p(b)` to be true or false. Constraint semantics gives a set of models, each for a different combination of true and false values of `p(c)` for different constants `c` such that the combination satisfies the given rule. This corresponds to what is often called open-world assumption and used in commonsense reasoning.

- **Complete.** Suppose `p` is declared complete but not closed. A completion rule is first added. The precise completion rule is:

  $\neg$ `p(x)` $\leftarrow$ `x` $\neq$ `a` $\vee$ `count` {`x: p(x)`} $\neq$ `1`

  Founded semantics does not infer `p(a)` to be true or false using the given rule or completion rule, because `count` {`x: p(x)`} $\neq$ `1` also cannot be determined to be true. Thus `p(a)` is undefined. Founded semantics infers `p(b)` for any constant `b` other than `a` to be false using the completion rule. Constraint semantics gives two models: one with `p(a)` being true, and `p(b)` being false for any constant `b` other than `a`; and one with `p(c)` being false for every constant `c`. This is the same as the two-model semantics per Kemp-Stuckey 1991 and Gelfond 2002.
- **Closed.** Supposed `p` is declared complete and closed. Both founded semantics and constraint semantics give only the second model above, i.e., `p(c)` is false for every constant `c`. They have `p(a)` being false because inferring `p(a)` to be true requires `p(a)` itself to be true. This is the same as the one-model semantics per Faber et al. 2011 and Gelfond-Zhang 2014–2019.

We see that simple binary declarations of the underlying assumptions, with simple inference following rules and taking rules as constraints, give the different desired semantics.

Relationship with Prior Semantics. Table 1 summarizes relationships between our unifying semantics and major prior semantics. With different predicate declarations capturing different underlying assumptions, founded semantics and constraint semantics for rules with aggregation extend different prior semantics for rules with negation uniformly, as shown in Table 1 left and middle columns. These extend the matching relationships proved for rules with negation in [41, 42]. All these relationships are when all predicates in a program have the same declarations, but our founded semantics and constraint semantics also allow different predicates to have different declarations.

Among many different prior semantics for rules with aggregations, there are even different extensions for the same prior semantics for rules with negation, as shown in the right column in Table 1. Unfortunately, most of them are defined for limited cases, or add some case-specific definitions. In particular, simple formal explanations for the disagreements, including among all different extensions for each of WFS and SMS, are completely missing. We are only aware of comparisons by examples or very restricted cases, even for disagreeing semantics by the same authors. However, for all such examples and cases we examined, we found that the desired results for them correspond to our semantics under some appropriate declarations for some predicates. These results are described in [44].

Table 1. Founded semantics and constraint semantics for rules with aggregation with different declarations (for all predicates in a program), extending prior semantics for rules with negation, and prior extensions.

Declarations	Semantics	Extending	Reference	Prior extensions
Certain	Founded, Constraint	Stratified (Perfect)	Van Gelder 1986 [62]	e.g., [49,53]
Uncertain, not complete	Founded	(None found)		(None found)
	Constraint	First-Order Logic		e.g., [33]
Uncertain, complete, not closed	Founded	Fitting (Kripke-Kleene)	Fitting 1985 [22]	Pelov et al. 2007 [50]
	Constraint	Supported	Apt et al. 1988 [7]	Pelov et al. 2007 [50]
Uncertain, complete, closed	Founded	WFS	Van Gelder et al. 1988 [65, 66]	Kemp-Stuckey 1991 [36]
				Van Gelder 1992 [63]
				Pelov et al. 2007 [50]
	Constraint	SMS	Gelfond-Lifschitz 1988 [26]	Kemp-Stuckey 1991 [36]
				Pelov et al. 2007 [50]
				Faber et al. 2011 [19]
				Gelfond-Zhang 2014–2019 [30]

3 Language

We consider Datalog rules extended with unrestricted negation, disjunction, quantification, aggregation, and comparison containing aggregation.

Domain. The *domain* of a program is the set of values that variables can be instantiated with. These values are called *constants*. The domain includes the values that appear in the program and a set *Num* of numbers. *Num* is a bounded range of numbers determined by a *numeric representation bound NRB* and a *numeric representation precision NRP*, i.e., *Num* contains all numbers in the range $[-NRB, NRB]$ with at most *NRP* decimal places. Numbers with more than *NRP* decimal places that appear in the program or arise during evaluation can be rounded to *NRP* decimal places, or a higher-precision representation can be used.

This rounding or increasing precision is not shown explicitly in the semantics, because the rule language in this paper does not include numeric operations that increase the number of decimal places. We use an *NRP* that is at least the maximum number of decimal places in numbers that appear in the program, so all numeric computations are exact. Our semantics detects and can report cases where an inference is blocked because it involves a value outside the range $[-NRB, NRB]$; for details, see the description of range-blocked inference in [44].

Datalog Rules with Unrestricted Negation. We first present a simple core form of rules and then describe additional constructs that can appear in rules. The *core form* of a rule is the following, where any P_i may be preceded with $\neg$:

$$Q(X_1, ..., X_a) \leftarrow P_1(X_{11}, ..., X_{1a_1}) \land ... \land P_h(X_{h1}, ..., X_{ha_h})$$

Q and P_i's are predicates, and each argument X_k and X_{ij} is a constant or a variable. In arguments of predicates in examples, we use numbers and quoted strings for constants and letters for variables.

If $h = 0$, there are no P_i's or X_{ij}'s, and each X_k must be a constant, in which case $Q(X_1, ..., X_a)$ is called a *fact*. For the rest of the paper, "rule" refers only to the case where $h \geq 1$, in which case the left side of $\leftarrow$ is called the *conclusion*, the right side is called the *body*, and each conjunct in the body is called a *hypothesis*. Note that we do not require variables in the conclusion to be in the hypotheses; it is not needed because rules are used with variables replaced by constants, and the domain of variables is finite.

Disjunction. In a rule body, hypotheses may be combined using disjunction as well as conjunction. Conjunction and disjunction may be nested arbitrarily.

Quantification. A hypothesis in a rule body can be an existential or universal quantification of the form

$$\exists\ X_1, ..., X_a \mid B \quad \text{existential quantification}$$
$$\forall\ X_1, ..., X_a \mid B \quad \text{universal quantification}$$

where each X_i is a variable that appears in B, and B has the same form as a rule body. Note that this recursive definition allows nested quantifications. Each quantified variable X_i ranges over the domain of the program. The quantifications return true iff for some or all, respectively, combinations of values of $X_1, ..., X_a$, the body B is true.

Aggregation and Comparison. A *set expression* has the form $\{X_1, ..., X_a : B\}$, where each X_i is a variable in B, and the body B has the same form as a rule body. The *arity* of this set expression is a. Each set expression body is first rewritten to have the same form as the body of a core-form rule, by introducing auxiliary predicates, e.g., `max` $\{$`Y:` $\exists$ `X | p(X,Y)`$\}$ $>$ `0` is rewritten to `max` $\{$`Y: q(Y)`$\}$ $>$ `0` together with `q(Y)` $\leftarrow$ $\exists$ `X | p(X,Y)`. Each auxiliary predicate has default declarations, except that it is declared closed if some predicate in the body of the rule defining the auxiliary predicate is declared closed.

An *aggregation* has the form *agg* S, where *agg* is an aggregation operator (`count`, `max`, `min`, or `sum`), and S is a set expression. The aggregation returns the result of applying the respective *agg* operation (cardinality, maximum, minimum, or sum) to the set value of S. `max` and `min` use the order on numbers, extended lexicographically to an order on tuples. `sum` is on numbers, and on tuples whose first components are numbers; in the latter case, the first components are summed. Note that `count` and `sum` applied to the empty set equal 0, while `max` and `min` applied to the empty set give an error.

A hypothesis of a rule may be a *comparison* of the form

$$agg\ S \odot k \qquad \text{or} \qquad agg\ S \odot agg'\ S'$$

where *agg* S and *agg'* S' are aggregations, the comparison operator $\odot$ is an equality ($=$) or inequality ($\neq, <\leq, >, \geq$), and k is a variable or numeric constant or, if the aggregation operator is `max` or `min`, a tuple of variables or numeric constants. Comparisons of the second form are first rewritten as two comparisons of the first form by introducing a fresh variable. For example, $agg\, S \neq agg'\, S'$ is rewritten as $agg\, S \neq V \wedge agg'\, S' = V$, and $agg\, S < agg'\, S'$ is rewritten as $agg\, S < V \wedge agg'\, S' \geq V$, where V is a fresh variable. The latter rewrite uses two inequalities, instead of an inequality and an equality, to increase the cases where occurrences of predicate atoms are positive (defined below).

Note that negation applied to comparisons can be eliminated by reversing the comparison operators; for example, the negation of a comparison using $\leq$ is a comparison using $>$.

The key idea here is that the value of a comparison (containing an aggregation) is undefined if there is not enough information about the predicates used to determine the value, or if applying the comparison (containing an aggregation) gives an error, such as a type error. Our principled approach can easily support additional aggregation and comparison functions, e.g., on other data types such as strings.

Programs, Atoms, and Literals. A *program* π is a set of rules and facts, plus declarations for predicates, described after dependencies are introduced next.

An *atom* of π is either a predicate symbol in π applied to constants in the domain of π and variables, or a comparison formed using predicate symbols in π, constants in the domain of π, and variables. These are called *predicate atoms* for P and *comparison atoms*, respectively.

A *literal* of π is either an atom of π or the negation of a predicate atom of π. These are called *positive literals* and *negative literals*, respectively. A literal containing a predicate atom or comparison atom is called a *predicate literal* or *comparison literal*, respectively. Note that negation of a comparison atom is not needed because the negation will be eliminated by reversing the comparison operator.

Dependency Graph. The dependency graph of a program characterizes dependencies between predicates induced by the rules, distinguishing positive from non-positive dependencies.

An occurrence A of a predicate atom in a hypothesis H is a *positive occurrence* if (1) H is A, which is a positive literal, (2) H is a quantification, and A is a positive literal in its body, (3) H is a comparison atom of the form $\texttt{count}\ S \geq k$, $\texttt{count}\ S > k$, $\texttt{max}\ S \geq k$, $\texttt{max}\ S > k$, $\texttt{min}\ S \leq k$, or $\texttt{min}\ S < k$, and A is in a positive literal in the set expression S, or (4) H is a comparison atom of the form $\texttt{count}\ S \leq k$, $\texttt{count}\ S < k$, $\texttt{max}\ S \leq k$, $\texttt{max}\ S < k$, $\texttt{min}\ S \geq k$, or $\texttt{min}\ S > k$, and A is in a negative literal in the set expression S. Otherwise, the occurrence is a *non-positive occurrence*.

This definition conservatively ensures that hypotheses are monotonic with respect to positive occurrences of predicate atoms, i.e., making a positive occurrence of a predicate atom in a hypothesis true cannot make the hypothesis change from true. This definition can be extended so that any occurrence A of a predicate atom in a hypothesis H is a *positive occurrence* if H can be determined to be monotonic with respective to A. For example, if predicate `p` holds for only non-negative numbers, then `p(x)` is a positive occurrence in `sum {x: p(x)} > k`.

The *dependency graph* $DG(\pi)$ of program π is a directed graph with a node for each predicate of π, and an edge from Q to P labeled positive (respectively, non-positive) if a rule whose conclusion contains Q has a hypothesis that contains a positive (respectively, non-positive) occurrence of an atom for P. If there is a path from Q to P in $DG(\pi)$, then Q *depends on* P in π. If the node for P is in a cycle containing a non-positive edge in $DG(\pi)$, then P has *circular non-positive dependency* in π.

Declarations. A predicate declared *certain* means that each assertion of the predicate has a unique true (*True*) or false (*False*) value. A predicate declared *uncertain* means that each assertion of the predicate has a unique true, false, or undefined (*Undef*) value. A predicate declared *complete* means that all rules with that predicate in the conclusion are given in the program. A predicate declared *closed* means that an assertion of the predicate is set to false, called *self-false*, if inferring it to be true using the given rules and facts requires assuming itself to be true.

A predicate must be declared uncertain if it has circular non-positive dependency, or depends on an uncertain predicate; otherwise, it may be declared certain or uncertain and is by default certain. A predicate may be declared complete or not only if it is uncertain, and it is by default complete. A predicate may be declared closed or not only if it is uncertain and complete, and it is by default not closed.

We do not give a syntax for predicate declarations, because it is straightforward, and most examples use default declarations. However, the language in [43,45] supports such declarations.

Notations. In presenting the semantics, in particular the completion rules, we allow negation in the conclusion of rules, and we allow hypotheses to be equalities ($=$) and negated equalities ($\neq$) between two variables or a variable and a constant.

4 Formal Semantics

This section extends the definitions of founded semantics and constraint semantics in [41,42] to handle aggregation and comparison. We introduce a new relation, namely, derivability of comparisons, and extend most of the foundational definitions, including the definitions of atom, literal, and positive occurrence in Sect. 3, and of complement, ground instance, truth value of a literal in an interpretation, completion rule, naming negation, unfounded set, and constraint model in this section. By carefully extending these foundational definitions, we are able to avoid explicit changes to the definitions of other terms and functions built on them, including the definition of completion and the definition of the least fixed point at the heart of the semantics, embodied mainly in the function *LFPbySCC*.

4.1 Interpretations and Derivability

Complements and Consistency. The predicate literals A and $\neg A$ are *complements* of each other. The following pairs of comparison literals are complements of each other: $agg\,S = k$ and $agg\,S \neq k$; $agg\,S \leq k$ and $agg\,S > k$; $agg\,S \geq k$ and $agg\,S < k$.

A set of predicate literals is *consistent* if it does not contain a literal and its complement.

Ground Instance. An occurrence of a variable X in a quantification Q is *bound* in Q if X is a variable to the left of the vertical bar in Q. An occurrence of a variable X in a set expression S is *bound* if X is a variable to the left of the colon in S. An occurrence of a variable in a rule R is *free* if it is not bound in a quantification or set expression in R.

A *ground atom* or *ground literal* is an atom or literal, respectively, not containing variables. A *ground instance* of a rule R in a program π is any rule obtained from R by expanding universal quantifications into conjunctions over all constants, instantiating existential quantifications with any constants, and instantiating the remaining free occurrences of variables with any constants (of course, all free occurrences of the same variable are replaced with the same constant). A *ground instance* of a comparison atom A is a comparison atom obtained from A by instantiating the free occurrences of variables in A with any constants. A *ground instance* of a set expression $\{X_1, ..., X_a : B\}$ is a pair $((X_1, ..., X_a), B)$ obtained by instantiating all variables in $X_1, ..., X_a$ and B with any constants.

Interpretations. An *interpretation* of a program π is a consistent set of ground predicate literals of π. Interpretations are generally 3-valued: a ground predicate literal is *true* (i.e., has truth value *True*) in interpretation I if it is in I, is *false* (i.e., has truth value *False*) in I if its complement is in I, and is *undefined* (i.e., has truth value *Undef*) in I if neither it nor its complement is in I. An interpretation of π is *2-valued* if it contains, for each ground predicate atom A of π, either A or its complement. Interpretations are ordered by set inclusion $\subseteq$.

Let $G(S)$ denote the set of ground instances of set expression S. For a set expression S, interpretation I, and truth value t, let

$$G(S, I, t) = \{x \mid (x, B) \in G(S) \wedge B \text{ has truth value } t \text{ in } I\}$$

That is, $G(S, I, t)$ is the set of combinations of constants for which the body of set expression S has truth value t in I.

Derivability of Comparisons. Informally, a ground comparison atom $agg\,S \odot k$ is *derivable* in interpretation I of π, denoted $\pi, I \vdash agg\,S \odot k$, if the comparison must be true in I, regardless of whether atoms with truth value *Undef* are true or false.

Precisely, founded semantics uses the *linear-time derivability relation* $\vdash_{\mathrm{L}}$ defined in Fig. 1 based on the aggregation operator and the comparison operator. It can be computed straightforwardly in linear time in $|G(S, I, \mathit{True})| + |G(S, I, \mathit{Undef})|$.

Derivability for each comparison in Fig. 1 has also a condition that the comparison does not give an error. It gives an error if the aggregation gives an error, or if there is a type error, i.e., either the aggregation is `count` or `sum`, or is `max` or `min` with arity of S being 1, and k is not a number, or the aggregation is `max`

$$\pi, I \vdash_{\mathrm{L}} \mathtt{count}\, S = k \Leftrightarrow |G(S,I,\mathit{True})| = k \wedge G(S,I,\mathit{Undef}) = \emptyset$$
$$\pi, I \vdash_{\mathrm{L}} \mathtt{count}\, S > k \Leftrightarrow |G(S,I,\mathit{True})| > k$$
$$\pi, I \vdash_{\mathrm{L}} \mathtt{count}\, S < k \Leftrightarrow |G(S,I,\mathit{True}) \cup G(S,I,\mathit{Undef})| < k$$
$$\pi, I \vdash_{\mathrm{L}} \mathtt{max}\, S = k \Leftrightarrow k \in G(S,I,\mathit{True}) \wedge \forall\, i \in G(S,I,\mathit{True}) \cup G(S,I,\mathit{Undef}) \mid i \leq k$$
$$\pi, I \vdash_{\mathrm{L}} \mathtt{max}\, S \neq k \Leftrightarrow k \notin G(S,I,\mathit{True}) \cup G(S,I,\mathit{Undef}) \vee \exists\, i \in G(S,I,\mathit{True}) \mid i > k$$
$$\pi, I \vdash_{\mathrm{L}} \mathtt{max}\, S > k \Leftrightarrow \exists\, i \in G(S,I,\mathit{True}) \mid i > k$$
$$\pi, I \vdash_{\mathrm{L}} \mathtt{max}\, S < k \Leftrightarrow \exists\, i \in G(S,I,\mathit{True}) \wedge \forall\, i \in G(S,I,\mathit{True}) \cup G(S,I,\mathit{Undef}) \mid i < k$$
$$\pi, I \vdash_{\mathrm{L}} \mathtt{sum}\, S = k \Leftrightarrow \mathrm{sum}\ G(S,I,\mathit{True}) = k \wedge \{\mathrm{first}(i) : i \in G(S,I,\mathit{Undef})\} \subseteq \{0\}$$
$$\pi, I \vdash_{\mathrm{L}} \mathtt{sum}\, S > k \Leftrightarrow \mathrm{sum}\ (G(S,I,\mathit{True}) \cup \{i \in G(S,I,\mathit{Undef}) : \mathrm{first}(i) < 0\}) > k$$
$$\pi, I \vdash_{\mathrm{L}} \mathtt{sum}\, S < k \Leftrightarrow \mathrm{sum}\ (G(S,I,\mathit{True}) \cup \{i \in G(S,I,\mathit{Undef}) : \mathrm{first}(i) > 0\}) < k$$

Fig. 1. Linear-time derivability relation for comparisons. first(i) returns the first component of i if i is a tuple, and returns i otherwise. Biconditionals ($\Leftrightarrow$) for derivability of other comparisons are obtained from those given as follows. (1) Biconditionals for deriving comparisons using `min` are obtained from those for `max` by replacing `max` with `min`, interchanging $\leq$ and $\geq$, and interchanging $<$ and $>$. (2) For aggregation operator *agg* being `count` or `sum`, the right side of the biconditional for deriving $agg\ S \neq k$ is the disjunction of the right sides of the biconditionals for deriving $agg\ S > k$ and $agg\ S < k$. (3) For each aggregation operator *agg*, biconditionals for deriving $agg\ S \geq k$ and $agg\ S \leq k$ are obtained from the given biconditionals for $agg\ S > k$ and $agg\ S < k$, respectively, by replacing $> k$ with $\geq k$ and replacing $< k$ with $\leq k$.

or `min` with arity a of S greater than 1, and k is not an a-tuple of numbers. The aggregation gives an error if it is `max` or `min` and $G(S, I, \mathit{True}) \cup G(S, I, \mathit{Undef})$ is empty, or if there is a type error, i.e., either it is `max` or `min` and $G(S, I, \mathit{True})$ or $G(S, I, \mathit{Undef})$ contains either a non-number or a tuple containing a non-number, or it is `sum` and S has arity 1 and $G(S, I, \mathit{True})$ or $G(S, I, \mathit{Undef})$ contains a non-number, or it is `sum` and S has arity greater than 1 and $G(S, I, \mathit{True})$ or $G(S, I, \mathit{Undef})$ contains a tuple whose first component is not a number. Comparisons that give errors can easily be detected and reported by checking these conditions.

This definition of derivability is relatively strict about errors, for example, it always makes a comparison give an error if the aggregation in it gives an error. One can be less strict about errors, for example, a comparison containing `max` or `min` applied to the empty set and using negated equality could be allowed to hold even if the aggregation in it gives an error, taking the view that an error is not equal to a value or a tuple of values in the domain. This generally yields more literals that are true or false, rather than undefined. Choices for error handling could also be specified using declarations.

An alternative to linear-time derivability is *exact derivability*, denoted $\vdash_{\mathrm{E}}$. Informally, $\pi, I \vdash_{\mathrm{E}} agg\ S \odot k$ holds iff (1) $agg\ S \odot k$ holds in all 2-valued interpretations I' that extend I and satisfy the part of π that S depends on, and (2) there is at least one such interpretation I'. Exact derivability is based on enumeration of interpretations and hence is less appropriate for founded semantics, which is designed to leave such enumeration for constraint semantics. Although exact derivability can be more precise in principle, linear-time derivability gives the same result as exact derivability for all examples we found in the literature.

Interpretations provide truth values for comparison literals similarly as for predicate literals. Let $DC(\pi, I)$ be the set of comparisons derivable for program π and interpretation I. A comparison literal A for π is *true* in I if it is in $DC(\pi, I)$, is *false* in I if its complement is in $DC(\pi, I)$, and is *undefined* in I otherwise.

Models. An interpretation I of a program π is a *model* of π if it (1) contains all facts in π, and (2) satisfies all rules of π, interpreted as formulas in 3-valued logic [22] (i.e., for each ground instance of each rule, if the body is true in I, then so is the conclusion).

One-Step Derivability The *one-step derivability* function T_π for program π performs one step of inference using rules of π. Formally, $A \in T_\pi(I)$ iff (1) A is a fact of π, or (2) there is a ground instance R of a rule of π with conclusion A such that the body of R is true in interpretation I.

4.2 Founded Semantics Without Closed Declarations

We first define a version of founded semantics, denoted $Founded_0$, that ignores declarations that predicates are closed. We then extend the definition to handle those declarations. Intuitively, the *founded model* of a program π ignoring closed-predicate declarations, denoted $Founded_0(\pi)$, is the least set of literals that are given as facts or can be inferred by repeatedly applying the rules. Formally, we define

$$Founded_0(\pi) = UnNameNeg(LFPbySCC(NameNeg(Cmpl(\pi)))),$$

where functions *Cmpl*, *NameNeg*, *LFPbySCC*, and *UnNameNeg* are defined as follows.

Completion. The completion function $Cmpl(\pi)$ returns the *completed program* of π. Formally, $Cmpl(\pi) = AddInv(Combine(\pi))$, where *Combine* and *AddInv* are defined as follows.

The function $Combine(\pi)$ returns the program obtained from π by replacing the facts and rules defining each uncertain complete predicate Q with a single *combined rule* for Q, defined as follows. First, transform the facts and rules defining Q so they all have the same conclusion $Q(V_1, ..., V_a)$, by replacing each fact or rule $Q(X_1, ..., X_a) \leftarrow B$ with

$$Q(V_1, ..., V_a) \leftarrow (\exists\, Y_1, ..., Y_k \mid V_1 = X_1 \wedge ... \wedge V_a = X_a \wedge B)$$

where $V_1, ..., V_a$ are fresh variables (i.e., not occurring in any given rule defining Q), and $Y_1, ..., Y_k$ are all variables occurring free in the original rule $Q(X_1, ..., X_a) \leftarrow B$. Then, combine the resulting rules for Q into a single rule defining Q whose body is the disjunction of the bodies of those rules. This combined rule for Q is logically equivalent to the original facts and rules for Q.

The function *AddInv*(π) returns the program obtained from π by adding, for each uncertain complete predicate Q, a *completion rule* that derives negative literals for Q. The completion rule for Q is obtained from the inverse of the combined rule defining Q (recall that the inverse of $A \leftarrow B$ is $\neg A \leftarrow \neg B$), by (1) putting the body of the rule in negation normal form, i.e., using laws of predicate logic to move negation inwards and eliminate double negations, and (2) eliminate negation applied to comparison atoms by reversing the comparison operators. As a result, in completion rules, negation is applied only to predicate atoms.

Similar completion rules but without aggregation are used in Clark's completion [14] and Fitting semantics [22].

Least Fixed Point. The least fixed point is preceded and followed by functions that introduce and remove, respectively, new predicates representing the negations of the original predicates.

The function *NameNeg*(π) returns the program obtained from π by replacing, except where $P(X_1,...,X_a)$ is a positive occurrence, $\neg P(X_1,...,X_a)$ with $\texttt{n.}P(X_1,...,X_a)$ and $P(X_1,...,X_a)$ not in $\neg P(X_1,...,X_a)$ with $\neg\,\texttt{n.}P(X_1,...,X_a)$. The new predicate $\texttt{n.}P$ represents the negation of predicate P. Since $P(X_1,...,X_a)$ and $\neg P(X_1,...,X_a)$ are complements of each other, we now also define $P(X_1,...,X_a)$ and $\texttt{n.}P(X_1,...,X_a)$ to be complements of each other.

Note that $\texttt{n.}P(X_1,...,X_a)$ is introduced to make the one-step derivability function explicitly monotonic, while maintaining consistency. We replace $\neg P(X_1,...,X_a)$ for any conclusion and any negative occurrence of $P(X_1,...,X_a)$ (where negative occurrence is defined symmetrically as positive occurrence) to allow negative conclusions to be derived and used as facts. We replace any negative occurrence of $P(X_1,...,X_a)$ not in $\neg P(X_1,...,X_a)$ with $\neg\,\texttt{n.}P(X_1,...,X_a)$ also to use these facts. Other occurrences, if any due to positive (and negative) occurrence being conservative, can be either replaced or left, with the result still being a model, because all derivation and use of $\texttt{n.}P(X_1,...,X_a)$ and $P(X_1,...,X_a)$ follow the one-step derivability. We have not seen any example that needs this, but one might obtain a more precise model, i.e., more atoms that are true or false, by trying all combinations of replacing and leaving. It is an open question whether some combination leads to a unique most precise model.

The function *LFPbySCC*(π) uses a least fixed point to infer facts for each strongly connected component (SCC) in the dependency graph of π, as follows. Let $C_1,...,C_n$ be a list of the SCCs in dependency order, so earlier SCCs do not depend on later ones; it is easy to show that any linearization of the dependency order leads to the same result for *LFPbySCC*. The *projection* of a program π onto an SCC C, denoted *Proj*(π, C), contains all facts of π whose predicates are in C and all rules of π whose conclusions contain predicates in C.

Define *LFPbySCC*(π) $=$ I_n, where I_0 $=$ $\emptyset$ and I_i $=$ *AddNeg*(*LFP*($T_{Proj(\pi,C_i)\cup I_{i-1}}$), C_i) for $i \in 1..n$. *LFP* is the least fixed point operator. *AddNeg*(I, C) returns the interpretation obtained from interpretation I by adding *completion facts* for the certain predicates in C to I; specifically, for each certain predicate P in C, for each combination of values $v_1,...,v_a$ of arguments of P, if

I does not contain $P(v_1, ..., v_a)$, then add $\mathtt{n}.P(v_1, ..., v_a)$. The least fixed point is well-defined, because the one-step derivability function $T_{Proj(\pi,C_i)\cup I_{i-1}}$ is monotonic with respect to $\subseteq$, i.e., for all interpretations J and J', $T_{Proj(\pi,C_i)\cup I_{i-1}}(J) \subseteq T_{Proj(\pi,C_i)\cup I_{i-1}}(J')$ whenever $J \subseteq J'$; the proof is straightforward [44].

The function $\mathit{UnNameNeg}(I)$ returns the interpretation obtained from interpretation I by replacing each atom $\mathtt{n}.P(X_1, ..., X_a)$ with $\neg P(X_1, ..., X_a)$.

4.3 Founded Semantics with Closed Declarations

Informally, when an uncertain complete predicate is declared closed, an atom A of the predicate is false in an interpretation I for a program π, called *self-false* in I, if every ground instance of a rule that concludes A has a hypothesis that is false in I or, recursively, is self-false in I. To simplify the formalization, we first transform ground instances of rules to eliminate disjunction, by putting the body of each ground instance R of a rule into disjunctive normal form (DNF) and then replacing R with multiple rules, one per disjunct of the DNF.

A set U of ground predicate atoms for closed predicates is an *unfounded set* of π with respect to an interpretation I of π iff U is disjoint from I and, for each atom A in U, for each ground instance R of a rule of π with conclusion A,

(1) some hypothesis of R is false in I,
(2) some positive predicate hypothesis of R is in U, or
(3) some comparison hypothesis H of R is false when all atoms in U are false, i.e., $\pi, I \cup \neg \cdot U \vdash_{\mathrm{L}} \neg H$,

where, for a set S of positive literals, $\neg \cdot S = \{\neg P(c_1, ..., c_a) \mid P(c_1, ..., c_a) \in S\}$, called the *element-wise negation* of S, and where $\neg H$ is implicitly simplified to eliminate negation applied to H by changing the comparison operator in H.

Note that this definition differs from the standard definition of unfounded set [66] in that we restricted the unfounded set to atoms for closed predicates, added clause (3), and added the disjointness condition. Because a comparison hypothesis depends non-conjunctively on the truth values of multiple literals for predicates used in the aggregation, and these literals may be spread across I and U, clause (3) checks whether H is false when all atoms in U are set to false in I. The explicit disjointness condition is not needed in WFS or founded semantics without aggregation, because one can prove in those settings that unfounded sets are disjoint from interpretations that arise in the semantics (e.g., see [66, Lemma 3.4]). The disjointness condition is needed here to ensure that the interpretation $I \cup \neg \cdot U$ in clause (3) is consistent and hence the meaning of the clause is well-defined.

The definition of unfounded set U ensures that extending I to make all atoms in U false is consistent with π, in the sense that no atom in U can be inferred to be true in the extended interpretation. We define $\mathit{SelfFalse}_\pi(I)$, the set of *self-false atoms* of π with respect to interpretation I, to be the greatest unfounded set of π with respect to I. Note that this set is empty when no predicate is declared closed.

The founded semantics is defined by repeatedly computing the semantics given by $Founded_0$ (founded semantics without closed declarations) and then setting self-false atoms to false, until a least fixed point is reached. Formally, the founded semantics is $Founded(\pi) = LFP(F_\pi)$, where $F_\pi(I) = Founded_0(\pi \cup I) \cup \neg \cdot SelfFalse_\pi(Founded_0(\pi \cup I))$.

4.4 Constraint Semantics

Constraint semantics is a set of 2-valued models based on founded semantics. A *constraint model* M of a program π is a 2-valued interpretation of π such that (1) $Founded(\pi) \subseteq M$, (2) M is a model of $Cmpl(\pi)$, and (3) if there are closed predicates, there is no non-empty subset S of $M \setminus Founded(\pi)$ such that S contains only positive literals for closed predicates and $S = SelfFalse_\pi(M \setminus S)$. Intuitively, condition (3) says that M should not contain a set S of positive literals for closed predicates that are not required to be true by the founded semantics and can be set to false.

We also require that an interpretation that leads to an error in a comparison is not a constraint model. Precisely, we require that for interpretation M to be a constraint model, no ground instance of a rule of π contains a comparison that gives an error in M. Errors are defined the same as in Sect. 4.1, but note that $G(S, I, Undef)$ is empty here. This definition of constraint models could be made less strict about errors.

Note that condition (3) differs from the corresponding condition in constraint semantics without aggregation [41,42], which is $\neg \cdot SelfFalse(M) \subseteq M$. The change is needed because of the new disjointness condition for unfounded sets. With the new disjointness condition, for any 2-valued interpretation M, $SelfFalse(M)$ must be empty, and hence $\neg \cdot SelfFalse(M) \subseteq M$ is vacuously true.

We define $Constraint(\pi)$ to be the set of constraint models of π. Constraint models can be computed by iterating over interpretations M that are supersets of $Founded(\pi)$, satisfying condition (1), and then checking whether the other conditions in the definition of constraint model are satisfied.

4.5 Properties of the Semantics

We briefly state several important properties of the semantics; detailed statements and proofs are in [44]. (1) *Consistency:* The founded model and constraint models of a program π are consistent. (2) *Correctness:* The founded model of a program π is a model of π and $Cmpl(\pi)$. The constraint models of π are 2-valued models of π and $Cmpl(\pi)$. (3) *Same SCC, same certainty:* All predicates in an SCC have the same certainty. (4) *Higher-order programming:* Founded semantics and constraint semantics are preserved by a transformation that facilitates higher-order programming by replacing a set S of compatible predicates with a single predicate `holds` whose first argument is the name of one of those predicates. (5) *Equivalent declarations:* Changing predicate declarations from uncertain, complete, and closed to certain when allowed, or vice versa, preserves founded and constraint semantics.

5 Examples: Company Control and Double Win

We discuss the well-known challenging company control problem [13,19,30,54] and an even more challenging game problem that generalizes the well-known win-not-win game [41,42].

5.1 Company Control—A Well-Known Challenge

This is Examples 1.1 and 2.13 in [19] and is also used in Example 12 in [30]. The problem was also discussed repeatedly before [36,49,50,54,63] and earlier [13]. It considers a set of facts of the form `company(c)`, denoting that `c` is a company, and a set of facts of the form `ownsStk(c1,c2,p)`, denoting the percentage `p` of shares of company `c2` that are owned by company `c1`. It defines that company `c1` controls company `c2`, denoted `controls(c1,c2)`, if the sum of the percentages of shares of `c2` that are owned either directly by `c1` or by companies controlled by `c1` is more than 50.

```
controlsStk(c1,c1,c2,p) ← ownsStk(c1,c2,p)
controlsStk(c1,c2,c3,p) ← company(c1)
        ∧ controls(c1,c2) ∧ ownsStk(c2,c3,p)
controls(c1,c3) ← company(c1) ∧ company(c3)
        ∧ sum {p,c2: controlsStk(c1,c2,c3,p)} > 50
```

It introduces `controlsStk(c1,c2,c3,p)`, denoting that company `c1` controls `p` percent of shares of company `c3` through company `c2`. It has become a most well-known challenging example for recursion with aggregation, because it involves aggregation in mutual recursion.

Founded semantics and constraint semantics are straightforward to compute. First, `company` and `ownsStk` as given are certain. Then, `controlsStk` and `controls` are certain by default, despite that `controlsStk` and `controls` are mutually recursive while involving aggregation, because `controlsStk(c1,c2,c3,p)` holds for only non-negative `p`, making the dependency through the comparison positive. Therefore, the semantics is simply a least fixed point using the given rules, giving the same result for founded semantics and constraint semantics. This is the desired result, same as in [19].

5.2 Double-Win Game—For Any Kind of Moves

Consider the following game, which we call the double-win game. Given a set of moves, the game uses the following single rule, called double-win rule, for winning:

```
dwin(x) ← count {y: move(x,y) ∧ ¬ dwin(y)} ≥ 2
```

It says that `x` is a winning position if the set of positions, `y`, such that there is a move from `x` to `y` and `y` is not a winning position, has at least two elements.

That is, x is a winning position if there are at least two positions to move to from x that are not winning positions.

We created the double-win game by generalizing the well-known win-not-win game [41,42], which has a single rule, stating that x is a winning position if there is a move from x to some position y and y is not a winning position:

```
win(x) ← move(x,y) ∧ ¬ win(y)
```

One could also rewrite the double-win rule using two explicit positions y1 and y2 and adding y1!=y2, but this approach does not scale when the count can be compared with any number, not just 2, and is not necessarily known in advance.

By default, move is certain, and dwin is uncertain but complete. First, add the completion rule:

```
¬ dwin(x) ← count {y: move(x,y) ∧ ¬ dwin(y)} < 2
```

Then, rename ¬dwin to n.dwin, in both the given rule and the completion rule, except the positive occurrence of dwin in the body of the completion rule, yielding:

```
dwin(x) ← count {y: move(x,y) ∧ n.dwin(y)} ≥ 2
n.dwin(x) ← count {y: move(x,y) ∧ ¬ dwin(y)} < 2
```

Now compute the least fixed point. Start with the base case, in the second rule, for positions x that have moves to fewer than 2 positions; this infers n.dwin(x) facts for those positions x. Then, the first rule infers dwin(x) facts for any position x that can move to 2 or more positions for which n.dwin is true.

This process iterates to infer more n.dwin and more dwin facts, until a fixed point is reached, where dwin gives winning positions, n.dwin gives losing positions, and the remaining positions are draw positions, corresponding to positions for which dwin is true, false, and undefined, respectively.

5.3 Experiments

We also performed experiments with our new semantics. We implemented straightforward and incremental least fixed-point computations for example problems in DistAlgo [46], an extension of Python. We also compared with results computed by three systems that support negation and aggregation in recursion: XSB [60], the most well-known such system that computes WFS, and clingo [4] and DLV [2,20], the most well-known such systems that compute SMS.

For the company control problem, our incremental program in DistAlgo (v.1.1.0b15 on Python 3.7) was the fastest; followed by clingo (v.5.4.0), about 7 times slower; followed by XSB (v.3.8.0), our straightforward program in DistAlgo, and DLV (https://www.dbai.tuwien.ac.at/proj/dlv/dlvRecAggr/

(accessed 2020-09-21))[1], each asymptotically and drastically slower than the preceding one. Most recent investigation found that changing the order of hypotheses in rules in XSB can improve the running times for this problem asymptotically.

For the double win problem, clingo and DLV cannot compute the desired 3-valued semantics, and XSB was found to compute incorrect results on some of our benchmarks. Most recent investigation found that SWI-Prolog [69] added support for computing WFS, but was found to compute incorrect results for this problem on some smallest inputs. Both SWI-Prolog and XSB have since found and fixed bugs that caused these incorrect results.

6 Related Work and Conclusion

The study of recursive rules with negation goes back at least to Russell's paradox, discovered over 120 years ago [35]. Many logic languages and disagreeing semantics have since been proposed, with significant complications and challenges described in various survey and overview articles, e.g., [8,23,52,61], and in works on relating and unifying different semantics, e.g., [10,17,18,34,37,42,51,55].

Recursive rules with aggregation have been a subject of study soon after rules with negation were used in programming. They received an even larger variety of disagreeing semantics in 20 years, e.g., [15,19–21,25,36,38,47,48,50,54,57–59,63], and even more intensive studies in the last few years, e.g., [1,3–6,11,12, 16,27–31,56,67,68,71–73], especially as they are needed in graph analysis and machine learning applications.

Major related works are as shown in Table 1, right column. They give disagreeing semantics with each other, without simple formal explanations for the disagreement, as explained there. More detailed comparisons with work by Kemp and Stuckey [36], Van Gelder [63], Pelov, Denecker, and Bruynooghe [50], Faber, Pfeifer, and Leone [19], Gelfond and Zhang [30], and Hella et al. [32,33] appear in [44]. Among all, Pelov et al.'s work [50], recently reworked for ASP [67], is notable for proposing a framework that can be instantiated to extend several prior semantics to handle aggregation. They develop several separate extended semantics. In contrast, our approach uses simple predicate declarations to capture different assumptions made by different semantics in a unifying single semantics.

Many other different semantics have been studied, all focused on restricted classes or issues. The survey by Ramakrishnan and Ullman [52] discusses some different semantics, optimization methods, and uses of recursive rules with aggregation in earlier projects. Ross and Sagiv [54] studies monotonic aggregation but not general aggregation. Beeri et al. [9] presents the valid model semantics for logic programs with negation, set expressions, and grouping, but not aggregation. Sudarshan et al. [59] extends the valid model semantics for aggregation, gives semantics for more programs than Van Gelder [63], and subsumes a class

[1] That version of DLV supports recursive aggregates, while the current release of DLV "does not yet contain a full implementation of recursive aggregates" according to http://www.dlvsystem.com/dlv/ (last accessed 2021-11-04).

of programs in Ganguly et al. [24], but it is only a 3-valued semantics. Hella et al. [32,33] study expressiveness of aggregation operators but without recursion. Liu et al. [38] give a semantics for logic programs with abstract constraints, which can represent aggregates, and show that, for positive programs, it agrees with one of Pelov et al.'s semantics [50]. A number of other works have followed Gelfond and Zhang's line of study for ASP [11,12,30].

Zaniolo et al. [16,24,31,70–72] study recursive rules with aggregation for database applications, especially including for big data analysis and machine learning applications in recent years. They study optimizations that exploit monotonicity as well as additional properties of the aggregation operators in computing the least fixed point, yielding superior performance and scalability necessary for these large applications. They discuss insight from their application experience as well as prior research for centering on fixed-point computation [72], which essentially corresponds to the assumption that predicates are certain.

Our founded semantics and constraint semantics for recursive rules with aggregation unify different previous semantics by allowing different underlying assumptions to be easily specified explicitly, and furthermore separately for each predicate if desired. Our semantics are also fully declarative, giving both a single 3-valued model from simply a least fixed-point computation and a set of 2-valued models from simply constraint solving.

The key enabling ideas of simple binary choices for expressing assumptions and simple lease fixed-point computation and constraint solving are taken from Liu and Stoller [41,42], where they present a simple unified semantics for recursive rules with negation and quantification.

Our semantics can be extended for rules with negation in the conclusion, in the same way as in [41]. It can also easily be extended for hypotheses that are equalities or negated equalities between variables and constants, because such hypotheses are already used in presenting the semantics.

There are many directions for future research, including additional language features, efficient implementation methods, and precise complexity guarantees [40] when possible.

Acknowledgement. We would like to thank David S. Warren and Jan Wielemaker for their excellent help with using XSB and SWI-Prolog.

References

1. Alviano, M.: Evaluating answer set programming with non-convex recursive aggregates. Fundamenta Informaticae **149**(1–2), 1–34 (2016)
2. Alviano, M., et al.: The ASP system DLV2. In: Balduccini, M., Janhunen, T. (eds.) LPNMR 2017. LNCS (LNAI), vol. 10377, pp. 215–221. Springer, Cham (2017). https://doi.org/10.1007/978-3-319-61660-5_19
3. Alviano, M., Dodaro, C., Maratea, M.: Shared aggregate sets in answer set programming. Theory Pract. Logic Program. **18**(3–4), 301–318 (2018)
4. Alviano, M., Faber, W., Gebser, M.: Rewriting recursive aggregates in answer set programming: back to monotonicity. Theory Pract. Logic Program. **15**(4–5), 559–573 (2015)

5. Alviano, M., Faber, W., Gebser, M.: From non-convex aggregates to monotone aggregates in ASP. In: Proceedings of the International Joint Conference on Artificial Intelligence, pp. 4100–4104 (2016)
6. Alviano, M., Leone, N.: Complexity and compilation of GZ-aggregates in answer set programming. Theory Pract. Logic Program. **15**(4–5), 574–587 (2015)
7. Apt, K.R., Blair, H.A., Walker, A.: Towards a theory of declarative knowledge. In: Foundations of Deductive Databases and Logic Programming, pp. 89–148. Morgan Kaufmann (1988)
8. Apt, K.R., Bol, R.N.: Logic programming and negation: a survey. J. Logic Program. **19**, 9–71 (1994)
9. Beeri, C., Ramakrishnan, R., Srivastava, D., Sudarshan, S.: The valid model semantics for logic programs. In: Proceedings of the 11th ACM SIGACT-SIGMOD-SIGART Symposium on Principles of Database Systems, pp. 91–104 (1992)
10. Bruynooghe, M., Denecker, M., Truszczynski, M.: First order logic with inductive definitions for model-based problem solving. AI Mag. **37**(3), 69–80 (2016)
11. Cabalar, P., Fandinno, J., Del Cerro, L.F., Pearce, D.: Functional ASP with intensional sets: application to Gelfond-Zhang aggregates. Theory Pract. Logic Program. **18**(3–4), 390–405 (2018)
12. Cabalar, P., Fandinno, J., Schaub, T., Schellhorn, S.: Gelfond-zhang aggregates as propositional formulas. Artif. Intell. **274**, 26–43 (2019)
13. Ceri, S., Gottlob, G., Tanca, L.: Logic Programming and Databases. Springer, Heidelberg (1990). https://doi.org/10.1007/978-3-642-83952-8
14. Clark, K.L.: Negation as failure. In: Gallaire, H., Minker, J. (eds.) Logic and Databases, pp. 293–322. Plenum Press, New York (1978)
15. Consens, M.P., Mendelzon, A.O.: Low-complexity aggregation in GraphLog and Datalog. Theor. Comput. Sci. **116**(1), 95–116 (1993)
16. Das, A., Li, Y., Wang, J., Li, M., Zaniolo, C.: Bigdata applications from graph analytics to machine learning by aggregates in recursion. In: Proceedings of the 35th International Conference on Logic Programming (Technical Communications), pp. 273–279 (2019)
17. Denecker, M., Ternovska, E.: A logic of nonmonotone inductive definitions. ACM Trans. Comput. Logic **9**(2), 14 (2008)
18. Dung, P.M.: On the relations between stable and well-founded semantics of logic programs. Theor. Comput. Sci. **105**(1), 7–25 (1992)
19. Faber, W., Pfeifer, G., Leone, N.: Semantics and complexity of recursive aggregates in answer set programming. Artif. Intell **175**(1), 278–298 (2011)
20. Faber, W., Pfeifer, G., Leone, N., Dell'Armi, T., Ielpa, G.: Design and implementation of aggregate functions in the DLV system. Theory Pract. Logic Program. **8**(5–6), 545–580 (2008)
21. Ferraris, P.: Logic programs with propositional connectives and aggregates. ACM Trans. Comput. Logic **12**(4), 1–40 (2011)
22. Fitting, M.: A Kripke-Kleene semantics for logic programs. J. Logic Program. **2**(4), 295–312 (1985)
23. Fitting, M.: Fixpoint semantics for logic programming: a survey. Theor. Comput. Sci. **278**(1), 25–51 (2002)
24. Ganguly, S., Greco, S., Zaniolo, C.: Minimum and maximum predicates in logic programming. In: Proceedings of the 10th ACM SIGACT-SIGMOD-SIGART Symposium on Principles of Database Systems, pp. 154–163 (1991)

25. Gelfond, M.: Representing knowledge in A-prolog. In: Kakas, A.C., Sadri, F. (eds.) Computational Logic: Logic Programming and Beyond. LNCS (LNAI), vol. 2408, pp. 413–451. Springer, Heidelberg (2002). https://doi.org/10.1007/3-540-45632-5_16
26. Gelfond, M., Lifschitz, V.: The stable model semantics for logic programming. In: Proceedings of the 5th International Conference and Symposium on Logic Programming, pp. 1070–1080. MIT Press (1988)
27. Gelfond, M., Zhang, Y.: Vicious circle principle and logic programs with aggregates. Theory Pract. Logic Program. **14**(4–5), 587–601 (2014)
28. Gelfond, M., Zhang, Y.: Vicious circle principle and formation of sets in ASP based languages. In: Balduccini, M., Janhunen, T. (eds.) LPNMR 2017. LNCS (LNAI), vol. 10377, pp. 146–159. Springer, Cham (2017). https://doi.org/10.1007/978-3-319-61660-5_14
29. Gelfond, M., Zhang, Y.: Vicious circle principle and logic programs with aggregates. Computing Research Repository (2018). cs.AI arXiv:1808.07050
30. Gelfond, M., Zhang, Y.: Vicious circle principle, aggregates, and formation of sets in ASP based languages. Artif. Intell. **275**, 28–77 (2019)
31. Gu, J., et al.: RaSQL: Greater power and performance for big data analytics with recursive-aggregate-SQL on Spark. In: Proceedings of the 2019 International Conference on Management of Data, pp. 467–484 (2019)
32. Hella, L., Libkin, L., Nurmonen, J., Wong, L.: Logics with aggregate operators. In: Proceedings of the 14th Annual IEEE Symposium on Logic in Computer Science, p. 35. IEEE Computer Society (1999)
33. Hella, L., Libkin, L., Nurmonen, J., Wong, L.: Logics with aggregate operators. J. ACM **48**(4), 880–907 (2001)
34. Hou, P., De Cat, B., Denecker, M.: FO(FD): extending classical logic with rule-based fixpoint definitions. Theory Pract. Logic Program. **10**(4–6), 581–596 (2010)
35. Irvine, A.D., Deutsch, H.: Russell's paradox. Stanford Encyclopedia of Philosophy (2020). https://plato.stanford.edu/entries/russell-paradox/, First published Fri Dec 8, 1995; substantive revision Mon Oct 12, 2020, Accessed 3 Jan 2021
36. Kemp, D.B., Stuckey, P.J.: Semantics of logic programs with aggregates. In: Proceedings of the International Symposium on Logic Programming, pp. 387–401 (1991)
37. Lin, F., Zhao, Y.: ASSAT: computing answer sets of a logic program by SAT solvers. Artif. Intell. **157**(1–2), 115–137 (2004)
38. Liu, L., Pontelli, E., Son, T.C., Truszczynski, M.: Logic programs with abstract constraint atoms: The role of computations. Artif. Intell. **174**(3), 295–315 (2010)
39. Liu, Y.A.: Logic programming applications: what are the abstractions and implementations? In: Kifer, M., Liu, Y.A. (eds.) Declarative Logic Programming: Theory, Systems, and Applications, chap. 10, pp. 519–557. ACM and Morgan & Claypool (2018)
40. Liu, Y.A., Stoller, S.D.: From datalog rules to efficient programs with time and space guarantees. ACM Trans. Program. Lang. Syst. **31**(6), 1–38 (2009)
41. Liu, Y.A., Stoller, S.D.: Founded semantics and constraint semantics of logic rules. In: Artemov, S., Nerode, A. (eds.) LFCS 2018. LNCS, vol. 10703, pp. 221–241. Springer, Cham (2018). https://doi.org/10.1007/978-3-319-72056-2_14
42. Liu, Y.A., Stoller, S.D.: Founded semantics and constraint semantics of logic rules. J. Logic Comput. **30**(8), 1609–1638 (2020). http://arxiv.org/abs/1606.06269
43. Liu, Y.A., Stoller, S.D.: Knowledge of uncertain worlds: programming with logical constraints. In: Artemov, S., Nerode, A. (eds.) LFCS 2020. LNCS, vol. 11972, pp. 111–127. Springer, Cham (2020). https://doi.org/10.1007/978-3-030-36755-8_8

44. Liu, Y.A., Stoller, S.D.: Recursive rules with aggregation: a simple unified semantics. Computing Research Repository (2020). cs.DB arXiv:2007.13053
45. Liu, Y.A., Stoller, S.D.: Knowledge of uncertain worlds: programming with logical constraints. J. Logic Comput. **31**(1), 193–212 (2021). https://arxiv.org/abs/1910.10346
46. Liu, Y.A., Stoller, S.D., Lin, B.: From clarity to efficiency for distributed algorithms. ACM Trans. Program. Lang. Syst. **39**(3), 12:1–12:41 (2017)
47. Marek, V.W., Remmel, J.B.: Set constraints in logic programming. In: Lifschitz, V., Niemelä, I. (eds.) LPNMR 2004. LNCS (LNAI), vol. 2923, pp. 167–179. Springer, Heidelberg (2003). https://doi.org/10.1007/978-3-540-24609-1_16
48. Marek, V.W., Truszczynski, M.: Logic programs with abstract constraint atoms. In: Proceedings of the 19th National Conference on Artificial Intelligence, 16th Conference on Innovative Applications of Artificial Intelligence, pp. 86–91. AAAI Press/The MIT Press (2004)
49. Mumick, I.S., Pirahesh, H., Ramakrishnan, R.: The magic of duplicates and aggregates. In: Proceedings of the 16th International Conference on Very Large Databases, pp. 264–277. Morgan Kaufmann (1990)
50. Pelov, N., Denecker, M., Bruynooghe, M.: Well-founded and stable semantics of logic programs with aggregates. Theory Pract. Logic Program. **7**(3), 301–353 (2007)
51. Przymusinski, T.C.: Well-founded and stationary models of logic programs. Ann. Math. Artif. Intell **12**(3), 141–187 (1994)
52. Ramakrishnan, R., Ullman, J.D.: A survey of deductive database systems. J. Logic Program. **23**(2), 125–149 (1995)
53. Ross, K.A., Sagiv, Y.: Monotonic aggregation in deductive databases. In: Proceedings of the 11th ACM SIGACT-SIGMOD-SIGART Symposium on Principles of Database Systems, pp. 114–126 (1992)
54. Ross, K.A., Sagiv, Y.: Monotonic aggregation in deductive databases. J. Comput. Syst. Sci. **54**(1), 79–97 (1997)
55. Schlipf, J.S.: The expressive powers of the logic programming semantics. J. Comput. Syst. Sci. **51**(1), 64–86 (1995)
56. Shkapsky, A., Yang, M., Zaniolo, C.: Optimizing recursive queries with monotonic aggregates in DeALS. In: Proceedings of the 2015 IEEE 31st International Conference on Data Engineering, pp. 867–878 (2015)
57. Simons, P., Niemelä, I., Soininen, T.: Extending and implementing the stable model semantics. Artif. Intell **138**(1–2), 181–234 (2002)
58. Son, T.C., Pontelli, E., Tu, P.H.: Answer sets for logic programs with arbitrary abstract constraint atoms. J. Artif. Intell. Res. **29**, 353–389 (2007)
59. Sudarshan, S., Srivastava, D., Ramakrishnan, R., Beeri, C.: Extending the well-founded and valid semantics for aggregation. In: Proceedings of the 1993 International Symposium on Logic programming, pp. 590–608 (1993)
60. Swift, T., et al.: The XSB System Version 3.8, x (2017). http://xsb.sourceforge.net
61. Truszczynski, M.: An introduction to the stable and well-founded semantics of logic programs. In: Kifer, M., Liu, Y.A. (eds.) Declarative Logic Programming: Theory, Systems, and Applications, pp. 121–177. ACM and Morgan & Claypool (2018)
62. Van Gelder, A.: Negation as failure using tight derivations for general logic programs. In: Proceedings of the 3rd IEEE-CS Symposium on Logic Programming, pp. 127–138 (1986)

63. Van Gelder, A.: The well-founded semantics of aggregation. In: Proceedings of the 11th ACM SIGACT-SIGMOD-SIGART Symposium on Principles of Database Systems, San Diego, California, 2–4 June 1992, pp. 127–138 (1992)
64. Van Gelder, A.: The alternating fixpoint of logic programs with negation. J. Comput. Syst. Sci. **47**(1), 185–221 (1993)
65. Van Gelder, A., Ross, K., Schlipf, J.S.: Unfounded sets and well-founded semantics for general logic programs. In: Proceedings of the 7th ACM SIGACT-SIGMOD-SIGART Symposium on Principles of Database Systems, pp. 221–230 (1988)
66. Van Gelder, A., Ross, K., Schlipf, J.S.: The well-founded semantics for general logic programs. J. ACM **38**(3), 620–650 (1991)
67. Vanbesien, L., Bruynooghe, M., Denecker, M.: Analyzing semantics of aggregate answer set programming using approximation fixpoint theory. Computing Research Repository (2021). cs.AI arXiv:2104.14789
68. Wang, Q., et al.: Automating incremental and asynchronous evaluation for recursive aggregate data processing. In: Proceedings of the 2020 ACM SIGMOD International Conference on Management of Data, pp. 2439–2454 (2020)
69. Wielemaker, J., Schrijvers, T., Triska, M., Lager, T.: SWI-Prolog. Theory Pract. Logic Program. **12**(1–2), 67–96 (2012)
70. Zaniolo, C., Arni, N., Ong, K.: Negation and aggregates in recursive rules: the LDL++ approach. In: Ceri, S., Tanaka, K., Tsur, S. (eds.) DOOD 1993. LNCS, vol. 760, pp. 204–221. Springer, Heidelberg (1993). https://doi.org/10.1007/3-540-57530-8_13
71. Zaniolo, C., Das, A., Gu, J., Li, Y., Li, M., Wang, J.: Monotonic properties of completed aggregates in recursive queries. Computing Research Repository (2019). cs.DB arXiv:1910.08888
72. Zaniolo, C., Yang, M., Das, A., Shkapsky, A., Condie, T., Interlandi, M.: Fixpoint semantics and optimization of recursive datalog programs with aggregates. Theory Pract. Logic Program. **17**(5–6), 1048–1065 (2017)
73. Zhang, Y., Rayatidamavandi, M.: A characterization of the semantics of logic programs with aggregates. In: Proceedings of the International Joint Conference on Artificial Intelligence, pp. 1338–1344 (2016)

Computational Properties of Partial Non-deterministic Matrices and Their Logics

Sérgio Marcelino, Carlos Caleiro, and Pedro Filipe(✉)

SQIG - Instituto de Telecomunicações, Departamento de Matemática, Instituto Superior Técnico, Universidade de Lisboa, Lisbon, Portugal
{smarcel,ccal}@math.tecnico.ulisboa.pt, pedro.g.filipe@tecnico.ulisboa.pt

Abstract. Incorporating non-determinism and partiality in the traditional notion of logical matrix semantics has proven to be decisive in a myriad of recent compositionality results in logic. However, several important properties which are known to be computable for finite matrices have not been studied in this wider context of partial non-deterministic matrices (PNmatrices).

This paper is dedicated to understanding how this generalization of the considered semantical structures affects the computational properties of basic problems regarding their induced logics, in particular their sets of theorems.

We will show that the landscape is quite rich, as some problems keep their computational status, for others the complexity increases, and for a few decidability is lost. Namely, we show that checking if the logics defined by two finite PNmatrices have the same theorems is undecidable. This latter result is obtained by reduction from the undecidable problem of checking universality of term-DAG-automata.

1 Introduction

Logical matrices are arguably the most widespread semantic structures for propositional logics [14,27]. After Łukasiewicz, a logical matrix consists in an underlying algebra, functionally interpreting logical connectives over a set of truth-values, together with a designated set of truth-values. The logical models are obtained by considering the homomorphisms from the free-algebra in the matrix similarity type into the algebra (valuations), and formulas that hold in the model are the ones that take designated values.

However, in recent years, it has become clear that there are advantages in departing from semantics based on logical matrices, by adopting a partial non-deterministic generalization of the standard notion. PNmatrices were introduced in [6], as a generalization of non-deterministic matrices (Nmatrices) that on

Research funded by FCT/MCTES through national funds and when applicable co-funded by EU under the project UIDB/50008/2020. The third author acknowledges the grant PD/BD/135513/2018 by FCT, under the LisMath Ph.D. programme.

S. Artemov and A. Nerode (Eds.): LFCS 2022, LNCS 13137, pp. 180–197, 2022.
https://doi.org/10.1007/978-3-030-93100-1_12

their turn were proposed in the beginning of this century by Avron and his collaborators [3,4]. In PNmatrices, the connectives are interpreted by (partial) multi-functions instead of functions. The central idea is that a connective can non-deterministically pick from a set of possible values instead of its value being completely determined by the input values. When for a certain input the interpretation function outputs the empty set it means that the input values are not compatible. Logical semantics based on PNmatrices are very malleable. The added expressiveness allows for finite characterizations of logics that do not admit finite semantics based on logical matrices [23,24] and provide valuable insight about proof theoretical properties of said logics [2]. It also allows for general recipes for various practical problems in logic, including procedures to constructively updating semantics when imposing new axioms [9,12], including language extensions; or effectively combining the semantics of two logics capturing the effect of joining their axiomatizations [8,21]. Recently, PNmatrices also provided new interpretations of quantum states as valuations [15]. Furthermore, finite PNmatrices still retain many useful properties of finite matrices. Namely, one can still smoothly produce analytic calculi for the induced logic, over which a series of reasoning activities in a purely symbolic fashion can be performed, including proof-search and countermodel generation [6,7,22]. However, naturally, certain properties of PNmatrices become more difficult to analyse. For instance, in [5], Zohar and Avron suggest investigating more deeply under which conditions two given Nmatrices define the same logic.

This paper is a first step towards understanding this question. We will consider simple standard problems regarding the evaluation of formulas, and the notion of theoremhood in the logics defined by different kinds of semantic structures. It is well known that checking whether a given formula is satisfiable, or valid, in a given finite logical matrix is decidable. Further (see [16,17]), the problems of checking the existence of theorems in the logic defined by a finite logical matrix, and of checking whether two finite matrices define logics with the same set of theorems, are also decidable. However, the computational status of such problems with respect to Nmatrices or PNmatrices has never been studied. Our aim is to analyse these richer problems, and assess how their computational hardness is affected by partiality, non-determinism, and the interaction between the two.

We will show that while some of these problems retain their computational properties in the wider context, namely determining the satisfiability or validity of a given formula, others become harder. Partiality, in particular, makes the problem of analysing possible valuations of a given formula non-local (with respect to the syntax of the formula). This fact by itself can be overcome, with limited computational cost, in the absence of non-determinism. However, in PNmatrices, there is an intrinsic added complexity to be accounted for in analysing theoremhood. Further, finite logical matrices are based on algebras, which allows us to compute a bound on the depth of the formulas that may be necessary to express all possible functions of the truth-values with a given arity [19]. In the presence of non-determinism, however, we will see that the analogous notion of clone of multi-functions does not behave well, and general upper-bounds cannot be obtained. Consequently, we will show that checking if the logics defined by two finite PNmatrices have the same theorems, or even just checking if some of the logics has any theorem, are both undecidable problems, by exploring a useful connection with *term-DAG-automata* [1,11].

The paper is organized as follows. In Sect. 2 we recall PNmatrices, their intrincacies, and associated logics, and introduce the computational properties we will analyse in the paper, along with some simple examples. As a warming up for the next sections, we show that the problems of satisfiability and validity are not essentially harder for PNmatrices. Section 3 addresses the problem of checking whether or not every formula is a theorem in the logic defined by a finite PNmatrix. We show that the problem is easily decidable, in constant time, in the absence of partiality. Further, we show that when having partiality without non-determinism the problem can still be decided in polynomial time, whereas the simultaneity of partiality and non-determinism turns the problem **NP-complete**. Then, in Sect. 4 we study the problem of checking the existence of theorems in the logic defined by a finite PNmatrix. We show that in the absence of non-determinism the problem can be decided in exponential time, and even in polynomial time in the presence of 0-ary connectives. However, allowing non-determinism makes the problem undecidable. To prove this result, we briefly recall term-DAG-automata, and show how an undecidable problem concerning term-DAG-automata can be reduced to the existence of theorems in a related PNmatrix. Section 5 is dedicated to our most interesting result, concerning the problem of determining whether two given PNmatrices determine logics with the same set of theorems. Namely, we show that the problem is decidable in double-exponential time in the absence of non-determinism, but again becomes undecidable in the presence of non-determinism. This undecidability result is a consequence of a simple reduction from the undecidability of the theorem existence problem of the previous section. We conclude, in Sect. 6, with a brief analysis of the results obtained and an outline of future research. We also attach an appendix containing the more technical details of some proofs.

2 Warming Up

We start by introducing partial non-deterministic matrices (PNmatrices), and their induced logics. We also define the (standard) computational problems we will study, and illustrate them with a few examples.

A signature Σ is a family of connectives indexed by their arity, $\Sigma = \{\Sigma^{(k)} : k \in \mathbb{N}\}$. The set of formulas over Σ based on a set of propositional variables P is denoted by $L_\Sigma(P)$. The set of subformulas (*resp.* variables) of a formula $\varphi \in L_\Sigma(P)$ is denoted by $\mathsf{sub}(\varphi)$ (*resp.* $\mathsf{var}(\varphi)$) and the size of a formula $\mathsf{size}(\varphi)$ is the cardinality of the set $\mathsf{sub}(\varphi)$ (that is, the size of the DAG representation of φ - we will get back to this in the end of this section). We say that formulas in $L_\Sigma(\emptyset)$ are *closed.* A *substitution* is a mapping $\sigma : P \to L_\Sigma(P)$, uniquely extendable into an endomorphism $\cdot^\sigma : L_\Sigma(P) \to L_\Sigma(P)$.

A logic over Σ is a Tarskian relation $\vdash \subseteq \wp(L_\Sigma(P)) \times L_\Sigma(P)$ satisfying

- $\Gamma \vdash \varphi$ if $\varphi \in \Gamma$ (*reflexivity*),
- $\Gamma \cup \Gamma' \vdash \varphi$ if $\Gamma \vdash \varphi$ (*monotonicity*),
- $\Gamma \vdash \varphi$ if $\Delta \vdash \varphi$ and $\Gamma \vdash \delta$ for every $\delta \in \Delta$ (*transitivity*)
- $\Gamma^\sigma \vdash \varphi^\sigma$ for every $\sigma : P \to L_\Sigma(P)$, if $\Gamma \vdash \varphi$ (*substitution invariance*)

We say that $\varphi \in L_\Sigma(P)$ is a *theorem* of $\vdash$ whenever $\emptyset \vdash \varphi$, and denote the set of theorems of $\vdash$ by $\mathsf{thm}(\vdash)$.

A Σ-PNmatrix, is a tuple $\mathbb{M} = \langle A, \cdot_\mathbb{M}, D\rangle$ where A is the set of truth-values, $D \subseteq A$ is the set of designated truth-values, and for each $© \in \Sigma^{(k)}$, the function $©_\mathbb{M} : A^k \to \wp(A)$ interprets the connective ©. We use $\mathsf{sig}(\mathbb{M})$ to denote the underlying signature of a PNmatrix $\mathbb{M}$. A PNmatrix $\mathbb{M}$ is finite if it contains only a finite number of truth-values and $\Sigma_\mathbb{M}$ is finite. We denote by PNMatr the class of all PNmatrices. There are 3 proper subclasses of PNMatr, that are particularly interesting: the class Matr of logical *matrices* is recovered when, for every $© \in \Sigma^{(k)}$ and $x_1, \ldots, x_k \in A$, $©_\mathbb{M}(x_1, \ldots, x_k)$ is a singleton; the class PMatr of *Pmatrices*, when for every $© \in \Sigma^{(k)}$ and $x_1, \ldots, x_k \in A$, $©_\mathbb{M}(x_1, \ldots, x_k)$ has at most one element; and the class NMatr of *Nmatrices*, when for every $© \in \Sigma^{(k)}$ and $x_1, \ldots, x_k \in A$, $©_\mathbb{M}(x_1, \ldots, x_k) \neq \emptyset$. Clearly, $\mathsf{Matr} \subsetneq \mathsf{PMatr} \cap \mathsf{NMatr}$ and $\mathsf{PMatr} \cup \mathsf{NMatr} \subsetneq \mathsf{PNMatr}$. The size of a PNmatrix is the size of the tables describing the interpretation of each connective which, for a fixed signature, is polynomial in the number of its truth-values.

Given $X \subseteq A$, $\mathbb{M}_X = \langle X, \cdot_{\mathbb{M}_X}, D \cap X\rangle$ is the sub-PNmatrix of $\mathbb{M}$ obtained by restricting it to values in X, that is, $©_{\mathbb{M}_X}(x_1, \ldots, x_k) = ©_\mathbb{M}(x_1, \ldots, x_k) \cap X$ for $© \in \Sigma^{(k)}$ and $x_1, \ldots, x_k \in X$. We say that $X \neq \emptyset$ is a (non-empty) *total component* of $\mathbb{M}$, or simply *total*, whenever $\mathbb{M}_X$ is an Nmatrix. We denote by $\mathsf{Tot}_\mathbb{M}$ the set of total components of $\mathbb{M}$. We say X is a *maximal* total component when $X \in \mathsf{Tot}_\mathbb{M}$ and $Y \notin \mathsf{Tot}_\mathbb{M}$ for all $Y \supsetneq X$. We say that a non-empty set of truth-values X is *compatible in* $\mathbb{M}$ if $X \subseteq Y$ for some $Y \in \mathsf{Tot}_\mathbb{M}$.

A *partial valuation* of $\mathbb{M} = \langle A, \cdot_\mathbb{M}, D\rangle$ is a function $v : \Gamma \to A$, where Γ is a set closed for subformulas, such that, for every $©(\varphi_1, \ldots, \varphi_n) \in \Gamma$, $v(©(\varphi_1, \ldots, \varphi_n)) \in ©_\mathbb{M}(v(\varphi_1), \ldots, v(\varphi_n))$. If $\Gamma = \mathbf{L}_\Sigma(P)$, then v is said to be a *total valuation*, or simply a *valuation*. We use $\mathsf{Val}(\mathbb{M})$ to denote the set of all valuations over $\mathbb{M}$. Partial valuations over Nmatrices are always extensible to valuations. However, when a PNmatrix is genuinely partial we additionally have to check if the image of the partial valuation is compatible.

Given a PNmatrix $\mathbb{M} = \langle A, \cdot_\mathbb{M}, D\rangle$, every formula $\varphi \in L_{\mathsf{sig}(\mathbb{M})}(P)$, where $\mathsf{var}(\varphi) = \{p_1, \ldots, p_k\}$, defines a multi-function $\varphi_\mathbb{M} : A^k \to \wp(A)$ such that $\varphi_\mathbb{M}(x_1, \ldots, x_k) = \{v(\varphi) : v \in \mathsf{Val}(\mathbb{M}), v(p_i) = x_i, 1 \leq i \leq k\}$.

The logic induced by a PNmatrix $\mathbb{M}$ is defined as $\Gamma \vdash_\mathbb{M} \varphi$ whenever, for every valuation over $\mathbb{M}$, if $v(\Gamma) \subseteq D$ then $v(\varphi) \in D$. This generalizes the usual semantics on matrices.

The main results of this paper are on the computational and complexity status of the following problems regarding the logics defined by finite PNmatrices. For $\mathcal{C} \subseteq \mathsf{PNMatr}$ we consider the following problems:

- $\mathsf{SAT}(\mathcal{C})$ – given finite $\mathbb{M} \in \mathcal{C}$ and $\varphi \in L_{\mathsf{sig}(\mathbb{M})}(P)$, check if $v(\varphi)$ is designated for some $v \in \mathsf{Val}(\mathbb{M})$,

- $\mathsf{VAL}(\mathcal{C})$ – given finite $\mathbb{M} \in \mathcal{C}$ and $\varphi \in L_{\mathsf{sig}(\mathbb{M})}(P)$, check if $v(\varphi)$ is designated for every $v \in \mathsf{Val}(\mathbb{M})$,
- $\forall\mathsf{Thm}(\mathcal{C})$ – given finite $\mathbb{M} \in \mathcal{C}$, check if $\mathsf{thm}(\vdash_{\mathbb{M}}) = L_\Sigma(P)$,
- $\exists\mathsf{Thm}(\mathcal{C})$ – given finite $\mathbb{M} \in \mathcal{C}$, check if $\mathsf{thm}(\vdash_{\mathbb{M}}) \neq \emptyset$,
- $\mathsf{EqThm}(\mathcal{C})$ – given two finite $\mathbb{M}_1, \mathbb{M}_2 \in \mathcal{C}$ s.t. $\mathsf{sig}(\mathbb{M}_1) = \mathsf{sig}(\mathbb{M}_2)$, check if $\mathsf{thm}(\vdash_{\mathbb{M}_1}) = \mathsf{thm}(\vdash_{\mathbb{M}_2})$

The problems SAT and VAL are the usual validity and satisfiability problems, now stated over (logics of) PNmatrices. The validity problem can of course be stated as checking if $\varphi \in \mathsf{thm}(\vdash_{\mathbb{M}})$. It is well known that the SAT and VAL problems are **NP-complete** and **coNP-complete**, respectively, already for logics induced by finite matrices. In the next example we recall the 3SAT problem for classical logic, a restricted version of SAT that is already **NP-complete**.

Example 1. Let $\mathbb{B} = \{\{0,1\}, \cdot_{\mathbb{B}}, \{1\}\}$ *be the Boolean matrix over a signature containing unary* $\neg$ *and binary* $\vee$ *and* $\wedge$*, defined by the tables*

$\wedge_{\mathbb{B}}$	0	1
0	0	0
1	0	1

$\vee_{\mathbb{B}}$	0	1
0	0	1
1	1	1

$\neg_{\mathbb{B}}$	
0	1
1	0

An instance of 3SAT *is a formula of the form* $\varphi = C_1 \wedge \cdots \wedge C_n$*, where each clause* C_i *is the disjunction of exactly three literals* $L_{i,1} \vee L_{i,2} \vee L_{i,3}$ *with each literal* $L_{i,j} \in P \cup \{\neg p : p \in P\}$*. The problem of determining the satisfability of instances of* 3SAT *over* $\mathbb{B}$ *is already **NP-complete** [13, 18]. Further ahead, we will make standard use of this fact in **NP-hardness** proofs.* △

On the other hand, it is known [6,7] that the complexity of these problems does not increase when ranging over the wider class of finite PNmatrices.

Proposition 2. *The problem* SAT *(*resp., VAL*) over finite PNmatrices is in* ***NP*** *(*resp., ***coNP****).*

Proof. Given a finite Σ-PNmatrix $\mathbb{M} = \langle A, \cdot_{\mathbb{M}}, D\rangle$ and $\varphi \in L_\Sigma(P)$ we guess $X \subseteq A$, together with a mapping $f : \mathsf{sub}(\varphi) \to X$. We check in linear time on the size of $\mathbb{M}$ if $\mathbb{M}_X$ is total and, if so, we check if f is a partial valuation, which can be done in linear time on the size of φ, and also if $f(\varphi) \in D$. Hence, $\mathsf{SAT}(\mathsf{PNMatr}) \in$ **NP**.

Of course, φ is a theorem of $\vdash_{\mathbb{M}}$ if and only if φ is not satisfiable in $\mathbb{M}' = \langle A, \cdot_{\mathbb{M}}, A \setminus D\rangle$, and thus $\mathsf{VAL}(\mathsf{PNMatr}) \in$ **coNP**. □

It will be useful to refer to the duals of the remaining problems, denoted by $\overline{\forall\mathsf{Thm}}$, $\overline{\exists\mathsf{Thm}}$ and $\overline{\mathsf{EqThm}}$, respectively. We are interested in understanding how instantiating $\mathcal{C}$ as Matr, PMatr, NMatr and PNMatr affects the hardness of the problems. This will be the subject of Sects. 3, 4, and 5. Before, let us look at some examples of (P)(N)matrices and their induced logics and theorems.

Example 3. Let $\mathbb{M}$ *be a PNmatrix. If* $\mathbb{M}$ *contains a single truth-value then either* $\mathbb{M}$ *is a matrix, or else* $\mathsf{Val}(\mathbb{M}) = \emptyset$ *and* $\mathsf{thm}(\vdash_{\mathbb{M}}) = L_{\mathsf{sig}(\mathbb{M})}(P)$*. Every formula is a theorem, also, in case every truth-value in* $\mathbb{M}$ *is designated. On the other hand, if* $\mathsf{Val}(\mathbb{M}) \neq \emptyset$ *and every truth-value of* $\mathbb{M}$ *is non-designated we necessarily have that* $\mathsf{thm}(\vdash_{\mathbb{M}}) = \emptyset$*. Let us look beyond these extreme cases.*

For simplicity, let us consider a signature with a single binary connective $\rightarrow$*, and the following five (Boolean)* $\rightarrow$*-Nmatrices* $2_{\mathsf{x}} = \langle\{0,1\}, \cdot_{\mathsf{x}}, \{1\}\rangle$*, defined for* $\mathsf{x} \in \{\mathsf{cl}, \mathsf{mp}, \mathsf{free}, \mathsf{empty}, \mathsf{false}\}$ *as follows.*

$\rightarrow_{\mathsf{cl}}$	0	1
0	1	1
1	0	1

$\rightarrow_{\mathsf{mp}}$	0	1
0	0,1	0,1
1	0	0,1

$\rightarrow_{\mathsf{in}}$	0	1
0	0,1	0,1
1	0,1	1

$\rightarrow_{\mathsf{free}}$	0	1
0	0,1	0,1
1	0,1	0,1

$\rightarrow_{\mathsf{empty}}$	0	1
0	1	1
1	0	$\emptyset$

$\rightarrow_{\mathsf{false}}$	0	1
0	0,1	$\emptyset$
1	0,1	0,1

Clearly, $2_{\mathsf{cl}} \in \mathsf{Matr}$*,* $2_{\mathsf{mp}}, 2_{\mathsf{in}}, 2_{\mathsf{free}} \in \mathsf{Nmatr}$*,* $2_{\mathsf{empty}} \in \mathsf{Pmatr}$*, and* $2_{\mathsf{false}} \in \mathsf{PNmatr}$*. All these PNmatrices define different sets of valuations. For each* $v : L_\Sigma(P) \rightarrow \{0,1\}$*,* $v \in \mathsf{Val}(2_{\mathsf{cl}})$ *precisely when* $v(\varphi \rightarrow \psi) = 0$ *iff* $v(\varphi) = 1$ *and* $v(\psi) = 0$*. Clearly,* $\mathsf{Val}(2_{\mathsf{cl}}) \subseteq \mathsf{Val}(2_{\mathsf{x}})$ *for* $\mathsf{x} \in \{\mathsf{mp}, \mathsf{in}, \mathsf{free}\}$*.* $\mathsf{Val}(2_{\mathsf{mp}})$ *is of course larger, as* $v \in \mathsf{Val}(2_{\mathsf{mp}})$ *if* $v(\varphi \rightarrow \psi) = 0$ *whenever* $v(\varphi) = 1$ *and* $v(\psi) = 0$*. Similarly,* $v \in \mathsf{Val}(2_{\mathsf{in}})$ *if* $v(\varphi \rightarrow \psi) = 1$ *whenever* $v(\varphi) = v(\psi) = 1$*. Note that non-determinism allows for new choices to be made every time a connective is applied to the values attributed to the subformulas. However, every occurence of a formula must be given the same value. Given* $v \in \mathsf{Val}(2_{\mathsf{in}})$ *we can have* $v(p \rightarrow p) = 0$ *or* $v(p \rightarrow p) = 1$*, but when determining the value of* $v((p \rightarrow p) \rightarrow (p \rightarrow p))$*, if* $v(p \rightarrow p) = 1$ *there are no choices to be made and we must have* $v((p \rightarrow p) \rightarrow (p \rightarrow p)) = 1$*, if* $v(p \rightarrow p) = 0$ *then we still might have* $v((p \rightarrow p) \rightarrow (p \rightarrow p))$ *take any value. Therefore,* $\mathsf{Val}(2_{\mathsf{free}})$ *collects all possible functions from the language to* $\{0,1\}$*, as the structure of* 2_{free} *imposes no restrictions. On the other hand, partiality destroys the locality of valuations. When looking for models we must guarantee that the partial valuations use only compatible values. Assume that* $v \in \mathsf{Val}(2_{\mathsf{empty}})$*, then* v *cannot take the value* 1 *as* $\rightarrow_{\mathsf{empty}} (1,1) = \emptyset$*, but then it cannot take the value* 0 *either, since* $\rightarrow_{\mathsf{empty}} (0,0) = \{1\}$*. Hence,* $\mathsf{Tot}_{2_{\mathsf{empty}}} = \emptyset$ *and* $\mathsf{Val}(2_{\mathsf{empty}}) = \emptyset$*. If* $v \in \mathsf{Val}(2_{\mathsf{false}})$ *then clearly* v *cannot mix* 0 *and* 1 *as* $\rightarrow_{\mathsf{false}} (0,1) = \emptyset$*. Thus,* $\mathsf{Tot}_{2_{\mathsf{false}}} = \{\{0\},\{1\}\}$ *and* $\mathsf{Val}(2_{\mathsf{false}}) = \{v_0, v_1\}$ *where* v_x *is the constant valuation with value* x*. So* 2_{false} *either satisfies every formula or falsifies every formula. It is well known [27] that removing a valuation that designates every value does not change the resulting logic and, therefore,* 2_{false} *has the same logic as the PNmatrix with no designated elements. Letting* $\vdash_{2_{\mathsf{x}}} = \vdash_{\mathsf{x}}$*:*

- $\vdash_{\mathsf{cl}}$ *is the logic of classical implication,*
- $\vdash_{\mathsf{mp}}$ *is the logic axiomatized by the single rule* $\frac{p,\, p\rightarrow q}{q}$ *(modus ponens),*
- $\vdash_{\mathsf{in}}$ *is axiomatized by the rule* $\frac{p,\, q}{p \rightarrow q}$*,*
- $\vdash_{\mathsf{free}}$ *is the smallest logic in* $\rightarrow$*, axiomatized by the empty set of rules,*

- $\vdash_{\mathsf{empty}}$ *is the largest (inconsistent) logic in* $\to$, *axiomatized by the rule* $\frac{}{p}$,
- $\vdash_{\mathsf{false}}$ *is the almost inconsistent logic, axiomatized by* $\frac{p}{q}$.

Hence, we have that $\mathsf{thm}(\vdash_{\mathsf{mp}}) = \mathsf{thm}(\vdash_{\mathsf{in}}) = \mathsf{thm}(\vdash_{\mathsf{free}}) = \mathsf{thm}(\vdash_{\mathsf{false}}) = \emptyset$ *and* $\mathsf{thm}(\vdash_{\mathsf{empty}}) = L_\Sigma(P)$. *The fact that completely different logics can have the same set of theorems was the initial reason for the definition of logic as a consequence relation. Also, the logic of classical implication contains theorems and non-theorems, thus* $\emptyset \neq \mathsf{thm}(\vdash_{\mathsf{cl}}) \neq L_\Sigma(P)$.

It is also relevant to note that, for every matrix $\mathbb{M}$, *if* $\varphi_{\mathbb{M}} = \psi_{\mathbb{M}}$ *then the two formulas are not only logically equivalent* ($\varphi \dashv\vdash_{\mathbb{M}} \psi$) *but also interchangeable in every context. With non-determinism the situation is dramatically different. Note that for* $\mathbb{M} \in \{2_{\mathsf{mp}}, 2_{\mathsf{free}}\}$ *and* $\varphi = (p \to p)$ *and* $\psi = (p \to p) \to p$ *we have that* $\varphi_{\mathbb{M}} = \psi_{\mathbb{M}}$ *(outputting the set* $\{0,1\}$ *for both inputs) but* $\varphi \not\vdash_{\mathbb{M}} \psi$ *and* $\psi \not\vdash_{\mathbb{M}} \varphi$. △

In the above example, Boolean partial non-deterministic matrices were enough to illustrate both the effects of partiality and non-determinism, and the problems we shall investigate. A more in-depth analysis of logics defined by Boolean partial non-deterministic matrices can be found in [20]. We should emphasize, though, that the full power of partiality and non-determinism is better appreciated when considering PNmatrices with more than two truth-values, for which we refer the reader to the literature, notwithstanding the few examples we shall present later in the paper.

3 Checking Theorem Universality

Given a PNmatrix $\mathbb{M}$, checking $\forall\mathsf{Thm}$ or, dually, if $\vdash_{\mathbb{M}}$ has a non-theorem is trivial in the absence of partiality since, when $\mathbb{M}$ is a (N)matrix, $\vdash_{\mathbb{M}}$ has a non-theorem if and only if $\mathbb{M}$ has non-designated elements, so that p is a non-theorem. Checking if the set of non-designated elements is non-empty can be done in at most linear time. In the presence of partiality, however, the problem is a little harder since, additionally, one needs to check if there is a total component containing a non-designated element. Before tackling the main results of this section we will show some computational intricacies of detecting total components in PNmatrices.

3.1 Computing Total Components

First, let us see how Pmatrices can host an exponential number (on the number of truth-values) of total components.

Example 4. Consider a signature containing a single binary connective g. *Fixed a set* I, *consider a* $\{g\}$*-Pmatrix* $\mathbb{I}$ *with domain* $A_I = \{0_i : i \in I\} \cup \{1_i : i \in I\}$ *and let*

$$g_{\mathbb{I}}(x,y) = \begin{cases} \emptyset & \textit{if } \{x,y\} = \{0_i, 1_i\} \textit{ for some } i \in I, \\ x & \textit{otherwise.} \end{cases}$$

Easily, $X \subseteq A_I$ is not compatible in $\mathbb{I}$ as long as $0_i, 1_i \in X$ for some $i \in I$. Consequently, we have that $\mathsf{Tot}_{\mathbb{I}} = \{X \subseteq A_I : \{0_i, 1_i\} \nsubseteq X \text{ for each } i \in I\}$. This means that if I has $n \in \mathbb{N}$ elements then $\mathbb{I}$ has 3^n total components, and 2^n maximal total components.

Note that all the matrices $\mathbb{I}_X$ corresponding to maximal total components X are isomorphic. However, by enriching the signature of $\mathbb{I}$ with more connectives, we could defined a new matrix $\mathbb{I}'$ in such a way that the submatrices $\mathbb{I}'_X$ would all induce different logics, whose intersection would still be induced by $\mathbb{I}'$.

Note also that if we changed the definition of $g_{\mathbb{I}}(x, y) = x$ to $g_{\mathbb{I}}(x, y) = A_I$ when the operation is non-empty, making $\mathbb{I}$ a PNmatrix, we would end up with the same set of total components. △

Hence, Pmatrices can host an exponential number of matrices and, similarly, PNmatrices can host an exponential number of Nmatrices. However, a crucial difference appears in the computational cost of checking if a set of truth-values is compatible when adding partiality and non-determinism, as the following two results show.

Proposition 5. *Let $\mathbb{M} = \langle A, \cdot_{\mathbb{M}}, D\rangle$ be a finite Pmatrix. Given $X \subseteq A$, the problem of deciding if X is compatible in $\mathbb{M}$ is in* ***P***.

Proof. This problem amounts to checking if the empty entries in the truth-tables of $\mathbb{M}$ can be avoided when starting with the truth-values in X. If $\mathbb{M}_X$ is total then clearly X itself is a total component. If there is an empty entry in $\mathbb{M}_X$ and that entry is already empty in $\mathbb{M}$ then X is not contained in any total component. If, instead, the empty entry in $\mathbb{M}_X$ corresponds to a non-empty entry in $\mathbb{M}$, then let Y collect those values. Clearly, $X \cap Y = \emptyset$. Now make $Z = X \cup Y$. Then X is contained in a total component if and only if Z is. Hence, we can repeat the strategy above for Z. We are bound to get a definite answer after at most $|A \setminus X|$ steps. □

The previous procedure only works in a deterministic setting, as we face no choices when building the set Z since the entries of $\mathbb{M}$ have at most one value. In the presence of non-determinism, it might be the case that an empty table entry can be filled with more than one value. In this case, every possible value needs to be checked, because it may be that only one of the available values leads to a total component.

Example 6. Consider the PNmatrix $\mathbb{M} = \langle A, \cdot_{\mathbb{M}}, D\rangle$ over a signature with a single connective g, with $D \subseteq A = \{a, b, c\}$ and

$g_{\mathbb{M}}$	a	b	c
a	b, c	c	A
b	A	A	$\emptyset$
c	c	A	A

In order to determine whether $X = \{a\}$ is compatible, as $g_{\mathbb{M}}(a, a) = \{b, c\}$ we need to consider two cases $Z = \{a, b\}$ and $Z = \{a, c\}$. However, only the choice

$\{a, c\}$ allows us to conclude that X is compatible, since $\{a, c\} \in \mathsf{Tot}_{\mathbb{M}}$. Choosing $\{a, b\}$ does not lead to a total component, since $g_{\mathbb{M}}(a, b) = \{c\}$ and $g_{\mathbb{M}}(b, c) = \emptyset$. △

We still know that the compatibility problem is in **NP** since, given a PNmatrix $\mathbb{M} = \langle A, \cdot_{\mathbb{M}}, D\rangle$ and $X \subseteq A$, we can guess Y such that $X \subseteq Y$ and check if $Y \in \mathsf{Tot}_{\mathbb{M}}$. Moreover, for PNmatrices checking if X is compatible is already **NP-complete**. In fact, this holds already for $X = \emptyset$ as the next theorem shows.

Theorem 7. *The problem of deciding if a given finite PNmatrix has a total component is **NP-complete**.*

Proof. Knowing that the problem is in **NP** we just have to show it is **NP-hard**. For that purpose, we will reduce 3SAT to it. Let $\varphi = C_1 \wedge \cdots \wedge C_n$ be an instance of 3SAT (recall Example 1). Let $\mathsf{var}(\varphi) = \{x_1, \ldots, x_k\}$ for some $k \in \mathbb{N}$. For each literal $L_{i,j}$, let $x_{i,j}$ be such that $\mathsf{var}(L_{i,j}) = \{x_{i,j}\}$. Let Σ be a signature containing one unary function symbol cycle, one binary function symbol cons and a ternary function symbol C_i, for every $1 \leq i \leq n$. Consider the Σ-PNmatrix $\mathbb{M} = \langle A, \cdot_{\mathbb{M}}, D\rangle$ where $A = \{x_r : 1 \leq r \leq k\} \cup \{\neg x_r : 1 \leq r \leq k\}$, $D \subseteq A$ (the choice of D is irrelevant), and

$$\mathsf{cons}_{\mathbb{M}}(y_1, y_2) = \begin{cases} \emptyset & \text{if } y_1 = x_r \text{ and } y_2 = \neg x_r \text{ for some } 1 \leq r \leq k, \\ A & \text{otherwise,} \end{cases}$$

$$\mathsf{cycle}_{\mathbb{M}}(y) = \begin{cases} \{x_{r+1}, \neg x_{r+1}\} & \text{if } y = x_r \text{ or } y = \neg x_r \text{ for some } 1 \leq r < k, \\ \{x_1, \neg x_1\} & \text{if } y = x_k \text{ or } y = \neg x_k, \end{cases}$$

$$(\mathsf{C}_i)_{\mathbb{M}}(y_1, y_2, y_3) = \begin{cases} \emptyset & \text{if } y_j \in \{x_{i,j}, \neg x_{i,j}\} \text{ and } y_j \neq L_{i,j} \text{ for all } j \in \{1, 2, 3\}, \\ A & \text{otherwise,} \end{cases}$$

for every $1 \leq i \leq n$.

We will now show that $\mathsf{Tot}(\mathbb{M}) \neq \emptyset$ if and only if φ is satisfiable. Notice that, if $X \in \mathsf{Tot}(\mathbb{M})$, then it must satisfy the following properties: (1) the definitions of $\mathsf{cons}_{\mathbb{M}}$ and $\mathsf{cycle}_{\mathbb{M}}$ guarantee that, for every $1 \leq r \leq k$, exactly one among x_r and $\neg x_r$ must be in X (2) for every $1 \leq i \leq n$, the definition of $(\mathsf{C}_i)_{\mathbb{M}}$ guarantees that $L_{i,j} \in X$ for some $j \in \{1, 2, 3\}$.

For the left to right direction, let $X \in \mathsf{Tot}(\mathbb{M})$, by property (1), we can define

$$v_X(x_r) = \begin{cases} 1 & \text{if } x_r \in X, \\ 0 & \text{if } \neg x_r \in X, \end{cases}$$

and, by property (2), we have that $v_X(\varphi) = 1$.

For the other direction, suppose v is a variable assignment such that $v(\varphi) = 1$. Consider the set

$$X_v = \{x_i : v(x_i) = 1\} \cup \{\neg x_i : v(x_i) = 0\}.$$

It is obvious that X_v satisfies the property (1), so $\mathsf{cons}_{\mathbb{M}}$ and $\mathsf{cycle}_{\mathbb{M}}$ are total when restricted to X_v. Furthermore, since $v(\varphi) = 1$, we have that for every $1 \leq i \leq n$, there is $j \in \{1, 2, 3\}$ such that $v(L_{i,j}) = 1$ and, consequently, $L_{i,j} \in X_v$. This ensures that $(\mathsf{C}_i)_{\mathbb{M}}$ is total as well, so $X_v \in \mathsf{Tot}(\mathbb{M})$. □

3.2 Determining the Existence of Non-theorems

As discussed in the begining of this section, theorem universality, or dually the existence of a formula that is not a theorem, can be decided in constant time for logical (N)matrices. On the other hand, a P(N)matrix has a non-theorem if and only if there is a total component containing a non-designated element. As the next theorem shows, partiality alone still allows for an efficient procedure to decide $\forall\mathsf{Thm}$.

Theorem 8. *The problem* $\forall\mathsf{Thm}$ *for finite Pmatrices is in* ***P***.

Proof. In order to decide $\forall\mathsf{Thm}(\mathsf{PMatr})$ we need to check, for a given Pmatrix $\mathbb{M} = \langle A, \cdot_{\mathbb{M}}, D\rangle$, if there is some $X \in \mathsf{Tot}_{\mathbb{M}}$ such that $X \not\subseteq D$. We can apply the polynomial procedure presented in Proposition 5 to every singleton containing a non-designated value. This procedure runs in polynomial time in the size of $\mathbb{M}$, so $\forall\mathsf{Thm}(\mathsf{PMatr}) \in \mathbf{P}$. □

Assuming $\mathbf{P} \neq \mathbf{NP}$, we conclude that the scenario changes radically when we allow for partiality together with non-determinism.

Theorem 9. *The problem* $\forall\mathsf{Thm}$ *for finite PNmatrices is* ***coNP-complete***.

Proof. We show that $\overline{\forall\mathsf{Thm}}(\mathsf{PNMatr}) \in$ **NP-complete**. This is equivalent to the problem of checking, for a given PNmatrix $\mathbb{M} = \langle A, \cdot_{\mathbb{M}}, D\rangle$, if there is $X \in \mathsf{Tot}_{\mathbb{M}}$ with $X \not\subseteq D$. The problem is clearly in **NP**, since we can guess $X \subseteq A$ with $X \not\subseteq D$, and check in polynomial time on the size of $\mathbb{M}$ if $X \in \mathsf{Tot}_{\mathbb{M}}$. **NP-hardness** of $\overline{\forall\mathsf{Thm}}(\mathsf{PNMatr})$ follows from Theorem 7 and the fact that $\mathbb{M}$ has a total component if and only if $\mathbb{M}' = \langle A, \cdot_{\mathbb{M}}, \emptyset\rangle$ has a total component containing a non-designated value (since every element of $\mathbb{M}'$ is non-designated). □

4 Checking Theorem Existence

We now address the problem $\exists\mathsf{Thm}$. Note that for any PNmatrix, by substitution invariance, there are theorems if and only if there are theorems in (at most) a single variable. Furthermore, if the signature has 0-ary connectives, then there are theorems if and only if there are closed theorems.

Theorem 10. *The problem* $\exists\mathsf{Thm}$ *for finite (P)matrices is in* ***EXPTIME****, and its restriction to (P)matrices over signatures containing at least one* 0*-ary connective is in* ***P***.

Proof. Let $\mathbb{M} = \langle A, \cdot_{\mathbb{M}}, D\rangle$ be a Σ-Pmatrix with $m = |\Sigma|$ and $n = |A|$. Of course, if $\Sigma^{(k)} = \emptyset$ for every k, then $\mathbb{M} \in \exists\mathsf{Thm}$ if and only if $A = D$, which can be checked in linear time. Henceforth we will assume $\Sigma^{(j)} \neq \emptyset$ for some j. Let k be the largest such that $\Sigma^{(k)} \neq \emptyset$.

Suppose $\Sigma^{(0)} \neq \emptyset$. If $©_{\mathbb{M}} = \emptyset$ for some $© \in \Sigma^{(0)}$, then $\mathsf{Val}(\mathbb{M}) = \emptyset$ and $\mathsf{Thm}(\mathbb{M}) = \mathbf{L}_\Sigma(P) \neq \emptyset$. Otherwise, $\vdash_{\mathbb{M}}$ has theorems if and only if it has closed theorems, i.e., if there is a designated element accessible from the set X_0 gathering the truth-values obtained by interpreting the 0-ary connectives. This can be checked in polynomial time on n. Start by constructing the set $X_0 = \bigcup_{©\in\Sigma^{(0)}} ©_{\mathbb{M}}$ (where each element is associated with some atomic formula $©$) and check if it is compatible. If not then $\mathsf{Thm}(\mathbb{M}) \neq \emptyset$. Otherwise, proceed to the next step. In step $i+1$ define

$$X_{i+1} = \bigcup\{©_{\mathbb{M}}(x_1,\dots,x_k) : x_1,\dots,x_k \in X_i, © \in \Sigma^k\} \cup X_i.$$

Each new element of X_{i+1} comes from some $x_1,\dots,x_k \in X_i$ and $© \in \Sigma^k$ and we associated it with the formula $©(\varphi_1,\dots,\varphi_k)$ whenever each φ_j was associated with x_j. If at any point a designated element is generated, the associated formula is a theorem. If the fixed point is reached without this happening then $\mathsf{Thm}(\mathbb{M}) = \emptyset$. Since $X_i \subseteq A$, for every i, we must have $X_{r+1} = X_r$ for some $r \leq |A \setminus X_0|$. Furthermore, in each step, the procedure evaluates, at most, $m \times n^k$ potentially new functions, which is polynomial in the size of $\mathbb{M}$. We conclude that $\exists\mathsf{Thm}$ over Pmatrices with $\Sigma^{(0)} \neq \emptyset$ is in $\mathbf{P}$.

Now suppose $\Sigma^{(0)} = \emptyset$. In order to deal with partiality, we extend the approach in [16], dealing with the total deterministic case. We only need to consider formulas with at most one variable, and for every $\varphi \in \mathbf{L}_\Sigma(\{p\})$, we have that $\varphi \in \mathsf{Thm}(\mathbb{M})$ if and only if $\varphi_{\mathbb{M}}(a) \in D$ for every a such that $\{a\}$ is compatible. Let $Y = \{a \in A : \{a\}$ is compatible$\}$, the problem is, therefore, equivalent to checking if there is an expressible 1-ary function f such that $f(Y) \subseteq D$ in which case the function is said to be *designated.* In a deterministic setting, the functions expressed by complex formulas are the composition of the interpretation function of the head connective with the functions expressed by the immediate subformulas. Therefore, the set of all 1-ary functions expressible in the underlying algebra can be algorithmically generated as it is usually done when calculating clones over finite algebras [19].

Start by constructing the set Y, which Proposition 5 guarantees that can be done in polynomial time. In the first step we make $X_0 = \{p_{\mathbb{M}}\}$, whose only element is the identity function, associated with the formula p. In step $i+1$, we make

$$X_{i+1} = \{©_{\mathbb{M}}(f_1,\dots,f_k) : f_1,\dots,f_k \in X_i, © \in \Sigma^{(k)}\} \cup X_i.$$

We associate each new function $f \in X_{i+1}$ with some formula: we pick some $© \in \Sigma^{(k)}$ and $f_1,\dots,f_k \in X_i$ such that $f = ©_{\mathbb{M}}(f_1,\dots,f_k)$ and associate f with the formula $©(\varphi_1,\dots,\varphi_k)$ whenever f_ℓ was associated with the formula φ_ℓ, for $1 \leq \ell \leq k$. If at any point a function f is generated such that $f(Y) \subseteq D$, then the associated formula is a theorem. If the fixed point is reached without this happening, then $\mathsf{Thm}(\mathbb{M}) = \emptyset$. Since there are at most n^n 1-ary functions over a set with n-elements, we conclude that a fixed point must be found in, at most, n^n steps. In each step, the procedure evaluates, at most,

$$m \times (n^n)^k = m \times n^{nk} = m \times 2^{\log(n)nk}$$

potentially new functions. We conclude that this procedure runs in exponential time and so $\exists\mathsf{Thm}(\mathsf{PMatr}) \in$ **EXPTIME**. $\square$

In the following example we give a hint at why a similar strategy may not work in the presence of non-determinism. A crucial difference from the deterministic case is that the (multi)-functions represented by formulas are sensitive to the syntax, as a repeated subformula cannot have a different value even if there are multiple choices. On a matrix $\mathbb{M}$ the function $©(\varphi_1, \dots, \varphi_k)_{\mathbb{M}}$ is the function obtained by composing the function $©_{\mathbb{M}}(p_1, \dots, p_k)$ with the functions $(\varphi_1)_{\mathbb{M}}, \dots, (\varphi_k)_{\mathbb{M}}$. On an Nmatrix $©(\varphi_1, \dots, \varphi_k)_{\mathbb{M}}$ is not a function but a multi-function. Crucially, the multi-function $©(\varphi_1, \dots, \varphi_k)_{\mathbb{M}}$ does not depend only on the multi-functions $©_{\mathbb{M}}$ and $(\varphi_1)_{\mathbb{M}}, \dots, (\varphi_k)_{\mathbb{M}}$.

Hence, contrary to what happens in the deterministic case, when generating the expressible multi-functions in an Nmatrix $\mathbb{M}$ (to find if there are theorems, or for any other purpose), we cannot just keep the information about the multi-functions themselves but also about the formulas that produce them. Otherwise we might generate a non-expressible function (as every occurrence of a subformula must have the same value) or miss some multi-functions that are still expressible.

Example 11. Consider $\mathbb{M} = \langle \{a, b, c\}, \cdot_{\mathbb{M}}, D\rangle$ *over a signature* Σ *with a single binary connective* g *with*

$g_{\mathbb{M}}$	a	b	c
a	c	a	b, c
b	b	c	a, c
c	b, c	a, c	c

For $\varphi \in L_\Sigma(\{p\})$, *let us identify the unary multi-function expressed by* φ *with the 3-tuple gathering the output set for each input,* $\varphi_{\mathbb{M}} = \langle \varphi_{\mathbb{M}}(a), \varphi_{\mathbb{M}}(b), \varphi_{\mathbb{M}}(c)\rangle$. *The formula* p *is associated to the identity multi-function* $p_{\mathbb{M}} = \langle \{a\}, \{b\}, \{c\}\rangle$ *and the formula* $g(p, p)$ *is associated to the constant multi-function* $g(p, p)_{\mathbb{M}} = \langle \{c\}, \{c\}, \{c\}\rangle$ *corresponding to the diagonal of* $g_{\mathbb{M}}$. *The formulas* $\varphi = g(g(p, p), p)$ *and* $\psi = g(p, g(p, p))$ *induce the same multi-function* $\langle \{b, c\}, \{a, c\}, \{c\}\rangle$. *However,* $g(\varphi, \varphi)_{\mathbb{M}} = g(\psi, \psi)_{\mathbb{M}} = g(p, p)_{\mathbb{M}}$ *but* $g(\varphi, \psi)_{\mathbb{M}} = g(\psi, \varphi)_{\mathbb{M}}$ *corresponds to the multi-function* $\langle \{a, c\}, \{b, c\}, \{c\}\rangle$, *thus showing the above mentioned sensitivity to the syntax.*

Note that $g_{\mathbb{M}}(\varphi_{\mathbb{M}}, \psi_{\mathbb{M}}) = \langle \{a, b, c\}, \{b, c\}, \{c\}\rangle \neq g(\varphi, \psi)_{\mathbb{M}}$. *That is, the multi-function expressed by* $g(\varphi, \psi)$ *is different from the composition* $g_{\mathbb{M}}(\varphi_{\mathbb{M}}, \psi_{\mathbb{M}})$ *of the involved multi-functions (as relations).* $\triangle$

As we shall see, the phenomenon illustrated in the previous example has a deep effect on the decidability of problem $\exists\mathsf{Thm}$ (and consequently also on problem EqThm).

4.1 A Bridge with (Term-DAG) Automata

To establish our undecidability results, we will take advantage of the undecidability results for term-DAG-automata in [1]. Recall that a DAG is a directed acyclic graph. A term-DAG is a DAG whose nodes correspond to the subformulas of some formula and whose arrows are numbered according to their arity pointing to the immediate subformulas in their corresponding position, as illustrated below. When discussing term-DAG-automata we will identify a formula with its term-DAG representation, and the nodes of the DAG with the corresponding subformulas. As an example, consider two term-DAGs

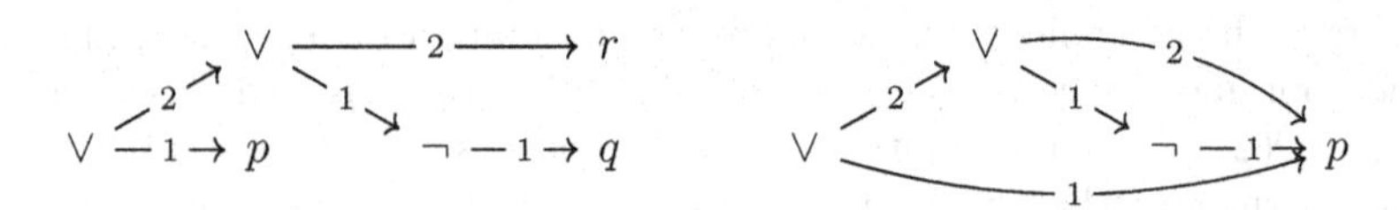

These represent, respectively, the clauses $p \vee (\neg q \vee r)$ and $p \vee (\neg p \vee p)$ (see Example 1).

A *Σ-(finite) term-DAG automaton* is a tuple $\mathcal{A} = \langle \Sigma, Q, F, \delta \rangle$ where Q is a finite non-empty set of states, $F \subseteq Q$ is the set of final (or accepting) states, and δ is a set of transition rules of the form $\langle f(q_1, \ldots, q_n), q \rangle$, where $f \in \Sigma$ is a function symbol of arity n and $q_1, \ldots, q_n, q \in Q$. A *run* of a $\mathcal{A}$ on a term-DAG φ is a mapping $r : \mathsf{sub}(\varphi) \to Q$ such that, for every subformula node u, if the connective labelling u and heading the subformulas is f of arity n, then $\langle f(r(u_1), \ldots, r(u_n)), r(u) \rangle \in \delta$, where $u_1, \ldots, u_n$ are the successor vertices of u, corresponding to its immediate subformulas, given in order. If $r(\varphi) \in F$ then the run is said to be *accepting*. A term-DAG automaton $\mathcal{A}$ is said to *accept* a term-DAG φ if there is an accepting run of $\mathcal{A}$ on φ. We denote by $\mathcal{L}(\mathcal{A})$ the *language* of the automaton $\mathcal{A}$, i.e., the set of all term-DAGs accepted by $\mathcal{A}$.

The universality problem for finite term-DAG is the problem of determining, for a given finite Σ-term-DAG automaton $\mathcal{A}$, if $\mathcal{L}(\mathcal{A}) = L_\Sigma(\emptyset)$. This problem was shown to be undecidable [1, Theorem 4] and it will help us establishing our undecidability results.

Given a Σ-term-DAG $\mathcal{A} = \langle \Sigma, Q, F, \delta \rangle$, consider the Σ-Nmatrix $\mathbb{M}_\mathcal{A} = \langle Q \cup \{*\}, \cdot_{\mathbb{M}_\mathcal{A}}, Q \setminus F \rangle$ where, for every $f \in \Sigma$ of arity n and $q_1, \ldots, q_n \in Q$, $f_{\mathbb{M}_\mathcal{A}}(q_1, \ldots, q_n) = \{*\}$ if there is no rule in δ with $f(q_1, \ldots, q_n)$ as its left side, and $f_{\mathbb{M}_\mathcal{A}}(q_1, \ldots, q_n) = \{q \in Q : \langle f(q_1, \ldots, q_n), q \rangle \in \delta\}$ otherwise.

Proposition 12. *Let $\mathcal{A}$ be a Σ-term-DAG, then $\mathcal{L}(\mathcal{A})$ is the set of closed non-theorems of $\mathbb{M}_\mathcal{A}$.*

Proof. As $\cdot_{\mathbb{M}_\mathcal{A}}$ codes the transitions of δ, given a closed formula φ and a partial valuation $v : \mathsf{sub}(\varphi) \to Q$ over $\mathbb{M}$, $v(\varphi)$ is non-designated, i.e., $v(\varphi) \in F$ precisely if v is an accepting run of φ in $\mathcal{A}$. □

Theorem 13. *The problem* $\exists\mathsf{Thm}$ *for finite PNmatrices is recursively enumerable but undecidable.*

Proof. The fact that $\exists\mathsf{Thm}(\mathsf{PNMatr})$ is recursively enumerable follows easily from the fact that we can enumerate all formulas with at most one variable and check whether each of them is a theorem of $\vdash_{\mathbb{M}}$. If there is a theorem, we will eventually find it.

As for the undecidability proof, we reduce the universality problem for term-DAG automata to $\overline{\exists\mathsf{Thm}}(\mathsf{PNMatr})$. For a given Σ-term-DAG automata $\mathcal{A}$, we pick the associated Σ-Nmatrix $\mathbb{M}_{\mathcal{A}}$ as per Proposition 12. Let $\mathbb{M}_{\mathcal{A}} = \langle A, \cdot_{\mathcal{A}}, D\rangle$ and $a \notin A$. Consider $\mathbb{M}^a_{\mathcal{A}} = \langle A \cup \{a\}, \cdot_a, D\rangle$ with

$$©_a(x_1, \dots, x_k) = \begin{cases} ©_{\mathbb{M}}(x_1, \dots, x_k) & \text{if } a \notin \{x_1, \dots, x_k\} \\ \{a\} & \text{otherwise.} \end{cases}$$

Clearly, the theorems of $\mathbb{M}^a_{\mathcal{A}}$ are the closed theorems of $\mathbb{M}_{\mathcal{A}}$, as formulas with variables can be falsified by sending the variables to a (see [10]). Hence, by Proposition 12, we conclude that $\mathcal{L}(\mathcal{A}) = L_{\Sigma}(\emptyset)$ if and only if $\mathsf{Thm}(\mathbb{M}^a_{\mathcal{A}}) = \emptyset$.

The undecidability of $\exists\mathsf{Thm}(\mathsf{PNMatr})$ follows directly from $\mathsf{NMatr} \subseteq \mathsf{PNmat}$. □

This theorem implies that the situation portrayed in Example 11 is not really avoidable and, in general, in the presence of non-determinism, there is not a bound on the size of the formulas that guarantees that we have covered all expressible multi-functions.

5 Deciding Equality of Theoremhood

We will see that just like in the case of checking the existence of theorems, when comparing the theorems induced by PNmatrices, the non-determinism spoils decidability.

Theorem 14. *The problem* EqThm *for finite (P)matrices is in* ***2-EXPTIME****.*

Proof. Let $\mathbb{M}_1$ and $\mathbb{M}_2$ be finite Pmatrices over a signature Σ. Just like in Theorem 10, when $\Sigma^{(k)} = \emptyset$ for all k, we have that either $\mathsf{Thm}(\mathbb{M}_i) = P$ when every truth-value of $\mathbb{M}_i$ is designated, or $\mathsf{Thm}(\mathbb{M}_i) = \emptyset$ otherwise. In which case we can check if $\mathsf{Thm}(\mathbb{M}_1) = \mathsf{Thm}(\mathbb{M}_2)$ in linear time. We now assume that $\Sigma^{(k)} \neq \emptyset$ for some k.

Let n be the maximum cardinality of the sets of truth-values of $\mathbb{M}_1$ and $\mathbb{M}_2$, $m = |\Sigma|$ and k be the largest such that $\Sigma^{(k)} \neq \emptyset$. The proof follows by extending the idea presented in [17] to deal with partiality.

In the absence of non-determinism, a valuation over $\mathbb{M}_i$ is completely determined by the values attributed to the variables, and as there are at most n possible values, we obtain that $\varphi \in \mathsf{thm}(\mathbb{M}_i)$ if and only if $\varphi^\sigma \in \mathsf{thm}(\mathbb{M}_i)$ for every $\sigma : P \to \{p_1, \dots, p_n\}$. Hence, $\mathsf{thm}(\mathbb{M}_1) = \mathsf{thm}(\mathbb{M}_2)$ if and only if $\mathsf{thm}(\mathbb{M}_1) \cap L_{\Sigma}(\{p_1, \dots, p_n\}) = \mathsf{thm}(\mathbb{M}_2) \cap L_{\Sigma}(\{p_1, \dots, p_n\})$.

Suppose $\mathbb{M}_1$ and $\mathbb{M}_2$ have, respectively, m_1 and m_2 maximal total components. For each $1 \leq i \leq m_1$ and $1 \leq j \leq m_2$ we denote by $\mathbb{M}_{1,i}$ and $\mathbb{M}_{2,j}$ each of the corresponding submatrices. Every formula $\varphi \in L_\Sigma(\{p_1,\dots,p_n\})$ induces functions $\varphi_{i,j} : \mathbb{M}^n_{i,j} \to \mathbb{M}_{i,j}$, for every choice of i and j. Of course, φ is a theorem of $\mathbb{M}_i$ if and only if $\varphi_{i,j}$ is a designated function for every $1 \leq j \leq m_i$.

Adapting the idea used in Theorem 10, we recursively generate the set of tuples $\bar{t} = \langle f_{1,1},\dots,f_{1,m_1}, f_{2,1},\dots,f_{2,m_2}\rangle$, where each of the $f_{i,j}$ correspond to the n-ary functions expressible in the various maximal total components of $\mathbb{M}_i$. For each variable $p_\ell \in \{p_1,\dots,p_n\}$ and each $\mathbb{M}_{i,j}$ we consider the function $(p_\ell)_{\mathbb{M}_{i,j}}(x_1,\dots,x_k) = \pi_{i,j,\ell}(x_1,\dots,x_n)$ where the $\pi_{i,j,\ell}(x_1,\dots,x_n) = x_\ell$ are the various ℓ-projections in n arguments for each of the maximal total components of $\mathbb{M}_1$ and $\mathbb{M}_2$. Given a k-vector $\boldsymbol{h} = \langle \bar{t}_1,\dots,\bar{t}_k\rangle$ of (m_1+m_2)-tuples

$$\bar{t}_\ell = \langle f_{1,1,\ell},\dots,f_{1,m_1,\ell}, f_{2,1,\ell},\dots,f_{2,m_2,\ell}\rangle,$$

define the function over $\mathbb{M}_{i,j}$ by $f^{©,\boldsymbol{h}}_{i,j} = ©_{\mathbb{M}_{i,j}}(f_{i,j,1},\dots,f_{i,j,k})$.

In the first step we consider the set

$$X_0 = \{\langle \pi_{1,1,\ell},\dots,\pi_{1,m_1,\ell},\pi_{2,1,\ell},\dots,\pi_{2,m_2,\ell}\rangle : 1 \leq \ell \leq n\},$$

where each tuple $\langle \pi_{1,1,\ell},\dots,\pi_{1,m_1,\ell},\pi_{2,1,\ell},\dots,\pi_{2,m_2,\ell}\rangle$ is associated with a variable $p_\ell \in \{p_1,\dots,p_n\}$.

In step $i+1$ we make

$$X_{i+1} = X_i \cup \{\langle f^{©,\boldsymbol{h}}_{1,1},\dots,f^{©,\boldsymbol{h}}_{1,m_1}, f^{©,\boldsymbol{h}}_{2,1},\dots,f^{©,\boldsymbol{h}}_{2,m_2}\rangle : © \in \Sigma^k, \boldsymbol{h} \in X_i^k\}.$$

Each new tuple in X_{i+1}, generated from a connective © and a vector $\boldsymbol{h} = \langle \bar{t}_1,\dots,\bar{t}_k\rangle$, is associated with the formula $©(\varphi_1,\dots,\varphi_k)$ whenever φ_ℓ was associated with the tuple $\bar{t}_\ell$.

An expressible function $f_{i,j}$ is said to be *designated* if it outputs a designated element for every possible input. We conclude that $\mathsf{thm}(\mathbb{M}_1) \neq \mathsf{thm}(\mathbb{M}_2)$, if at any point we generate a tuple $\bar{t} = \langle f_{1,1},\dots,f_{1,m_1}, f_{2,1},\dots,f_{2,m_2}\rangle$ such that all $f_{1,1},\dots,f_{1,m_1}$ are designated and some among $f_{2,1},\dots,f_{2,m_2}$ is not, or all $f_{2,1},\dots,f_{2,m_2}$ are designated and some among $f_{1,1},\dots,f_{1,m_1}$ is not. Furthermore, the formula associated with that tuple testifies the difference between $\mathsf{thm}(\mathbb{M}_1)$ and $\mathsf{thm}(\mathbb{M}_2)$. Otherwise, we continue to the next step until a fixed point is reached, in which case we conclude that $\mathsf{thm}(\mathbb{M}_1) = \mathsf{thm}(\mathbb{M}_2)$.

Since there are, at most, n^{n^n} of n-ary expressible functions over a set with n elements, we have that $|X_i| \leq (n^{n^n})^{m_1+m_2} \leq (n^{n^n})^{2^{n+1}}$, and the fixed point $X_{r+1} = X_r$ must be reached for

$$r \leq (n^{n^n})^{2^{n+1}} = n^{2^{n+1}\times n^n} \leq n^{2^{n^2+n+1}} \leq 2^{n\times 2^{n^2+n+1}} \leq 2^{2^n\times 2^{n^2+n+1}} = 2^{2^{n^2+2n+1}}.$$

Furthermore, in each step i, it takes $m \times |X_i|^k \times (m_1+m_2) \leq m \times 2^{2^{n^2+2n+1}} \times 2^{n+1}$ operations to construct the set X_{i+1}, so the procedure runs in double-exponential time and $\mathsf{EqThm}(\mathsf{PMatr}) \in$ **2-EXPTIME**. □

In the presence of non-determinism and the consequent lack of a bound in the depth of formulas that may express new multi-functions, we have again a negative result for testing the equality of the sets of theorems in the logics defined by (P)Nmatrices.

Theorem 15. *The problem* EqThm *for finite PNmatrices is undecidable and co-recursively enumerable.*

Proof. It is clear that $\overline{\mathsf{EqThm}}(\mathsf{PNMatr})$ is recursively enumerable. Given $\mathbb{M}_1$ and $\mathbb{M}_2$ and an enumeration of their formulas, for each we check if it is a theorem of $\vdash_{\mathbb{M}_1}$ and not $\vdash_{\mathbb{M}_2}$, and vice-versa. If so, the procedure terminates and the matrices have a different set of theorems. If their sets of theorems are distinct we are bound to find the difference at some point.

However, there is no bound allowing us to conclude in general that the sets of theorems of $\vdash_{\mathbb{M}_1}$ and $\vdash_{\mathbb{M}_2}$ are the same, and the problem $\mathsf{EqThm}(\mathsf{PNMatr})$ is in fact undecidable. To see that let us reduce the problem $\overline{\exists\mathsf{Thm}}(\mathsf{PNMatr})$ (already proven to be undecidable in Theorem 13) to $\mathsf{EqThm}(\mathsf{PNMatr})$.

For a given signature Σ, consider the Nmatrix $\mathbb{M}_{\mathsf{free}} = \langle\{0,1\}, \cdot_{\mathsf{free}}, \{1\}\rangle$ with $©_{\mathsf{free}}(x_1,\ldots,x_k) = \{0,1\}$ for every $© \in \Sigma^k$ and $x_1,\ldots,x_k \in \{0,1\}$. Then, given a Σ-PNmatrix $\mathbb{M}$, we have $\mathsf{thm}(\vdash_{\mathbb{M}}) = \emptyset$ if and only if $\mathsf{thm}(\vdash_{\mathbb{M}}) = \mathsf{thm}(\vdash_{\mathbb{M}_{\mathsf{free}}})$. □

6 Conclusions and Further Work

In this paper we have shown that the computational properties of PNmatrices are, in many cases, harder than those of logical matrices. Even worse, properties like the existence of a theorem, or equality of the sets of theorems, become undecidable. Our results are summarized in the table below. Recall that Σ_1^0 and Π_1^0 correspond, in the arithmetical hierarchy, precisely to the classes of undecidable problems which are nevertheless recursively enumerable and co-recursively enumerable, respectively.

$\mathcal{C}$	SAT($\mathcal{C}$)	VAL($\mathcal{C}$)	∀Thm($\mathcal{C}$)	∃Thm($\mathcal{C}$)	EqThm($\mathcal{C}$)
Matr	**NP**	**coNP**	**P**	**EXPTIME**	**2-EXPTIME**
PMatr	**NP**	**coNP**	**P**	**EXPTIME**	**2-EXPTIME**
NMatr	**NP**	**coNP**	**P**	Σ_1^0	Π_1^0
PNMatr	**NP**	**coNP**	**coNP**	Σ_1^0	Π_1^0

Notwithstanding, these facts do not hinder the considerable usefulness of using partiality and non-determinism in semantical approaches to logic, as explained in the introduction, and the search for necessary and/or sufficient conditions for such properties to hold, namely by means of rexpansion homomorphisms and generalized algebraic techniques.

The results in this paper are not definitive, though, as they do not settle the initial motivating problem posed by Zohar and Avron, of checking whether two Nmatrices define the same logic. In future work we aim at applying similar ideas and techniques to address that problem, both in the Set × Fmla setting of Tarskian logics as considered in this paper, and in the more general Set × Set setting of multiple-conclusion logics [25,26]. We also want to study the computational import of other crucial properties, namely monadicity [7,22], which is instrumental for the automated treatment of the underlying logics using analytic calculi.

Concerning the techniques used in this paper, it is certainly important to further explore the relationship between PNmatrices and term-DAG-automata, and understand it at the light of the introduction of infectious values [10] as used in the proof of Theorem 13. Further, note that the computational characterizations shown in the table are all tight, though completeness is not mentioned, for simplicity, with the possible exception of the **EXPTIME** and **2-EXPTIME** entries. We expect that also here, a careful reduction from the bounded halting problem for counter machines may help settle the **EXPTIME-completeness** of $\exists\mathsf{Thm}(\mathsf{PMatr})$.

References

1. Anantharaman, S., Narendran, P., Rusinowitch, M.: Closure properties and decision problems of dag automata. Inf. Process. Lett. **94**(5), 231–240 (2005). https://doi.org/10.1016/j.ipl.2005.02.004
2. Avron, A.: Non-deterministic semantics for families of paraconsistent logics. In: Handbook of Paraconsistency, Studies in Logic, vol. 9. College Publications (2007)
3. Avron, A., Lev, I.: Non-deterministic multiple-valued structures. J. Log. Comput. **15**(3), 241–261 (2005). https://doi.org/10.1093/logcom/exi001
4. Avron, A., Zamansky, A.: Non-deterministic semantics for logical systems. In: Gabbay, D., Guenthner, F. (eds.) Handbook of Philosophical Logic. Handbook of Philosophical Logic, vol. 16. Springer, Dordrecht (2011). https://doi.org/10.1007/978-94-007-0479-4_4
5. Avron, A., Zohar, Y.: Rexpansions of non-deterministic matrices and their applications. Rev. Symb. Log. **12**(1), 173–200 (2019). https://doi.org/10.1017/S1755020318000321
6. Baaz, M., Lahav, O., Zamansky, A.: Finite-valued semantics for canonical labelled calculi. J. Autom. Reason. **51**(4), 401–430 (2013). https://doi.org/10.1007/s10817-013-9273-x
7. Caleiro, C., Marcelino, S.: Analytic calculi for monadic PNmatrices. In: Iemhoff, R., Moortgat, M., de Queiroz, R. (eds.) WoLLIC 2019. LNCS, vol. 11541, pp. 84–98. Springer, Heidelberg (2019). https://doi.org/10.1007/978-3-662-59533-6_6
8. Caleiro, C., Marcelino, S.: Modular semantics for combined many-valued logics (2021, submitted)
9. Caleiro, C., Marcelino, S.: On axioms and rexpansions. In: Arieli, O., Zamansky, A. (eds.) Arnon Avron on Semantics and Proof Theory of Non-Classical Logics. OCL, vol. 21, pp. 39–69. Springer, Cham (2021). https://doi.org/10.1007/978-3-030-71258-7_3
10. Caleiro, C., Marcelino, S., Filipe, P.: Infectious semantics and analytic calculi for even more inclusion logics. In: 2020 IEEE 50th International Symposium on Multiple-Valued Logic, pp. 224–229 (2020). https://doi.org/10.1109/ISMVL49045.2020.000-1
11. Charatonik, W.: Automata on DAG representations of finite trees. Technical report, MPI-I-1999-2-001, Max-Planck-Institut für Informatik, Saarbrücken, Germany (1999)
12. Ciabattoni, A., Lahav, O., Spendier, L., Zamansky, A.: Taming paraconsistent (and other) logics: an algorithmic approach. ACM Trans. Comput. Log. **16**(1), 5:1–5:16 (2014). https://doi.org/10.1145/2661636

13. Cook, S.: The complexity of theorem-proving procedures. In: Proceedings of the 3rd Annual ACM Symposium on Theory of Computing, STOC 1971, pp. 151–158. Association for Computing Machinery (1971). https://doi.org/10.1145/800157.805047
14. Font, J.: Abstract Algebraic Logic. Mathematical Logic and Foundations, vol. 60. College Publications (2016)
15. Jorge, J.P., Holik, F.: Non-deterministic semantics for quantum states. Entropy **22**(2), 156 (2020). https://doi.org/10.3390/e22020156
16. Kalicki, J.: A test for the existence of tautologies according to many-valued truth-tables. J. Symb. Log. **15**(3), 182–184 (1950). https://doi.org/10.2307/2266783
17. Kalicki, J.: A test for the equality of truth-tables. J. Symb. Log. **17**(3), 161–163 (1952). https://doi.org/10.2307/2267687
18. Karp, R.M.: Reducibility among combinatorial problems. In: Miller, R.E., Thatcher, J.W., Bohlinger, J.D. (eds.) Complexity of Computer Computations. The IBM Research Symposia Series. Springer, Boston, MA (1972). https://doi.org/10.1007/978-1-4684-2001-2_9
19. Lau, D.: Function Algebras on Finite Sets. A Basic Course on Many-Valued Logic and Clone Theory. Springer Monographs in Mathematics. Springer, Heidelberg (2006). https://doi.org/10.1007/3-540-36023-9
20. Marcelino, S.: An unexpected Boolean connective. Logica Universalis (2021). https://doi.org/10.1007/s11787-021-00280-7
21. Marcelino, S., Caleiro, C.: Disjoint fibring of non-deterministic matrices. In: Kennedy, J., de Queiroz, R.J.G.B. (eds.) WoLLIC 2017. LNCS, vol. 10388, pp. 242–255. Springer, Heidelberg (2017). https://doi.org/10.1007/978-3-662-55386-2_17
22. Marcelino, S., Caleiro, C.: Axiomatizing non-deterministic many-valued generalized consequence relations. Synthese **198**(22), 5373–5390 (2019). https://doi.org/10.1007/s11229-019-02142-8
23. Marcos, J.: What is a non-truth-functional Logic? Stud. Logica. **92**(2), 215–240 (2009). https://doi.org/10.1007/s11225-009-9196-z
24. Omori, H., Skurt, D.: Untruth, falsity and non-deterministic semantics. In: 2021 IEEE 51th International Symposium on Multiple-Valued Logic, pp. 74–80 (2021). https://doi.org/10.1109/ISMVL51352.2021.00022
25. Scott, D.: Completeness and axiomatizability in many-valued logic. In: Henkin, L., Addison, J., Chang, C., Craig, W., Scott, D., Vaught, R. (eds.) Proceedings of the Tarski Symposium. Proceedings of Symposia in Pure Mathematics, vol. XXV, pp. 411–435. American Mathematical Society (1974). https://doi.org/10.1007/978-3-0346-0145-0_24
26. Shoesmith, D., Smiley, T.: Multiple-Conclusion Logic. Cambridge University Press (1978). https://doi.org/10.1017/CBO9780511565687
27. Wójcicki, R.: Theory of Logical Calculi. Basic Theory of Consequence Operations. Synthese Library, vol. 199. Kluwer (1998). https://doi.org/10.1007/978-94-015-6942-2

Soundness and Completeness Results for LEA and Probability Semantics

Eoin Moore(✉)

City University of New York Graduate Center, New York, NY 10016, USA
emoore@gradcenter.cuny.edu

Abstract. In [2], a logical system called the *logic of evidence aggregation* (LEA) was introduced, along with an intended semantics for it called *probability semantics*. The goal was to describe probabilistic evidence aggregation in the setting of formal logic. However, as noted in that paper, LEA is not complete with respect to probability semantics. This leaves open the tasks to find sound and complete semantics for LEA and a proper axiomatization for probability semantics. In this paper we do both. We define a class of basic models called deductive basic models. We show LEA is sound and complete with respect to the class of deductive basic models. We also define an axiomatic system LEA_+ extending LEA and show it is sound and complete with respect to probability semantics.

Keywords: Justification logic · Probability logic

1 Introduction

1.1 Overview of the Logic of Evidence Aggregation

Suppose there is a database of sentences Γ, from which follows the sentence A. Suppose, in addition, that the sentences in Γ are said not to hold certainly, but only with some individual probabilities. That is, we are now considering sentences as events in a probabilistic sense. How can we best estimate the probability for A to occur, given the probability estimates for Γ?

This is a central question in probability logic. See [1,4,6,7] for alternative approaches to this problem and the related problem of evidence aggregation. In [2], an elegant solution was proposed, where the power of the justification logic format was applied to the probabilistic setting. Let us recall the main ideas of that paper, for it is the axiom system and semantics introduced there which we are interested in developing here.

For each event $C_i \in \Gamma$, let there be some event u_i with known probability $P(u_i)$, such that C_i will occur if u_i occurs. In a justification logic format, we write

$$u_i{:}\, C_i.$$

A will follow if a specific configuration of u_i's occurs, written as

$$u_1{:}\, C_1, \ldots u_n{:}\, C_n \vdash t(u_1, \ldots, u_n){:}\, A.$$

S. Artemov and A. Nerode (Eds.): LFCS 2022, LNCS 13137, pp. 198–216, 2022.
https://doi.org/10.1007/978-3-030-93100-1_13

The maximal event t, (maximal in the sense of most assured to occur, minimal in the sense of requiring the weakest assumptions,) will be a symbolic representation of the probability we should assign to A given Γ. This justification term is called the *aggregated evidence* for A given Γ, written $AE^{\Gamma}(A)$.

For example, suppose $\Gamma = \{B, B \to A, A, X\}$, with X irrelevant to A. We would like to calculate the best estimate for A to occur, given Γ. Symbolically, we compute the maximal t such that

$$r{:}\, B, s{:}\, (B \to A), u{:}\, A, v{:}\, X \vdash t{:}\, A.$$

A will occur if both r and s occur, or if u occurs. It may occur if *additionally* v also occurs, but that adds an additional assumption, weakening the evidence. Therefore, the strongest evidence for A given Γ is $(r \cap s) \cup u$, where $\cap$ represents that both events occur, and $\cup$ represents that either event occurs. That is $AE^{\Gamma}(A) = (r \cap s) \cup u$, and

$$r{:}\, B, s{:}\, (B \to A), u{:}\, A, v{:}\, X \vdash (r \cap s) \cup u{:}\, A.$$

The formal system which allows us to make such deductions is called the *logic of evidence aggregation*, or LEA. To bring the problem back to the original question of probabilities, we employ a semantics for LEA called *probability semantics*. Each probabilistic interpretation $\circ$ maps justification terms to events in a probability space, which itself comes equipped with a probability function P. To give the "best estimate" for A to occur given Γ, for a particular interpretation $\circ$, we may take the probability of the event corresponding to the strongest possible evidence for A given Γ. That is,

$$P((AE^{\Gamma}(A))^{\circ}).$$

Although LEA is sufficient for the task of formalizing the process of evidence aggregation, and probabilistic semantics is sufficient for interpreting evidence probabilistically, the formal system and its semantics are not sound and complete with respect to each other, as shown in [2]. The contribution of this paper is to give a sound and complete semantics for LEA, and a sound and complete axiomatization for probability semantics.

Additional attempts to combine justification logic with probability may be found in [8,10]. LEA was discussed in [9]. That paper introduced a novel type of justification logic semantics called *subset models*. LEA was shown to be sound, but not shown to be complete, for a particular class of subset models.

1.2 LEA Definition

We first define the language and then the proof system for LEA.

Definition 1 (Justification Term). *Justification terms or simply* terms *are defined according to the following grammar:*

$$t := \quad e_i \quad | \quad \mathbf{1} \quad | \quad \mathbf{0} \quad | \quad t \cup t \quad | \quad t \cap t\,.$$

1 and **0** are justification constants intended to represent empty and total justification specifications, respectively. Justification variables are $e_i \in \{e_1, \ldots, e_n\}$. We may abbreviate $s \cap t$ as st.

The set of all terms is denoted $Term$ and is isomorphic to the bounded free distributive lattice on n-generators. Top and bottom elements are $\mathbf{1}, \mathbf{0}$, respectively. Generators are $e_1, \ldots, e_n$. Meet and join are $\cap, \cup$, respectively. The order on lattice terms is $\leq$. Since each n corresponds to a unique set of terms, to each n there is a unique LEA language, and therefore a unique logic. In practice, however, the n will not be consequential. Therefore, we will not distinguish between the different logics corresponding between different n in our notation.

Definition 2 (LEA **Formula**). LEA *formulas are defined inductively in the usual way for justification logics, according to the following grammar:*

$$A := \quad p \quad | \quad \neg A \quad | \quad A \vee A \quad | \quad A \wedge A \quad | \quad A \rightarrow A \quad | \quad t{:}\, A$$

where p ranges over a countably infinite set of propositional variables, and t is a justification term. $A \leftrightarrow B$ is shorthand notation for $(A \rightarrow B) \wedge (B \rightarrow A)$. The set of all formulas is denoted as $Form$. (We will use "$Form$" for other logics besides LEA. *In all contexts, $Form$ will be understood as the set of all formulas in the logic under current consideration.)*

A special feature of our logical system is that LEA comes equipped with an order $\leq$ on terms, which is not part of the language, yet we reference in the axioms.

Definition 3 (LEA **Proof System**). *Any uniform substitution of* LEA *formulas into the following formulas is an axiom of* LEA. *Modus Ponens is the only rule of inference for* LEA.

1. *The axioms and rules of classical logic in the language of* LEA
2. $s{:}\,(A \rightarrow B) \rightarrow (t{:}\, A \rightarrow st{:}\, B)$
3. $(s{:}\, A \wedge t{:}\, A) \rightarrow s \cup t{:}\, A$
4. $s{:}\, A \rightarrow t{:}\, A$ *for any evidence terms s and t such that $t \leq s$*
5. $\mathbf{1}{:}\, C$ *where C is any axiom*
6. $\mathbf{0}{:}\, A$ *where A is any formula.*

1.3 Alternative Formulation of LEA

In [2], the following axiom scheme is used for LEA:

$$s{:}\, A \rightarrow t{:}\, A \text{ if } t \leq s.$$

LEA is already interesting simply for introducing an order on justification terms. In forthcoming work, we study in more detail the proof theory and semantics of justification logics with an order imposed on their justification terms. However, in the case of LEA, due to the underlying lattice structure of the terms, we can do without this axiom. We choose here to use an equivalent axiom scheme, because defining basic models in the presence of an existing order invites subtleties that we do not need to address here.

Moreover, the presentation that we now give will provide more transparency on the behavior of LEA and its related systems we will study in this paper. This equivalent axiom system we will call LEA′. It has the same language as LEA.

Definition 4 (LEA′ Proof System). *Any uniform substitution of* LEA *formulas into the following formulas is an axiom of* LEA′*. Modus Ponens is the only rule of inference for* LEA′*.*

1. *The axioms classical logic in the language of* LEA
2. $t\colon (A \to B) \to (t\colon A \to t\colon B)$, *(terms are closed under Modus Ponens)*
3. *(a)* $s \cup t\colon A \to (s\colon A \wedge t\colon A)$
   (b) $(s\colon A \wedge t\colon A) \to s \cup t\colon A$
4. $(s\colon A \vee t\colon A) \to st\colon A$
5. $\mathbf{1}\colon C$ *where* C *is any axiom*
6. $\mathbf{0}\colon A$ *where* A *is any formula*
7. $s\colon A \leftrightarrow t\colon A$ *if* $s = t$ *in the lattice ordering.*

Lemma 1. LEA′ *is equivalent to* LEA*. That is, for all sets of formulas* $\Gamma \cup \{A\} \subset Form$,

$$\Gamma \vdash_{\mathsf{LEA}} A \textit{ iff } \Gamma \vdash_{\mathsf{LEA'}} A.$$

Proof. Since the only rule of inference in both systems is Modus Ponens, and since they share the same language, it is enough to show that an axiom from one system is provable in the other. Through liberal use of LEA Axiom 4, one can show that LEA contains LEA′. The reverse direction is not much more difficult. To give an example of how this is done, we will give a proof in LEA′ of LEA Axiom 2, and a proof in LEA of LEA′ Axiom 3(a).

Working in LEA′, let us derive $s\colon (A \to B) \to (t\colon A \to st\colon B)$. From LEA′ Axiom 4, LEA′ proves $s\colon (A \to B) \to st\colon (A \to B)$. From Axiom 4 again, LEA′ proves $t\colon A \to st\colon A$. Then LEA′ proves $(s\colon (A \to B) \wedge t\colon A) \to (st\colon (A \to B) \wedge st\colon A)$. Then, using Axiom 2, we get $s\colon (A \to B) \to (t\colon A \to st\colon B)$.

Working in LEA, let us derive $s \cup t\colon A \to (s\colon A \wedge t\colon A)$. In the lattice order we have $s \leq s \cup t$, so $s \cup t\colon A \to s\colon A$ is an instance of LEA Axiom 4. Similarly, $t \leq s \cup t$, so again $s \cup t\colon A \to t\colon A$ is an instance of Axiom 4. Together they give $s \cup t\colon A \to (s\colon A \wedge t\colon A)$.

Since the two systems are equivalent, we will no longer distinguish between them, and will refer to the latter LEA′ as LEA. However, we may still reference the order $\leq$ when convenient to do so.

1.4 Probability Semantics Definition

Definition 5 (Probability Semantics).
A probabilistic interpretation $\circ$ *consists of:*

- *a probability space* $(\Omega, \mathcal{F}, P)$*, where* Ω *is a set of outcomes,* $\mathcal{F}$ *is a sigma-algebra of measurable events, and* P *is a probability measure on* $\mathcal{F}$
- *a mapping* $\circ$ *which takes terms to elements of* $\mathcal{F}$ *and formulas to subsets of* Ω*. (*$\circ$ *is overloaded to mean both the interpretation its mapping.)*
- $\circ$ *satisfies the conditions below for formulas* A, B *and terms* $s, t, \mathbf{1}$*, and* $\mathbf{0}$*:*

Terms	*Formulas*
$\mathbf{0}^\circ = \emptyset$	$(A \wedge B)^\circ = A^\circ \cap B^\circ$
$\mathbf{1}^\circ = \Omega$	$(A \vee B)^\circ = A^\circ \cup B^\circ$
$(st)^\circ = s^\circ \cap t^\circ$	$(\neg A)^\circ = \overline{A^\circ}$
$(s \cup t)^\circ = s^\circ \cup t^\circ$	$(A \rightarrow B)^\circ = \overline{A^\circ} \cup B^\circ$
	$(t{:}\, A)^\circ = \overline{t^\circ} \cup A^\circ.$

In order to define an entailment relation, we extend $\circ$ to sets of formulas, as

$$\Gamma^\circ = \bigcap_{C \in \Gamma} C^\circ.$$

In particular, notice $\emptyset^\circ = \Omega$.

Definition 6 (Entailment in Probability Semantics). *An entailment relation* $\Vdash$ *is defined as*

$$\Gamma \Vdash A \textit{ iff } \Gamma^\circ \subset A^\circ, \textit{ for all probabilistic interpretations } \circ.$$

LEA is sound with respect to probability semantics, but it is not complete, as shown in [2]. This can be seen by noting that $\mathbf{1}{:}\, A \rightarrow A$ is valid in probability semantics, but is not provable in LEA.

In Sect. 2 we give sound and complete semantics for LEA. In Sect. 3 we give a sound and complete axiomatization of probability semantics. In Sect. 4 we discuss issues of decidability and possible avenues for future research.

2 Sound and Complete Semantics for LEA

2.1 Basic Models

We will define a class of models which are sound and complete with respect to LEA. The models will be a class of *basic models.* We recall here the definition of basic models. For a more detailed exposition, see for example, [3].

Definition 7 (Basic Model). *A basic model* $*$ *is a mapping of justification terms and formulas, which takes formulas to truth values* 1 *or* 0 *– representing true or false, respectively – takes justification terms to sets of formulas, and satisfies the following conditions:*

- $\bot^* = 0$
- $(A \wedge B)^* = 1$ *iff* $A^* = 1$ *and* $B^* = 1$
- $(A \vee B)^* = 1$ *iff* $A^* = 1$ *or* $B^* = 1$

- $(\neg A)^* = 1$ *iff* $A^* = 0$
- $(A \to B)^* = 1$ *iff* $A^* = 0$ *or* $B^* = 1$
- $(t{:}\, A)^* = 1$ *iff* $A \in t^*$.

Definition 8 (Entailment in a Basic Model). *For an entailment relation* $\models$, *we write* $\Gamma \models_* A$ *iff* $A^* = 1$ *or there exists* $C \in \Gamma$ *such that* $C^* = 0$. *We write* $\Gamma \models A$ *iff, for all basic models* $*$ *(of the basic model class under consideration)* $\Gamma \models_* A$. *We may write* $* \models A$ *or* $\models_* A$ *both to mean* $\emptyset \models_* A$.

Definition 9 ($\Gamma \vdash_{\mathsf{CL}} A$). *For* $\Gamma \cup \{A\} \subset Form$, *we write* $\Gamma \vdash_{\mathsf{CL}} A$ *iff* A *can be derived from assumptions in* Γ *using the axioms of classical logic, treating justification formulas* $t{:}\, A$ *as distinct propositional variables.*

Definition 10 (Class of Basic Models). *For* $\Gamma \subset Form$, $BM(\Gamma)$ *is the class of all basic models which satisfy all formulas in* Γ. *That is,* $* \in BM(\Gamma)$ *iff* $* \models C$ *for all* $C \in \Gamma$. *If* $\mathcal{J}$ *is an axiomatic system with axioms* $Ax(\mathcal{J})$, *then abusing notation we write* $BM(\mathcal{J})$ *to denote* $BM(Ax(\mathcal{J}))$.

Theorem 1 (Generic Completeness). *Each set of formulas* Γ *is sound and complete with respect to its class of basic models. That is* $\Gamma \vdash_{\mathsf{CL}} A$ *iff* A *is true in all basic models of* Γ.

Proof. See, for example, [3].

Corollary 1. *Let a justification logic* $\mathcal{J}$ *contain Modus Ponens as its only rule of inference, and contain as axioms the axioms of classical logic in the language of* $\mathcal{J}$. *Then* $\Gamma \vdash_{\mathcal{J}} A$ *iff* A *is true in all basic models of* Γ.

This allows us to apply the generic completeness theorem to any justification logic containing classical logic with Modus Ponens as its only rule of inference. In particular, all the justification logics studied in this paper fall into this class.

2.2 Deductive Basic Models

Now we define the class of basic models for which LEA is sound and complete.

Definition 11 (Deductive Basic Model). *A deductive basic model is a basic model* $*$ *which satisfies the following conditions:*

- t^* *is a deductively closed set of* LEA *formulas, for all terms* t
- $(st)^* \supset s^* \cup t^*$
- $(s \cup t)^* = s^* \cap t^*$
- $\mathbf{1}^* \supset Taut$
- $\mathbf{0}^* = Form$

where $Taut$ *is the set of all* LEA *tautologies.*

Lemma 2. *If* $* \in BM(\mathsf{LEA})$ *and* $s = t$ *in the lattice order, then* $s^* = t^*$.

Proof. $* \models s{:}A \leftrightarrow t{:}A$ is an axiom when $s = t$ in the lattice order. Using this fact along with the definition of satisfaction in a basic model, we have the following chain of inferences.

$$A \in s^* \iff * \models s{:}A \iff * \models t{:}A \iff A \in t^*.$$

Lemma 3. *If* $* \in BM(\mathsf{LEA})$ *then* $*$ *is a deductive basic model.*

Proof. Check the axioms.

Lemma 4. *If* $*$ *is a deductive basic model, then* $* \in BM(\mathsf{LEA})$.

Proof. We check the axioms one-by-one.

1. Since $*$ is a basic model, it satisfies the axioms of classical logic and is closed under Modus Ponens.
2. Since t^* is deductively closed, then $* \models t{:}(A \to B) \to (t{:}A \to t{:}B)$.
3. Since $(s \cup t)^* = s^* \cap t^*$ then $* \models (s{:}A \wedge t{:}A) \leftrightarrow s \cup t{:}A$.
4. Since $(st)^* \supset (s^* \cup t^*)$, then $* \models (s{:}A \wedge t{:}A) \to st{:}A$.
5. Since $Taut \subset \mathbf{1}^*$, then $* \models \mathbf{1}{:}C$ for any axiom C.
6. Since $\mathbf{0}^* = Form$, then $* \models \mathbf{0}{:}A$ for any formula A.
7. By Lemma 2, $* \models s{:}A \leftrightarrow t{:}A$ if $s = t$ in the lattice ordering.

Corollary 2. $BM(\mathsf{LEA})$ *equals the class of all deductive basic models.*

Corollary 3. LEA *is sound and complete with respect to the class of all deductive basic models.*

3 Sound and Complete Axiomatization of Probability Semantics

We will now work towards an axiom system which is sound and complete with respect to probability semantics. It turns out that an extension of LEA, which we will call LEA_+, is sound and complete with respect to probability semantics. Before investigating LEA_+, however, we turn our attention to a reduct of the language of LEA, with an axiomatic system we will call LEA_-. As a language, LEA_- is very simple – it contains no operations on justification terms, only justification variables. Studying LEA_- will shed light on the probability semantics and simplify our completeness proof for LEA_+.

3.1 LEA_- Definition and Models

Definition 12 (LEA_- Formula). *The terms of the* LEA_- *language consist of a countably infinite number of atomic evidence terms* $e_1, e_2, \ldots, e_n, \ldots$. *There are no constants or term operations. Formulas are built up using this set of terms as usual.*

Definition 13 (LEA$_-$ Proof System). *Any uniform substitution of* LEA$_-$ *formulas into the following formulas is an axiom of* LEA$_-$. *Modus Ponens is the only rule of inference.*

1. The axioms of classical logic in the language of LEA$_-$
2. $A \to t\colon A$
3. $t\colon (A \to B) \to (t\colon A \to t\colon B)$.

Let us write $\vdash_-$ to denote $\vdash_{\mathsf{LEA}_-}$.

Axiom 2 is cofactivity. It says that if A is true, then any evidence justifies it. Axiom 3 tells us that justification terms are closed under Modus Ponens.

Lemma 5. LEA$_-$ *is consistent.*

Proof. Here is a model of LEA$_-$. Consider the basic model $*$, where $t^* = Form$ for all terms t. All formulas $t\colon A$ will evaluate to 1 (true), and so will LEA$_-$ Axioms 2 and 3. Since it is a basic model, $*$ satisfies the axioms of classical logic in the language of LEA$_-$, as well as Modus Ponens.

Definition 14 (Two-model). *A basic model* $*$ *is a two-model iff*

$$t^* \in \{True^*, Form\}$$

for all terms t, *where* $True^* = \{A \mid * \models A\}$.

Theorem 2 (Soundness and Completeness). $BM(\mathsf{LEA}_-)$ *is the class of all two-models.*

Proof. From Lemma 5 we know $BM(\mathsf{LEA}_-)$ is nonempty.

For one direction of inclusion, let $*$ be a basic model of LEA$_-$. Due to Axiom 3, we have that t^* must be a deductively closed set. Due to Axiom 2, we have that $True^* \subset t^*$. From this it follows that $t^* \in \{True^*, Form\}$, for if $B \notin True^*$ then the deductive closure of $(True^* \cup B)$ is in fact $Form$.

For the other direction, let $*$ be a two-model. Since $*$ is a basic model, it satisfies the axioms of classical logic. Axiom 2 holds since $True^* \subset t^*$. For Axiom 3, suppose $\{A \to B, A\} \subset t^*$. Since both $True^*$ and $Form$ are deductively closed sets, $B \in t^*$. It follows that Axiom 3 holds in $*$.

We now work towards showing a soundness and completeness result for LEA$_-$ with respect to probability semantics. First, we show that to each two-model $*$, we can associate a probability model $\circ$ that validates the same formulas.

Definition 15 (Probability Model $\circ$ Corresponding to Two-model $*$). *Let* $*$ *be a two-model. Define a probability model* $\circ$ *corresponding to* $*$ *as follows. The underlying probability space for* $\circ$ *is* $(\Omega, \mathcal{F}, P)$, *where* $\Omega = \mathbb{1} = \{\emptyset\}$; $\mathcal{F} = \{\emptyset, \mathbb{1}\}$; $P(\emptyset) = 0$ *and* $P(\mathbb{1}) = 1$. *Define* $p^\circ = \mathbb{1}$ *if* $p^* = 1$, *and* $p^\circ = \emptyset$ *if* $p^* = 0$. *For atomic evidence terms, define* $e_i^\circ = \emptyset$ *if* $e_i^* = Form$, *and* $e_i^\circ = \mathbb{1}$ *if* $e^* = True^*$.

Proposition 1. *For all* LEA_- *formulas* A, $A^\circ = \mathbb{1}$ *iff* $A^* = 1$; $A^\circ = \emptyset$ *iff* $A^* = 0$.

Proof. Argue by induction on the complexity of A.

- The claim is true for propositional atoms by definition of $\circ$.
- For the Boolean connectives, the proof is standard. We give the proof for negation. Suppose $(\neg A)^\circ = \mathbb{1}$. This is true iff $A^\circ = \emptyset$, iff (by the induction hypothesis) $A^* = 0$, iff $(\neg A)^* = 1$.
- For the justification case, $(e_i\colon A)^\circ = \overline{e_i^\circ} \cup A^\circ = \mathbb{1}$ iff $e_i^\circ = \emptyset$ or $A^\circ = \mathbb{1}$. This holds iff $e_i^* = Form$ (by definition of $\circ$) or $A^* = 1$ (by induction hypothesis.) This holds iff $e_i^* = Form$ or $A \in True^*$ (since if $A^* = 1$ then A is true in the model $*$). This holds iff $A \in e_i^*$ (for if A is false in the model and $A \in e_i^*$, then $e_i^* = Form$.) This holds iff $(e_i\colon A)^* = 1$.

Theorem 3 (Soundness and Completeness). *For any set of* LEA_- *formulas* $\Gamma \cup \{A\}$,

$$\Gamma \vdash_- A \text{ iff } \Gamma \Vdash A.$$

Proof. For soundness, the axioms of LEA_- are clearly true in any probability model. We can see this, for example, by noticing that in any probability model, $t\colon A$ would have the same interpretation as $t \to A$, if only the latter were actually a well formed LEA_- formula. Then the axiom $A \to t\colon A$ corresponds to the classical tautology $A \to (t \to A)$ and similarly for Axiom 3.

For completeness, suppose $\Gamma \nvdash_- A$. Since LEA_- is sound and complete with respect to two-models, there exists is a two-model $*$ such that $C^* = 1$ for all $C \in \Gamma$, and $A^* = 0$. Then there exists a corresponding probability model $\circ$, with underlying probability space $\Omega = \mathbb{1} = \{\emptyset\}$, such that $C^\circ = \mathbb{1}$ for all $C \in \Gamma$, and $A^\circ = \emptyset$. Then $\cap_{C\in\Gamma} C^\circ = \mathbb{1} \not\subset A^\circ = \emptyset$, so $\Gamma \not\Vdash A$.

Corollary 4. *Classical logic is sound and complete with respect to probability semantics.*

Proof. Examining the axioms of any standard formulation of classical logic shows it to be sound. The completeness proof exactly mirrors that for LEA_-. For each Boolean assignment $*$ with truth values 1 and 0, there exists a probability model $\circ$, with its underlying probability space the same as in Definition 15, such that for all formulas A, $A^* = 1$ iff $A^\circ = \mathbb{1}$ and $A^* = 0$ iff $A^\circ = \emptyset$. The rest of the proof follows the steps in Theorem 3.

3.2 More LEA_- Results

Proposition 2 (Substitution). $\vdash_- (A \to B) \to (t\colon A \to t\colon B)$

Proof. $(A \to B) \to t\colon (A \to B)$ is an instance of Axiom 2. Applying Axiom 3 proves the result.

Proposition 3. $\vdash_- \neg t\colon A \to t\colon (\neg A)$

Proof. $\neg t\colon A \to \neg A$ is the contrapositive of Axiom 2. $\neg A \to t\colon(\neg A)$ is an instance of Axiom 2. Combined they give $\neg t\colon A \to t\colon(\neg A)$.

Proposition 4. $\vdash_{-} t\colon(A \vee B) \leftrightarrow (t\colon A \vee t\colon B)$

Proof. Argue inside LEA_{-}. Suppose $t\colon(A \vee B)$ holds. Then, by Proposition 3.5, so does $t\colon(\neg A \to B)$. If $t\colon\neg A$ holds, then $t\colon B$ holds, by Axiom 3. If, on the other hand, $\neg t\colon\neg A$ holds, then from Proposition 3.6, we have $t\colon A$. In either case, $t\colon A \vee t\colon B$ holds.

For the other direction, suppose $t\colon A$ holds. Since $A \to A \vee B$ is a tautology, by Axiom 2, $t\colon(A \to A \vee B)$. Applying Axiom 3, we get $t\colon(A \vee B)$. We similarly get $t\colon(A \vee B)$ if $t\colon B$ holds.

Proposition 5. $\vdash_{-} t\colon(A \wedge B) \leftrightarrow (t\colon A \wedge t\colon B)$

Proof. Argue inside LEA_{-}. Suppose $t\colon(A \wedge B)$ holds. $A \wedge B \to A$ is a tautology, so by Axiom 2, $t\colon(A \wedge B \to A)$. Applying Axiom 3, we derive $t\colon A$. We may similarly derive $t\colon B$. Therefore, we get $t\colon A \wedge t\colon B$.

For the other direction, $t\colon(A \to (B \to A \wedge B))$ holds from Axiom 2. Suppose $t\colon A$ and $t\colon B$ hold. Using Axiom 3 twice, we first apply $t\colon A$ to $t\colon(A \to (B \to (A \wedge B))$, and then $t\colon B$ to that operation's result. Doing so, we derive $t\colon(A \wedge B)$.

Definition 16 (Justification Literal). *A class of formulas called justification literals are defined inductively. p and $\neg p$ are justification literals, for any propositional variable p. If A is a justification literal, so is $e_i\colon A$ and $\neg e_i\colon A$, for a justification variable e_i.*

Definition 17 (Disjunctive Justified Normal Form). *For a sentence A, we say A is in disjunctive justified normal form (djnf) iff A is in the form of a disjunction of clauses, where a clause is a conjunction of justification literals.*

Theorem 4. *Any LEA_{-} sentence A is provably equivalent in LEA_{-} to a sentence A^n which is in disjunctive justified normal form.*

We omit the proof here, but notice it essentially follows by breaking justification formulas $t\colon A$ into "simpler" formulas, using Propositions 4 and 5.

Notice, this says something about the meaning of LEA_{-} formulas. The informational content of justification formulas $t\colon X$ is located entirely in justification literals. Let us consider propositional variables as atomic concepts. Evidence for or against these concepts are also atomic concepts. These would be interpreted as $e_i\colon p$ or $e_i\colon\neg p$. Inductively, we may have evidence for evidence, or evidence against evidence, $e_j\colon e_i\colon p$, $e_j\colon\neg e_i\colon p$, etc., which are also considered atomic concepts. By that, we mean that all other sentences can be build up from them, using conjunction, disjunction, and negation. Everything boils down in an inductive chain to evidence for or evidence against propositional variables.

3.3 LEA$_+$ Definition and Basic Results

Let us now expand the language and axiomatic system of LEA$_-$. Our new system, LEA$_+$, will have the same language as LEA. That means it includes the justification constants **1** and **0**, and includes justification term operations $\cap$ and $\cup$.

Definition 18 (LEA$_+$ Proof System). *Any uniform substitution of* LEA$_+$ *formulas into the following formulas is an axiom of* LEA$_+$*. Modus Ponens is the only rule of inference for* LEA$_+$*.*

1. *The axioms classical logic in the language of* LEA
2. $A \to t{:}\, A$
3. $t{:}\,(A \to B) \to (t{:}\, A \to t{:}\, B)$
4. (a) $st{:}\, A \to (s{:}\, A \vee t{:}\, A)$
   (b) $(s{:}\, A \vee t{:}\, A) \to st{:}\, A$
5. (a) $s \cup t{:}\, A \to (s{:}\, A \wedge t{:}\, A)$
   (b) $(s{:}\, A \wedge t{:}\, A) \to s \cup t{:}\, A$
6. $\mathbf{1}{:}\, A \to A$
7. $\mathbf{0}{:}\, A$.

Let us write $\vdash_+$ to denote $\vdash_{\mathsf{LEA}_+}$.

Lemma 6. *$\vdash_+ s{:}\, A \leftrightarrow t{:}\, A$ if $s = t$ in the lattice ordering.*

Proof. Define an equivalence relation on terms as

$$s \sim t \text{ iff } \vdash_+ s{:}\, A \leftrightarrow t{:}\, A \text{ for all formulas } A.$$

Let $[s] = \{t \mid s \sim t\}$. We wish to define $[s] \cup [t] = [s \cup t]$ and $[s] \cap [t] = [s \cap t]$. These operators on $\sim$-equivalence classes will be defined if, for all terms s_1, t_1, s_2, t_2,

$$\text{if } [s_1] = [s_2] \text{ and } [t_1] = [t_2], \text{ then } [s_1 \cup t_1] = [s_2 \cup t_2] \text{ and } [s_1 \cap t_1] = [s_2 \cap t_2].$$

This is indeed the situation. We prove it below for $\cup$, but omit the $\cap$ case, which is similar. Let A be an arbitrary LEA$_+$ formula.

$\vdash_+ s_1 \cup t_1{:}\, A \leftrightarrow (s_1{:}\, A \wedge t_1{:}\, A)$	Axioms 5(a), 5(b)
$\vdash_+ s_1{:}\, A \leftrightarrow s_2{:}\, A$	since $[s_1] = [s_2]$
$\vdash_+ t_1{:}\, A \leftrightarrow t_2{:}\, A$	since $[t_1] = [t_2]$
$\vdash_+ s_1 \cup t_1{:}\, A \leftrightarrow (s_2{:}\, A \wedge t_2{:}\, A)$	substitution
$\vdash_+ s_1 \cup t_1{:}\, A \leftrightarrow s_2 \cup t_2{:}\, A$	Axioms 5(a), 5(b)

so $[s_1 \cup t_1] = [s_2 \cup t_2]$. Again, we state, but do not prove, that $[s_1 \cap t_1] = [s_2 \cap t_2]$. Therefore, the $\sim$-equivalence class operations $\cap, \cup$ are well defined.

As might be expected, $< Term/\sim, \cup, \cap >$ is a lattice, and the natural quotient map $i : Term \longrightarrow Term/\sim$ given by $i(s) = [s]$ is a lattice homomorphism. For details, see [5].

To finish the proof, if $s = t$, then since i is a lattice homomorphism, in particular a well defined function, $[s] = i(s) = i(t) = [t]$.

Therefore $\vdash_+ s\colon A \leftrightarrow t\colon A$ for all A, by definition of the equivalence class.

Proposition 6 (Substitution). $\vdash_+ (A \to B) \to (t\colon A \to t\colon B)$

Proposition 7. $\vdash_+ \neg t\colon A \to t\colon (\neg A)$

Proposition 8. $\vdash_+ t\colon (A \vee B) \leftrightarrow (t\colon A \vee t\colon B)$

Proposition 9. $\vdash_+ t\colon (A \wedge B) \leftrightarrow (t\colon A \wedge t\colon B)$

Proof. The proofs of the above propositions are the same as in the LEA_- cases.

Next we will show that for each LEA_+ formula A, there exists a LEA_- formula A^- such that $\vdash_+ A \leftrightarrow A^-$.

Definition 19 (LEA_- Translation). *For a* LEA *formula A, we define the* LEA_- *translation of A, written A^-. Here is the inductive definition.*

- $p^- = p$ *for propositional variable* p
- $(\neg A)^- = \neg(A^-)$
- $(A \vee B)^- = A^- \vee B^-$
- $(A \wedge B)^- = A^- \wedge B^-$
- $(A \to B)^- = A^- \to B^-$
- $(e_i\colon A)^- = e_i\colon (A^-)$
- $(st\colon A)^- = (s\colon A)^- \vee (t\colon A)^-$
- $(s \cup t\colon A)^- = (s\colon A)^- \wedge (t\colon A)^-$
- $(1\colon A)^- = A^-$
- $(0\colon A)^- = \top$

Proposition 10. *For each* LEA_+ *formula A, A^- is a well defined* LEA_- *formula.*

Proof. Argue on the complexity of A.

Lemma 7. *For each* LEA_+ *formula A, $\vdash_+ A \leftrightarrow A^-$.*

Proof. Argue by induction on the complexity of A. The base case when A is a propositional variable holds trivially. The cases for Boolean connectives are standard. If $A = t\colon B$, then perform subinduction on the complexity of t. In particular, from Proposition 6 and the induction hypothesis, we have $\vdash_+ e_i\colon B \leftrightarrow e_i\colon (B^-)$. The cases for more complex terms and term constants – $st, s \cup t, \mathbf{1}, \mathbf{0}$ – follow directly from the application of the induction hypothesis to the corresponding axioms.

Corollary 5. *Each* LEA_+ *formula is provably equivalent in* LEA_+ *to a formula in disjunctive justified normal form.*

Proof. Each LEA_+ formula A is provably equivalent to its LEA_- translation A^-, which in turn is provably equivalent to a djnf formula $(A^-)^n$.

3.4 Basic Models of LEA_+

The basic models of LEA_+ are essentially two-models, with additional clauses to define the term operations and constants $\cap, \cup, \mathbf{1}, \mathbf{0}$. We call them *trivial-lattice models.*

Definition 20 (Trivial-lattice Model). *A trivial-lattice model is a basic model satisfying the following conditions:*

- $t^* \in \{Form, True^*\}$ for all terms t
- $(t \cap s)^* = t^* \cup s^*$
- $(t \cup s)^* = t^* \cap s^*$
- $\mathbf{1}^* = True^*$
- $\mathbf{0}^* = Form.$

It is clear from the definitions that a trivial-lattice model is a deductive model. It is also clear that to any two-model, there is an associated trivial-lattice model resulting from defining the interpretation of justification terms in accordance with the definitions given above. These models will agree on the truth value of LEA_- formulas.

Note, we now have similar semantics for two distinct systems LEA_- and LEA_+. When it is clear from the context which one we are working in, we may simply write $\models$ or $\Vdash$ to designate, respectively, basic model semantics or probability semantics. When we need to be precise regarding which language we are interpreting, we may write $\models_-$, $\Vdash_-$ when interpreting LEA_- formulas, and $\models_+$, $\Vdash_+$ when interpreting LEA_+ formulas.

Theorem 5. LEA_+ *is sound and complete with respect to trivial-lattice models.*

Proof. Simple inspection of the axioms and semantics shows that LEA_+ is sound with respect to trivial-lattice models.

To show completeness, suppose $\Gamma \nvdash_+ A$. By Lemma 7, $\vdash_+ A \leftrightarrow A^-$. Since A^- is a LEA_- formula, and since LEA_+ is an extension of LEA_-, we have $\Gamma \nvdash_- A^-$. By the completeness theorem for LEA_-, we have $\Gamma \not\models_- A^-$. Then we have a two-model $*$ which validates all the formulas in Γ and which falsifies A^-. From $*$ we can produce a trivial-lattice model $\tilde{*}$ which agrees with $*$ on all LEA_- formulas. Thus we have $\tilde{*} \models_+ C$ for all $C \in \Gamma$ and $\tilde{*} \not\models_+ A^-$, i.e. $\Gamma \not\models_+ A^-$. Since $\models_+ A \rightarrow A^-$, then $\Gamma \not\models_+ A$.

Lemma 8. LEA_+ *is a conservative extension of* LEA_-.

Proof. It is clear that LEA_+ is an extension of LEA_-. Suppose $\vdash_+ A$ for some LEA_- formula A. By the completeness theorem for LEA_+, A holds in every trivial-lattice model. If A yet fails in some two-model, it fails in its associated trivial-lattice mode, which is a contradiction. Therefore A holds in every two-model. By the completeness theorem for LEA_- we obtain $\vdash_- A$.

Note, since LEA_+ is a conservative extension of LEA_-, and since each LEA_+ formula is equivalent to a LEA_- formula in disjunctive justified normal form which is produced by "decomposing" formulas in a uniform way, we may draw the same conclusion for LEA_+ as we did for LEA_-. That is, the informational content of formulas $t{:}X$ are entirely contained in justification literals. Moreover, LEA_+ essentially has the same expressive power as LEA_-. The terms operators and constants $\sqcap, \sqcup, \mathbf{1}, \mathbf{0}$ are merely convenient shorthand to rewrite djnf formulas.

Lemma 9. LEA_+ *is consistent.*

Proof. LEA_+ is a conservative extension of LEA_-, and LEA_- is consistent. Conservative extensions of consistent theories are consistent.

3.5 Probability Semantics for LEA_+

Theorem 6 (Soundness). LEA_+ *is sound with respect to probability semantics. That is, $\Gamma \vdash_+ A$ implies $\Gamma \Vdash_+ A$.*

Proof. All the axioms of LEA_+ are true in any probability model. Probability models respect Modus Ponens, which is the only rule of inference for LEA_+.

Theorem 7 (Completeness). LEA_+ *is complete with respect to probability semantics. That is, $\Gamma \Vdash_+ A$ implies $\Gamma \vdash_+ A$.*

Proof. Let $\Gamma^- = \{C^- \mid C \in \Gamma\}$.

Suppose $\Gamma \nvdash_+ A$. Then $\Gamma \nvdash_+ A^-$ by Lemma 7. Then also $\Gamma^- \nvdash_+ A^-$, since $\vdash_+ C \to C^-$ for all $C \in \Gamma$. Then $\Gamma^- \nvdash_- A^-$ since LEA_+ is an extension of LEA_-. Then $\Gamma^- \nVdash_- A^-$ by the completeness theorem for LEA_-. Then $\Gamma^- \nVdash_+ A^-$ since $\Vdash_+$ and $\Vdash_-$ agree on LEA_- formulas. Then $\Gamma \nVdash_+ A^-$ since every probabilistic model of Γ is a model of Γ^-. Then $\Gamma \nVdash_+ A$, since $\Vdash_+ A \to A^-$.

4 Further Discussion

4.1 Decidability Results

In Corollary 4 we showed that classical logic is sound and complete with respect to probability semantics. We may extend classical logic with n unique atomic formulae $e_1, \ldots, e_n$ to a system we call CL_n, and the same proof will show CL_n to also be sound and complete with respect to probability semantics. We then give a *Boolean translation* b from LEA_+ formulas to CL_n formulas, such that A is provable in LEA_+ iff A^b is provable in LEA_+. It follows that, to decide if A is valid in LEA_+, it is enough to decide if A^b is valid in CL_n. The decision problem for LEA_+ reduces to the decision problem for classical logic.

Definition 21 (CL_n)**.** *Classical logic with n justification variables, or CL_n, is an extension of the language of classical propositional logic. The axioms are the same as for classical logic. The language contains a countable set of propositional* *variables* $\{p_i \mid i < \omega\}$*, and Boolean connectives* $\wedge, \vee, \neg, \rightarrow$*. In addition* CL_n *contains* n *unique* justification variables $E_1, \ldots, E_n$ *which are considered atomic formulae, and constants* $\top, \bot$ *which are also atomic formulae. We will assume that* $\top, \bot$ *were not in the language of our formulation of classical logic.*

Let $\Vdash_{\mathsf{c}}$ denote the probability semantics applied to CL_n formulas, $\models_{\mathsf{c}}$ the usual Boolean semantics applied to CL_n formulas, and $\vdash_{\mathsf{c}}$ denote provable in CL_n.

Lemma 10. CL_n *is sound and complete with respect to probability semantics.*

Proof. The proof is the same as in Corollary 4.

Next we give the translation from LEA_+ formulas to CL_n formulas.

Definition 22 (Boolean Translation).
Given a LEA_+ *formula* A*, we define the Boolean translation of* A*, written* A^b*, inductively as follows:*

- $e_i^b = E_i$
- $\mathbf{1}^b = \top$
- $\mathbf{0}^b = \bot$
- $(st)^b = s^b \wedge t^b$
- $(s \cup t)^b = s^b \vee t^b$
- $(p)^b = p$ *for propositional variable* p
- $(A \wedge B)^b = A^b \wedge B^b$ *and similar for other Boolean connectives*
- $(s\colon A)^b = (s)^b \rightarrow (A)^b$.

Proposition 11. *For all* LEA_+ *formulas* A, $\vdash_+ A$ *iff* $\vdash_{\mathsf{c}} A^b$.

Proof. From Lemma 10, we have that $\vdash_{\mathsf{c}} A^b$ iff $\Vdash_{\mathsf{c}} A^b$. Since LEA_+ is sound and with respect to probability semantics, it is therefore enough to show $\Vdash_{\mathsf{c}} A^b$ iff $\Vdash_+ A$. This is the case, since for all A, and all probabilistic interpretations $\circ$, $A^\circ = (A^b)^\circ$. This latter claim we may show by induction on the complexity of A.

Theorem 8. LEA_+ *is decidable.*

Proof. The decision algorithm for LEA_- consists of translating A to A^b, then using a Boolean decision algorithm on A^b. The translation from A to A^b is linear in length; A^b results from A by replacing each instance of e_i with E_i, $\mathbf{1}$ with $\top$, $\mathbf{0}$ with $\bot$, $\cap$ with $\wedge$, $\cup$ with $\vee$, and **:** with $\rightarrow$.

Corollary 6. LEA_- *is decidable.*

Theorem 9. *The satisfiability problems for* LEA_+ *and* LEA_- *are in NP.*

Proof. To cover both cases, it is enough to show that the satisfiability problem for LEA_+ is in NP. First, we show that A is satisfiable in LEA_+ iff A^b is satisfiable in CL_n.

Following the proof as in Corollary 4, given a trivial-lattice model $*$, define a corresponding Boolean evaluation for CL_n $\tilde{*}$, such that if $e_i^* = Form$ then $E_i^{\tilde{*}} = 0$; if $e_i^* = True^*$ then $E_i^{\tilde{*}} = 1$; and $p_i^* = p_i^{\tilde{*}}$. It will follow that $* \models_+ A$ iff $\tilde{*} \models_{\mathsf{c}} A^b$. In fact, we see that the mapping $* \mapsto \tilde{*}$ provides a one-to-one correspondence between trivial-lattice models for LEA_+ and Boolean evaluations for CL_n. Therefore, it stands that A is satisfiable in LEA_+ iff A^b is satisfiable in CL_n.

Therefore, an algorithm to verify if A is satisfiable in LEA_+ is as follows. First, translate A to A^b. This is done in deterministically in polynomial time. Then run a nondeterministic polynomial time algorithm to determine if A^b is satisfiable in CL_n.

4.2 Justification Logic as Propositional Logic

The relationship between LEA_+ and CL_n explains why LEA_+ is the axiomatization of probability semantics. Probability semantics is fundamentally a classical semantics. To interpret the LEA_+ language in this classical format, we treat ":" like classical implication $\rightarrow$. In this way, using the Boolean translation, all LEA_+ formulas can be interpreted classically. However, not all classical formulas have an LEA_+ counterpart.

Due to the soundness and completeness of probability semantics with respect to CL_n and with respect to LEA_+ we can use the Boolean translation to embed LEA_+ into CL_n in the following sense. If A is a LEA_+ formula, then

$$\vdash_+ A \text{ iff } \vdash_{\mathsf{c}} A^b.$$

In words, the theorems of LEA_+ are exactly those sentences whose Boolean translation are theorems in classical logic, (more precisely, in CL_n).

CL_n itself can be thought of as simply classical logic, where we distinguish a finite set of propositional variables $E_1, \ldots, E_n$ for our attention, and add the constants $\top, \bot$. In this way, we can identify LEA_+ with a fragment of CL_n, which we will call LEA_+^b.

$$\mathsf{LEA}_+^b = \{A^b \mid A \in \mathcal{L}(\mathsf{LEA}_+)\}$$

How can we characterize LEA_+^b? Essentially, it is the fragment of CL_n where the image of justification terms appear only in the antecedent of an implication. They may appear alone, or with other justification terms in a combination of disjunctions and conjunctions. This can be defined rigorously if one desires.

From this characterization we see how LEA_+^b – and therefore LEA_+ – straddles the line between justification logic and classical propositional logic.

Terms are propositional in character, simply because they map to classical propositional formulas. Formulas $t{:}\,A$ are propositional, because they map to an implication.

Yet, LEA_+ is still different than classical logic. LEA_+ retains its justificational character because there is a restriction on where terms can be mapped to. Terms always map (in prescribed ways) to the antecedent of an implication. They have no life of their own, but are only used to justify other propositions. This latter fact is true for all justification logics. A term t is never considered by itself, but only in relation to other formulas in the format $t{:}A$. We can say nothing about the fact of t, but only about how it relates as evidence for some proposition. Typical propositional logics do not have this kind of distinction between propositions, whereby some propositions are used only to justify others. This distinction should be considered a useful feature of LEA_+ as compared to classical propositional logic.

One may ask if all justification logics behave like LEA_+. That is, do we need special justification terms and operations, or can we do without them, interpreting terms as propositions and : as implication? The answer in general is no.

The typical situation for justification logics is that justification variables and constants are allowed to justify arbitrary collections of formulas. Yet, if : were interpreted as implication, and terms as propositions, then it would follow that if $t{:}A$ and $t{:}B$ hold then $t{:}(A \wedge B)$ holds. Similarly, if $t{:}A$ and $A \rightarrow B$ hold, then $t{:}B$ holds. In general, this is not the case for justification logics. That it is so for the logics studied in this paper points to the peculiarity of these systems.

Another related feature that justification logics typically have, which the ones in this paper do not have, is hyperintensionality at the level of justification. Conceptually speaking, this means that t may be evidence for A, and A may be logically equivalent to B, yet t is not evidence for B. Formally, $t{:}A$ and $A \leftrightarrow B$ may hold, yet $t{:}B$ may not hold. In any justification logic with hyperintensional justifications, : will therefore not behave like $\rightarrow$. Conversely, any justification logic where terms are interpreted as deductively closed sets will not be hyperintensional. That is the situation here. Mathematically, the basic model interpretation of a term will be closed under Modus Ponens whenever application is idempotent.

The increased flexibility of standard justification logics is an asset, not a weakness. We are more free to create complicated relationships between justifications and propositions, including the phenomenon of hyperintensionality. When justifications are treated as propositions and : as implication, a great simplification and flattening naturally occurs. The $\mathsf{LEA}_{+/-}$ framework may be regarded as a clean mathematical answer to the question of justification logic with propositional evidence and material implication as :.

4.3 Future Research

There are a few different directions that appear for further research.

First, we would like to continue research into the computational complexities of the systems discussed in this paper. Discovering if LEA is decidable or not is a natural goal.

Towards the goal of determining if the LEA is decidable, it could be fruitful to develop the proof theory of these three systems, in particular Gentzen-style sequent systems. Also, we should examine the issue of cut and cut-elimination in each system. Observe the resemblance between cut and the axiom $t{:}\,A \rightarrow s{:}\,A$ if $s \leq t$.

$$\frac{\Gamma \vdash B \qquad B \vdash A}{\Gamma \vdash A} \quad \frac{s \leq t \qquad t{:}\,A}{s{:}\,A}$$

The deduction on the right is valid in any deductive model, which thereby includes LEA and LEA_+. Ideologically $t{:}\,A$ was intended to mean that A follows from t. This justifies reading as $t{:}\,A$ as $t \vdash A$. Moreover, we have $s^b \vdash_{\mathsf{c}} t^b$ whenever $s \leq t$ in our lattice structure. This justifies reading $s \leq t$ as $s \vdash t$.

The elimination of references to $\leq$ in our formulation of LEA seems to bear resemblance to the elimination of the cut rule in a sequent system. In further generalizations of LEA-like logics, we may ask for criteria to determine when so-called $\leq$-elimination may take place, just as one looks for cut-elimination theorems in sequent calculi. The parallels between $\leq$, :, and cut should be elucidated and examined formally in future work.

This leads us towards generalizing the types of evidence lattices we work with, for, in the cases where the lattices are finite and distributive, $\leq$-elimination seems feasible, perhaps inevitable. We might generalize to the cases of infinite lattices and lattices without distributivity, for example, and study the behavior of these systems. This could be fruitful for building connections with other areas of logic and formal reasoning, such as argumentation theory, [11]. Moreover, the interplay between the axioms and the lattice structure for terms should be further investigated.

As a further area for research, we can look into alternative interpretations of :. In probability semantics, : is treated like $\rightarrow_{cl}$, where $\rightarrow_{cl}$ is classical (material) implication. One might argue to change this on philosophical grounds. For example, this reading requires the strange validity $A \rightarrow (t{:}\,A)$, which we read as "either A does not happen, or event t will secure event A." If we read $t{:}\,A$ as t is evidence for A, then, since t is arbitrary, we have that if A happens then *anything* is evidence for A. This triviality is reflected in the degenerate nature of two-models and trivial-lattice models. Conceptually, this shows the principal limitations of reading evidence as propositions and $t{:}\,A$ as a classical implication $t \rightarrow_{cl} A$.

To get around this, we might consider alternative interpretations of $t{:}\,A$. For example, one may believe that the processes of evidence collection and aggregation should be understood in an intuitionistic sense, or perhaps in the sense of relevant logic. Yet, one may believe that the physical world operates under classical laws. Then one may have a semantics similar to probability semantics, with $(A \rightarrow B)^\circ = (A \rightarrow_{cl} B)^\circ$, while $(t{:}\,A)^\circ = (t \rightarrow_{int} A)^\circ$ or $(t{:}\,A)^\circ = (t \rightarrow_{rel} A)^\circ$, where $\rightarrow_{int}$, $\rightarrow_{rel}$, are intuitionistic implication and relevant implication, respectively.

Acknowledgements. The author wishes to thank Sergei Artemov, participants of the Computational Logic seminar at the CUNY Graduate Center, and anonymous reviewers for helpful and fruitful comments, suggestions, and ideas.

References

1. Adams, E.W.: A Primer of Probability Logic. CSLI Publications, Stanfort (1998)
2. Artemov, S.: On aggregating probabilistic evidence. **30**(1), 61–76. https://doi.org/10.1007/978-3-319-27683-0_3
3. Artemov, S., Fitting, M.: Justification Logic Reasoning with Reasons. Cambridge University Press, Cambridge (2019)
4. Clemen, R., Winkler, R.: Aggregating probability distributions. In: Edwards, W., Miles, R., von Winterfeldt, D. (eds.) Advances in Decision Analysis: From Foundations to Applications, pp. 154–176. Cambridge University Press (2007)
5. Davey, B.A., Priestley, H.A.: Introduction to Lattices and Order, 2 edn. Cambridge University Press, Cambridge (2002)
6. Hailperin, T.: Sentential Probability Logic. Lehigh University Press, London (1996)
7. Halpern, J.: Reasoning About Uncertainty. MIT Press, Cambridge (2003)
8. Kokkinis, I., Ognjanović, Z., Studer, T.: Probabilistic justification logic. In: Artemov, S., Nerode, A. (eds.) LFCS 2016. LNCS, vol. 9537, pp. 174–186. Springer, Cham (2016). https://doi.org/10.1007/978-3-319-27683-0_13
9. Lehmann, E., Studer, T.: Subset models for justification logic. In: Logic, Language, Information and Computation (WoLLIC) (2019), pp. 433–449 (2019)
10. Lurie, J.: Probabilistic justification logic. In: Philosophies 3.1 (2018). https://doi.org/10.3390/philosophies3010002
11. Van Eemeren, F.H., Grootendorst, R. , Kruiger, T.: Handbook of Argumentation Theory. De Gruyter Mouton (2019)

On Inverse Operators in Dynamic Epistemic Logic

Shota Motoura[1(✉)] and Shin-ya Katsumata[2]

[1] NEC Corporation, Kawasaki 211-8666, Japan
motoura@nec.com
[2] National Institute of Informatics, Tokyo 101-8430, Japan
s-katsumata@nii.ac.jp

Abstract. We extend Dynamic Epistemic Logic with inverse operators $\langle\!\langle \alpha^{-1} \rangle\!\rangle$ of an action α along the line of tense logics. The meaning of the formula $\langle\!\langle \alpha^{-1} \rangle\!\rangle \varphi$ is 'φ is the case before an action α'. This augmentation of expressivity enables us to capture important aspects of communication actions. We also propose its semantics using model transition systems provided in our previous work, which are a suitable framework for interpreting inverse operators. In this framework, we give several soundness/completeness correspondences, which lead to modular proofs of completeness of public announcement logic and epistemic action logic of Baltag-Moss-Solecki extended with inverse operators with respect to suitable classes of MTSs.

Keywords: Dynamic epistemic logic · Inverse operators · General framework · Category theory · Tense logic

1 Introduction

Dynamic Epistemic Logic (*DEL*) is a branch of modal logic for reasoning about knowledge changes or belief revisions caused by communication [10,16]. Up until today, many DELs for various kinds of actions of communications have been proposed and studied: *Public Announcement Logic* (*PAL*) [17], *Epistemic Action Logic* (*EA*) [3], *Preference Upgrade* [7], *etc.* Their methodologies of formulation are basically common. From an external point of view, an epistemic situation of involved *agents* is expressed as a pointed Kripke model, and a formula $\langle\!\langle \alpha \rangle\!\rangle \varphi$ ('after an action α, φ is the case') is true when φ holds after transforming the model according to action α. This is called a *model transformation.* To illustrate this methodology, let us see a quick example.

This work was mainly done while the authors were at Research Institute for Mathematical Sciences, Kyoto University, Kyoto, Japan.

S. Artemov and A. Nerode (Eds.): LFCS 2022, LNCS 13137, pp. 217–235, 2022.
https://doi.org/10.1007/978-3-030-93100-1_14

Example 1. Let us quote a sentence from [5]:

> *Sending an email message to someone with a list of people under the cc button is like making a public announcement in the total group, assuming the computers work perfectly... .*

Following this idea, let us think of sending an email message with content p and with cc list $Ag = \{a, b\}$ as public announcement $!p$ to Ag. This knowledge situation may be expressed as in the left Kripke model ('before') below.

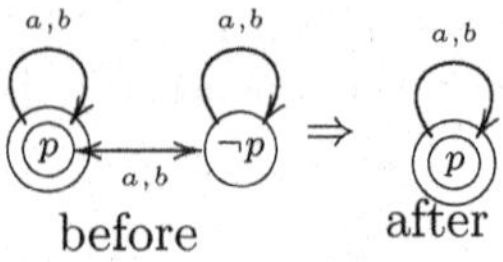

Here, the arrows express uncertainty: for example, the a-labelled arrow from the p-world to the $\neg p$-world indicates that agent a cannot distinguish, on the basis of his knowledge, the $\neg p$-world from the p-world. The double circle indicates the actual state. Conversely, since agent a cannot distinguish the p-world from the $\neg p$-world in the actual world, we may say that agent a does not know p, which we write as $\neg[a]p$. Similarly, we may also say that b does not know p: $\neg[b]p$. Let us next consider sending the email, which makes a public announcement. This is expressed by the elimination of the contradicting possible worlds, namely the $\neg p$-world, resulting in the model on the right. Then, both of agents a and b turn to know p. Thus, we can say that it is true at the actual world before the announcement that, after a truthful public announcement p, both a and b know p is true. This is expressed as $[[!p]][x]p$ for $x \in \{a, b\}$ in syntax. Moreover, $[[!p]][x][y]p$ is also true for any $x, y \in \{a, b\}$.

Backward Operators. As the above example of an email indicates, intended applications of DEL include not only philosophical but also practical ones, such as artificial intelligence and distant communication. Therefore, it is reasonable to extend DELs with more and more practical facilities. In particular, since communication is closely associated with the concept of time, a natural improvement is to give enough expressivity about time by, for example, adding backward operators like 'φ was the case yesterday'. Actually, several studies already exist on backward operators in specific DELs. For example, the yesterday operator $Y\varphi$ is introduced in [18–21]. Notably, the paper [20] also proposes to use an inverse operator $\langle (p!)^{-1} \rangle \top$ in PAL to express that an announcement of p took place before. Their semantics employs a sequence of Kripke models satisfying certain constraints, which is called a *history*. Even the completeness and decidability of EA augmented by inverse operators are discussed in [21]. However, there has not been any common framework that can encompass backward operators in variations of DELs.

Contributions of this Paper. To express various backward operators in a common framework, we introduce its syntax and semantics. Then, we give several soundness/completeness correspondences on inverse operators in the proposed framework. More precisely, our contributions are the following.

1. We introduce inverse operators $\langle\!\langle \alpha^{-1} \rangle\!\rangle \varphi$ ('φ is the case before an event α') in a generic way independent of specific DELs and along the line of tense logics. Inverse operators enable us to express a happened event itself $\langle\!\langle \alpha^{-1} \rangle\!\rangle \top$ as well as what is the case before the event $\langle\!\langle \alpha^{-1} \rangle\!\rangle \varphi$ and can be used to capture many important aspects of actions that ordinary languages cannot.
2. We also propose their semantics by using a modified *update universe* [6]. The original update universe is a big collection of Kripke models linked with each other by *update relations*. In our previous work [15], we have introduced a variant, called *model transition systems* (*MTSs*), where the main difference is that transition relations are labelled by arbitrary actions α rather than by extensions $[\![\varphi]\!]$ of formulae. This provides us with a suitable framework to interpret inverse operators.
3. We give several soundness/completeness correspondences in the proposed framework, which leads to modular proofs of the completeness of PAL and EA with inverse operators.

Let us elaborate the contribution 1 in the following example.

Example 2 (Continued from Example 1). The fact that the email has been sent can be expressed by $\langle\!\langle !\varphi^{-1} \rangle\!\rangle \top$, under the assumption that the past is totally determined (cf. [20]). Since logic PAL intends to capture *truthful* announcement, φ should be true before the announcement took place. This is accounted for by the validity of $\langle\!\langle !\varphi^{-1} \rangle\!\rangle \top \to \langle\!\langle !\varphi^{-1} \rangle\!\rangle \varphi$ in the PAL-specialised MTSs. Moreover, once the message has been sent, both a and b should know that it has been sent, as explained by the validity of $\langle\!\langle !\varphi^{-1} \rangle\!\rangle \top \to [x]\langle\!\langle !\varphi^{-1} \rangle\!\rangle \top$ for $x \in \{a, b\}$, which in turn implies $\langle\!\langle !\varphi^{-1} \rangle\!\rangle \top \to [x][y]\langle\!\langle !\varphi^{-1} \rangle\!\rangle \top$ for any $x, y \in \{a, b\}$. Thus, sending an email with cc implies that it is common knowledge that it has been sent. In this sense, a public announcement is always 'successful'. Inverse operators can be used to capture such important aspects of actions. ⊣

Organisation. In Sect. 2, we recall the general framework for DELs—languages, MTSs and logics—introduced in our previous study [15]. Several notions, including the languages, are slightly modified. In Sect. 3, we add inverse operators to the languages and give several soundness/completeness correspondences. This leads to modular proofs of the completeness of PAL and EA with inverse operators. We also show that the extensions of PAL and EA are conservative and properly increase the expressivity. In Sect. 4 we give a categorical account of MTSs through categorical constructions, which sheds new light on the duality between MTSs and algebraic MTSs discussed in [15].

2 A General Framework for DEL

In this section, we recall the general framework for DEL proposed in [15].

2.1 Syntax

We recall the general framework for DEL in [15]. Let us begin with a generic language:

Definition 1. *Let $\mathcal{P}$ be a set of atomic propositions, $\mathcal{E}$ a set of epistemic expressions and $\mathcal{A}$ a set of action expressions. We define a DEL language $\mathcal{L}(\mathcal{E},\mathcal{A})$ by the following rule:*

$$\varphi ::= \top \mid p \mid \neg\varphi \mid \varphi \vee \psi \mid \langle e\rangle\varphi \mid \mathrm{E}\varphi \mid \langle\langle\alpha\rangle\rangle\varphi$$

where p ranges over $\mathcal{P}$, e over $\mathcal{E}$ and α over $\mathcal{A}$. ⊣

Here, E is the *global operator*, which is included in the language for a technical reason. $\langle e\rangle$ and $\langle\langle\alpha\rangle\rangle$ are called an *epistemic operator* and an *action operator*, respectively. Boolean connectives $\wedge$, $\rightarrow$ and $\leftrightarrow$ are defined as usual, and dual operators $[e]\varphi$, $[[\alpha]]\varphi$ and $\mathrm{A}\varphi$ are defined to be $\neg\langle e\rangle\neg\varphi$, $\neg\langle\langle\alpha\rangle\rangle\neg\varphi$ and $\neg\mathrm{E}\neg\varphi$, respectively.

The languages of *Public Announcement Logic* (PAL) [17] and *Epistemic Action* (EA) [3] can be seen as instances of $\mathcal{L}(\mathcal{E},\mathcal{A})$:

Example 3. Let Ag be a given set of *agents*.

PAL. The language of PAL can be expressed as $\mathcal{L}_{\mathsf{PAL}} = \mathcal{L}(Ag, \mathcal{A}_{\mathsf{PAL}})$, where $\mathcal{A}_{\mathsf{PAL}} = \{!\varphi \mid \varphi \in \mathcal{L}_{\mathsf{PAL}}\}$. As $\mathcal{A}_{\mathsf{PAL}}$ depends on $\mathcal{L}_{\mathsf{PAL}}$, these two sets are actually defined by simultaneous induction; however, the resulting language fits the pattern of $\mathcal{L}(\mathcal{E},\mathcal{A})$. The same remark applies to the language of EA below. The intended meaning of $[n]\varphi$ is that agent n knows φ, while $[[!\varphi]]\psi$ means that ψ holds after a truthful public announcement of φ.

EA. The language of EA is $\mathcal{L}_{\mathsf{EA}} = \mathcal{L}(Ag, \mathcal{A}_{\mathsf{EA}})$ where $\mathcal{A}_{\mathsf{EA}}$ is the set of *action models* [3]. An action model (U, s) consists of a (non-empty) finite Kripke frame $U = (S, (\rightarrow_n)_{n\in Ag}, \mathrm{Pre})$ with a *precondition function* $\mathrm{Pre} : S \rightarrow \mathcal{L}_{\mathsf{EA}}$ and s is an element of S. When we write $x \in U$, we mean $x \in S$. A formula $[[(U, s)]]\varphi$ is read as 'φ holds after an epistemic action (U, s)' and $[n]\varphi$ is as in PAL. ⊣

For the language $\mathcal{L}(\mathcal{E},\mathcal{A})$, we consider the multimodal logic $\mathsf{K_g}$ with the global operator (without the substitution rule) as the base logic; $\mathsf{K_g}$ is defined to be the multimodal logic K together with the following axiom schemata (cf. [8]):

$$\varphi \rightarrow \mathrm{E}\varphi, \quad \varphi \rightarrow \mathrm{AE}\varphi, \quad \mathrm{EE}\varphi \rightarrow \mathrm{E}\varphi, \quad \langle e\rangle\varphi \rightarrow \mathrm{E}\varphi.$$

A specific system of DEL is obtained by adding axioms to $\mathsf{K_g}$, which are often called *reduction axioms*.

Example 4. Some of them can be expressed in any $\mathcal{L}(\mathcal{E},\mathcal{A})$. Other reduction axioms depend on specific choice of $\mathcal{E}$ and $\mathcal{A}$; see Example 5 below.

$$\mathsf{R_P} : \langle\langle\alpha\rangle\rangle p \leftrightarrow \langle\langle\alpha\rangle\rangle\top \wedge p \qquad \mathsf{R_N} : \langle\langle\alpha\rangle\rangle\neg\varphi \leftrightarrow \langle\langle\alpha\rangle\rangle\top \wedge \neg\langle\langle\alpha\rangle\rangle\varphi$$
$$\mathsf{R_E} : \langle\langle\alpha\rangle\rangle\langle e\rangle\varphi \leftrightarrow \langle\langle\alpha\rangle\rangle\top \wedge \langle e\rangle\langle\langle\alpha\rangle\rangle\varphi \quad \mathsf{R_G} : \langle\langle\alpha\rangle\rangle\mathrm{E}\varphi \leftrightarrow \langle\langle\alpha\rangle\rangle\top \wedge \mathrm{E}\langle\langle\alpha\rangle\rangle\varphi$$

Observe that these axioms allow us to reduce the scope of a dynamic operator $\langle\!\langle \alpha \rangle\!\rangle$ to a simpler formula. Notice also that the rule $\mathsf{R_P}$ refers to an atomic proposition p. Hence, logics involving $\mathsf{R_P}$ are not closed under substitution.

Example 5. The proof systems PAL and EA are obtained as follows:

1. PAL is defined to be $\mathsf{K_g} \oplus \mathsf{R_P R_N R_E R_G R_T}$, where $\mathsf{R_T}$ is $\langle\!\langle !\varphi \rangle\!\rangle \top \leftrightarrow \varphi$.
2. EA is defined to be $\mathsf{K_g} \oplus \mathsf{R_P R_N E_{EA} G_{EA} Pre}$, where
   - $\mathsf{E_{EA}} : \langle\!\langle (U,s) \rangle\!\rangle \langle n \rangle \varphi \leftrightarrow \langle\!\langle (U,s) \rangle\!\rangle \top \wedge \bigvee \{ \langle n \rangle \langle\!\langle (U,t) \rangle\!\rangle \varphi \mid s \to_n t \}$,
   - $\mathsf{G_{EA}} : \langle\!\langle (U,s) \rangle\!\rangle \mathrm{E} \varphi \leftrightarrow \langle\!\langle (U,s) \rangle\!\rangle \top \wedge \bigvee \{ \mathrm{E} \langle\!\langle (U,t) \rangle\!\rangle \varphi \mid t \in U \}$,
   - $\mathsf{Pre} : \langle\!\langle (U,s) \rangle\!\rangle \top \leftrightarrow \mathrm{Pre}(s)$. ⊣

2.2 Semantics

Models: Model Transition Systems. In [6], van Benthem introduced a semantic framework for DELs, called an *update universe.* It consists of a family of Kripke models linked by update relations. In our previous work [15], the first author introduced a modification of update universe called *model transition systems* (*MTS* for short), which better fits our parameterised DEL $\mathcal{L}(\mathcal{E},\mathcal{A})$. In Sect. 4 we will derive the category of MTSs through a combination of categorical constructions. Let us first recall the definition of MTS.

Definition 2 (Model Transition Systems[1] [15]). *A* model transition system (MTS) *for $\mathcal{L}(\mathcal{E},\mathcal{A})$ is a triple $\mathsf{M} = (I, (M_i)_{i \in I}, T)$ such that:*

1. *I is a set whose elements are called* indices*;*
2. *M_i for each $i \in I$ is a (possibly empty) Kripke model $M_i = (W_i, (R_i^e)_{e \in \mathcal{E}}, V_i)$; and*
3. *T is a pair $((t^\alpha)_{\alpha \in \mathcal{A}}, (T_i^\alpha)_{\alpha \in \mathcal{A}, i \in I})$ such that*
   (a) $t^\alpha : I \to I$ is a function and
   (b) $T_i^\alpha \subseteq W_i \times W_i^\alpha$ is a binary relation, where W_i^α is the carrier set of $M_{t^\alpha(i)}$.

The relations T_i^α are called action relations.

We also use the notation M_i^α to denote $M_{t^\alpha(i)}$. Notations M_i^α, W_i^α, T_i^α and t^α are naturally extended for a sequence of actions $\gamma = \alpha_1 \cdots \alpha_n \in \mathcal{A}^*$: M_i^γ, W_i^γ, T_i^γ and t^γ. A *frame transition system* (*FTS*) is defined analogously by replacing Kripke models with Kripke frames. The *underlying FTS* of an MTS $\mathsf{M} = (I, (M_i)_{i \in I}, T)$ is obtained by replacing each M_i with its underlying frame. It is denoted by $U(\mathsf{M})$.

An MTS expresses model transformations as follows. A Kripke model M_i is transformed into M_i^α by an action α, where the target M_i^α is determined by the function $t^\alpha : I \to I$. Kripke models M_i and M_i^α are linked by the binary relation T_i^α, and $w T_i^\alpha v$ expresses that state w in M_i is sent to state v in M_i^α. As a result, the MTS expresses the transformation of a pointed Kripke model (M_i, w) into (M_i^α, v) that satisfies $w T_i^\alpha v$.

A bounded morphism between MTSs is also defined:

[1] This notion is called a *functional MTS* in [16] since any t^α in an MTS $(I, (M_i)_{i \in I}, T)$ with $T = ((t^\alpha)_{\alpha \in \mathcal{A}}, (T_i^\alpha)_{\alpha \in \mathcal{A}, i \in I})$ is a function. Inverse operators in the more general settings where each t^α is a relation is a part of our future work.

Definition 3 (Bounded Morphisms between MTSs [15]**).** *For MTSs* $\mathsf{M} = (I, (M_i)_{i \in I}, T)$ *with* $T = ((t^\alpha)_{\alpha \in \mathcal{A}}, (T_i^\alpha)_{\alpha \in \mathcal{A}, i \in I})$ *and* $\mathsf{M}' = (I', (M'_{i'})_{i' \in I'}, T')$ $T' = ((t'^\alpha)_{\alpha \in \mathcal{A}}, (T'^\alpha_{i'})_{\alpha \in \mathcal{A}, i' \in I'})$*, a* bounded morphism $(h, (H_i)_{i \in I}) : \mathsf{M} \to \mathsf{M}'$ *is a pair of:*

1. *a function* $h : I \to I'$ *such that* $h \circ t^\alpha = t'^\alpha \circ h$ *for all* α *in* $\mathcal{A}$ *and*
2. *an* I*-indexed family of bounded morphisms* $H_i : M_i \to M'_{h(i)}$ *that additionally satisfies the following conditions for any* α *in* $\mathcal{A}$ *(here* H_i^α *denotes* $H_{t^\alpha(i)}$*):*
   (a) $wT_i^\alpha v$ *implies* $H_i(w) T'^\alpha_{h(i)} H_i^\alpha(v)$*, and*
   (b) $H_i(w) T'^\alpha_{h(i)} v'$ *implies* $wT_i^\alpha v$ *and* $H_i^\alpha(v) = v'$ *for some* $v \in M_i^\alpha$*.*

We remark that $H_i^\alpha : M_i^\alpha \to M'^\alpha_{h(i)}$ *since* $M_{t^\alpha(i)} = M_i^\alpha$ *and* $M'_{h(t^\alpha(i))} = M'_{t'^\alpha(h(i))} = M'^\alpha_{h(i)}$ *by condition 1.*

Conditions (a) and (b) in the definition correspond to the homomorphic condition and the back condition in the definition of ordinary bounded morphism. Condition 1 is their precondition.

Interpretation. Formulae in the language $\mathcal{L}(\mathcal{E}, \mathcal{A})$ are interpreted in an MTS as follows:

Definition 4 (Interpretation [15]**).** *Suppose that* $\mathsf{M} = (I, (M_i)_{i \in I}, T)$ *is an MTS and that* $M_i = (W_i, (R_i^e)_{e \in \mathcal{E}}, V_i)$ *is its* i*-th Kripke model. We inductively define the notion of a formula* φ *being satisfied at index* $i \in I$ *and state* $w \in W_i$ *in* M *(notation:* $\mathsf{M}, i, w \models \varphi$*) as follows:*

$$
\begin{array}{lll}
\mathsf{M}, i, w \models \top & & \textit{always} \\
\mathsf{M}, i, w \models p & \iff & w \in V_i(p) \\
\mathsf{M}, i, w \models \neg\varphi & \iff & \mathsf{M}, i, w \not\models \varphi \\
\mathsf{M}, i, w \models \varphi \vee \psi & \iff & \mathsf{M}, i, w \models \varphi \textit{ or } \mathsf{M}, i, w \models \psi \\
\mathsf{M}, i, w \models \langle e \rangle \varphi & \iff & \mathsf{M}, i, v \models \varphi \textit{ for some } v \in W_i \textit{ with } wR_i^e v \\
\mathsf{M}, i, w \models \mathrm{E}\varphi & \iff & \mathsf{M}, i, v \models \varphi \textit{ for some } v \in W_i \\
\mathsf{M}, i, w \models \langle\!\langle \alpha \rangle\!\rangle \varphi & \iff & \mathsf{M}, t^\alpha(i), v \models \varphi \textit{ for some } v \in W_i^\alpha \textit{ with } wT_i^\alpha v
\end{array}
$$

We also denote $\{w \in W_i \mid \mathsf{M}, i, w \models \varphi\}$ by $[\![\varphi]\!]_{\mathsf{M},i}$. We say that M *validates* φ if $\mathsf{M}, i, w \models \varphi$ for all i in I and $w \in W_i$. A formula φ is *valid in an FTS* F if any MTS M with $U(\mathsf{M}) = \mathsf{F}$ validates the formula φ.

A bounded morphism, defined above, preserves the truth values:

Proposition 1. *Let* $\mathsf{M} = (I, (M_i)_{i \in I}, T)$ *and* $\mathsf{M}' = (I', (M'_{i'})_{i' \in I'}, T')$ *be MTSs and* $(h, (H_i)_{i \in I}) : \mathsf{M} \to \mathsf{M}'$ *be a bounded morphism such that each* H_i *is surjective. Then, for any world* w *in* M_i*,* w *and* $H_i(w)$ *satisfy exactly the same formulae in* $\mathcal{L}(\mathcal{E}, \mathcal{A})$*.*

Note that surjectivity is needed for the global operator.

Examples of MTSs. By using an MTS and the interpretation above, we can express several types of model transformations up to isomorphism.

Example 6.

1. **PAL.** Informally speaking, the model transformation caused by an announcement $!\varphi$ changes a Kripke model M_i into its submodel $M_i|_\varphi$ whose carrier set is $[\![\varphi]\!]_{\mathsf{M},i}$. Formally, a *PAL-transition system* $(I, (M_i)_{i\in I}, T)$ is an MTS such that for any i in I, the relation $\{(w,v) \mid wT_i^{!\varphi}v\}$ gives rise to an isomorphism between $M_i|_\varphi$ and $M_i^{!\varphi}$ (the left diagram below).
2. **EA.** An action (U, s) with $U = (S, (\rightarrow_n)_{n\in Ag}, \mathrm{Pre})$ causes the model transformation that changes a Kripke model $M_i = (W_i, (R_i^n)_{n\in Ag}, V_i)$ into the Kripke model $M_i \otimes U = (W', (R'_n)_{n\in Ag}, V')$ given by

$$W' = \{(w,s) \in W_i \times S \mid w \in [\![\mathrm{Pre}(s)]\!]_{\mathsf{M},i}\},$$
$$(w,s)R'_n(v,t) \Leftrightarrow wR_nv \text{ and } s{\rightarrow_n}t, \quad (w,s) \in V'(p) \Leftrightarrow w \in V(p).$$

   We therefore define an *EA-transition system* to be an MTS $(I, (M_i)_{i\in I}, T)$ such that (for any index i in I and action model U):
   - $t^{(U,s)}(i) = t^{(U,t)}(i)$ for any s and t in U,
   - $\bigcup_{s\in U}\{((w,s),v) \mid wT_i^{(U,s)}v\}$ gives rise to an isomorphism between $M_i\otimes U$ and $M_i^{(U,s)}$ (the right diagram below).

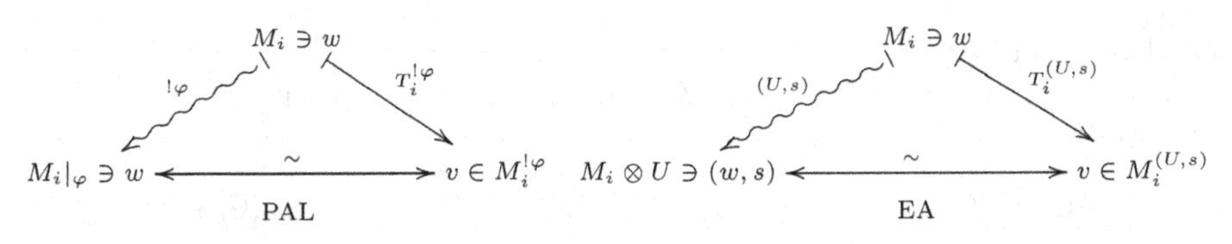

⊣

The proof systems PAL and EA are sound and strongly complete with respect to the class of PAL-transition systems and that of EA-transition systems, respectively, as established in [15] (adding the global operator is easy).

3 Inverse Operators

In the previous section, we have recalled the general framework, a generic language and MTSs, for DEL proposed in [15]. We next employ MTSs to extend DELs with inverse operators. We then discuss completeness, conservativity and properness of extended DELs.

3.1 Introduction of Inverse Operators

We define a *DEL language with inverse operators* $\mathcal{L}^+(\mathcal{E},\mathcal{A})$ to be $\mathcal{L}(\mathcal{E},\mathcal{A})$ augmented with the inverse operator $\langle\!\langle\alpha^{-1}\rangle\!\rangle\varphi$ of α. Specifically, $\mathcal{L}^+(\mathcal{E},\mathcal{A})$ is defined by the following BNF:

$$\varphi ::= \top \mid p \mid \neg\varphi \mid \varphi\vee\psi \mid \langle e\rangle\varphi \mid \mathrm{E}\varphi \mid \langle\!\langle\alpha\rangle\!\rangle\varphi \mid \langle\!\langle\alpha^{-1}\rangle\!\rangle\varphi$$

where p ranges over $\mathcal{P}$, e over $\mathcal{E}$ and α over $\mathcal{A}$. $\langle\!\langle\alpha^{-1}\rangle\!\rangle\varphi$ is called the inverse operator of α. A dual operator $[[\alpha^{-1}]]\varphi$ is defined to be $\neg\langle\!\langle\alpha^{-1}\rangle\!\rangle\neg\varphi$. An inverse operator is interpreted by an MTS $\mathsf{M} = (I,(M_i)_{i\in I},T)$ as follows: $\mathsf{M},i,w \models \langle\!\langle\alpha^{-1}\rangle\!\rangle\varphi$ iff $\mathsf{M},j,v\models\varphi$ for some $j\in I$ and $v\in W_j$ with $t^\alpha(j)=i$ and $vT^\alpha_j w$. The above generic definition specialises to the inverse operators for PAL and EA:

Example 7.

PAL. We denote the language $\mathcal{L}^+(\mathcal{E},\mathcal{A}_{\mathsf{PAL}})$ by $\mathcal{L}^+_{\mathsf{PAL}}$. The intended meaning of $[[!\varphi^{-1}]]\psi$ is that ψ always holds before a public announcement of φ.

EA The language $\mathcal{L}^+_{\mathsf{EA}}$ is defined to be $\mathcal{L}^+(\mathcal{E},\mathcal{A}_{\mathsf{EA}})$. A formula $[[(U,s)^{-1}]]\varphi$ means that φ always holds before an epistemic action (U,s). ⊣

Obviously, the inverse operators are related to the past operators in tense logics. Hence, it is natural to extend $\mathsf{K_g}$ with the following *converse axioms* (cf. [8]):

$$\varphi\to[[\alpha]]\langle\!\langle\alpha^{-1}\rangle\!\rangle\varphi,\quad \varphi\to[[\alpha^{-1}]]\langle\!\langle\alpha\rangle\!\rangle\varphi.$$

We denote this logic by K^+.[2]

The proof systems PAL^+ and EA^+ are obtained by replacing the base logic $\mathsf{K_g}$ with K^+:

$$\mathsf{PAL}^+ = \mathsf{K}^+\oplus \mathsf{R_P R_N R_E R_G R_T} \qquad \mathsf{EA}^+ = \mathsf{K}^+\oplus\mathsf{R_P R_N E_{EA} G_{EA} Pre}.$$

Then, PAL^+ and EA^+ are indeed sound with respect to the class of PAL-transition systems and that of EA-transition systems, respectively. The next subsection discusses their completeness.

Remark 1. Inverse operators $\langle(p!)^{-1}\rangle\top$ in [20] assert that either p or an equivalent expression was just announced. Therefore, $\langle(p!)^{-1}\rangle\top$ and $\langle(\neg\neg p!)^{-1}\rangle\top$ assert the same. On the other hand, our inverse operators $\langle\!\langle !p^{-1}\rangle\!\rangle\top$ and $\langle\!\langle !\neg\neg p^{-1}\rangle\!\rangle\top$ does not necessarily assert the same. This is because $T^{!p}_i$ and $T^{!\neg\neg p}_i$ in an MTS $(I,(M_i)_{i\in I},T)$ do not necessarily coincide. ⊣

[2] The axioms can be rephrased by adjunction (residuation) inference rules $\varphi\to[[\alpha]]\psi \Leftrightarrow \langle\!\langle\alpha^{-1}\rangle\!\rangle\varphi\to\psi$. Hence, adding inverse operators is a natural thing to do from an algebraic point of view. Also quite naturally, a display calculus has been proposed for EA with adjoint operators [11]. In *loc. cit.*, an example of a Kripke model is given to show that the residuation rules with any interpretation of the adjoint operators to $[[\alpha]]$ do not hold in some model. On the other hand, our residuation rules above hold since ours are based on global satisfaction: i.e. $\varphi\to[[\alpha]]\psi$ holds for any model iff $\langle\!\langle\alpha^{-1}\rangle\!\rangle\varphi\to\psi$ holds for any model.

3.2 Completeness

We now address completeness issues[3]. Our ultimate targets are completeness theorems for PAL and EA with inverse operators. Towards this goal, we incrementally add reduction axioms starting from K^+, and see whether completeness (partly established in [15]) can be maintained by adding of inverse operators. It will turn out that assumption of backward determinism is essential in many cases. We introduce the following additional axioms:

- $\mathsf{G} : \mathrm{E}\langle\!\langle\alpha\rangle\!\rangle\varphi \to [[\alpha]]\mathrm{E}\varphi$,
- $\mathsf{R}_\mathsf{N}^{-1} : \langle\!\langle\alpha^{-1}\rangle\!\rangle\neg\varphi \leftrightarrow \langle\!\langle\alpha^{-1}\rangle\!\rangle\top \wedge \neg\langle\!\langle\alpha^{-1}\rangle\!\rangle\varphi$,
- $\mathsf{B_d} : \{\langle\!\langle\alpha^{-1}\rangle\!\rangle\top \to \neg\langle\!\langle\beta^{-1}\rangle\!\rangle\top \mid \alpha, \beta \in \mathcal{A} \text{ and } \alpha \neq \beta\}$.

Let us begin with the base case, to which we give a short proof:

Proposition 2. $\mathsf{K}^+ \oplus \mathsf{G}$ *is sound and strongly complete with respect to the class of MTSs.*

Proof. Let $\mathcal{M} = (W, (R_e)_{e\in\mathcal{E}}, (R_\alpha)_{\alpha\in\mathcal{A}}, R_\mathrm{E}, V)$ be the canonical Kripke model for $\mathsf{K}^+ \oplus \mathsf{G}$ seen as an ordinary multimodal logic and $\mathcal{M}^-$ the Kripke model $(W, (R_e)_{e\in\mathcal{E}}, V)$. The relation R_E for the global operator E is an equivalence relation that includes R_e for all $e \in \mathcal{E}$ due to the four axioms added to K to obtain $\mathsf{K_g}$. We define an MTS $\mathsf{M} = (I \cup \{*\}, (M_i)_{i\in I\cup\{*\}}, T)$: I is the set of equivalence classes i for R_E; M_i is the submodel of $\mathcal{M}^-$ whose carrier set W_i is the equivalence class i and M_* is the empty structure; t^α is defined by $t^\alpha(i) = j$ if the restricted relation $R_\alpha \cap (W_i \times W_j) \neq \emptyset$ and otherwise $t^\alpha(i) = *$; T_i^α is the relation $R_\alpha \cap (W_i \times W_{t^\alpha(i)})$. Function t^α is well-defined due to the axiom G, which ensures that the equivalence relation R_E is preserved by action α. Corresponding two states $(\mathcal{M}, w)$ and (M, i, w) with $w \in i$ satisfy exactly the same formulae in $\mathcal{L}^+(\mathcal{E}, \mathcal{A})$. □

The proof above is easy: we just transform the canonical Kripke model into an MTS. We call this MTS the *canonical MTS for* $\mathsf{K}^+ \oplus \mathsf{G}$. Observe how the global operator, as well as axiom G, helps us cut the canonical Kripke model into relevant pieces which constitute the canonical MTS. This construction quite useful and, indeed is essential when proving Proposition 6, *infra.* All propositions in this subsection are proved by a similar construction of canonical MTSs.[4]

Determinism. We next consider actions in MTSs which are deterministic in forward or backward directions.

Definition 5. *Let* $\mathsf{M} = (I, (M_i)_{i\in I}, T)$ *be an MTS.*

1. M *is called* forward deterministic *if for any action* α*,* i *in* I *and* w *in* M_i*,* $wT_i^\alpha v$ *and* $wT_i^\alpha v'$ *imply* $v = v'$*.*

[3] Similar results *not* involving inverse operators are shown in [6,15,16,23,24].

[4] A previous work [16] studies this construction in more general settings without the assumption of the functionality of MTSs in the ordinary language without inverse operators.

2. M *is called* backward deterministic *if for any action α, any i in I and w in M_i, $vT_j^\alpha w$ and $v'T_{j'}^\alpha w$ imply $j = j'$ and $v = v'$ for all j and j' in I.*

These two types of determinism can be axiomatised by $\mathsf{R_N}$ and $\mathsf{R_N^{-1}}$:

Proposition 3. *Soundness and strong completeness hold for each of the following pairs:*

1. $\mathsf{K^+ \oplus GR_N}$ *and the class of forward deterministic MTSs;*
2. $\mathsf{K^+ \oplus GR_N^{-1}}$ *and the class of backward deterministic MTSs.*

This is due to the fact that $\mathsf{R_N}$ and $\mathsf{R_N^{-1}}$ are equivalent to $\langle\langle\alpha\rangle\rangle\varphi \to [[\alpha]]\varphi$ and $\langle\langle\alpha^{-1}\rangle\rangle\varphi \to [[\alpha^{-1}]]\varphi$ with the substitution of $\neg\varphi$ for φ, and they naturally capture determinism. Note that $\mathsf{K^+ \oplus GR_N}$ is equivalent to $\mathsf{K^+ \oplus R_G^L R_N}$ where $\mathsf{R_G^L}$ is the left direction of $\mathsf{R_G}$. In addition to the forward and the backward determinism, we introduce the notion of strict backward determinism: an MTS $\mathsf{M} = (I, (M_i)_{i\in I}, T)$ is *strictly backward deterministic* if for any i in I and w in M_i, $vT_j^\alpha w$ and $v'T_{j'}^{\alpha'} w$ imply $\alpha = \alpha'$, $j = j'$ and $v = v'$. Intuitively, strict backward determinism means that each state is a result of a unique action from a unique previous state. This property also has an axiomatisation:

Proposition 4. $\mathsf{K^+ \oplus GR_N^{-1}B_d}$ *is sound and strongly complete with respect to the class of strictly backward deterministic MTSs.*

The notion of strict backward determinism helps when it comes to the completeness of $\mathsf{EA^+}$.

Preservation of Facts. An action of communication itself does not change the facts: for example, communication does not change the weather. This nature is expressed as follows:

Definition 6 (Epistemic MTSs [15]). *An MTS* $\mathsf{M} = (I, (M_i)_{i\in I}, T)$ *is* epistemic *if for any action α in $\mathcal{A}$, index i in I, v and w with $vT_i^\alpha w$ satisfy exactly the same atomic propositions.*

As already established in [15] for the ordinary language without inverse operators, this notion corresponds to axiom schema $\mathsf{R_P}$ under the condition $\mathsf{R_N}$:

Proposition 5. $\mathsf{K^+ \oplus GR_N R_P}$ *is sound and strongly complete with respect to the class of forward deterministic and epistemic MTSs.*

Eliminativity. We now consider a class of actions whose only effect is to eliminate some of possible worlds.

Definition 7 (Eliminative MTSs [15]). *An MTS* $\mathsf{M} = (I, (M_i)_{i\in I}, T)$ *is called* eliminative *if for any index i in I and action α in $\mathcal{A}$, M_i^α is a submodel of M_i and the inverse relation $(T_i^\alpha)^{-1}$ is an embedding of M_i^α into M_i. An* eliminative FTS *is defined analogously.*

This kind of update appears in [4] and [17]. Eliminativity is axiomatised by $\mathsf{R_N R_E R_G}$ under the assumption of backward determinism:

Proposition 6. *Soundness and strong completeness hold for the following pairs:*

1. $\mathsf{K}^+ \oplus \mathsf{R}_\mathsf{N}^{-1}\mathsf{R}_\mathsf{N}\mathsf{R}_\mathsf{E}\mathsf{R}_\mathsf{G}$ *and the class of backward deterministic and eliminative FTSs;*
2. $\mathsf{K}^+ \oplus \mathsf{R}_\mathsf{N}^{-1}\mathsf{R}_\mathsf{N}\mathsf{R}_\mathsf{E}\mathsf{R}_\mathsf{G}\mathsf{R}_\mathsf{P}$ *and the class of backward deterministic eliminative MTSs.*

Proof. As before, it is proved by constructing a canonical MTS (or a canonical FTS) $\mathcal{M}$. Each action relation T_i^α in $\mathcal{M}$ indeed satisfies the desired properties:

- R_N ensures functionality: $wT_i^\alpha v$ and $wT_i^\alpha v'$ imply $v = v'$;
- $\mathsf{R}_\mathsf{N}^{-1}$ ensures injectivity: $wT_i^\alpha v$ and $w'T_i^\alpha v$ imply $w = w'$;
- R_E ensures surjectivity: for any w in M_i^α, there exists v in M_i such that $vT_i^\alpha w$;
- $\mathsf{R}_\mathsf{E}\mathsf{R}_\mathsf{N}$ ensures the forward homomorphic condition: $wT_i^\alpha v$, $w'T_i^\alpha v'$ and $wR_i^e w'$ imply $vR_{t^\alpha(i)}^e v'$;
- $\mathsf{R}_\mathsf{E}\mathsf{R}_\mathsf{N}\mathsf{R}_\mathsf{N}^{-1}$ ensures the backward homomorphic condition: $wT_i^\alpha v$, $w'T_i^\alpha v'$ and $vR_{t^\alpha(i)}^e v'$ imply $wR_i^e w'$ (Note that backward determinism is crucial here. $\mathsf{R}_\mathsf{E}\mathsf{R}_\mathsf{N}$ only guarantees that there is w'' such that $wR_i^e w''$ and $w''T_i^\alpha v'$.);
- $\mathsf{R}_\mathsf{P}\mathsf{R}_\mathsf{N}$ ensures the preservation of valuation: $wT_i^\alpha v$ implies that $w \in V_i(p)$ and $v \in V_{t^\alpha(i)}(p)$ are equivalent.

We thus conclude that $\mathcal{M}$ is eliminative. □

PAL and EA. The results so far extend to PAL (resp. EA) with inverse operators under the assumption of backward (resp. strict backward) determinism.

Proposition 7. *For each of the following pairs, soundness and strong completeness hold:*

1. $\mathsf{PAL}^+ \oplus \mathsf{R}_\mathsf{N}^{-1}$ *and the class of backward deterministic PAL-transition systems.*
2. $\mathsf{EA}^+ \oplus \mathsf{R}_\mathsf{N}^{-1}\mathsf{B}_\mathsf{d}$ *and the class of strictly backward deterministic EA-transition systems.*

For the latter result, the canonical MTS has to be slightly modified so that it meets the first condition of the definition of EA-transition system (Example 6). It is open whether completeness holds without the determinism assumptions.

Remark 2. We proved the completeness in a modular and plain manner. In many DELs, we often use the following fact to prove their completeness: any formula in $\mathcal{L}(\mathcal{E}, \mathcal{A})$ has a provably equivalent formula not containing any dynamic operators. This technique may be called *reduction*, which is why the axioms in Example 4 are called reduction axioms. However, we could not use this technique for the reason discussed in the next subsection.

3.3 Irreducibility

It is well known [10,17] that any formula in $\mathcal{L}_\mathsf{PAL}$ or $\mathcal{L}_\mathsf{EA}$ can be reduced to an equivalent formula not containing any dynamic operators; however, this is no longer the case for $\mathcal{L}_\mathsf{PAL}^+$ and $\mathcal{L}_\mathsf{EA}^+$. In this subsection we prove this fact.

We first introduce an *unravelling construction*. Intuitively, the unravelling of an MTS $(I, (M_i)_{i \in I}, T)$ around $i \in I$ is a tree-like MTS with root M_i, which consists of (copies of) its offspring. This part is sufficient for interpreting formulae in the original language $\mathcal{L}(\mathcal{E}, \mathcal{A})$.

Definition 8 (Unravellings of MTSs (cf. [16])). *Let $\mathsf{M} = (I, (M_i)_{i \in I}, T)$ be an MTS. The unravelled MTS $(\mathcal{A}^*, ({M'}_\gamma)_{\gamma \in \mathcal{A}^*}, T')$ of M around i in I is given as follows:*

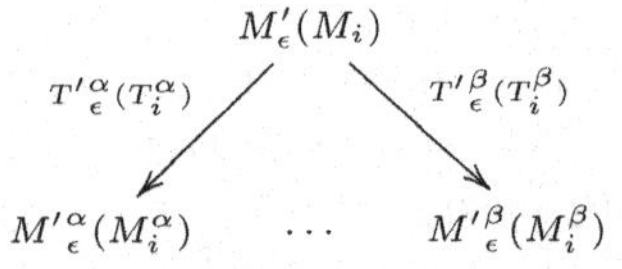

1. *the index set is the Kleene closure $\mathcal{A}^*$ of $\mathcal{A}$;*
2. *M'_γ is M_i^γ for each γ in $\mathcal{A}^*$;*
3. *T' is $(({t'}^\alpha)_{\alpha \in \mathcal{A}}, ({T'}^\alpha_\gamma)_{\alpha \in \mathcal{A}, \gamma \in \mathcal{A}^*})$ such that*
   (a) ${t'}^\alpha(\gamma) = \gamma\alpha$ for any γ in $\mathcal{A}^*$ and
   (b) ${T'}^\alpha_\gamma \subseteq W'_\gamma \times {W'}^\alpha_\gamma$ is $T^\alpha_{t^\gamma(i)} \subseteq W_i^\gamma \times W_i^{\gamma\alpha}$.

$M'_\epsilon(M_i)$

${T'}^\alpha_\epsilon(T_i^\alpha)$ ↙ ↘ ${T'}^\beta_\epsilon(T_i^\beta)$

${M'}^\alpha_\epsilon(M_i^\alpha)$ $\cdots$ ${M'}^\beta_\epsilon(M_i^\beta)$

If we think of $t^\alpha : I \to I$ and ${t'}^\alpha : \mathcal{A}^* \to \mathcal{A}^*$ as binary relations, $(\mathcal{A}^*, ({t'}^\alpha)_{\alpha \in \mathcal{A}})$ is nothing but the unravelled Kripke frame of $(I, (t^\alpha)_{\alpha \in \mathcal{A}})$ around i (cf. [8]). The unravelled MTS is obtained by naturally enriching $(\mathcal{A}^*, ({t'}^\alpha)_{\alpha \in \mathcal{A}})$ on the basis of $(I, (M_i)_{i \in I}, T)$.

Remark 3. A model used in [20] is called a *history*. A history is, intuitively, a sequence $(M_0, \cdots, M_n)$ of Kripke models by which model transformations $M_0 \overset{\alpha_1}{\mapsto} \cdots \overset{\alpha_n}{\mapsto} M_n$ in PAL or EA are recorded. An unravelled PAL- or EA-transition system starting from M_0 records all chains of transformations from M_0 as the index $\alpha_1 \cdots \alpha_n$. In this sense, the notion of MTS is an extension of that of history. ⊣

Proposition 8. *Let $(I, (M_i)_{i \in I}, T)$ and $(\mathcal{A}^*, ({M'}_\gamma)_{\gamma \in \mathcal{A}^*}, T')$ be an MTS and its unravelled one around i in I. Then, $(h, (H_\gamma)_{\gamma \in \mathcal{A}^*}) : (\mathcal{A}^*, ({M'}_\gamma)_{\gamma \in \mathcal{A}^*}, T') \to (I, (M_i)_{i \in I}, T)$ is a bounded morphism, where, for each $\gamma \in \mathcal{A}^*$, $h(\gamma)$ is defined to be $t^\gamma(i)$ and H_γ to be the identity mapping from M'_γ to M_i^γ.*

The irreducibility is proved as follows:

Proposition 9.

1. *Any formula of the form $\langle\!\langle !\varphi^{-1} \rangle\!\rangle \top$ in $\mathcal{L}^+_{\mathsf{PAL}}$ does not have any $\mathsf{PAL}^+ \oplus \mathsf{R}_\mathsf{N}^{-1}$-equivalent formula in $\mathcal{L}_{\mathsf{PAL}}$ unless it is equivalent to $\bot$.*
2. *Any formula $\langle\!\langle (U, s)^{-1} \rangle\!\rangle \top$ in $\mathcal{L}^+_{\mathsf{EA}}$ does not have any $\mathsf{EA}^+ \oplus \mathsf{R}_\mathsf{N}^{-1}\mathsf{B}_\mathsf{d}$-equivalent formula in $\mathcal{L}_{\mathsf{EA}}$ unless it is equivalent to $\bot$.*

Proof. We here prove (1) only. Let us suppose, for the sake of contradiction, that there is an equivalent formula ψ in $\mathcal{L}_{\mathsf{PAL}}$. As $\langle\!\langle !\varphi^{-1} \rangle\!\rangle \top$ is $\mathsf{PAL}^+ \oplus \mathsf{R}_\mathsf{N}^{-1}$-consistent by assumption, there is a backward deterministic PAL-transition system $\mathsf{M} = (I, (M_i)_{i \in I}, T)$ such that $\mathsf{M}, i, w \models \langle\!\langle !\varphi^{-1} \rangle\!\rangle \top$ for some i in I and $w \in M_i$. Namely, w is a result of action $!\varphi$ from a previous state. Note that we also have $\mathsf{M}, i, w \models \psi$. We next take the unravelled MTS M' of M around the index i. Since M' is a backward deterministic PAL-transition system the equivalence should still hold at any state in M'. However, $\mathsf{M}', \epsilon, w \models \psi$ holds due to Propositions 1 and 8 while $\mathsf{M}', \epsilon, w \models \langle\!\langle !\varphi^{-1} \rangle\!\rangle \top$ does not since w it is no more a result of any action. This is a contradiction. □

3.4 Conservativity

We conclude this section by showing that adding inverse operators and some axioms in Sects. 3.1 and 3.2 conservatively extends base logics.

We call an axiom schema A in $\mathcal{L}(\mathcal{E}, \mathcal{A})$ *canonical* if for any logic $\Lambda \supseteq \mathsf{K_g} \oplus \mathsf{A}$, its canonical frame $\mathcal{F}_\Lambda = (W, (R_e)_{e \in \mathcal{E}}, (R_\alpha)_{\alpha \in \mathcal{A}}, R_{\mathrm{E}})$ validates A.[5,6] Using the notion of canonicity, we have the following:

Proposition 10. *$\mathsf{K}^+ \oplus \mathsf{A}$ is conservative over $\mathsf{K_g} \oplus \mathsf{A}$ for any canonical schema A in $\mathcal{L}(\mathcal{E}, \mathcal{A})$.*

Proof. The canonical Kripke frame for $\mathsf{K_g} \oplus \mathsf{A}$ validates A thus all instances of A in $\mathcal{L}^+(\mathcal{E}, \mathcal{A})$. In particular, the canonical Kripke model validates A in $\mathcal{L}^+(\mathcal{E}, \mathcal{A})$. □

Axioms $\mathsf{R_G}$ and $\mathsf{R_E}$ in conjunction with $\mathsf{R_N}$ and many static axioms in the language $\mathcal{L}(\mathcal{E}, \emptyset)$ are canonical. For instance, Proposition 10 applies to $\mathsf{K}^+ \oplus \mathsf{R_N R_G R_E}$ and $\mathsf{K_g} \oplus \mathsf{R_N R_G R_E}$.

Now that we have a basic conservativity result, let us next consider the cases of $\mathsf{PAL}^+ \oplus \mathsf{R_N^{-1}}$ and $\mathsf{EA}^+ \oplus \mathsf{R_N^{-1} B_d}$. Notice that $\mathsf{R_P}$ is not closed under substitution and that $\mathsf{R_N^{-1}}$ and $\mathsf{B_d}$ contain inverse operators. To prove conservativity we use the following fact:

Fact 1. *Suppose that a theory Λ in $\mathcal{L}(\mathcal{E}, \mathcal{A})$ is (weakly) complete with respect to a class of MTSs and that an extension Λ' in $\mathcal{L}^+(\mathcal{E}, \mathcal{A})$ of Λ is sound with respect to the same class. Then, Λ' is conservative over Λ.*

Now a crucial observation is the following:

Lemma 1. *Let $\mathsf{M} = (I, (M_i)_{i \in I}, T)$ be an MTS. If the action relation T_i^α for each index $i \in I$ and action $\alpha \in \mathcal{A}$ is injective, then the unravelled MTS N of M around any index in I is strictly backward deterministic.*

By using this lemma, we obtain the following completeness results (in the case of EA, a minor modification is required to satisfy the condition on t^α).

Lemma 2. *For each of the following pairs, soundness and strong completeness hold:*

1. *PAL and the class of backward deterministic PAL-transition systems;*
2. *EA and the class of strictly backward deterministic EA-transition systems.*

Proof. We prove (1) only. For any consistent set Γ, there is a PAL-transition system $\mathsf{M} = (I, (M_i)_{i \in I}, T)$ satisfying Γ at some $i \in I$ and $w \in W_i$ by the completeness of PAL w.r.t. the class of all PAL-transition systems (Proposition 7 in [15]). Notice that each T_i^α is injective. Hence, by unravelling M around i, we obtain the theorem. □

[5] We here consider addition of all the instances of the schema unlike $\mathsf{R_P}$.

[6] A similar notion of canonicity is introduced in [16] to study the conservativity of DELs extended by the global operators.

By Lemma 2, Fact 1 and Proposition 7, we obtain the conservativity results:

Proposition 11. $\mathsf{PAL}^+ \oplus \mathsf{R}_\mathsf{N}^{-1}$ *and* $\mathsf{EA}^+ \oplus \mathsf{R}_\mathsf{N}^{-1}\mathsf{B}_\mathsf{d}$ *are conservative over* PAL *and* EA*, respectively.*

To other cases, we may apply the following fact:[7]

Fact 2. *Suppose that* $\mathsf{M} = (I, (M_i)_{i \in I}, T)$ *with* $M_i = (W_i, ({R_i}^e)_{e \in \mathcal{E}}, V_i)$ *for* $i \in I$ *is an MTS whose action relations* T_i^α *are functional and surjective. For any* $i' \in I$*, define an MTS* $\mathsf{M}' = (\mathcal{A}^*, ({M'}_\gamma)_{\gamma \in \mathcal{A}^*}, T')$ *with* $M'_\gamma = (W'_\gamma, ({R'_\gamma}^e)_{e \in \mathcal{E}}, V'_\gamma)$ *for* $\gamma \in \mathcal{A}^*$ *as blow:*

$$W'_\gamma = \{w \in W_{i'} \mid w R^\gamma_{i'} v \text{ for some } v\} \text{ (denote such state } v \text{ by } \gamma(w)),$$
$${R'_\gamma}^e = \{(w, w') \mid \gamma(w) R^e_{r^\gamma(i')} \gamma(w')\}, \quad V'_\gamma(p) = \{w \mid \gamma(w) \in V_{r^\gamma(i')}(p)\},$$
$$t'^\alpha(\gamma) = \gamma\alpha, \quad {T'_\gamma}^\alpha = \{(w, w) \mid w \in W'_\gamma \text{ and } w \in W'_{\gamma\alpha}\}.$$

Then, each action relation $T_i'^\alpha$ *of* M' *is injective and we have that*

$$\mathsf{M}, t^\gamma(i), \gamma(w) \models \varphi \text{ iff } \mathsf{M}', \gamma, w \models \varphi$$

for any $\gamma \in \mathcal{A}^*$*,* $w \in W'_\gamma$ *and* $\varphi \in \mathcal{L}(\mathcal{E}, \mathcal{A})$*.*

Note that functionality and surjectivity correspond to R_N and the right direction of R_G. Therefore, using this fact, we can see that $\mathsf{R}_\mathsf{N}^{-1}$ and $\mathsf{R}_\mathsf{N}^{-1}\mathsf{B}_\mathsf{d}$ conservatively extend other logics (e.g. from $\mathsf{K_g} \oplus \mathsf{R_N R_G R_E R_P}$ to $\mathsf{K}^+ \oplus \mathsf{R_N R_G R_E R_P R_N^{-1}}$ and $\mathsf{K}^+ \oplus \mathsf{R_N R_G R_E R_P R_N^{-1} B_d}$) by modifying canonical MTSs and using Lemma 1 carefully to ensure that modified MTSs are in desired classes.

4 Categorical Construction of Model Transition Systems

In Sect. 2.2 we have introduced model transition systems and bounded morphisms between them, which naturally form a category **MTS**. One might wonder how the definition of MTS arises; a part of the definition refers to the category $\mathbf{KM}_b$ of Kripke models and bounded morphisms, while other parts consist of an intricate indexing system with binary relations between them. To understand this category better, we introduce *MTS construction* that generalizes the category **MTS** and its algebraic counterpart studied in [15]. We first introduce three sub-constructions.

Indexed Coalgebras. For a set $\mathcal{A}$ and an endofunctor $F : \mathbb{C} \to \mathbb{C}$, we define the category $\mathbf{Coalg}(\mathcal{A}, F)$ of $\mathcal{A}$-indexed family of F-coalgebras by the following data. An object is a pair of $X \in \mathbb{C}$ and a family $(x^\alpha : X \to FX)_{\alpha \in \mathcal{A}}$ of F-coalgebras on X. A morphism from $(X, (x^\alpha)_{\alpha \in \mathcal{A}})$ to $(Y, (y^\alpha)_{\alpha \in \mathcal{A}})$ is a morphism $h : X \to Y$ such that it is a coalgebra morphism from (X, x^α) to (Y, y^α) for all $\alpha \in \mathcal{A}$. It comes with the evident forgetful functor, which we name $K_{\mathcal{A},F}$: $\mathbf{Coalg}(\mathcal{A}, F) \to \mathbb{C}$.

[7] This construction is a simpler version of that used in the proof of Proposition 8 in [15].

Family Construction. For a category $\mathbb{C}$, we define the category $\mathbf{Fam}(\mathbb{C})$ by the following data [12]. An object is pair of a set I and an I-indexed family $(X_i)_{i\in I}$ of $\mathbb{C}$-objects. A morphism from $(I,(X_i)_{i\in I})$ to $(J,(Y_j)_{j\in J})$ is a pair of a function $f : I \to J$ and an I-indexed family $(f_i : X_i \to Y_{f(i)})_{i\in I}$ of $\mathbb{C}$-morphisms. The composition of $(f,x) : (I,(X_i)_{i\in I}) \to (J,(Y_j)_{j\in J})$ and $(g,y) : (J,(Y_j)_{j\in J}) \to (K,(Z_k)_{k\in K})$ is given by $(g \circ f, (y_{f(i)} \circ x_i)_{i\in I})$. We then extend the **Fam** construction to functors and natural transformations as follows. This makes **Fam** a 2-endofunctor on **CAT**.

$$\mathbf{Fam}(F)(I,(X_i)_{i\in I}) = (I,(FX_i)_{i\in I}), \quad \mathbf{Fam}(F)(f,(x_i)_{i\in I}) = (f,(Fx_i)_{i\in I})$$
$$\mathbf{Fam}(\alpha)_{(I,(X_i)_{i\in I})} = (\mathrm{id}_I,(\alpha_{X_i})_{i\in I}).$$

Pullback. For two functors $\mathbb{C} \xrightarrow{F} \mathbb{D} \xleftarrow{G} \mathbb{E}$, we define the pullback category $\mathbf{PB}(F,G)$ by the following data. An object is a pair (C,E) of objects $C \in \mathbb{C}$ and $E \in \mathbb{E}$ such that $FC = GE$. A morphism from (C,E) to (C',E') is a pair of morphisms $f : C \to C'$ and $h : E \to E'$ such that $Ff = Gh$. Composition is defined componentwisely. The category $\mathbf{PB}(F,G)$ has evident projection functors into $\mathbb{C}$ and $\mathbb{E}$.

We now introduce MTS construction. It takes a set $\mathcal{A}$ and functors $\mathbb{C} \xrightarrow{U} \mathbb{D} \xrightarrow{F} \mathbb{D}$ as parameters. We first apply **Fam** construction to U and F to obtain functors $\mathbf{Fam}(U)$ and $\mathbf{Fam}(F)$ respectively. We then take the pullback of $\mathbf{Fam}(U)$ along the forgetful functor $K_{\mathcal{A},\mathbf{Fam}(F)}$:

$$\begin{array}{ccc}
\mathbf{PB}(K_{\mathcal{A},\mathbf{Fam}(F)},\mathbf{Fam}(U)) & \longrightarrow & \mathbf{Fam}(\mathbb{C}) \\
\big\downarrow & & \big\downarrow {\scriptstyle \mathbf{Fam}(U)} \\
\mathbf{Coalg}(\mathcal{A},\mathbf{Fam}(F)) & \xrightarrow{K_{\mathcal{A},\mathbf{Fam}(F)}} & \mathbf{Fam}(\mathbb{D})
\end{array}$$

The vertex category, which we name $\mathbf{MTS}(\mathcal{A}, \mathbb{C} \xrightarrow{U} \mathbb{D} \xrightarrow{F} \mathbb{D})$ (or $\mathbf{MTS}(\mathcal{A},U,F)$ for short), is the result of MTS construction. Its concrete description is given as follows.

- An object consists of a set I, I-indexed family of $\mathbb{C}$-objects $(M_i)_{i\in I}$, $\mathcal{A}$-indexed family of functions $(t^\alpha : I \to I)_{\alpha\in\mathcal{A}}$ and $I\times\mathcal{A}$-indexed family of $\mathbb{D}$-morphisms $(T_i^\alpha : UM_i \to FUM_{t^\alpha(i)})_{(i,\alpha)\in I\times\mathcal{A}}$.
- A morphism from $(I,(M_i),(t^\alpha),(T_i^\alpha))$ to $(J,(N_j),(s^\alpha),(S_j^\alpha))$ is a pair of a function $h : I \to J$ and an I-indexed family of $\mathbb{C}$-morphisms $(H_i : M_i \to N_{h(i)})_{i\in I}$ making the following diagrams commute for any $\alpha \in \mathcal{A}, i \in I$.

$$\begin{array}{ccc}
I & \xrightarrow{h} & J \\
{\scriptstyle t^\alpha}\big\downarrow & & \big\downarrow{\scriptstyle s^\alpha} \\
I & \xrightarrow[h]{} & J
\end{array}
\qquad
\begin{array}{ccc}
UM_i & \xrightarrow{UH_i} & UN_{h(i)} \\
{\scriptstyle UT_i^\alpha}\big\downarrow & & \big\downarrow{\scriptstyle US_{h(i)}^\alpha} \\
FUM_{t^\alpha(i)} & \xrightarrow[FUH_{t^\alpha(i)}]{} & FUN_{s^\alpha(h(i))}
\end{array}$$

The category **MTS** can be constructed by the MTS construction. We consider the category $\mathbf{KM}_b$ of Kripke models, including the empty structure, and bounded morphisms between them. It has the forgetful functor $U_{\mathbf{KM}_b} : \mathbf{KM}_b \to \mathbf{Set}$ extracting the underlying set. Also let $P : \mathbf{Set} \to \mathbf{Set}$ be the covariant powerset functor.

Theorem 3. *The category* $\mathbf{MTS}(\mathcal{A}, \mathbf{KM}_b \xrightarrow{U_{\mathbf{KM}_b}} \mathbf{Set} \xrightarrow{P} \mathbf{Set})$ *is isomorphic to* **MTS**.

One may replace $U_{\mathbf{KM}_b}$ with the forgetful functor from the category $\mathbf{KF}_b$ of Kripke *frames* and bounded morphisms between them. Then $\mathbf{MTS}(\mathcal{A}, U_{\mathbf{KF}_b}, P)$ is isomorphic to the category **FTS** of frame transition systems.

We next use MTS construction to recover the algebraic counterpart of MTSs called *algebraic model transition systems* [15] . Let **pAM** be the category of algebraic models whose underlying BAOs are *perfect* (i.e. complete atomic boolean algebras with operators preserving all joins [22, Definition 5.1]). Morphisms of **pAM** are complete homomorphisms between them. Note that $\mathbf{pAM}^{\mathrm{op}}$ and $\mathbf{KM}_b$ are equivalent [22, Theorem 5.8]. Next, let **CABA** be the category of complete atomic boolean algebras and complete homomorphisms between them. We also let $U_{\mathbf{pAM}} : \mathbf{pAM} \to \mathbf{CABA}$ be the evident forgetful functor, and $H : \mathbf{CABA}^{\mathrm{op}} \to \mathbf{CABA}^{\mathrm{op}}$ be the functor obtained by transferring the covariant powerset functor P on **Set** to $\mathbf{CABA}^{\mathrm{op}}$ by the equivalence $\mathbf{Set} \equiv \mathbf{CABA}^{\mathrm{op}}$.

One easily finds that $\mathbf{MTS}(\mathcal{A}, U^{\mathrm{op}}_{\mathbf{pAM}}, H)$ is a subcategory of the category of *algebraic model transition system* (*AMTS*) in [15]. We also have the duality result in the spirit of [15, Proposition 18]:

Theorem 4. *The category* $\mathbf{MTS}(\mathcal{A}, \mathbf{pAM}^{\mathrm{op}} \xrightarrow{U^{\mathrm{op}}_{\mathbf{pAM}}} \mathbf{CABA}^{\mathrm{op}} \xrightarrow{H} \mathbf{CABA}^{\mathrm{op}})$ *is equivalent to* **MTS**.

5 Related Work

As already referred to in footnote 2, Greco *et al.* [11] propose the adjoint operators to $[[\alpha]]$ in EA. They also propose using the final coalgebra to interpret the language of EA extended by the adjoint operators. This interpretation is justified by the invariance of EA-formulae under bisimulation, which holds due to the fact that reduction axioms are true for all connectives in EA. In contrast, our semantics does not require a 'full' set of reduction axioms and thus it may encompass more varieties of DEL extended with inverse operators.

Other studies on inverse operators include [1,2]. Aucher and Herzig [1] propose inverse operators for EA by using a 'large' Kripke model wherein the effect of an action α is expressed by its corresponding accessibility relation R_α on the set of possible worlds. In this sense, their inverse operators are introduced on the basis of study on how to express the effect of actions without model transformation, while ours are on that of study on how to express model transformations without model transformations. On the other hand, Balbiani *et al.* [2] propose inverse operators for mono-agent PAL. Its semantics uses an 'initial' set and a set of its subsets. It is sufficient for interpretation since the logic is mono-agent PAL. The main difference from ours is that our semantics does not require the 'initial' set.

Regarding the categorical reformulation of MTSs, in [9,13], Cîrstea and Sadrzadeh give a coalgebraic model of modal logic for actions and agents. They take $TX = P_\kappa(X)^{\mathcal{E}} \times (1+X)^{\mathcal{A}} \times \wp(\mathcal{P})$, where κ is a regular cardinal, as the coalgebra functor (over **Set**). A T-coalgebra determines an *extended Kripke model* proposed in [23] and [24] for EA and PAL, respectively. Such an extended Kripke model further determines an MTS whose index set is singleton.

The main feature of the semantics of DEL in an MTS (Definition 4), which was originally given in [15], is the use of indexed Kripke models and the interpretation of actions as index updates. The categorical reformulation of MTS in Sect. 4 focuses on capturing this indexing mechanism by **Fam** construction. Categorical semantics of DEL in [9,13,14] does not have such indexing systems. On the other hand, the first author extends index update functions (t^α component of MTS) to relations in [16]. This extension is beyond the categorical framework in Sect. 4, and its categorical understanding is left as a future work.

6 Conclusion and Future Work

Conclusion. In this paper, we have introduced inverse operators $\langle\!\langle \alpha^{-1} \rangle\!\rangle$ of an action α in Dynamic Epistemic Logic (DEL) in a way independent of specific DELs and along the line of tense logics. We have also provided its semantics using model transition systems (MTSs), which expresses model transformations in one model. In this syntax and semantics, we have proved several soundness and completeness results of well-known axioms and, as results, have given modular proofs of completeness of public announcement logic and epistemic action logic extended with inverse operators with respect to suitable classes of MTSs under the assumptions of 'backward determinism'. In addition, we have also analysed categorical structures behind the MTSs.

Future Work. There are some problems left. Firstly, our completeness results rest on the assumption of backward determinism, whose necessity is unclear. Secondly, decidability of PAL^+ and EA^+ should be addressed. Thirdly, for the generalised notion of MTS proposed in [16], whose model transitions t^α are not necessarily functional, we shall study its categorical reformulation and whether analogous results regarding inverse operators given in this paper hold. Lastly, the MTS construction in Sect. 4 is independent from the inverse operators studied in Sect. 3, and their integration is desirable.

Acknowledgements. We would like to thank Kazushige Terui and Manuela Antoniu for discussion and many helpful comments. The second author is supported by ERATO HASUO Metamathematics for Systems Design Project (No. JPMJER1603), JST. Thanks are also due to all anonymous referees who gave valuable comments on earlier versions of this article.

References

1. Aucher, G., Herzig, A.: Exploring the power of converse events. In: Girard, P., Roy, O., Marion, M. (eds.) Dynamic Formal Epistemology, pp. 51–74. Springer, Dordrecht (2011). https://doi.org/10.1007/978-94-007-0074-1_4
2. Balbiani, P., van Ditmarsch, H., Herzig, A.: Before announcement (regular paper). In: Advances in Modal Logic (AiML 2016), Budapest, Hungary, 30 August 2016–02 Sept 2016, pp. 58–77. College Publications (2016). http://www.collegepublications.co.uk/, http://oatao.univ-toulouse.fr/19192/
3. Baltag, A., Moss, L.S., Solecki, S.: The logic of public announcements, common knowledge, and private suspicions. In: Proceedings of the 7th Conference on Theoretical Aspects of Rationality and Knowledge, TARK '98, p. 43–56. Morgan Kaufmann Publishers Inc., San Francisco (1998)
4. van Benthem, J.: Dynamic logic for belief revision. J. Appl. Non-Classic. Logics **17**(2), 129–155 (2007). https://doi.org/10.3166/jancl.17.129-155
5. van Benthem, J.: Logical Dynamics of Information and Interaction. Cambridge University Press, Cambridge (2011). https://doi.org/10.1017/CBO9780511974533
6. van Benthem, J.: Two logical faces of belief revision. In: Trypuz, R. (ed.) Krister Segerberg on Logic of Actions. OCL, vol. 1, pp. 281–300. Springer, Dordrecht (2014). https://doi.org/10.1007/978-94-007-7046-1_13
7. van Benthem, J., Liu, F.: Dynamic logic of preference upgrade. J. Appl. Non-Classic. Logics **17**(2), 157–182 (2007). https://doi.org/10.3166/jancl.17.157-182
8. Blackburn, P., Rijke, M.d., Venema, Y.: Modal Logic. Cambridge Tracts in Theoretical Computer Science. Cambridge University Press, Cambridge (2001). https://doi.org/10.1017/CBO9781107050884
9. Cîrstea, C., Sadrzadeh, M.: Coalgebraic epistemic update without change of model. In: Mossakowski, T., Montanari, U., Haveraaen, M. (eds.) CALCO 2007. LNCS, vol. 4624, pp. 158–172. Springer, Heidelberg (2007). https://doi.org/10.1007/978-3-540-73859-6_11
10. van Ditmarsch, H., van der Hoek, W., Kooi, B.: Dynamic Epistemic Logic, 1st edn. Springer, Dordrecht (2008). https://doi.org/10.1007/978-1-4020-5839-4
11. Greco, G., Kurz, A., Palmigiano, A.: Dynamic epistemic logic displayed. In: Grossi, D., Roy, O., Huang, H. (eds.) LORI 2013. LNCS, vol. 8196, pp. 135–148. Springer, Heidelberg (2013). https://doi.org/10.1007/978-3-642-40948-6_11
12. Jacobs, B.: Categorical Logic and Type Theory. No. 141 in Studies in Logic and the Foundations of Mathematics. North Holland, Amsterdam (1999)
13. Kishida, K.: Categories for dynamic epistemic logic. In: Proceedings of the 16th Conference on Theoretical Aspects of Rationality and Knowledge, TARK 2017, pp. 353–372. Open Publishing Association (2017). https://doi.org/10.4204/eptcs.251.26
14. Kurz, A., Palmigiano, A.: Epistemic updates on algebras. Logical Methods Comput. Sci. **9**(4) (2013). https://doi.org/10.2168/LMCS-9(4:17)2013
15. Motoura, S.: A general framework for modal correspondence in dynamic epistemic logic. In: van der Hoek, W., Holliday, W.H., Wang, W. (eds.) LORI 2015. LNCS, vol. 9394, pp. 282–294. Springer, Heidelberg (2015). https://doi.org/10.1007/978-3-662-48561-3_23
16. Motoura, S.: A general framework for dynamic epistemic logic: towards canonical correspondences. J. Appl. Non-Classic. Logics **27**(1–2), 50–89 (2017). https://doi.org/10.1080/11663081.2017.1370663

17. Plaza, J.: Logics of public communications. Synthese **158**(2), 165–179 (2007). https://doi.org/10.1007/s11229-007-9168-7
18. Renne, B., Sack, J., Yap, A.: Dynamic epistemic temporal logic. In: He, X., Horty, J., Pacuit, E. (eds.) LORI 2009. LNCS (LNAI), vol. 5834, pp. 263–277. Springer, Heidelberg (2009). https://doi.org/10.1007/978-3-642-04893-7_21
19. Renne, B., Sack, J., Yap, A.: Logics of temporal-epistemic actions. Synthese **193**(3), 813–849 (2016). https://doi.org/10.1007/s11229-015-0773-6
20. Sack, J.: Temporal languages for epistemic programs. J. Logic Lang. Inf. **17**(2), 183–216 (2008). https://doi.org/10.1007/s10849-007-9054-1
21. Sack, J.: Logic for update products and steps into the past. Ann. Pure Appl. Logic **161**(12), 1431–1461 (2010). https://doi.org/10.1016/j.apal.2010.04.011, https://www.sciencedirect.com/science/article/pii/S0168007210000527
22. Venema, Y.: 6 algebras and coalgebras. In: Blackburn, P., Van Benthem, J., Wolter, F. (eds.) Handbook of Modal Logic, Studies in Logic and Practical Reasoning, vol. 3, pp. 331–426. Elsevier (2007). https://doi.org/10.1016/S1570-2464(07)80009-7, https://www.sciencedirect.com/science/article/pii/
23. Wang, Y., Aucher, G.: An alternative axiomatization of del and its applications. In: Proceedings of the Twenty-Third International Joint Conference on Artificial Intelligence, IJCAI '13, pp. 1139–1146. AAAI Press (2013)
24. Wang, Y., Cao, Q.: On axiomatizations of public announcement logic. Synthese **190**(1), 103–134 (2013). https://doi.org/10.1007/s11229-012-0233-5

Betwixt Turing and Kleene

Dag Normann[1] and Sam Sanders[2](✉)

[1] Department of Mathematics, The University of Oslo, P.O. Box 1053, Blindern, 0316 Oslo, Norway
dnormann@math.uio.no

[2] Department of Philosophy II, RUB Bochum, Universitätsstrasse 150, 44780 Bochum, Germany
sam.sanders@rub.de

Abstract. Turing's famous 'machine' model constitutes the first intuitively convincing framework for *computing with real numbers.* Kleene's computation schemes S1–S9 extend Turing's approach and provide a framework for *computing with objects of any finite type.* Various research programs have been proposed in which higher-order objects, like functions on the real numbers, are *represented/coded* as real numbers, so as to make them amenable to the Turing framework. It is then a natural question whether there is any significant difference between the Kleene approach or the Turing-approach-via-codes. Continuous functions being well-studied in this context, we study *functions of bounded variation*, which have **at most countably** many points of discontinuity. A central result is the *Jordan decomposition theorem* that a function of bounded variation on $[0, 1]$ equals the difference of two monotone functions. We show that for this theorem and related results, the difference between the Kleene approach and the Turing-approach-via-codes is *huge*, in that full second-order arithmetic readily comes to the fore in Kleene's approach, in the guise of Kleene's quantifier $\exists^3$.

Keywords: Representations · Computability theory · Kleene S1–S9 · Bounded variation

1 Introduction: Jordan, Turing, and Kleene

In a nutshell, we study the computational properties of the *Jordan decomposition theorem* as in Theorem 1 and other results on *functions of bounded variation*, establishing the huge differences between the *Turing and Kleene approaches* to computability theory. For the rest of this section, we introduce the above italicised notions and sketch the contents of this paper in more detail. All technical notions are introduced in Sect. 2 while our main results are in Sect. 3.

First of all, Turing's famous 'machine' model, introduced in [45], constitutes the first intuitively convincing framework for *computing with real numbers.* Kleene's computation schemes S1–S9, introduced in [18] extend Turing's framework and provide a framework for *computing with objects of any finite type.* Now,

S. Artemov and A. Nerode (Eds.): LFCS 2022, LNCS 13137, pp. 236–252, 2022.
https://doi.org/10.1007/978-3-030-93100-1_15

various[1] research programs have been proposed in which higher-order objects are *represented/coded* as real numbers or similar second-order representations, so as to make them amenable to the Turing framework. It is then a natural question whether there is any significant difference[2] between the Kleene approach and the Turing-approach-via-codes. Continuous functions being well-studied (see footnote 2) in this context, we investigate *functions of bounded variation*, which have **at most** countably many points of discontinuity.

Secondly, the notion of *bounded variation* was first introduced by Jordan around 1881 [17] yielding a generalisation of Dirichlet's convergence theorems for Fourier series. Indeed, Dirichlet's convergence results are restricted to functions that are continuous except at a finite number of points, while functions of bounded variation can have (at most) countable many points of discontinuity, as also shown by Jordan, namely in [17, p. 230]. The fundamental theorem about functions of bounded variation is as follows and can be found in [17, p. 229].

Theorem 1 (Jordan decomposition theorem). *A function $f : [0,1] \to \mathbb{R}$ of bounded variation can be written as the difference of two non-decreasing functions $g, h : [0,1] \to \mathbb{R}$.*

The computational properties of Theorem 1 have been studied extensively via second-order representations, namely in e.g. [23,28,29,48]. The same holds for constructive analysis by [5,7,14,39], involving different (but related) constructive enrichments. Now, finite iterations of the Turing jump suffice to compute g, h from Theorem 1 in terms of *represented* functions f of bounded variation by [23, Cor. 10].

Thirdly, in light of the previous, it is a natural question what the computational properties of Theorem 1 are in Kleene's framework. In particular, the following question is central to this paper.

How hard is it to compute (S1-S9) from $f : [0,1] \to \mathbb{R}$ of bounded variation, two monotone functions g, h such that $f = g - h$ on $[0,1]$?

A functional that can perform this computational task will be called a *Jordan realiser* (see Definition 6). A related and *natural* computational task is as follows.

How hard is it to compute (S1-S9) from $f : [0,1] \to \mathbb{R}$ of bounded variation, the supremum $\mathsf{sup}_{x\in[0,1]} f(x)$?

[1] Examples of such frameworks include: reverse mathematics [43,44], constructive analysis ([2, I.13], [3]), predicative analysis [12], and computable analysis [47]. Note that Bishop's constructive analysis is not based on Turing computability *directly*, but one of its 'intended models' is however (constructive) recursive mathematics, as discussed in [8]. One aim of Feferman's predicative analysis is to capture constructive reasoning in the sense of Bishop.

[2] The *fan functional* constitutes an early *natural* example of this difference: it has a computable code but is not S1–S9 computable (but S1–S9 computable in Kleene's $\exists^2$ from Sect. 2.2). The fan functional computes a modulus of uniform continuity for continuous functions on Cantor space; details may be found in [26].

A functional that can perform this computational task will be called a sup-*realiser* (see Definition 6). This task restricted to *continuous* functions is well-studied, and rather weak by [21, Footnote 6]. In light of the above, the following computational task is also natural:

How hard is it to compute S1–S9 from $f : [0,1] \to \mathbb{R}$ of bounded variation, a sequence $(x_n)_{n\in\mathbb{N}}$ listing the points of discontinuity of f?

By way of a robustness result, we show that the above three tasks are *the same* modulo Kleene's $\exists^2$ from Sect. 2.2. Moreover, we show that Jordan realisers are *hard* to compute: no type two functional, in particular the functionals S_k^2 which decide Π_k^1-formulas (see Sect. 2.2), can compute a Jordan realiser. We also show that Jordan realisers are *powerful*: when combined with other natural functionals, one can go all the way up to Kleene's quantifier $\exists^3$ which yields full second-order arithmetic (see again Sect. 2.2 for the definition of $\exists^3$). We also study special cases of Jordan realisers, which connects to the computational tasks associated to the *uncountability of* $\mathbb{R}$.

Finally, our main results are obtained in Sect. 3 while some preliminary notions, including some essential parts of Kleene's higher-order computability theory, can be found in Sect. 2.

2 Preliminaries

2.1 Kleene's Higher-Order Computability Theory

We first make our notion of 'computability' precise as follows.

(I) We adopt ZFC, i.e. Zermelo-Fraenkel set theory with the Axiom of Choice, as the official metatheory for all results, unless explicitly stated otherwise.
(II) We adopt Kleene's notion of *higher-order computation* as given by his nine clauses S1–S9 (see [26, Ch. 5] or [18]) as our official notion of 'computable'.

We refer to [26] for a thorough overview of higher-order computability theory. We do mention the distinction between 'normal' and 'non-normal' functionals based on the following definition from [26, §5.4].

Definition 2. For $n \geq 2$, a functional of type n is called *normal* if it computes Kleene's $\exists^n$ following S1–S9, and *non-normal* otherwise.

We only make use of $\exists^n$ for $n = 2, 3$, as defined in Sect. 2.2.

It is a historical fact that higher-order computability theory based on S1–S9 has focused primarily on *normal* functionals (see [26, §5.4] for this opinion). We have previously studied the computational properties of new *non-normal* functionals, namely those that compute the objects claimed to exist by:

- the Heine-Borel and Vitali covering theorems [30,32,33],
- the Baire category theorem [35],
- local-global principles like *Pincherle's theorem* [36],

- the uncountability of $\mathbb{R}$ and the Bolzano-Weierstrass theorem for countable sets in Cantor space [34,38],
- weak fragments of the Axiom of (countable) Choice [37].

In this paper, we continue this study for the Jordan decomposition theorem and other basic properties of functions of bounded variation. Next, we introduce some required higher-order notions in Sect. 2.2.

2.2 Some Higher-Order Notions

Some Higher-Order Functionals. We introduce a number of comprehension functionals from the literature. We are dealing with *conventional* comprehension, i.e. only parameters over $\mathbb{N}$ and $\mathbb{N}^{\mathbb{N}}$ are allowed in formula classes like Π^1_k and related notions.

First of all, the functional[3] φ as in $(\exists^2)$ is clearly discontinuous at $f = 11\ldots$; in fact, $(\exists^2)$ is equivalent to the existence of $F : \mathbb{R} \to \mathbb{R}$ such that $F(x) = 1$ if $x > 0$, and 0 otherwise by [21, §3].

$$(\exists \varphi^2 \leq_2 1)(\forall f^1)\big[(\exists n^0)(f(n) = 0) \leftrightarrow \varphi(f) = 0\big]. \tag{$\exists^2$}$$

Intuitively speaking, the functional φ from $(\exists^2)$ can decide the truth of any Σ^0_1-formula in its (Kleene) normal form. Related to $(\exists^2)$, the functional μ^2 in (μ^2) is also called *Feferman's* μ [1], defined as follows:

$$(\exists \mu^2)(\forall f^1)\begin{bmatrix}(\exists n^0)(f(n) = 0) \to [f(\mu(f)) = 0 \wedge (\forall i < \mu(f))(f(i) \neq 0)] \\ \wedge \\ (\forall n^0)(f(n) \neq 0) \to (\mu(f) = 0)\end{bmatrix}. \tag{1}$$

We have $(\exists^2) \leftrightarrow (\mu^2)$ over a weak system by [20, Prop. 3.4 and Cor. 3.5]) while μ^2 is readily computed from φ^2 in $(\exists^2)$. The third-order functional from $(\exists^2)$ is called 'Kleene's quantifier $\exists^2$'; we use the same convention for other functionals.

Secondly, S^2 as in (S^2) is called (see footnote 3) *the Suslin functional* [1,21]:

$$(\exists \mathsf{S}^2 \leq_2 1)(\forall f^1)\big[(\exists g^1)(\forall n^0)(f(\overline{g}n) = 0) \leftrightarrow \mathsf{S}(f) = 0\big]. \tag{S^2}$$

Intuitively, the Suslin functional S^2 can decide the truth of any Σ^1_1-formula in its normal form. We similarly define the functional S^2_k which decides the truth or falsity of Σ^1_k-formulas (again in normal form). We note that the operators ν_n from [10, p. 129] are essentially S^2_n strengthened to return a witness to the Σ^1_n-formula at hand. As suggested by its name, ν_k is the restriction of Hilbert-Bernays' ν from [15, p. 495] to Σ^1_k-formulas. We sometimes use the special cases S^2_0 and S^2_1 to denote the functionals $\exists^2$ and S^2.

Thirdly, second-order arithmetic is readily derived from (see footnote 3) the following:

$$(\exists E^3 \leq_3 1)(\forall Y^2)\big[(\exists f^1)(Y(f) = 0) \leftrightarrow E(Y) = 0\big]. \tag{$\exists^3$}$$

[3] The notation '$\varphi^2 \leq_2 1$' means that $\varphi(f) \leq 1$ for all $f \in \mathbb{N}^{\mathbb{N}}$ and guarantees uniqueness. The same holds for S^2, S^2_k, and $\exists^3$ below.

The functional from $(\exists^3)$ is also called 'Kleene's quantifier $\exists^3$'. Hilbert-Bernays' ν from [15, p. 495] trivially computes $\exists^3$.

Finally, the functionals S_k^2 are defined using the usual formula class Π_k^1, i.e. only allowing first- and second-order parameters. We have dubbed this the *conventional approach* and the associated functionals are captured by the umbrella term *conventional comprehension.* Comprehension involving third-order parameters has previously (only) been studied in [12,19], to the best of our knowledge.

Some Higher-Order Definitions. We introduce some required definitions.

First of all, a fruitful and faithful approach is the representation of sets by characteristic functions (see e.g. [24,32,35,37,40–42]), well-known from e.g. measure and probability theory. We shall use this approach, always assuming $\exists^2$ to make sure open sets represented by countable unions of basic opens are indeed sets in our sense.

Secondly, we now turn to *countable* sets. Of course, the notion of 'countable set' can be formalised in various ways, as follows.

Definition 3 [Enumerable set]. A set $A \subset \mathbb{R}$ is *enumerable* if there is a sequence $(x_n)_{n\in\mathbb{N}}$ such that $(\forall x \in \mathbb{R})(x \in A \leftrightarrow (\exists n \in \mathbb{N})(x = x_n))$.

Definition 3 reflects the notion of 'countable set' from reverse mathematics [43, V.4.2]. Our definition of 'countable set' is as follows.

Definition 4 [Countable set]. A set $A \subset \mathbb{R}$ is *countable* if there is $Y : \mathbb{R} \to \mathbb{N}$ with

$$(\forall x, y \in A)(Y(x) =_0 Y(y) \to x = y). \tag{2}$$

The functional Y as in (2) is called *injective* on A or *an injection* on A. If $Y : \mathbb{R} \to \mathbb{N}$ is also *surjective*, i.e. $(\forall n \in \mathbb{N})(\exists x \in A)(Y(x) = n)$, we call A *strongly countable.* In this case, Y is called *bijective* on A or *a bijection* on A.

The first part of Definition 4 is from Kunen's set theory textbook [25, p. 63] and the second part is taken from Hrbacek-Jech's set theory textbook [16] (where the term 'countable' is used instead of 'strongly countable'). According to Veldman [46, p. 292], Brouwer studies set theory based on injections in [9]. Hereinafter, 'strongly countable' and 'countable' shall exclusively refer to Definition 4.

3 Main Results

In this section, we shall obtain our main results as follows. Recall that a *Jordan realiser* outputs the monotone functions claimed to exist by the Jordan decomposition theorem as in Theorem 1.

- We introduce Jordan realisers and other functionals witnessing basic properties of functions of bounded variation; we show that three of these are computationally equivalent (Sect. 3.1).
- We show that Jordan realisers are *hard* to compute based on results from [34] (Sect. 3.2).

- We show that Kleene's $\exists^3$ can be computed from $\exists^2$, a Jordan realiser, and a well-ordering of $[0, 1]$ (Sect. 3.3).
- We show that Jordan realisers remain hard to compute even if we severely restrict the output (Sect. 3.4).

3.1 Jordan Realisers and Equivalent Formulations

We introduce functionals witnessing basic properties of functions of bounded variation, including the Jordan decomposition theorem (Theorem 1). We show that three of these are computationally equivalent given $\exists^2$.

As noted above, we always assume $\exists^2$ but specify the use when essential. This means that we can use the concept of Kleene-computability over $\mathbb{R}$ or $[0, 1]$ without focusing on how these spaces are represented.

First of all, as to definitions, the *total variation* of a function $f : [a, b] \to \mathbb{R}$ is (nowadays) defined as follows:

$$V_a^b(f) := \sup_{a \le x_0 < \cdots < x_n \le b} \sum_{i=0}^n |f(x_i) - f(x_{i+1})|. \tag{3}$$

If this quantity exists and is finite, one says that f has bounded variation on $[a, b]$. Now, the notion of bounded variation is defined in [28] *without* mentioning the supremum in (3); this approach can also be found in [5,7,23]. Hence, we shall hereafter distinguish between the following two notions. As it happens, Jordan seems to use item (a) of Definition 5 in [17, pp. 228–229], providing further motivation for the functionals introduced in Definition 6.

Definition 5 [Variations on variation]

a. The function $f : [a, b] \to \mathbb{R}$ *has bounded variation* on $[a, b]$ if there is $k_0 \in \mathbb{N}$ such that $k_0 \ge \sum_{i=0}^n |f(x_i) - f(x_{i+1})|$ for any partition $a \le x_0 < x_1 < \cdots < x_{n-1} < x_n \le b$.
b. The function $f : [a, b] \to \mathbb{R}$ *has a variation* on $[a, b]$ if the supremum in (3) exists and is finite.

We can now introduce the following notion of 'realiser' for the Jordan decomposition theorem and related functionals.

Definition 6

- A *Jordan realiser* is a partial functional $\mathcal{J}$ of type 3 taking as input $f : [0, 1] \to \mathbb{R}$ of bounded variation (item (a) in Definition 5), and providing a pair (g, h) of non-decreasing $g, h : [0, 1] \to \mathbb{R}$ with $f = g - h$ on $[0, 1]$.
- A *weak Jordan realiser* is a partial functional $\mathcal{J}_{\mathsf{w}}$ of type 3 taking as inputs a function $f : [0, 1] \to \mathbb{R}$ and its bounded variation $V_0^1(f)$ (item (b) in Definition 5), and providing a pair (g, h) of increasing functions g and h such that $f = g - h$ on $[0, 1]$.
- A sup-*realiser* is a partial functional $\mathcal{S}$ of type 3 taking as input a function $f : [0, 1] \to \mathbb{R}$ which has bounded variation (item (a) in Definition 5), and providing the supremum $\mathsf{sup}_{x \in [0,1]} f(x)$.

- A *continuity-realiser* is a partial functional $\mathcal{L}$ of type 3 taking as input a function $f : [0,1] \to \mathbb{R}$ which has bounded variation (item (a) in Definition 5), and providing a sequence $(x_n)_{n\in\mathbb{N}}$ which lists all points of discontinuity of f on $[0,1]$.

Next, we need the following lemma. The use of $\exists^2$ is perhaps superfluous in light of the constructive proof in [6], but the latter seems to make essential use of the Axiom of (countable) Choice.

Lemma 7. *There is a functional $\mathcal{D}$, computable in $\exists^2$, such that if $f : [0,1] \to \mathbb{R}$ is increasing (decreasing), then $\mathcal{D}(f)$ enumerates all points of discontinuity of f on $[0,1]$.*

Proof. Let $\{q_i\}_{i\in\mathbb{N}}$ be an enumeration of $\mathbb{Q} \cap [0,1]$, let $f : [0,1] \to \mathbb{R}$ be monotone, and define $a_i := f(q_i)$. Let A be the set of pairs of rationals $p < r$ such that there is no i with $p < a_i < r$. For $(p,r) \in A$, define

$$x_{p,r} := \sup\{q_i : a_i \leq p\} = \inf\{q_j : r \leq a_j\}. \tag{4}$$

It is easy to see that the reals in (4) are equal; indeed, the existence of a rational q_i between them, together with assuming that $(p,r) \in A$, leads to a contradiction.

Then all discontinuities of f will be among the elements $x_{p,r}$ for $(p,r) \in A$. Clearly, $i \mapsto a_i$, A and $(p,r) \mapsto x_{p,r}$ for $(p,r) \in A$ are computable in f and $\exists^2$. This ends the proof. □

Finally, we show that three 'non-weak' realisers from Definition 6 are in fact one and the same, in part based on Lemma 7.

Theorem 8. *Assuming $\exists^2$, Jordan realisers,* sup*-realisers, and continuity realisers are computationally equivalent.*

Proof. We first show that a continuity realiser computes a sup-realiser. To this end, let f be of bounded variation on $[0,1]$ and let $\mathcal{L}(f) = (x_n)_{n\in\mathbb{N}}$ be a list of its points of discontinuity. From $\mathcal{L}(f)$ we can find a list $(y_j)_{j\in\mathbb{N}}$ containing both the points of discontinuity and all rational numbers in $[0,1]$. Then we can compute

$$\mathcal{S}(f) = \sup\{f(x) : x \in [0,1]\} = \sup\{f(y_j) : j \in \mathbb{N}\}, \tag{5}$$

since $\sup\{f(y_j) : j \in \mathbb{N}\}$ is computable from $\mathcal{L}(f)$, f, and $\exists^2$.

Secondly, that a Jordan realiser computes a continuity realiser, assuming $\exists^2$, is immediate from Lemma 7.

Thirdly, we show that a sup-realiser $\mathcal{S}$ computes a Jordan realiser. To this end, let a and b be such that $0 \leq a < b \leq 1$ and define

$$\mathcal{S}^+_{a,b}(f) := \sup_{x\in[a,b]} f(x) \text{ and } \mathcal{S}^-_{a,b}(f) := \inf_{x\in[a,b]} f(x). \tag{6}$$

These functionals are clearly computable in $\mathcal{S}$, for $f : [0,1] \to \mathbb{R}$ a function of bounded variation. Now let $\mathbf{Var}(P,f)$ be the sum $\sum_{i=0}^{n-1} |f(x_{i+1}) - f(x_i)|$ for a partition $P = \{0 \leq x_0 < \cdots < x_n \leq 1\}$ of $[0,1]$, while $\mathbf{Var}^+(P,f)$ is the sum of all *positive* differences $f(x_{i+1}) - f(x_i)$.

Claim 9. *To compute a Jordan realiser, it suffices to compute the following number* $\Delta(f) := \sup_P \mathbf{Var}^+(P, f)$*, where* P *varies over all partitions of* $[0,1]$.

To prove Claim 9, we compute increasing functions f^+ and f^- such that $f = f^+ - f^-$. Without loss of generality, we may assume that $f(0) = 0$, and by symmetry it suffices to compute f^+. We can define $f^+(0) = 0$ and $f^+(x) = \Delta(f_x)$ for $x > 0$, where $f_x(y) = f(\frac{y}{x})$. This ends the proof of Claim 9.

We now employ the functionals $\mathcal{S}^+_{a,b}$ and $\mathcal{S}^-_{a,b}$ from (6) as follows.

Definition 10. Let f be of bounded variation on $[0,1]$ and let $n \in \mathbb{N}$. An *n, f-trail* is a sequence $k_0, \ldots, k_{2m-1}$ such that $0 \le k_0 < \ldots < k_{2m-1} < n$ and such that when $0 \le j < m$ we have

$$\mathcal{S}^-_{\frac{k_{2j}}{n}, \frac{k_{2j}+1}{n}}(f) < \mathcal{S}^+_{\frac{k_{2j+1}}{n}, \frac{k_{2j+1}+1}{n}}(f).$$

Define the positive *n-move* $\bar{M}(n, f)$ as the maximal value of the sum

$$\sum_{j=0}^{m-1} \left(\mathcal{S}^+_{\frac{k_{2j+1}}{n}, \frac{k_{2j+1}+1}{n}}(f) - \mathcal{S}^-_{\frac{k_{2j}}{n}, \frac{k_{2j}+1}{n}}(f) \right),$$

where the finite sequence $k_0, \ldots, k_{2m-1}$ varies over all n, f-trails.

Claim 11. *For each* $n \in \mathbb{N}$*, we have that* $\bar{M}(n, f) \le \Delta(f)$.

To prove Claim 11, it suffices to show that $\bar{M}(n, f) \le \Delta(f) + \epsilon$ for each $\epsilon > 0$. Let $k_0, \ldots, k_{2m-1}$ be an n, f-trail giving the value of $\bar{M}(n, f)$. For each $j < m$, select $x_j \in \left[\frac{k_{2j+1}}{n}, \frac{k_{2j+1}+1}{n}\right]$ such that $f(x_k) > \mathcal{S}^+_{\frac{k_{2j+1}}{n}, \frac{k_{2j+1}+1}{n}}(f) - \frac{\epsilon}{2m}$ and $y_j \in \left[\frac{k_{2j}}{n}, \frac{k_{2j}+1}{n}\right]$ such that $f(y_k) < \mathcal{S}^-_{\frac{k_{2j}}{n}, \frac{k_{2j}+1}{n}}(f) + \frac{\epsilon}{2m}$. For any partition P containing all points x_j and y_j we have that $\mathbf{Var}^+(P, f) > \bar{M}(n, f) - \epsilon$. As a result, we obtain

$$\bar{M}(n, f) < \mathbf{Var}^+(P, f) + \epsilon \le \Delta(f) + \epsilon,$$

which ends the proof of Claim 11.

Claim 12. *Let* f *be of bounded variation on* $[0,1]$ *and let* P *be a partition of* $[0,1]$*. Then there is* $n \in \mathbb{N}$ *such that* $\mathbf{Var}^+(P, f) \le \bar{M}(n, f)$.

To prove Claim 12, fix $P = \{0 \le s_0 < \cdots < s_{m'} \le 1\}$. Without loss of generality, we may assume that $f(s_0) < f(s_1) > f(s_2) < \cdots < f(s_{m'})$, i.e. the values $f(s_i)$ alternate between going up and going down, so m' is an odd number $2m-1$. Let n be such that each $[\frac{k}{n}, \frac{k+1}{n}]$, for $k < n$, contains at most one s_i and each s_i is contained in exactly one $[\frac{k_i}{n}, \frac{k_i+1}{n}]$. Then $k_0, \ldots, k_{2m-1}$ is an n, f-trail witnessing that $\mathbf{Var}^+(P, f) \le \bar{M}(n, f)$. This ends the proof of Claim 12.

Finally, by the above three claims, we have $\Delta(f) = \sup_{n \in \mathbb{N}} \bar{M}(n, f)$. Since we can compute the latter from $\mathcal{S}$ and $\exists^2$, the former is likewise computable. As we can compute a Jordan realiser from Δ, the proof of Theorem 8 is complete. □

In conclusion, as expected in light of (5), rather effective and pointwise approximation results exist for functions of bounded variation *at points of continuity* (see e.g. [11, p. 261]). For points of discontinuity, it seems one only approximates the average of the left and right limits, i.e. not the function value itself.

3.2 Jordan Realisers and Countable Sets

We show that Jordan realisers are *hard* to compute by connecting them to computability theoretic results pertaining to *countable sets* from [34]. Recall the definitions from Sect. 2.2 pertaining to the latter notion.

Now, the most fundamental property of countable sets is that they can be enumerated, i.e. listed as a sequence, as explicitly noted by e.g. Borel in [4] in his early discussions of the Heine-Borel theorem. Next, we show that Jordan realisers can indeed enumerate countable sets as in Definition 4.

Theorem 13. *Together with $\exists^2$, a Jordan realiser $\mathcal{J}$ can perform the following computational procedures.*

- *Given a set $A \subset [0,1]$ and $Y : [0,1] \to \mathbb{N}$ injective on A, produce a sequence $(x_n)_{n\in\mathbb{N}}$ listing exactly the elements of A.*
- *Given $F : [0,1] \to [0,1]$, $A \subset [0,1]$, and $Y : [0,1] \to \mathbb{N}$ injective on A, produce $\sup_{x\in A} F(x)$.*

Proof. We only need to establish the first item, as the second item readily follows from the first one using $\exists^2$. Let $A \subset [0,1]$ be countable, i.e. there is $Y : [0,1] \to \mathbb{N}$ which is injective on A. Use $\exists^2$ to define the function $f : \mathbb{R} \to \mathbb{R}$ defined as follows:

$$f(x) := \begin{cases} \frac{1}{2^{Y(x)+1}} & x \in A \\ 0 & \text{otherwise} \end{cases}. \tag{7}$$

Following item (a) in Definition 5, the function f has bounded variation on $[0,1]$ as any sum $\sum_{i=0}^{n} |f(x_i) - f(x_{i+1})|$ is at most $\sum_{i=0}^{n} \frac{1}{2^{i+1}}$ for x_i in $[0,1]$ and $i \le n+1$. Now let $\mathcal{J}(f) = (g,h)$ be such that $f = g - h$ on $[0,1]$ and recall $\mathcal{D}$ from Lemma 7. Use $\exists^2$ to define the sequence $(x_n)_{n\in\mathbb{N}}$ as all the reals in $\mathcal{D}(g)$ and $\mathcal{D}(h)$. Now consider the following formula for any $x \in [0,1]$:

$$\big[(\exists n \in \mathbb{N})(x = x_n) \wedge g(x) \neq h(x)\big] \leftrightarrow x \in A. \tag{8}$$

The forward direction in (8) is immediate as $g(x) \neq h(x)$ for $x \in [0,1]$ implies $f(x) > 0$, and hence $x \in A$ by definition. For the reverse direction, fix $x \in A$ and note that $0 < f(x) = g(x) - h(x)$ by the definition of f in (7), i.e. $g(x) \neq h(x)$ holds. Moreover, in case $(\forall n \in \mathbb{N})(x \neq x_n)$, then g and h are continuous at x, by the definition of $(x_n)_{n\in\mathbb{N}}$. Hence, f is continuous at x, which is only possible if $f(x) = 0$, but the latter implies $x \notin A$, by (7), a contradiction. In this way, we obtain (8) and we may enumerate A by removing from $(x_n)_{n\in\mathbb{N}}$ all elements not in A, which can be done using $\exists^2$. □

Secondly, weak Jordan realisers can enumerate *strongly* countable sets (Definition 4).

Corollary 14. *Together with $\exists^2$, a weak Jordan realiser $\mathcal{J}_{\mathsf{w}}$ can perform the following computational procedures.*

- *Given a set $A \subset [0,1]$ and $Y : [0,1] \to \mathbb{N}$* ***bijective*** *on A, produce a sequence $(x_n)_{n\in\mathbb{N}}$ listing exactly the elements of A.*
- *Given $F : [0,1] \to [0,1]$, $A \subset [0,1]$, and $Y : [0,1] \to \mathbb{N}$* ***bijective*** *on A, produce $\sup_{x\in A} F(x)$.*

Proof. Following item (b) in Definition 5, the function f in (7) has total variation exactly 1 in case Y is additionally a bijection. □

A weak Jordan realiser cannot compute a Jordan realiser; this remains true if we combine the former with an arbitrary type 2 functional. Since the proof of this claim is rather lengthy, we have omitted the former from this paper.

Thirdly, the functional Ω_{BW} introduced and studied in [34, §4], performs the computational procedure from the second item in Theorem 13, leading to the following corollary.

Corollary 15. *Together with the Suslin functional S^2, a Jordan realiser $\mathcal{J}$ computes S^2_2, i.e. a realiser for Π^1_2-CA_0.*

Proof. By [34, Theorem 4.6.(b)], $\Omega_{\mathsf{BW}} + \mathsf{S}^2$ computes S^2_2, while a Jordan realiser $\mathcal{J}$ computes Ω_{BW} by Theorem 13. □

Finally, under the additional set-theoretic hypothesis $\mathsf{V} = \mathsf{L}$, the combination $\Omega_{\mathsf{BW}} + \mathsf{S}^2$ even computes $\exists^3$ by [34, Theorem 4.6.(c)]. An obvious question is whether a similar result can be obtained within ZFC, which is the topic of the following section.

3.3 Computing Kleene's $\exists^3$ from Jordan Realisers

We show that Kleene's quantifier $\exists^3$ is computable in the combination of:

- Kleene's quantifier $\exists^2$,
- any Jordan realiser $\mathcal{J}$ (or: Ω_{BW} from Sect. 3.2),
- a well-ordering $\prec$ of $[0,1]$.

We note that the third item exists by the Axiom of Choice. Assuming $\prec$ and $\preceq$ are the irreflexive and reflexive versions of the same well-ordering of $[0,1]$, they are computable in each other and $\exists^2$.

Theorem 16. *Let $\mathcal{J}$ be a Jordan realiser. Then $\exists^3$ is Kleene-computable in $\mathcal{J}$, $\prec$, and $\exists^2$.*

Proof. We actually prove a slightly stronger result. Let Ω be a partial functional of type 3 such that $\Omega(X)$ terminates whenever $X \subset \mathbb{R}$ has at most one element, and $\Omega(X) \in X$ whenever X contains exactly one element. One readily[4] shows that Ω is computationally equivalent to Ω_{BW}, given $\exists^2$. We now show that $\exists^3$ is computable in Ω, $\prec$, and $\exists^2$.

We let x, y vary over $[0,1]$ and we fix $h : [0,1] \to \{0,1\}$. We aim to compute $\exists^3(h)$ by deciding the truth of the formula $(\exists x \in [0,1])(h(x) = 1)$. To this end, consider the functionals $E_{\prec x}$ and $E_{\preceq x}$ defined as:

$$E_{\prec x}(h) = 1 \leftrightarrow (\exists y \prec x)(h(y) = 1) \text{ and } E_{\preceq x}(h) = 1 \leftrightarrow (\exists y \preceq x)(h(y) = 1).$$

We shall show that these are computable in Ω, $\prec$, and $\exists^2$, uniformly in x. Note that $E_{\preceq x}$ is trivially computable in $E_{\prec x}$, uniformly in x, so we settle for computing $E_{\prec x}(h)$. The argument will be by the recursion theorem, so we give the algorithm for computing $E_{\prec x}(h)$ using x, h, Ω and $E_{\prec y}$ for $y \prec x$. Now let x and h be fixed and define h_x as:

$$h_x(y) := \begin{cases} 0 & \text{if } x \preceq y \\ h(y) & \text{if } y \prec x \wedge (\forall z \prec y)(h(z) = 0) \\ 0 & \text{otherwise} \end{cases},$$

where we use $E_{\prec y}$ to decide whether the second case holds. Then h_x is constant zero if $(\forall y \prec x)(h(y) = 0)$, and if not, h_x takes the value 1 in exactly the *least* point $y \prec x$ where $h(y) = 1$. Hence, we have

$$(\exists y \prec x)(h(y) = 1) \leftrightarrow [\Omega(h_x) \prec x \wedge h(\Omega(h_x)) = 1].$$

We now apply the recursion theorem to find an index $e \in \mathbb{N}$ such that for all $x \in [0,1]$ and $h : [0,1] \to \{0,1\}$ (and well-orderings $\preceq$), we have

$$\{e\}(\Omega, h, x, \preceq) \simeq E_{\preceq x}(h).$$

Then we use transfinite induction over $\preceq$ to prove that $e \in \mathbb{N}$ defines a total functional doing what it is supposed to do. For readers not familiar with this use of the recursion theorem, what we do is defining $\{e_0\}(d, \Omega, h, x, \preceq)$ as in the construction, but replacing all uses of $E_{\preceq y}(h')$ by $\{d\}(\Omega, h', y, \preceq)$; we then use the fact that there is an index e such that $\{e\}(\dots) \simeq \{e_0\}(e, \dots)$, where '...' are the other parameters.

Having established the computability of each $E_{\prec x}$ from x, we can use the same trick to compute $\exists^3(h)$ as follows: construct from h a function h' that takes the value 1 in at most one place, namely the $\prec$-least x where $h(x) = 1$, in case such exists. □

[4] To obtain an enumeration of $A \subset [0,1]$ given $Y : [0,1] \to \mathbb{R}$ injective on A, define $E_n := \{x \in A : Y(x) = n\}$ and define $x_n := \Omega(E_n)$ in case the latter is in E_n, and 0 otherwise.

Finally, we note that Kleene's computation scheme S9 is essentially a 'hard-coded' version of the recursion theorem for S1–S9, while S1–S8 merely define (higher-order) primitive recursion. In this way, the recursion theorem is central to S1–S9, although we have previously witnessed S1–S9 computations via primitive recursive terms.

3.4 Jordan Realisers and the Uncountability of $\mathbb{R}$

We show that a number of interesting functionals, including 'heavily restricted' Jordan realisers, are (still) quite hard to compute, based on the computational properties of the uncountability of $\mathbb{R}$ pioneered in [34].

First of all, in more detail, Theorem 8 implies that a Jordan realiser can enumerate all points of discontinuity of a function of bounded variation. It is then a natural question whether Jordan realisers remain hard to compute if we only require the output to be e.g. *one* point of continuity. By way of an answer, Theorem 19 lists a number of interesting functionals -including the aforementioned 'one-point' Jordan realisers- that compute functionals witnessing the uncountability of $\mathbb{R}$. Functionals related to the uncountability of $\mathbb{R}$ are special in the following sense, as was first studied in [34, §4].

By the previous, Jordan realisers have surprising properties and are a nice addition to the pantheon of interesting non-normal functionals stemming from ordinary mathematics (see [30–38] or Sect. 2.1 for other examples). It is then a natural question what the *weakest* such functional is; a candidate is provided by the *uncountability of* $\mathbb{R}$, which can be formulated in various guises as follows.

- Cantor's theorem: there is no surjection from $\mathbb{N}$ to $\mathbb{R}$.
- NIN: there is no injection from $[0,1]$ to $\mathbb{N}$.
- NBI: there is no bijection from $[0,1]$ to $\mathbb{N}$.

Cantor's theorem is provable in constructive and computable mathematics [3,43], while there is even an efficient algorithm to compute from a sequence of reals, a real not in that sequence [13]. As explored in [34], NIN and NBI are hard to prove in terms of conventional comprehension. We will not study NBI in this paper while Cantor's theorem and NIN give rise to the following specifications.

Definition 17 [Realisers for the uncountability of $\mathbb{R}$].

- A *Cantor functional/realiser* takes as input $A \subset [0,1]$ and $Y : [0,1] \to \mathbb{N}$ such that Y is injective on A, and outputs $x \notin A$.
- A ***weak*** *Cantor realiser* takes as input $A \subset [0,1]$ and $Y : [0,1] \to \mathbb{N}$ such that Y is **bijective** on A, and outputs $x \notin A$.
- A NIN-*realiser* takes as input $Y : [0,1] \to \mathbb{N}$ and outputs $x, y \in [0,1]$ with $x \neq y \wedge Y(x) = Y(y)$.

As explored in [34], NIN-realisers are among the weakest non-normal functionals originating from ordinary mathematics we have studied. Moreover, one readily[5]

[5] Let N be a NIN-realiser and let $A \subset [0,1]$ and $Y : [0,1] \to \mathbb{N}$ be such that Y is injective on A. Define $Z : [0,1] \to \mathbb{N}$ as follows: $Z(x) := Y(x) + 1$ in case $x \in A$, and 0 otherwise. Clearly, $N(Z)(1) \notin A$ or $N(Z)(2) \notin A$ as required.

proves that a NIN-realiser computes a Cantor realiser, while the latter are still hard to compute as follows.

Theorem 18. *No type 2 functional computes a weak Cantor realiser.*

Proof. Fix some functional F^2 and assume wlog that F computes $\exists^2$. Assume there is a Cantor realiser $\mathcal{C}$ computable in F. Now let A be the set of reals computable in F and define $Y : [0,1] \to \mathbb{N}$ as follows using *Gandy selection* (see [26,37] for an introduction): for the first part, define $Y(x)$ as an index for computing x from F in case $x \in A$; we put $Y(x) := 0$ in case $x \notin A$. By assumption, $\mathcal{C}(A, Y)$ terminates as Y is injective on A. Since the restriction of Y to A is partially computable in F, all oracle calls of the form '$x \in A$' will be answered with yes, since x then is computable in F. Hence, all oracle calls for the value $Y(x)$ can be answered *computably in* F. In this way, $\mathcal{C}(A, Y)$ is computable in F, which also follows from [34, Lemma 2.15]. Hence, $\mathcal{C}(A, Y) \in A$ by definition, a contradiction. The proof remains valid if we extend A to some $B \subset [0,1]$ and extend Y to $Z : [0,1] \to \mathbb{N}$ bijective on B. □

Secondly, it is readily proved (classically) that there is no *continuous* injection from $[0,1]$ to $\mathbb{Q}$, based on the intermediate value theorem. Now consider the following, which expresses a *very special case* of the uncountability of $\mathbb{R}$.

- $\mathsf{NIN}_{\mathsf{BV}}$: there is no injection from $[0,1]$ to $\mathbb{Q}$ that has *bounded variation* as in item (a) in Definition 5.

One readily establishes the equivalence $\mathsf{NIN} \leftrightarrow \mathsf{NIN}_{\mathsf{BV}}$ over a weak system, following the proof of Theorem 19. By the latter and Theorem 18, while NIN-realisers are defined for all $Y : [0,1] \to \mathbb{N}$, the restriction to functions of bounded variation, which only have countably many points of discontinuity, is (still) hard to compute *and* intermediate between Cantor and NIN-realisers.

Thirdly, we have the following theorem where the functional $\mathfrak{L}$ defined as

$$\mathfrak{L}(f)(s) := \int_0^{+\infty} e^{-st} f(t)\, dt$$

is the *Laplace transform* of $f : \mathbb{R} \to \mathbb{R}$. Since we restrict to functions of bounded variation, we interpret $\mathfrak{L}(f)$ as the limit of Riemann integrals, if the latter exists. It is well-known that if $\mathfrak{L}(f)$ and $\mathfrak{L}(g)$ exists and are equal everywhere, f and g are equal almost everywhere, inspiring the final -considerably weaker- item in Theorem 19. In the below items, 'bounded variation' refers to item (a) of Definition 5.

Theorem 19. *Assuming $\exists^2$, a Cantor realiser can be computed from a functional performing any of the following tasks.*

- *For $f : [0,1] \to \mathbb{Q}$ which has bounded variation, find $x, y \in [0,1]$ such that $x \neq y$ and $f(x) = f(y)$.*
- *For $f : [0,1] \to \mathbb{R}$ which has bounded variation, find a point of continuity in $[0,1]$.*

– *If* $f : [0,1] \to [0,1]$ *is Riemann integrable (or has bounded variation) with* $\int_0^1 f(x)\,dx = 0$, *find* $y \in [0,1]$ *with* $f(y) = 0$.
– *If* $f, g : \mathbb{R} \to \mathbb{R}$ *satisfy the following:*
  - f, g *have bounded variation on* $[0,a]$ *for any* $a \in \mathbb{R}^+$,
  - $\mathfrak{L}(f)$ *and* $\mathfrak{L}(g)$ *exists and are equal on* $[0,+\infty)$,

  find $x \in (0,\infty)$ *with* $f(x) = g(x)$.

Any NIN*-realiser computes a functional as in the first item. Assuming* $\exists^2$, *a* ***weak*** *Cantor realiser can be computed from a functional performing any of the above tasks restricted as in item* (b) *of Definition 5.*

Proof. The penultimate sentence is immediate as $\mathbb{Q}$ and $\mathbb{N}$ are bijective. Now fix some countable set $A \subset [0,1]$ and $Y : [0,1] \to \mathbb{N}$ injective on A. Consider $f : [0,1] \to \mathbb{R}$ as in (7) and recall it has bounded variation. For the second item, if $x_0 \in [0,1]$ is a point of continuity of f, we must have $f(x_0) = 0$ by continuity. Then $x_0 \notin A$ by definition, as required. For the first item, in case $x \neq y$ and $f(x) = f(y)$, we must have $x \notin A$ or $y \notin A$, in light of (7).

For the third item, consider $f : [0,1] \to \mathbb{R}$ as in (7); that $\int_0^1 f(x)\,dx$ exists and equals 0, follows from the usual ε-δ-definition of Riemann integrability. Indeed, fix $\varepsilon_0 > 0$ and find $k_0 \in \mathbb{N}$ such that $\frac{1}{2^{k_0}} < \varepsilon_0$. Let P be a partition $x_0 := 0, x_1, \ldots, x_n, x_{n+1} := 1$ of $[0,1]$ with $t_i \in [x_i, x_{i+1}]$ for $i \leq n$ and with mesh $\|P\| := \max_{i \leq n}(x_{i+1} - x_i)$ at most $\frac{1}{2^{k_0}}$. Then the Riemann sum $S(f,P) := \sum_{i \leq n} f(t_i)(x_{i+1} - x_i)$ satisfies

$$S(f,P) \leq \tfrac{1}{2^{k_0}} \textstyle\sum_{i \leq n} f(t_i) \leq \tfrac{1}{2^{k_0}} \sum_{i \leq n} \tfrac{1}{2^{i+1}} \leq \tfrac{1}{2^{k_0}},$$

as Y is injective on A and f is zero outside of A. Hence, $\int_0^1 f(x)\,dx = 0$ and any $y \in [0,1]$ with $f(y) = 0$ yields $y \notin A$ by (7).

For the final item, the tangent and arctangent functions provide bijections between $(0,1)$ and $\mathbb{R}$. Hence, we may work with $A \subset \mathbb{R}$ and $Y : \mathbb{R} \to \mathbb{N}$ injective on A. We again consider f as in (7), now as an $\mathbb{R} \to \mathbb{R}$-function. This function f has bounded variation on any interval $[0,a]$ for $a > 0$, in the same way as in the proof of Theorem 13. Since $e^{-z} \leq 1$ for $z \geq 0$, the previous paragraph yields $\int_0^N e^{-st} f(t)\,dt = 0$ for any $N \in \mathbb{N}, s \geq 0$, and hence $\mathfrak{L}(f)(s) = 0$ for all $s \geq 0$. For $g : \mathbb{R} \to \mathbb{R}$ the zero everywhere function, we trivially have $\mathfrak{L}(g)(s) = 0$ for any $s \geq 0$. Clearly, any $x \in [0,+\infty)$ such that $f(x) = g(x) = 0$ also satisfies $x \notin A$, yielding a Cantor functional. The final sentence now follows in light of (the proof of) Corollary 14. □

We note that theorem also goes through if we formulate the second item using the much weaker notions of quasi[6] or cliquish (see footnote 6) continuity in at least one point in $[0,1]$. These notions are found in e.g. [22,27].

[6] A function $f : X \to \mathbb{R}$ is *quasi-continuous* (resp. *cliquish*) at $x \in X$ if for any $\epsilon > 0$ and any open neighbourhood U of x, there is a non-empty open ball $G \subset U$ with $(\forall y \in G)(|f(x) - f(y)| < \varepsilon)$ (resp. $(\forall y, z \in G)(|f(z) - f(y)| < \varepsilon)$).

Acknowledgements. We thank Anil Nerode for his most helpful advise. Our research was kindly supported by the *Deutsche Forschungsgemeinschaft* via the DFG grant SA3418/1-1 for Sam Sanders. We thank the anonymous referees for their suggestions that have greatly improved this paper.

References

1. Avigad, J., Feferman, S.: Gödel's functional ("Dialectica") interpretation. In: Handbook of Proof Theory, pp. 337–405 (1998)
2. Beeson, M.J. Foundations of Constructive Mathematics: Metamathematical Studies. Ergebnisse der Mathematik und ihrer Grenzgebiete, vol. 6. Springer, Heidelberg (1985). https://doi.org/10.1007/978-3-642-68952-9
3. Bishop, E.: Foundations of Constructive Analysis. McGraw-Hill (1967)
4. Borel, E.: Leçons sur la théorie des fonctions. Gauthier-Villars, Paris (1898)
5. Bridges, D.: A constructive look at functions of bounded variation. Bull. Lond. Math. Soc. **32**(3), 316–324 (2000)
6. Bridges, D.S.: Constructive continuity of increasing functions. In: Kosheleva, O., Shary, S.P., Xiang, G., Zapatrin, R. (eds.) Beyond Traditional Probabilistic Data Processing Techniques: Interval, Fuzzy etc. Methods and Their Applications. SCI, vol. 835, pp. 9–19. Springer, Cham (2020). https://doi.org/10.1007/978-3-030-31041-7_2
7. Bridges, D., Mahalanobis, A.: Bounded variation implies regulated: a constructive proof. J. Symb. Log. **66**(4), 1695–1700 (2001)
8. Bridges, D., Richman, F.: Varieties of Constructive Mathematics. London Mathematical Society Lecture Note Series, vol. 97. Cambridge University Press, Cambridge (1987)
9. Brouwer, L.E.J.: Begründung der mengenlehre unabhängig vom logischen satz vom ausgeschlossenen dritten. erster teil: Allgemeine mengenlehre. Koninklijke Nederlandsche Akademie van Wetenschappen, Verhandelingen, 1ste sectie, vol. 12, no. 5, p. 43 (1918)
10. Buchholz, W., Feferman, S., Pohlers, W., Sieg, W.: Iterated Inductive Definitions and Subsystems of Analysis: Recent Proof-Theoretical Studies. LNM, vol. 897. Springer, Heidelberg (1981). https://doi.org/10.1007/BFb0091894
11. Hua, C.F.: On the rate of convergence of Bernstein polynomials of functions of bounded variation. J. Approx. Theor. **39**(3), 259–274 (1983)
12. Feferman, S.: How a Little Bit goes a Long Way: Predicative Foundations of Analysis (2013). Unpublished notes from 1977–1981 with updated introduction. https://math.stanford.edu/~feferman/papers/pfa(1).pdf
13. Gray, R.: Georg cantor and transcendental numbers. Amer. Math. Mon. **101**(9), 819–832 (1994)
14. Heyting, A.: Recent progress in intuitionistic analysis. In: Intuitionism and Proof Theory, Proceedings of the Summer Conference, Buffalo, N.Y., pp. 95–100 (1970)
15. Hilbert, D., Bernays, P.: Grundlagen der Mathematik. II. Zweite Auflage. Die Grundlehren der mathematischen Wissenschaften, vol. 50. Springer, Heidelberg (1970). https://doi.org/10.1007/978-3-642-86896-2
16. Hrbacek, K., Jech, T.: Introduction to Set Theory, 3rd edn. Monographs and Textbooks in Pure and Applied Mathematics, vol. 220. Marcel Dekker Inc., New York (1999)
17. Jordan, C.: Sur la série de fourier. Comptes rendus de l'Académie des Sciences, Paris, Gauthier-Villars, vol. 92, pp. 228–230 (1881)

18. Kleene, S.C.: Recursive functionals and quantifiers of finite types. i. Trans. Amer. Math. Soc. **91**, 1–52 (1959)
19. Kohlenbach, U.: Foundational and mathematical uses of higher types. In: Lecture Notes in Logic, vol. 15, pp. 92–116. ASL (2002)
20. Kohlenbach, U.: On uniform weak König's lemma. Ann. Pure Appl. Log. **114**(1–3), 103–116 (2002). Commemorative Symposium Dedicated to Anne S. Troelstra (Noordwijkerhout, 1999)
21. Kohlenbach, U.: Higher order reverse mathematics. In: Reverse Mathematics 2001, pp. 281–295. ASL (2005)
22. Kowalewski, M., Maliszewski, A.: Separating sets by cliquish functions. Topology Appl. **191**, 10–15 (2015)
23. Kreuzer, A.P.: Bounded variation and the strength of Helly's selection theorem. Log. Meth. Comput. Sci. **10**(4:16), 1–23 (2014)
24. Kreuzer, A.P.: Measure theory and higher order arithmetic. Proc. Amer. Math. Soc. **143**(12), 5411–5425 (2015)
25. Kunen, K.: Set Theory. Studies in Logic, vol. 34. College Publications, London (2011)
26. Longley, J., Normann, D.: Higher-Order Computability. Theory and Applications of Computability. Springer, Heidelberg (2015). https://doi.org/10.1007/978-3-662-47992-6
27. Neubrunn, T.: Quasi-continuity. Real Anal. Exch. **14**(2), 259–306 (1988/89)
28. Nies, A., Triplett, M.A., Yokoyama, K.: The reverse mathematics of theorems of Jordan and Lebesgue. J. Symb. Log., 1–18 (2021)
29. Greenberg, N., Miller, J.S., Nies, A.: Highness properties close to PA completeness. Isr. J. Math. **244**, 419–465 (2021). https://doi.org/10.1007/s11856-021-2200-7
30. Normann, D., Sanders, S.: Nonstandard analysis, computability theory, and their connections. J. Symb. Log. **84**(4), 1422–1465 (2019)
31. Normann, D., Sanders, S.: On the mathematical and foundational significance of the uncountable. J. Math. Log. **19**(01), 1950001 (2019). https://doi.org/10.1142/S0219061319500016
32. Normann, D., Sanders, S.: Representations in measure theory. arXiv arXiv:1902.02756 (2019)
33. Normann, D., Sanders, S.: The strength of compactness in computability theory and nonstandard analysis. Ann. Pure Appl. Log. **170**(11), 102710 (2019)
34. Normann, D., Sanders, S.: On the uncountability of $\mathbb{R}$, p. 37. arXiv arXiv:2007.07560 (2020)
35. Normann, D., Sanders, S.: Open sets in reverse mathematics and computability theory. J. Log. Comput. **30**(8), 40 (2020)
36. Normann, D., Sanders, S.: Pincherle's theorem in reverse mathematics and computability theory. Ann. Pure Appl. Log. **171**(5), 102788 (2020)
37. Normann, D., Sanders, S.: The axiom of choice in computability theory and reverse mathematics. J. Log. Comput. **31**(1), 297–325 (2021)
38. Normann, D., Sanders, S.: On robust theorems due to Bolzano, Weierstrass, and Cantor in reverse mathematics, p. 30. arXiv arXiv:2102.04787 (2021)
39. Richman, F.: Omniscience principles and functions of bounded variation. Math. Log. Q. **48**, 111–116 (2002)
40. Sanders, S.: Nets and reverse mathematics. In: Manea, F., Martin, B., Paulusma, D., Primiero, G. (eds.) CiE 2019. LNCS, vol. 11558, pp. 253–264. Springer, Cham (2019). https://doi.org/10.1007/978-3-030-22996-2_22

41. Sanders, S.: Reverse mathematics and computability theory of domain theory. In: Iemhoff, R., Moortgat, M., de Queiroz, R. (eds.) WoLLIC 2019. LNCS, vol. 11541, pp. 550–568. Springer, Heidelberg (2019). https://doi.org/10.1007/978-3-662-59533-6_33
42. Sanders, S.: Nets and reverse mathematics: a pilot study. Computability **10**(1), 31–62 (2021)
43. Simpson, S.G.: Subsystems of Second Order Arithmetic. Perspectives in Logic, 2nd edn. Cambridge University Press (2009)
44. Stillwell, J.: Reverse Mathematics, Proofs from the Inside Out. Princeton University Press (2018)
45. Turing, A.: On computable numbers, with an application to the Entscheidungsproblem. Proc. Lond. Math. Soc. **42**, 230–265 (1936)
46. Veldman, W.: Understanding and using Brouwer's continuity principle. In: Schuster, P., Berger, U., Osswald, H. (eds.) Reuniting the Antipodes – Constructive and Nonstandard Views of the Continuum. Synthese Library (Studies in Epistemology, Logic, Methodology, and Philosophy of Science), vol. 306, pp 285–302. Springer, Dordrecht (2001). https://doi.org/10.1007/978-94-015-9757-9_24
47. Weihrauch, K.: Computable Analysis. TTCSAES, Springer, Heidelberg (2000). https://doi.org/10.1007/978-3-642-56999-9
48. Zheng, X., Rettinger, R.: Effective Jordan decomposition. Theor. Comput. Syst. **38**(2), 189–209 (2005)

Computability Models over Categories and Presheaves

Iosif Petrakis[(✉)]

University of Munich, Munich, Germany
petrakis@math.lmu.de

Abstract. Generalising slightly the notions of a strict computability model and of a simulation between them, which were elaborated by Longley and Normann in [9], we define canonical computability models over certain categories and appropriate presheaves on them. We study the canonical total computability model over a category $\mathcal{C}$ and a covariant presheaf on $\mathcal{C}$, and the canonical partial computability model over a category $\mathcal{C}$ with pullbacks and a pullback preserving, covariant presheaf on $\mathcal{C}$. These computability models are shown to be special cases of a computability model over a category $\mathcal{C}$ with a so-called base of computability and a pullback preserving, covariant presheaf on $\mathcal{C}$. In this way Rosolini's theory of dominions is connected with the theory of computability models. All our notions and results are dualised by considering certain (contravariant) presheaves on appropriate categories.

1 Introduction

In [9] Longley and Normann not only give a comprehensive introduction to Higher-Order Computability (HOC) presenting the various approaches to HOC, but they also fit them into a coherent and unifying framework, making their comparison possible. Their main tool is a general notion of computability model[1] and of an appropriate concept of simulation of one computability model in another. Turing machines, programming languages, λ-calculus etc., are shown to be computability models, not necessarily in a unique way.

As it is remarked in [2], p. 3, computability theory has "still not received the level of categorical attention it deserves". It seems though, that the "categorical spirit" of the notions of a computability model and a simulation between them is behind the remarkable success of the framework of Longley and Normann. E.g., their, completely categorical in nature, notion of equivalence of computability models (see Definition 2) is deeper than the one suggested by the standard presentations of computability theory, according to which different notions of computability are equivalent if they generate the same class of partial computable

[1] This notion is rooted in previous work of Longley in [6–8], and is influenced by the work of Cockett and Hofstra in [3] and [4] (see [9], p. 52).

S. Artemov and A. Nerode (Eds.): LFCS 2022, LNCS 13137, pp. 253–265, 2022.
https://doi.org/10.1007/978-3-030-93100-1_16

functions from $\mathbb{N}$ to $\mathbb{N}$ (see [9], Section 1.1.5). Longley and Normann associated in a canonical way to a computability model[2] $\boldsymbol{C}$ its category of assemblies $\mathcal{A}sm(\boldsymbol{C})$, "the world of all datatypes that can be represented in" $\boldsymbol{C}$. They also showed that the computability models $\boldsymbol{C}$ and $\boldsymbol{D}$ are equivalent if and only if the categories of assemblies $\mathcal{A}sm(\boldsymbol{C})$ and $\mathcal{A}sm(\boldsymbol{D})$ are equivalent. The category of computability models and the corresponding functor $\boldsymbol{C} \mapsto \mathcal{A}sm(\boldsymbol{C})$ are studied extensively in [8].

In this paper we associate in a canonical way a computability model to a category $\mathcal{C}$, given an appropriate presheaf on $\mathcal{C}$. For that we slightly generalise the definition of a computability model given in [9], allowing the classes of type names T and of partial functions $\boldsymbol{C}[\sigma, \tau]$ in a computability model to be proper classes. The computability models considered in [9] are called *small* in Definition 1, and working solely with them we can only define the computability model of a small category. The notion of a simulation between computability models is also slightly generalised, as the function between the corresponding classes of type names is, generally, a class-function. Although Longley and Normann usually work with *lax* computability models, to our needs the so-called *strict* computability models suit best[3].

In Sect. 2 we include all basic notions and facts necessary to the rest of this paper. In Sect. 3 we study the canonical, total computability model over a category $\mathcal{C}$ and a covariant presheaf on $\mathcal{C}$. In Sect. 4 we study the canonical (partial) computability model over a category $\mathcal{C}$ with pullbacks and a covariant presheaf on $\mathcal{C}$ that preserves pullbacks. In Sect. 5 we define the notion of a base of computability for a category $\mathcal{C}$, and the canonical (partial) computability model over a category $\mathcal{C}$ with a base of computability and a pullback preserving, covariant presheaf on $\mathcal{C}$. The first two constructions are shown to be special cases of the last one. All these models can be dualised. E.g., the dual of the last model is the notion of a computability model over a category $\mathcal{C}$ with a cobase of computability and a (contravariant) preshaef on $\mathcal{C}$ that sends pushouts to pullbacks. For all categorical notions and facts mentioned here without explanation or proof we refer to [1] and [14]. Due to lack of space we omit here some proofs.

2 Computability Models

Definition 1. *Let T be a class, the elements of which are called type names. A computability model $\boldsymbol{C}$ over T is a pair*

$$\boldsymbol{C} = \big(\big(\boldsymbol{C}(\tau)\big)_{\tau \in T}, \big(\boldsymbol{C}[\sigma, \tau]\big)_{(\sigma,\tau) \in T \times T}\big),$$

where $|\boldsymbol{C}| = \big(\boldsymbol{C}(\tau)\big)_{\tau \in T}$ is a family of sets $\boldsymbol{C}(\tau)$ over T, called the datatypes of $\boldsymbol{C}$, and $\big(\boldsymbol{C}[\sigma, \tau]\big)_{(\sigma,\tau) \in T \times T}$ is a family of classes $\boldsymbol{C}[\sigma, \tau]$ of partial functions of type $\boldsymbol{C}(\sigma) \rightharpoonup \boldsymbol{C}(\tau)$ over $T \times T$, such that the following conditions hold:

[2] In [9] this category is defined for lax computability models, but, as it is remarked in [9], p. 91, the definition makes sense for an arbitrary computability model.

[3] For a discussion on "strict vs. lax" see [8], Section 2.1.

$(\mathtt{CM}_1)$ *For every* $\tau \in T$ $\mathrm{id}_{\boldsymbol{C}(\tau)} : \boldsymbol{C}(\tau) \to \boldsymbol{C}(\tau)$ *is in* $\boldsymbol{C}[\tau, \tau]$.
$(\mathtt{CM}_2)$ *For every* $\rho, \sigma, \tau \in T$, *if* $f \in \boldsymbol{C}[\rho, \sigma]$ *and* $g \in \boldsymbol{C}[\sigma, \tau]$, *the composite partial function* $g \circ f$ *is in* $\boldsymbol{C}[\rho, \tau]$.
If the classes $\boldsymbol{C}[\sigma, \tau]$ *are sets, for every* $\sigma, \tau \in T$, *we call* $\boldsymbol{C}$ *locally small. If* T *is also a set, we call* $\boldsymbol{C}$ *small. If every element of* $\boldsymbol{C}[\sigma, \tau]$ *is a total function, for every* $\sigma, \tau \in T$, *then* $\boldsymbol{C}$ *is called total. We say that* $\boldsymbol{C}$ *contains the constants, if every constant function* $\boldsymbol{C}(\sigma) \to \boldsymbol{C}(\tau)$ *is in* $\boldsymbol{C}[\sigma, \tau]$, *for every* $\sigma, \tau \in T$.

E.g., Kleene's first model K_1 is a computability model over $T = \{0\}$, $\boldsymbol{C}(0) = \mathbb{N}$, and

$$\boldsymbol{C}[0,0] = \{f : \mathbb{N} \to \mathbb{N} \mid f \text{ is Turing computable}\}.$$

For the rest of this paper T, U, W are classes, and $\boldsymbol{C}, \boldsymbol{D}, \boldsymbol{E}$ are computability models over T, U, W, respectively.

Definition 2. *A simulation* $\boldsymbol{\gamma} \colon \boldsymbol{C} \twoheadrightarrow \boldsymbol{D}$ *of* $\boldsymbol{C}$ *in* $\boldsymbol{D}$ *is a pair*

$$\boldsymbol{\gamma} = \big(\gamma, (\Vdash^{\gamma}_{\tau})_{\tau \in T}\big),$$

where $\gamma \colon T \to U$ *is a class function, and* $\Vdash^{\gamma}_{\tau} \subseteq \boldsymbol{D}(\gamma(\tau)) \times \boldsymbol{C}(\tau)$, *for every* $\tau \in T$, *such that the following conditions hold:*

$$(\mathtt{Siml}_1) \qquad \forall_{\tau \in T} \forall_{x \in \boldsymbol{C}(\tau)} \exists_{x' \in \boldsymbol{D}(\gamma(\tau))} \big(x' \Vdash^{\gamma}_{\tau} x\big),$$

$$(\mathtt{Siml}_2) \qquad \forall_{\sigma, \tau \in T} \forall_{f \in \boldsymbol{C}[\sigma, \tau]} \exists_{f' \in \boldsymbol{D}[\gamma(\sigma), \gamma(\tau)]} \big(f' \Vdash^{\gamma}_{(\sigma,\tau)} f\big),$$

where the relation "f is tracked by f' through $\boldsymbol{\gamma}$" is defined by

$$f' \Vdash^{\gamma}_{(\sigma,\tau)} f :\Leftrightarrow \forall_{x \in \boldsymbol{C}(\sigma)} \big(x \in \mathrm{dom}(f) \Rightarrow$$

$$\forall_{x' \in \boldsymbol{D}(\gamma(\sigma))} \big(x' \Vdash^{\gamma}_{\sigma} x \;\&\; x' \in \mathrm{dom}(f') \Rightarrow f'(x') \Vdash^{\gamma}_{\tau} f(x)\big)\big).$$

The identity simulation $\boldsymbol{\iota_C} : \boldsymbol{C} \twoheadrightarrow \boldsymbol{C}$ *is the pair* $\big(\mathrm{id}_T, (\Vdash^{\iota_C}_{\tau})_{\tau \in T}\big)$, *where* $x' \Vdash^{\iota_C}_{\tau} \Leftrightarrow x' = x$, *for every* $x', x \in \boldsymbol{C}(\tau)$. *If* $\boldsymbol{\delta} \colon \boldsymbol{D} \twoheadrightarrow \boldsymbol{E}$, *the composite simulation* $\boldsymbol{\delta} \circ \boldsymbol{\gamma} \colon \boldsymbol{C} \twoheadrightarrow \boldsymbol{E}$ *is the pair* $\big(\delta \circ \gamma, (\Vdash^{\delta \circ \gamma}_{\tau})_{\tau \in T}\big)$, *where the relation* $\Vdash^{\delta \circ \gamma}_{\tau} \subseteq \boldsymbol{E}\big(\delta(\gamma(\tau))\big) \times \boldsymbol{C}(\tau)$ *is defined by*

$$z \Vdash^{\delta \circ \gamma}_{\tau} x \Leftrightarrow \exists_{y \in \boldsymbol{D}(\gamma(\tau))} \big(z \Vdash^{\delta}_{\gamma(\tau)} y \;\&\; y \Vdash^{\gamma}_{\tau} x\big).$$

Let $\mathtt{CompMod}$ *be the category of computability models with simulations. If* $\boldsymbol{\gamma}, \boldsymbol{\delta} \colon \boldsymbol{C} \twoheadrightarrow \boldsymbol{D}$, *then* $\boldsymbol{\gamma}$ *is transformable to* $\boldsymbol{\delta}$, *in symbols* $\boldsymbol{\gamma} \preceq \boldsymbol{\delta}$, *if for every* $\tau \in T$ *there is* $f \in \boldsymbol{D}[\gamma(\tau), \delta(\tau)]$ *such that*

$$\forall_{x \in \boldsymbol{C}(\tau)} \forall_{x' \in \boldsymbol{D}(\gamma(\tau))} \big(x' \Vdash^{\gamma}_{\tau} x \;\&\; x' \in \mathrm{dom}(f) \Rightarrow f(x') \Vdash^{\delta}_{\tau} x\big).$$

Let $\boldsymbol{\gamma} \sim \boldsymbol{\delta}$, *if* $\boldsymbol{\gamma} \preceq \boldsymbol{\delta}$ *and* $\boldsymbol{\delta} \preceq \boldsymbol{\gamma}$. *The computational models* $\boldsymbol{C}$ *and* $\boldsymbol{D}$ *are equivalent, if there are simulations* $\boldsymbol{\gamma} \colon \boldsymbol{C} \twoheadrightarrow \boldsymbol{D}$ *and* $\boldsymbol{\delta} \colon \boldsymbol{D} \twoheadrightarrow \boldsymbol{C}$ *such that* $\boldsymbol{\delta} \circ \boldsymbol{\gamma} \sim \boldsymbol{\iota_C}$ *and* $\boldsymbol{\gamma} \circ \boldsymbol{\delta} \sim \boldsymbol{\iota_D}$.

If $f' \Vdash^{\gamma}_{(\rho,\sigma)} f$ and $g' \Vdash^{\gamma}_{(\sigma,\tau)} g$, then $g' \circ f' \Vdash^{\gamma}_{(\rho,\tau)} g \circ f$, and if $\gamma \preceq \gamma'$ and $\delta \preceq \delta'$, then $\delta \circ \gamma \preceq \delta' \circ \gamma'$, where $\gamma, \gamma' : \boldsymbol{C} \twoheadrightarrow \boldsymbol{D}$ and $\delta, \delta' : \boldsymbol{D} \twoheadrightarrow \boldsymbol{E}$. Next we include the definition of a computability model with weak products, given in [9], p. 53. It is a characteristic example of a categorical notion translated appropriately in the framework of partiality within computability models.

Definition 3. *A computability model $\boldsymbol{C}$ has weak products, if for every $\sigma, \tau \in T$, there is some $\rho \in T$ and computable projections $\mathtt{pr}_\sigma \in \boldsymbol{C}[\rho, \sigma]$ and $\mathtt{pr}_\tau \in \boldsymbol{C}[\rho, \tau]$ such that for every $v \in T$, and for every $f \in \boldsymbol{C}[v, \sigma]$ and $g \in \boldsymbol{C}[v, \tau]$, there is (a not necessarily unique) partial function $\langle\langle f, g\rangle\rangle \in \boldsymbol{C}[v, \rho]$*

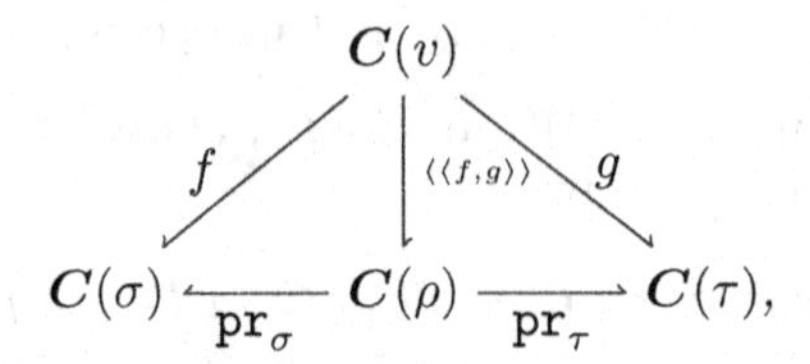

such that for every $x \in \boldsymbol{C}(v)$ the following conditions hold:

$(\mathtt{WP}_1)$ $x \in \mathrm{dom}(\langle\langle f, g\rangle\rangle) \Leftrightarrow x \in \mathrm{dom}(f)$ & $x \in \mathrm{dom}(g)$.

$(\mathtt{WP}_2)$ *If* $x \in \mathrm{dom}(\langle\langle f, g\rangle\rangle)$, *then* $\mathtt{pr}_\sigma((\langle\langle f, g\rangle\rangle)(x)) = f(x)$ *and* $\mathtt{pr}_\tau((\langle\langle f, g\rangle\rangle)(x)) = g(x)$. *A computability model $\boldsymbol{C}$ has standard products, if $\boldsymbol{C}(\rho) = \boldsymbol{C}(\sigma) \times \boldsymbol{C}(\tau)$ and $\mathtt{pr}_\sigma, \mathtt{pr}_\tau$ are the standard (total) projection functions.*

In $(\mathtt{WP}_2)$ $(\langle\langle f, g\rangle\rangle)(x)$ is in the domain of $\mathtt{pr}_\sigma$ and of $\mathtt{pr}_\tau$[4].

Definition 4. *The category of assemblies $\mathcal{A}sm(\boldsymbol{C})$ over $\boldsymbol{C}$ has objects triplets $(X, \tau_X, \Vdash_X)$, where X is a set, $\tau_X \in T$, and $\Vdash_X \subseteq \boldsymbol{C}(\tau_X) \times X$ such that*

$$\forall_{x \in X} \exists_{x' \in \boldsymbol{C}(\tau_X)} (x' \Vdash_X x).$$

An arrow $f : (X, \tau_X, \Vdash_X) \to (Y, \tau_Y, \Vdash_Y)$ is a function $f : X \to Y$, such that there is $\bar{f} \in \boldsymbol{C}[\tau_X, \tau_Y]$ that "tracks" f, in symbols $\bar{f} \Vdash^Y_X f$. The tracking relation $\bar{f} \Vdash^Y_X f$ is defined by the following condition:

$$\forall_{x \in X} \forall_{y \in \boldsymbol{C}(\tau_X)} (y \Vdash_X x \ \& \ y \in \mathrm{dom}(\bar{f}) \Rightarrow \bar{f}(y) \Vdash_Y f(x)).$$

$1_{(X, \tau_X, \Vdash_X)} = \mathrm{id}_X$, *which is tracked by* $\mathrm{id}_{\boldsymbol{C}(\tau_X)}$. *If $g : (Y, \tau_Y, \Vdash_Y) \to (Z, \tau_Z, \Vdash_Z)$ is tracked by $\bar{g} \in \boldsymbol{C}[\tau_Y, \tau_Z]$, then $g \circ f : (X, \tau_X, \Vdash_X) \to (Z, \tau_Z, \Vdash_Z)$ is tracked by $\bar{g} \circ \bar{f}$. As usual, the forgetful functor*

$$\mathrm{Frg}^{\boldsymbol{C}} : \mathcal{A}sm(\boldsymbol{C}) \to \mathtt{Set}$$

is defined by $\mathrm{Frg}^{\boldsymbol{C}}_0(X, \tau_X, \Vdash_X) = X$ *and* $\mathrm{Frg}^{\boldsymbol{C}}_1(f) = f$.

---

[4] If we write $(\mathtt{WP}_2)$ as the implication: if $x \in \mathrm{dom}(\langle\langle f, g\rangle\rangle)$, then $[\mathtt{pr}_\sigma \circ (\langle\langle f, g\rangle\rangle)](x) = f(x)$ and $[\mathtt{pr}_\tau \circ (\langle\langle f, g\rangle\rangle)](x) = g(x)$, we need to use the definition of composition of partial functions. As it is noted in [9], p. 53, a computability model with weak products is equivalent to one with standard products.

Clearly, $\mathrm{Frg}^{\boldsymbol{C}}$ is injective on arrows. In [9], p. 92, it is shown that if $\boldsymbol{C}$ has weak products, then $\mathcal{A}sm(\boldsymbol{C})$ has products. In this case $\mathcal{A}sm(\boldsymbol{C})$ also has pullbacks.

Proposition 1. (i) *If* $\boldsymbol{C}$ *has standard products, then* $\mathcal{A}sm(\boldsymbol{C})$ *has (products and) pullbacks. Moreover,* $\mathrm{Frg}^{\boldsymbol{C}}$ *preserves pullbacks.*
(ii) *If* $\boldsymbol{C}$ *has weak products, if every datatype* $\boldsymbol{C}(\tau)$ *is inhabited, and if* $\boldsymbol{C}$ *contains the constants, then* $\mathcal{A}sm(\boldsymbol{C})$ *has pullbacks.*

For the rest of this paper $\mathcal{C}, \mathcal{D}$ are categories, **Cat** is the category of categories, $\mathtt{Set}$ is the category of sets, $[\mathcal{C}, \mathtt{Set}]$ is the category of functors from $\mathcal{C}$ to $\mathtt{Set}$, and $\mathbf{Cat}/\mathtt{Set}$ is the slice category of **Cat** over $\mathtt{Set}$. Let C_0, C_1 be the classes of objects and arrows in $\mathcal{C}$, respectively. If $a, b \in C_0$, let $C_1(a, b)$ be the class of arrows f in C_1 with $\mathrm{dom}(f) = a$ and $\mathrm{cod}(f) = b$.

A central functor defined in [9] is $F \colon \mathtt{CompMod} \to \mathbf{Cat}/\mathtt{Set}$, where $F_0(\boldsymbol{C}) = \mathrm{Frg}^{\boldsymbol{C}} : \mathcal{A}sm(\boldsymbol{C}) \to \mathtt{Set}$.

3 Total Computability Models over Categories

Definition 5. *Let* $S \colon \mathcal{C} \to \mathtt{Set}$ *be a (covariant) functor. The total computability model* $\mathbf{CM}^{\mathrm{tot}}(\mathcal{C}; S)$ *over* $\mathcal{C}$ *and* S *is the pair*

$$\mathbf{CM}^{\mathrm{tot}}(\mathcal{C}; S) = \big((S_0(a))_{a \in C_0}, (S^{\mathrm{tot}}[a, b])_{(a,b) \in C_0 \times C_0}\big),$$

$$S^{\mathrm{tot}}[a, b] = \{S_1(f) \mid f \in C_1(a, b)\}.$$

If $T \colon \mathcal{C}^{\mathrm{op}} \to \mathtt{Set}$ *is contravariant, the total computability model* $\mathbf{CM}^{\mathrm{tot}}(\mathcal{C}; T)$ *over* $\mathcal{C}$ *and* T *is defined dually i.e.,*

$$T^{\mathrm{tot}}[a, b] = \{T_1(f) \mid f \in C_1(b, a)\}.$$

Let $(\mathbb{N}, \leq)$ be seen as a thin category, and $S \colon \mathbb{N} \to \mathtt{Set}$. If $n \leq m$, $S^{\mathrm{tot}}[n, m] = \{S_1(f)\}$, where $f \colon n \to m$ is the unique arrow in $Hom(n, m)$. Clearly, $\mathbf{CM}^{\mathrm{tot}}(\mathcal{C}; S)$ satisfies conditions (CM_1) and (CM_2). With replacement, if $\mathcal{C}$ is locally small, then $\mathbf{CM}^{\mathrm{tot}}(\mathcal{C}; S)$ is locally small, and if $\mathcal{C}$ is small, then $\mathbf{CM}^{\mathrm{tot}}(\mathcal{C}; S)$ is small. Clearly, if $\mathcal{C}$ has (fixed) products, the total computability model $\mathbf{CM}^{\mathrm{tot}}(\mathcal{C}; S)$ has weak products. Next we describe an "external" and an "internal" functor in the full subcategory $\mathtt{TotCompMod}$ of total computability models. The corresponding proofs are omitted, as these are similar to the included proofs of Propositions 7 and 8.

Proposition 2. *There is a functor*

$$\mathrm{CM}^{\mathrm{tot}} \colon \mathbf{Cat}/\mathtt{Set} \to \mathtt{TotCompMod},$$

defined by

$$\mathrm{CM}_0^{\mathrm{tot}}(\mathcal{C}, S) = \mathbf{CM}^{\mathrm{tot}}(\mathcal{C}; S),$$

$$\mathrm{CM}_1^{\mathrm{tot}}\big(F \colon (\mathcal{C}, S) \to (\mathcal{D}, T)\big) : \mathbf{CM}^{\mathrm{tot}}(\mathcal{C}; S) \twoheadrightarrow \mathbf{CM}^{\mathrm{tot}}(\mathcal{D}; T),$$

$$\mathrm{CM}_1^{\mathrm{tot}}(F) = \gamma_F = \big(F_0, (\Vdash_a^{\gamma_F})_{a \in C_0}\big),$$

$$\Vdash_a^{\gamma_F} \subseteq T_0(F_0(a)) \times S_0(a), \quad y \Vdash_a^{\gamma_F} x \Leftrightarrow y = x.$$

If F^{tot} is the restriction of F (mentioned in the end of Sect. 2), to the subcategory `TotCompMod`, we get the pair of adjoint functors

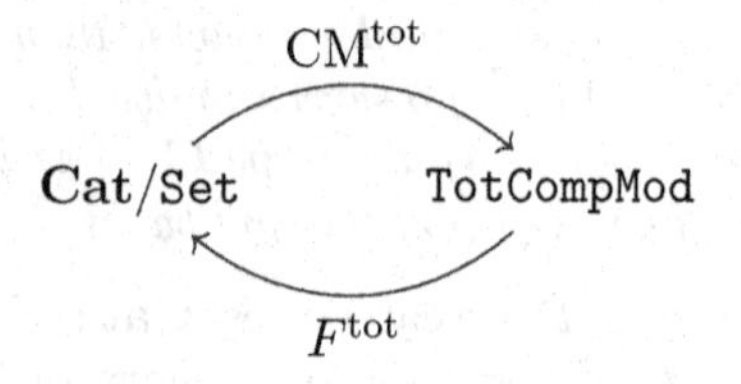

namely $\mathrm{CM}^{\mathrm{tot}} \dashv F^{\mathrm{tot}}$, as there is a natural isomorphism

$$Hom\big(\mathbf{CM}^{\mathrm{tot}}(\mathcal{C};S),\boldsymbol{D}\big) \cong Hom\big(S,\mathrm{Frg}^{\boldsymbol{D}} : \mathcal{A}sm(\boldsymbol{D}) \to \mathtt{Set}\big).$$

Proposition 3. *The functor* ${}^{\mathrm{tot}}\mathrm{CM}^{\mathcal{C}} \colon [\mathcal{C},\mathtt{Set}] \to \mathtt{TotCompMod}$, *with*

$${}^{\mathrm{tot}}\mathrm{CM}_0^{\mathcal{C}}(S) = \mathbf{CM}^{\mathrm{tot}}(\mathcal{C};S),$$

$${}^{\mathrm{tot}}\mathrm{CM}_1^{\mathcal{C}}\big(\eta\colon S \Rightarrow T\big) : \mathbf{CM}^{\mathrm{tot}}(\mathcal{C};S) \to \mathbf{CM}^{\mathrm{tot}}(\mathcal{C};T),$$

$${}^{\mathrm{tot}}\mathrm{CM}_1^{\mathcal{C}}(\eta) = \gamma_\eta = \big(\mathrm{id}_{C_0}, (\Vdash_a^{\gamma_\eta})_{a\in C_0}\big),$$

$$\Vdash_a^{\gamma_\eta} \subseteq T_0(a) \times S_0(a), \qquad y \Vdash_a^{\gamma_\eta} x \Leftrightarrow y = \eta_a(x),$$

is faithful.

By Definition 4 the category $\mathcal{A}sm\big(\mathbf{CM}^{\mathrm{tot}}(\mathcal{C};S)\big)$ of assemblies of the total model $\mathbf{CM}^{\mathrm{tot}}(\mathcal{C};S)$ has objects $(X,a_X,\Vdash_X)$, where X is a set, $a_X \in C_0$, and $\Vdash_X \subseteq S_0(a_X) \times X$ such that $\forall_{x\in X}\exists_{x'\in S_0(a_X)}\big(x' \Vdash_X x\big)$. An arrow $(X,a_X,\Vdash_X) \to (Y,a_Y,\Vdash_Y)$ is a function $f\colon X \to Y$, such that there is $\overline{f} \in S[a_X,a_Y]$ i.e., $g\colon a_X \to a_Y \in C_1$ and $\overline{f} = S_1(g)\colon S_0(a_X) \to S_0(a_Y)$ tracks f, where in this case $\overline{f} \Vdash_X^Y f$ if and only if $\forall_{x\in X}\forall_{y\in S_0(a_X)}\big(y \Vdash_X x \Rightarrow [S_1(g)](y) \Vdash_Y f(x)\big)$.

Proposition 4. *Let* $F^{\mathrm{tot}}\colon \mathcal{C} \to \mathcal{A}sm\big(\mathbf{CM}^{\mathrm{tot}}(\mathcal{C};S)\big)$ *be defined by*

$$F_0^{\mathrm{tot}}(a) = \big(S_0(a), a, \Vdash_{S_0(a)}\big), \quad y \Vdash_{S_0(a)} x \Leftrightarrow y = x; \qquad y,x \in S_0(a),$$

$$F_1^{\mathrm{tot}}(f\colon a \to b) = S_1(f).$$

(i) F^{tot} *is a full functor, injective on objects.*
(ii) *If S is injective on arrows, then F^{tot} is an embedding.*

Proof. (i) $S_1(f)$ tracks itself and is an arrow in $\mathcal{A}sm\big(\mathbf{CM}^{\mathrm{tot}}(\mathcal{C};S)\big)$. F^{tot} is a functor, injective on objects. Let $g\colon \big(S_0(a), a \Vdash_{S_0(a)}\big) \to \big(S_0(b), b \Vdash_{S_0(b)}\big)$ in $\mathcal{A}sm\big(\mathbf{CM}^{\mathrm{tot}}(\mathcal{C};S)\big)$ i.e., $g\colon S_0(a) \to S_0(b)$ such that there is $\overline{g} = S_1(h)$, for some $h\colon a \to b$ in C_1, with $S_1(h) \Vdash_{S_0(a)}^{S_0(b)} g$. Consequently, $\forall_{x\in S_0(a)}\forall_{y\in S_0(a)}\big(y = x \Rightarrow [S_1(h)](y) = g(x)\big)$, hence $\forall_{x\in S_0(a)}\big([S_1(h)](x) = g(x)\big)$ i.e., $F_1^{+}(h) = S_1(h) = g$ and F^{tot} is full.
(ii) If S is injective on arrows, F^{tot} is injective on arrows, hence faithful, and since it is full and injective on objects, it is an embedding.

Let $\boldsymbol{C}$ be a small, and $\mathrm{Frg}^{\boldsymbol{C}}$ the forgetful functor on $\mathcal{A}sm(\boldsymbol{C})$. If $\overline{X} = (X, \tau_X, \Vdash_X)$ and $\overline{Y} = (Y, \tau_Y, \Vdash_Y)$, the computability model $\mathbf{CM}^{\mathrm{tot}}(\mathcal{A}sm(\boldsymbol{C}); \mathrm{Frg}^{\boldsymbol{C}})$ over $\mathcal{A}sm(\boldsymbol{C})$ and $\mathrm{Frg}^{\boldsymbol{C}}$ is the pair

$$\left((X)_{\overline{X} \in \mathcal{A}sm(\boldsymbol{C})_0}, (\mathrm{Frg}^{\boldsymbol{C}}[\overline{X}, \overline{Y}])_{\overline{X}, \overline{Y} \in \mathcal{A}sm(\boldsymbol{C})_0}\right)$$

$$\mathrm{Frg}^{\boldsymbol{C}}[\overline{X}, \overline{Y}] = \{f : X \to Y \mid \exists_{\overline{f} : C(\tau_x) \rightharpoonup C(\tau_Y)} (\overline{f} \Vdash_X^Y f)\}.$$

I.e., from a small, and, in general, partial model $\boldsymbol{C}$, we get the locally small and total computability model $\mathbf{CM}^{\mathrm{tot}}(\mathcal{A}sm(\boldsymbol{C}); \mathrm{Frg}^{\boldsymbol{C}})$.

Proposition 5. (i) *If $\boldsymbol{C}$ is small, there is a simulation*

$$\boldsymbol{\delta}^{\mathrm{tot}} : \mathbf{CM}^{\mathrm{tot}}(\mathcal{A}sm(\boldsymbol{C}); \mathrm{Frg}^{\boldsymbol{C}}) \twoheadrightarrow \boldsymbol{C},$$

where $\boldsymbol{\delta}^{\mathrm{tot}} = (\delta^{\mathrm{tot}}, (\Vdash_{\overline{X}}^{\boldsymbol{\delta}^{\mathrm{tot}}})_{\overline{X} \in \mathcal{A}sm(\boldsymbol{C})_0})$, *with* $\delta^{\mathrm{tot}} : \mathcal{A}sm(\boldsymbol{C})_0 \to T$, *defined by* $\delta^{\mathrm{tot}}(\overline{X}) = \tau_X$, *and* $\Vdash_{\overline{X}}^{\boldsymbol{\delta}^{\mathrm{tot}}} \subseteq C(\tau_X) \times X$ *by* $\Vdash_{\overline{X}}^{\boldsymbol{\delta}^{\mathrm{tot}}} = \Vdash_X$.

(ii) *If $\boldsymbol{C}$ is total and small, there is a simulation*

$$\boldsymbol{\gamma}^{\mathrm{tot}} : \boldsymbol{C} \twoheadrightarrow \mathbf{CM}^{\mathrm{tot}}(\mathcal{A}sm(\boldsymbol{C}); \mathrm{Frg}^{\boldsymbol{C}}),$$

where $\boldsymbol{\gamma}^{\mathrm{tot}} = (\gamma^{\mathrm{tot}}, (\Vdash_{\tau}^{\boldsymbol{\gamma}^{\mathrm{tot}}})_{\tau \in T})$, *with* $\gamma^{\mathrm{tot}} : T \to \mathcal{A}sm(\boldsymbol{C})_0$, *defined by* $\gamma^{\mathrm{tot}}(\tau) = (C(\tau), \tau, \Vdash_{C(\tau)})$ *and* $\Vdash_{C(\tau)} \subseteq C(\tau) \times C(\tau)$ *is the equality on* $C(\tau)$, *and* $\Vdash_{\tau}^{\boldsymbol{\gamma}^{\mathrm{tot}}} \subseteq C(\tau) \times C(\tau)$ *is also the equality on* $C(\tau)$.

(iii) *If $\boldsymbol{C}$ is total and small,* $\boldsymbol{\delta}^{\mathrm{tot}} \circ \boldsymbol{\gamma}^{\mathrm{tot}} = \iota_{\boldsymbol{C}}$ *and* $\boldsymbol{\gamma}^{\mathrm{tot}} \circ \boldsymbol{\delta}^{\mathrm{tot}} \sim \iota_{\mathbf{CM}^{\mathrm{tot}}(\mathcal{A}sm(\boldsymbol{C}); \mathrm{Frg}^{\boldsymbol{C}})}$.

(iv) *If* $\mathcal{A}^{\mathrm{tot}} = \mathcal{A}sm(\mathbf{CM}^{\mathrm{tot}}(\mathcal{C}; S))$, *the total models* $\mathrm{CM}^{\mathrm{tot}}(\mathcal{C}; S)$ *and* $\mathrm{CM}^{\mathrm{tot}}(\mathcal{A}^{\mathrm{tot}}, \mathrm{Frg}^{\mathcal{A}^{\mathrm{tot}}})$ *are equivalent.*

4 Partial Models over Categories with Pullbacks

Definition 6. *Let* $C_1(a \hookrightarrow)[C_1(\hookrightarrow a)]$ *be the subclass of monos in* C_1 *with domain [codomain]* a. *If* $\mathcal{C}$ *has pullbacks, a partial arrow*[5] *is a pair* $(i, f) : a \rightharpoonup b$, *where* $i \in C_1(\hookrightarrow a)$ *and* $f \in C_1(\mathrm{dom}(i), b)$

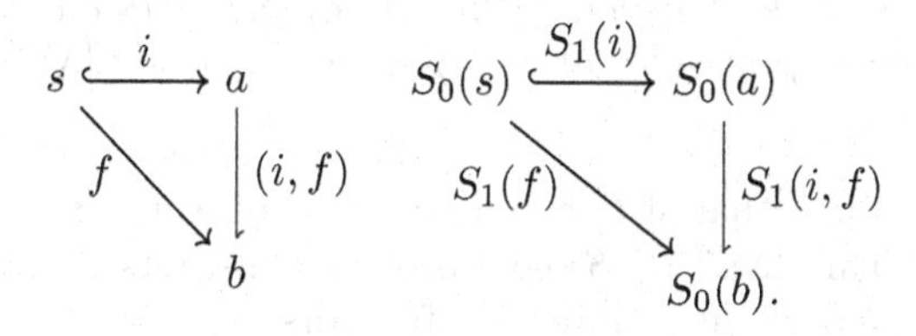

[5] As we don't define categories of partial arrows, we avoid the equivalence relation between them. A partial function between sets is a partial arrow in `Set`, also in accordance with the notion of partial function in Bishop set theory (see [10, 11]).

If $(j, g): b \rightharpoonup c$, where $j: t \hookrightarrow b$ and $g: t \to c$, their composition is the partial arrow $(i \circ i', g \circ f'): a \rightharpoonup c$, where $i': s \times_b t \hookrightarrow s$ and $f': s \times_b t \to t$ is determined by the corresponding pullback diagram

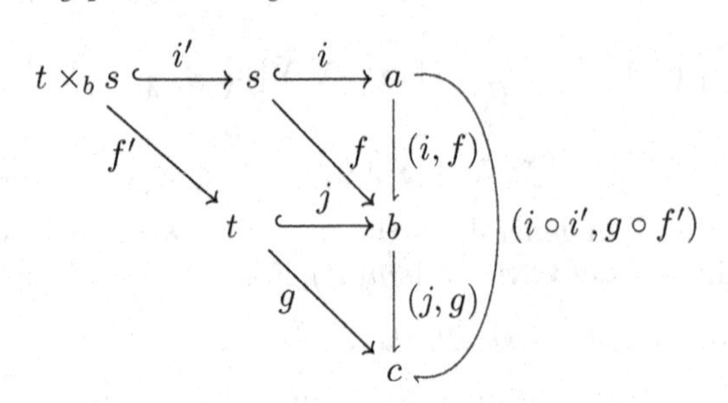

If $S: \mathcal{C} \to$ `Set` preserves pullbacks, hence monos, and if $(i, f): a \rightharpoonup b$, let $S_1(i, f) = (S_1(i), S_1(f)): S_0(a) \rightharpoonup S_0(b)$. Let also

$$\mathrm{dom}(S_1(i, f)) = \big\{[S_1(i)](x) \mid x \in S_0(s)\big\},$$

$$[S_1(i, f)](y) = [S_1(f)](x); \qquad y = [S_1(i)](x) \in \mathrm{dom}(S_1(i, f)).$$

Definition 7. *If $\mathcal{C}$ has pullbacks, and $S: \mathcal{C} \to$ `Set` preserves pullbacks, the computability model $\mathbf{CM}^{\mathrm{prt}}(\mathcal{C}; S)$ over $\mathcal{C}$ and S is the pair*

$$\mathbf{CM}^{\mathrm{prt}}(\mathcal{C}; S) = \big((S_0(a))_{a \in C_0}, (S^{\mathrm{prt}}[a, b])_{(a,b) \in C_0 \times C_0}\big),$$

$$S^{\mathrm{prt}}[a, b] = \{S_1(i, f) \mid i \in C_1(\hookrightarrow a) \ \&\ f \in C_1(\mathrm{dom}(i), b)\}.$$

If $\mathcal{D}$ has pushouts and $T: \mathcal{C}^{\mathrm{op}} \to$ `Set` sends pushouts to pullbacks, the computability model $\mathbf{CM}^{\mathrm{prt}}(\mathcal{D}; T)$ over $\mathcal{D}$ and T is defined dually i.e., $T^{\mathrm{prt}}[a, b] = \{T_1(i, f) \mid i \in C_1(a \hookrightarrow) \ \&\ f \in C_1(b, \mathrm{cod}(i))\}$.

Let again the thin category $(\mathbb{N}, \leq)$, which has pullbacks (the minimum $\wedge$). If $S: \mathbb{N} \to$ `Set` preserves pullbacks, it is easy to see that $S^{\mathrm{prt}}[n, m]$ has $(n \wedge m)$-number of elements. The fact that $\mathbf{CM}^{\mathrm{prt}}(\mathcal{C}; S)$ is a computability model depends on the preservation of pullbacks by S. If $\mathcal{C}$ is small, $\mathbf{CM}^{\mathrm{prt}}(\mathcal{C}; S)$ is small, but if $\mathcal{C}$ is locally small, $\mathbf{CM}^{\mathrm{prt}}(\mathcal{C}; S)$ need not be locally small. The proof of the next proposition is found in [12].

Proposition 6. *If $\mathcal{C}$ is a category with (fixed) products and pullbacks, and if $S: \mathcal{C} \to$ `Set` preserves pullbacks, the computability model $\mathbf{CM}^{\mathrm{prt}}(\mathcal{C}; S)$ has weak products.*

One can also show that if $\mathcal{C}$ has a terminal object 1 and if S is also full and preserves 1, then $\mathbf{CM}^{\mathrm{prt}}(\mathcal{C}; S)$ contains the constants. Next we describe the corresponding "external" and "internal" functors.

Proposition 7. *Let $\mathbf{Cat}^{\mathrm{pull}}$ be the category of categories with pullbacks and pullback preserving functors, and let $\mathbf{Cat}^{\mathrm{pull}}/$`Set` its slice category over `Set`. If*

$\mathbf{Cat}^{\mathrm{pull}}_{\hookrightarrow}/\mathtt{Set}$ *is the subcategory of* $\mathbf{Cat}^{\mathrm{pull}}/\mathtt{Set}$ *with arrows the monos preserving*[6] *arrows in* $\mathbf{Cat}^{\mathrm{pull}}/\mathtt{Set}$*, then*

$$\mathrm{CM}^{\mathrm{prt}} \colon \mathbf{Cat}^{\mathrm{pull}}_{\hookrightarrow}/\mathtt{Set} \to \mathtt{CompMod}$$

is a functor, where $\mathrm{CM}^{\mathrm{prt}}_0(\mathcal{C}, S) = \mathbf{CM}^{\mathrm{prt}}(\mathcal{C}; S)$*, and*

$$\mathrm{CM}^{\mathrm{prt}}_1(F) : \mathbf{CM}^{\mathrm{prt}}(\mathcal{C}; S) \twoheadrightarrow \mathbf{CM}^{\mathrm{prt}}(\mathcal{D}; T)$$

is defined as the simulation $\mathrm{CM}^{\mathrm{tot}}_1(F) = \gamma_F$ *in Proposition 2.*

Proof. We show that $\gamma_F = \big(F_0, (\Vdash^{\gamma_F}_a)_{a \in C_0}\big)$, where

$$\Vdash^{\gamma_F}_a \subseteq T_0(F_0(a)) \times S_0(a), \quad y \Vdash^{\gamma_F}_a x \Leftrightarrow y = x,$$

is a simulation $\mathbf{CM}^{\mathrm{prt}}(\mathcal{C}; S) \twoheadrightarrow \mathbf{CM}^{\mathrm{prt}}(\mathcal{D}; T)$. First we show ($\mathtt{Siml}_1$). If $a \in C_0$ and $x \in S_0(a)$, then $S_0(a) = T_0(F_0(a))$ and $x \Vdash^{\gamma_F}_a x$. To show ($\mathtt{Siml}_2$), let $a, b \in C_0$ and $(S_1(i), S_1(f)) \colon S_0(a) \rightharpoonup S_0(b) \in S[a, b]$, for some $s \in C_0$. We find an element of $T[F_0(a), F_0(b)]$ that tracks $(S_1(i), S_1(f))$. Let

$$(T_1 \circ F_1)(i, f) = \big(T_1(F_1(i)), T_1(F_1(f))\big) \colon T_0(F_0(a)) \rightharpoonup T_0(F_0(b)),$$

as F preserves monos. If $x \in \mathrm{dom}(S_1(i, f))$ i.e., there is a (unique) $x' \in S_0(s)$ with $x = [S_1(i)](x')$, then if $y \in T_0(F_0(a) = S_0(a)$ such that $y \Vdash^{\gamma_F}_a x$, then $y = x$. We show that $x \in \mathrm{dom}\big((T_1 \circ F_1)(i, f)\big)$, which by Definition 6 means that $x = [T_1(F_1(i))](x'')$, for some $x' \in T_0(F_0(s)) = S_0(s)$. Clearly, $x'' = x'$ works. By Definition 6 the equality obtained unfolding the relation

$$[(T_1 \circ F_1(i, f)](x) \Vdash^{\gamma_F}_b [S_1(i, f)](x)$$

follows immediately. Clearly,

$$\mathrm{CM}^{\mathrm{prt}}_1(1_{(\mathcal{C}, S)}) = \iota_{\mathbf{CM}^{\mathrm{prt}}(\mathcal{C}; S)},$$

and if $F \colon \mathcal{C} \to \mathcal{D}$, $G \colon \mathcal{D} \to \mathcal{E}$, then $\gamma_{G \circ F} = \gamma_G \circ \gamma_F$.

Proposition 8. *If* $\mathcal{C}$ *has pullbacks and* $[\mathcal{C}, \mathtt{Set}]^{\mathrm{pull}}$ *is the category of pullback preserving, covariant presheaves on* $\mathcal{C}$*, then the functor*

$$^{\mathrm{prt}}\mathrm{CM}^{\mathcal{C}} \colon [\mathcal{C}, \mathtt{Set}]^{\mathrm{pull}} \to \mathtt{CompMod}$$

is faithful, where $^{\mathrm{prt}}\mathrm{CM}^{\mathcal{C}}_0(S) = \mathbf{CM}^{\mathrm{prt}}(\mathcal{C}; S)$*, and the simulation*

$$^{\mathrm{prt}}\mathrm{CM}^{\mathcal{C}}_1(\eta) : \mathbf{CM}^{\mathrm{prt}}(\mathcal{C}; S) \twoheadrightarrow \mathbf{CM}^{\mathrm{prt}}(\mathcal{C}; T)$$

is defined as $^{\mathrm{tot}}\mathrm{CM}^{\mathcal{C}}_1(\eta) = \gamma_\eta$ *in Proposition 3.*

[6] If $F \colon \mathcal{C} \to \mathcal{D}$ such that $T \circ F = S$, where $S \colon \mathcal{C} \to \mathtt{Set}$ and $T \colon \mathcal{D} \to \mathtt{Set}$, we cannot show, in general, that F preserves monos. It does, if, e.g., T is injective on arrows.

Proof. Let $\gamma_\eta = \big(\mathrm{id}_{C_0}, (\Vdash_a^{\gamma_\eta})_{a \in C_0}\big)$ with $\Vdash_a^{\gamma_\eta} \subseteq T_0(a) \times S_0(a)$ is defined by $y \Vdash_a^{\gamma_\eta} x \Leftrightarrow y = \eta_a(x)$. To show $(\mathtt{Siml}_1)$, if $a \in C_0$ and $x \in S_0(a)$, then $T_0(a) \ni y = \eta_a(x) \Vdash_a^{\gamma_\eta} x$. To show $(\mathtt{Siml}_2)$, if $a, b \in C_0$ and $(S_1(i), S_1(f))\colon S_0(a) \rightharpoonup S_0(b) \in S[a,b]$, for some $s \in C_0$, we find an element of $T[a,b]$ that tracks $(S_1(i), S_1(f))$. Let $T_1(i,f) = \big(T_1(i), T_1(f)\big) : T_0(a) \rightharpoonup T_0(b)$. We show that

$$\big(T_1(i), T_1(f)\big) \Vdash_{(a,b)}^{\gamma_\eta} \big(S_1(i), S_1(f)\big).$$

Let $x \in S_0(a)$ such that $x \in \mathrm{dom}(S_1(i,f))$ i.e., there is a (unique) $x' \in S_0(s)$ with $x = [S_1(i)](x')$. Let $y \in T_0(a)$ such that $y \Vdash_a^{\gamma_\eta} x \Leftrightarrow y = \eta_a(x)$. First we show that $y \in \mathrm{dom}(T_1(i,f))$ i.e., $y = [T_1(i)](y')$, for some $y' \in T_0(s)$. If $y' = \eta_s(x')$, then by the commutativity of the following left diagram we have that

$$\begin{array}{ccc} S_0(s) & \xrightarrow{S_1(i)} & S_0(a) \\ {\scriptstyle \eta_s}\downarrow & & \downarrow{\scriptstyle \eta_a} \\ T_0(s) & \xrightarrow[T_1(i)]{} & T_0(a) \end{array} \qquad\qquad \begin{array}{ccc} S_0(s) & \xrightarrow{S_1(f)} & S_0(b) \\ {\scriptstyle \eta_s}\downarrow & & \downarrow{\scriptstyle \eta_b} \\ T_0(s) & \xrightarrow[T_1(f)]{} & T_0(b) \end{array}$$

$$[T_1(i)](y') = [T_1(i)](\eta_s(x')) = \eta_a\big([S_1(i)](x')\big) = \eta_a(x) = y.$$

To show $[T_1(i,f)](y) \Vdash_b^{\gamma_\eta} [S_1(i,f)](x)$ i.e., $[T_1(f)](y') \Vdash_b^{\gamma_\eta} [S_1(f)](x')$, we use the commutativity of the above right diagram:

$$\eta_b\big([S_1(f)](x')\big) = [T_1(f)]\big(\eta_s(x')\big) = [T_1(f)](y').$$

It is straightforward to show that ${}^{\mathrm{prt}}\mathrm{CM}_1(1_S\colon S \Rightarrow S) = \iota_{\mathbf{CM}^{\mathrm{prt}}(\mathcal{C};S)}$, and if $\tau\colon S \Rightarrow T$, $\eta\colon T \Rightarrow U$, then $\gamma_{\eta \circ \tau} = \gamma_\eta \circ \gamma_\tau$. To show that ${}^{\mathrm{prt}}\mathrm{CM}^{\mathcal{C}}$ is faithful, let $\eta, \theta\colon S \Rightarrow T$ such that $\gamma_\eta = \gamma_\theta$ i.e., $\Vdash_a^{\gamma_\eta} = \Vdash_a^{\gamma_\theta}$, for every $a \in C_0$. If $a \in C_0$, let $y \in T_0(a)$ and $x \in S_0(a)$. Then $y \Vdash_a^{\gamma_\eta} x \Leftrightarrow y \Vdash_a^{\gamma_\theta} x$ i.e., $y = \eta_a(x) \Leftrightarrow y = \theta_a(x)$, hence $\eta_a = \theta_a$, and since $a \in C_0$ is arbitrary, we conclude that $\eta = \theta$.

The category of assemblies $\mathcal{A}sm\big(\mathbf{CM}^{\mathrm{prt}}(\mathcal{C};S)\big)$ has objects the triplets $(X, a_X, \Vdash_X)$, and an arrow $(X, a_X, \Vdash_X) \to (Y, a_Y, \Vdash_Y)$ is a function $f\colon X \to Y$, such that there is $\bar{f} \in S[a_X, a_Y]$ i.e., there is $s \in C_0$ and $i\colon s \hookrightarrow a_X, f\colon s \to a_Y$ such that $\bar{f} = \big(S_1(i), S_1(f)\big)$ and $\bar{f} \Vdash_X^Y f$. One defines $F^{\mathrm{prt}}\colon \mathcal{C} \to \mathcal{A}sm\big(\mathbf{CM}^{\mathrm{prt}}(\mathcal{C};S)\big)$, as in Proposition 4, where $F_1^{\mathrm{prt}}(f\colon a \to b) = S_1(f)$ is tracked by the partial function $\overline{S_1(f)} = \big(S_1(1_a), S_1(f)\big)$. This functor F^{prt} though, is not, in general, full. If $\mathcal{A}^{\mathrm{prt}} = \mathcal{A}sm\big(\mathbf{CM}^{\mathrm{prt}}(\mathcal{C};S)\big)$ one can study the total model $\mathbf{CM}^{\mathrm{tot}}\big(\mathcal{A}^{\mathrm{prt}}; \mathrm{Frg}^{\mathcal{A}^{\mathrm{prt}}}\big)$ and the partial model $\mathbf{CM}^{\mathrm{prt}}\big(\mathcal{A}^{\mathrm{prt}}; \mathrm{Frg}^{\mathcal{A}^{\mathrm{prt}}}\big)$, and relate the latter to $\mathbf{CM}^{\mathrm{prt}}(\mathcal{C};S)$.

5 Categories with a Base of Computability

We introduce the notion of a base of computability, which resembles Rosolini's notion of dominion (see [13], Section 2.1), and its dual.

Definition 8. *A base of computability is a family $B = (B(a))_{a \in C_0}$, where $B(a)$ is a subclass of the class $C_1(\hookrightarrow a)$ of monos having codomain the object a, for every $a \in C_0$, such that the following conditions are satisfied:*

($\texttt{Base}_1$) *For every $a \in C_0$, we have that $1_a \in B(a)$.*

($\texttt{Base}_2$) *For every $i \colon s \hookrightarrow a \in B(a)$, for every $b \in C_0$, for every $f \colon s \to b$ and for every $j \colon t \hookrightarrow b \in B(b)$ a pullback $s \times_b t$ exists and $i \circ i' \colon s \times_b t \hookrightarrow a \in B(a)$*

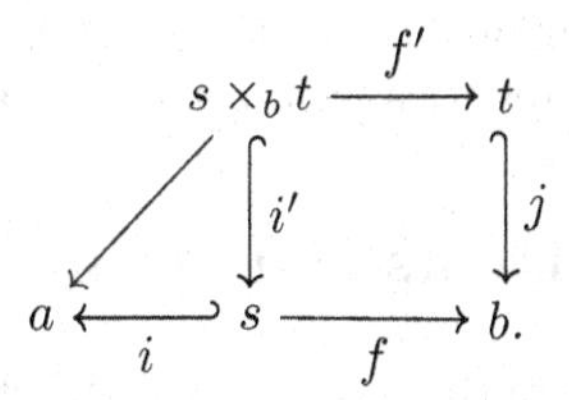

A cobase of computability for $\mathcal{C}$ is a family $C = (C(a))_{a \in C_0}$, where $C(a)$ is a subclass of the class $C_1(a \hookrightarrow)$ of monos with domain a such that ($\texttt{Base}_1$) *and the dual of* ($\texttt{Base}_2$) *are satisfied.*

Proposition 9. (i) *If $I(a) = \{1_a \colon a \hookrightarrow a\}$, for every $a \in C_0$, then $I = (I(a))_{a \in C_0}$ is a base and a cobase of computability for $\mathcal{C}$.*

(ii) *If* $\mathrm{Iso}(a) = \{i \in C_1(\hookrightarrow a) \mid i$ is an iso$\}$, *for every $a \in C_0$, then* $\mathrm{Iso} = (\mathrm{Iso}(a))_{a \in C_0}$ *is a base of computability for $\mathcal{C}$.*

(iii) *If $\mathcal{C}$ has pullbacks, and $B(a) = C_1(\hookrightarrow a)$, for every $a \in C_0$, then $B^{\hookrightarrow} = (B(a))_{a \in C_0}$ is a base of computability for $\mathcal{C}$.*

(iv) *If $\mathcal{C}$ has pushouts, and $C(a) = C_1(a \hookrightarrow)$, for every $a \in C_0$, then $C^{\hookleftarrow} = (C(a))_{a \in C_0}$ is a cobase of computability for $\mathcal{C}$.*

Proof. (i) If $a, b \in C_0$, and $f \colon a \to b$, then $I(b) - \{1_b \colon b \hookrightarrow b\}$ and the following is a pullback square with $1_a \circ 1_a = 1_a \in I(a)$

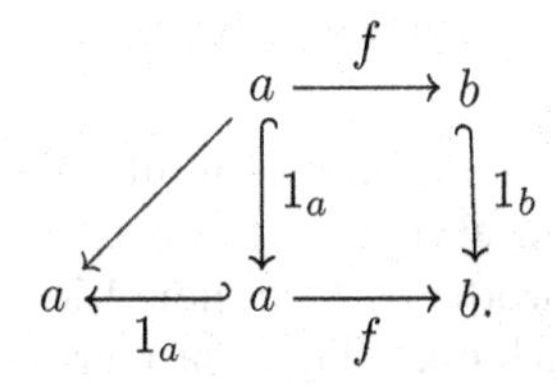

(ii) In this case $a = s \times_b t$. Cases (iii) and (iv) are trivial.

Definition 9. *If $B = (B(a))_{a \in C_0}$ is a base of computability for $\mathcal{C}$ and $S \colon \mathcal{C} \to$* $\texttt{Set}$ *preserves pullbacks, the computability model $\mathbf{CM}^B(\mathcal{C}; S)$ over $\mathcal{C}$ and S with respect to B is the pair*

$$\mathbf{CM}^B(\mathcal{C}; S) = \big((S_0(a))_{a \in C_0}, (S[a, b])_{(a,b) \in C_0 \times C_0}\big),$$

$$S[a, b] = \{S_1(i, f) \mid i \in B(a) \ \& \ f \in C_1(\mathrm{dom}(i), b)\}.$$

If $C = (C(a))_{a \in C_0}$ is a cobase of computability for $\mathcal{C}$ and $T \colon \mathcal{C}^{\mathrm{op}} \to \mathtt{Set}$ sends pushouts to pullbacks, the computability model $\mathbf{CM}^B(\mathcal{C}; T)$ over $\mathcal{D}$ and T is defined dually i.e.,

$$T[a, b] = \{T_1(i, f) \mid i \in C(a) \ \& \ f \in C_1(b, \mathrm{cod}(i))\}.$$

If $B(a)$ is a set, for every $a \in C_0$, and $\mathcal{C}$ is locally small, then $\mathbf{CM}^B(\mathcal{C}; S)$ is locally small. By Proposition 9, if $B = I$, we have that $\mathbf{CM}^I(\mathcal{C}; S) = \mathbf{CM}^{\mathrm{tot}}(\mathcal{C}; S)$. Actually, every covariant presheaf preserves the pullbacks related to the base I. If $B = B^{\hookrightarrow}$, then $\mathbf{CM}^{B^{\hookrightarrow}}(\mathcal{C}; S) = \mathbf{CM}^{\mathrm{prt}}(\mathcal{C}; S)$. Consequently, the results shown in the previous two sections can be seen as special cases of the corresponding results for categories with a base (or a cobase) of computability.

6 Concluding Comments and Future Work

The results presented here are the first steps in the study of computability models over categories and presheaves. While the notions of a computability model and simulation defined in [9] permitted the transition from computability models to categories, their slight generalisation presented here facilitated the transition from categories (and presheaves on them) to computability models, allowing new connections between the two subjects. Next we include some topics that we plan to investigate in a future extension of our work.

1. The further study of the notion of a category with a (co)base of computability and the possible influence of Rosolini's theory of dominions in the theory of computability models over such a category.

2. The study the computability models over a locally small category with pullbacks and the pullback preserving, covariant representable functors $Hom(a, \_ \,)$.

3. The formulation of conditions on a model $\boldsymbol{C}$ such that its category of assemblies $\mathcal{A}sm(\boldsymbol{C})$ has a base of computability B, and the study of the relation between $\boldsymbol{C}$ and $\mathbf{CM}^B(\mathcal{A}sm(\boldsymbol{C}); \mathrm{Frg}^{\boldsymbol{C}})$.

4. The effect of various constructions in categories to their canonical computability models.

5. The generalisation of the notions of a computability model and of a simulation between them so that the datatypes $C(\tau)$ are not necessarily in $\mathtt{Set}$, but in a category $\mathcal{S}$ with pullbacks, products, terminal object and factorisation of arrows to compositions of monos with epis. These properties of $\mathcal{S}$ allow the formulation of all set-based notions of Sect. 1 within $\mathcal{S}$ itself. For example, $\Vdash_\tau^\gamma$ is a relation between the objects $\boldsymbol{D}(\gamma(\tau))$ and $\boldsymbol{C}(\tau)$ of $\mathcal{S}$ i.e., an object of $\mathcal{S}$ and a pair of jointly monic arrows in $\mathcal{S}$ from $\Vdash_\tau^\gamma$ to $\boldsymbol{D}(\gamma(\tau))$ and $\boldsymbol{C}(\tau)$. The general theory of relations in categories (see [5]) can be used to formulate in abstract categorical terms all related notions and to study $\mathcal{S}$-computability models over a category $\mathcal{C}$ and an $\mathcal{S}$-valued functor on $\mathcal{C}$.

6. The study of relations between the theory of computability models over categories and presheaves on them with interesting computability models within the theory of Higher-Order Computability, as this is developed in [9].

Acknowledgment. Our research was supported by LMUexcellent, funded by the Federal Ministry of Education and Research (BMBF) and the Free State of Bavaria under the Excellence Strategy of the Federal Government and the Länder.

References

1. Carboni, A., Pedicchio, M.C., Rosolini, G. (eds.): Category Theory. LNM, vol. 1488. Springer, Heidelberg (1991). https://doi.org/10.1007/BFb0084207
2. Cockett, R.: Categories and Computability, Lecture Notes (2014)
3. Cockett, R., Hofstra, P.: Introduction to turing categories. Ann. Pure Appl. Log. **156**(2–3), 183–209 (2008). https://doi.org/10.1016/j.apal.2008.04.005
4. Cockett, R., Hofstra, P.: Categorical simulations. J. Pure Appl. Algebra **214**(10), 1835–1853 (2010). https://doi.org/10.1016/j.jpaa.2009.12.028
5. Klein, A.: Relations in categories. Illinois J. Math. **14**(4), 536–550 (1970)
6. Longley, J.: Realizability Toposes and language semantics. Ph.D. thesis ECS-LFCS-95-332 University of Edinbourgh (1995)
7. Longley, J.: On the ubiquity of certain total type structures. Math. Struct. Comput. Sci. **17**(5), 841–953 (2007). https://doi.org/10.1016/j.entcs.2004.08004
8. Longley, J.: Computability structures, simulations and realizability. Math. Struct. Comput. Sci. **24**(2), E240201 (2014). https://doi.org/10.1017/S0960129513000182
9. Longley, J., Normann, D.: Higher-Order Computability. THEOAPPLCOM. Springer, Heidelberg (2015). https://doi.org/10.1007/978-3-662-47992-6
10. I. Petrakis: Dependent sums and dependent products in Bishop's set theory. In: Dybjer, P., et al. (eds.) TYPES 2018. LIPIcs, vol. 130 (2019). https://doi.org/10.4230/LIPIcs.TYPES.2018.3. Article no. 3
11. Petrakis, I.: Families of sets in bishop set theory. Habilitation thesis, Ludwig-Maximilians-Universität (2020)
12. Petrakis, I.: Computability models over categories. arXiv:2105.06933v1 (2021)
13. Rosolini, G.: Continuity and effectiveness in topoi. Ph.D. thesis, University of Oxford (1986)
14. Riehl, E.: Category Theory in Context. Dover Publications Inc., Mineola (2016)

Reducts of Relation Algebras: The Aspects of Axiomatisability and Finite Representability

Daniel Rogozin(✉)

Institute for Information Transmission Problems, Russian Academy of Sciences, Moscow, Russia
daniel.rogozin@serokell.io

Abstract. In this paper, we show that the class of representable residuated semigroups has the finite representation property. That is, every finite representable residuated semigroup is representable over a finite base. This result gives a positive solution to Problem 19.17 from the monograph by Hirsch and Hodkinson [13].

We also show that the class of representable join semilattice-ordered semigroups is pseudo-universal and it has a recursively enumerable axiomatisation. For this purpose, we introduce representability games for join semilattice-ordered semigroups.

Keywords: Algebraic logic · Relation algebras · Finite representation property · Residuated semigroups · Join semilattice-ordered semigroups

1 Introduction

Relation algebras are a kind of Boolean algebras with operators that provide algebraisation of binary relations [20]. The class of all relation algebras, denoted as **RA**, consists of algebras of the signature $\{0, 1, +, -, ;, ^{\smile}, \mathbf{1'}\}$, and all those algebras obey certain axioms. The class of representable relation algebras, **RRA**, consists of algebras isomorphic to set relation algebras. **RRA** is a subclass of **RA**, but the converse inclusion does not hold. That is, there exist non-representable relation algebras [22]. Moreover, the class **RRA** is not a finitely axiomatisable variety [24] with neither Sahlqvist [29] nor canonical axiomatisation [19]. The problem of determining whether a given finite relation algebra is representable is undecidable, see [12].

For this reason, we are interested in reducts since one may extract more positive results in the aspects of decidability, representability, and finite axiomatisability. There are several results on reducts of relation algebras that have no finite axiomatisation. The examples of non-finitely axiomatisable classes are ordered monoids [10], distributive residuated lattices [1], join semilattice-ordered semigroups [2], meet semilattice-ordered semigroups with converses [18], etc. On the

The research is supported by the project MK-1184.2021.1.1.

S. Artemov and A. Nerode (Eds.): LFCS 2022, LNCS 13137, pp. 266–280, 2022.
https://doi.org/10.1007/978-3-030-93100-1_17

other hand, such classes as representable residuated semigroups [1] and ordered domain algebras [15] are finitely axiomatisable. There are also subsignatures for which the question of finite axiomatisability remains open, see, e.g., [2].

The other direction we discuss is related to finite representability. A finite algebra of relations has the finite representation property if it is isomorphic to some algebra of relations over a finite base. The investigation of this problem is of interest to study such aspects as decidability of membership of $\mathbf{R}(\tau)$ for finite structures. The finite representation property also implies recursivity of the class of all finite representable τ-structures [9], if the whole class is finitely axiomatisable. Here, τ is a subsignature of operations and predicates definable in $\{0, 1, +, -, ;, \smile, \mathbf{1}\}$. The examples of the class having the finite representation property are some classes of algebras [9,15,23], the subsignature of which contains the domain and range operators. The other kind of algebras of binary relations having the finite representation property is semigroups with so-called demonic refinement has been recently studied by Hirsch and Šemrl [16], but the same authors have recently shown that semigroups with demonic joins fail to have the finite representation property [8].

There are subsignatures τ such that the class $\mathbf{R}(\tau)$ of representable reducts fails to have the finite representation property, for example, $\{;, \cdot\}$, see [16, Theorem 4.1]. In general, (un)decidability of determining whether a finite relation algebra has a finite representation is an open question [13, Problem 18.18].

In this paper, we consider reducts of relation algebras the signature of which consists of composition, residuals, and the binary relation symbol that denotes partial ordering. That is, we study the class of representable residuated semigroups. We show that $\mathbf{R}(;, \backslash, /, \leq)$ has the finite representation property. As a result, Problem 19.17 of [13] has a positive solution. The solution is based on the Dedekind-MacNeille completions and relational representations of quantales. We embed a finite residuated semigroup into a finite quantale by mapping every element to its lower cone. After that, we apply the relational representation for quantales. As a result, the original finite residuated semigroup has a Zaretski-style representation [30] and this satisfies the finite base requirement.

In the final section, we study the class of representable join semilattice-ordered semigroups, denoted as $\mathbf{R}(;, +)$. It is already known that this class is not finitely axiomatisable [2]. We show that $\mathbf{R}(;, +)$ has a recursively enumerable axiomatisation. For that, we define networks and representability games. This class is axiomatised with the axioms of join semilattice-ordered semigroups plus the countable set of universal formulas claiming that $\exists$ has a winning strategy on every finite step. The question of finite representability for this class remains open, see [27, Problem 2].

2 Definitions

2.1 Relation Algebras and Their Reducts

Let us introduce some basic definitions related to relation algebras. See [13, Section 3] to have more details.

Definition 1. *A relation algebra is an algebra* $\mathcal{R} = \langle R, 0, 1, +, -, ;, {}^{\smile}, \mathbf{1}' \rangle$ *such that* $\langle R, 0, 1, +, - \rangle$ *is a Boolean algebra,* $\langle R, ;, \mathbf{1}' \rangle$ *is a monoid, and the following equations hold, for all* $a, b, c \in R$*:*

1. $(a + b); c = (a; c) + (b; c)$,
2. $a^{\smile\smile} = a$,
3. $(a + b)^{\smile} = a^{\smile} + b^{\smile}$,
4. $(a; b)^{\smile} = b^{\smile}; a^{\smile}$,
5. $a^{\smile}; (-(a; b)) \leq -b$.

where $a \leq b$ *is defined usually as* $a + b = b$. **RA** *is the class of all relation algebras.*

Definition 2. *A proper relation algebra (or, a set relation algebra) is an algebra* $\mathcal{R} = \langle R, 0, 1, \cup, -, ;, {}^{\smile}, \mathbf{1}' \rangle$ *such that* $R \subseteq \mathcal{P}(W)$*, where* X *is a base set,* $W \subseteq X \times X$ *is an equivalence relation,* $0 = \emptyset$*,* $1 = W$*,* $\cup$ *and* $-$ *are set-theoretic union and complement respectively,* $;$ *is relation composition,* ${}^{\smile}$ *is relation converse,* $\mathbf{1}'$ *is the identity relation restricted to* W*, that is:*

1. $a; b = \{(x, z) \in W \mid \exists y \, (x, y) \in a \,\&\, (y, z) \in b\}$
2. $a^{\smile} = \{(x, y) \in W \mid (y, x) \in a\}$
3. $\mathbf{1}' = \{(x, y) \in W \mid x = y\}$

PRA *is the class of all proper relation algebras.* **RRA** *is the class of all representable relation algebras, that is, the closure of* **PRA** *under isomorphic copies.*

Let τ be a subset of operations and predicates definable in **RA**. $\mathbf{R}(\tau)$ is the class of subalgebras of τ-subreducts of algebras belonging to **RRA**. We also assume that $\mathbf{R}(\tau)$ is closed under isomorphic copies. A τ-structure is *representable* if it is isomorphic to some algebra of relations of this signature. A representable finite τ-structure has a *finite representation over a finite base* if it is isomoprhic to some finite representable over a finite base. $\mathbf{R}(\tau)$ has the finite representation property if every $\mathcal{A} \in \mathbf{R}(\tau)$ has a finite representation over a finite base.

2.2 Residuated Semigroups

A *residuated semigroup* is a structure $\mathcal{A} = \langle A, ;, \leq, \backslash, / \rangle$ such that, for all $a, b, c \in \mathcal{A}$:

1. $\leq$ is reflexive, antisymmetric, and transitive.
2. $a; (b; c) = (a; b); c$.
3. $a \leq b \Rightarrow a; c \leq b; c$ and $a \leq b \Rightarrow c; a \leq c; b$.
4. $b \leq a \backslash c \Leftrightarrow a; b \leq c \Leftrightarrow a \leq c \,/\, b$.

We can express residuals in every $\mathcal{R} \in$ **RA** using Boolean negation, inversion, and composition as follows:

1. $a \backslash b = -(a^{\smile}; -b)$
2. $a \,/\, b = -(-a; b^{\smile})$

These residuals have the following explicit definition in $\mathcal{R} \in \mathbf{PRA}$:

1. $a \setminus b = \{(x, y) \mid \forall z \, (z, x) \in a \Rightarrow (z, y) \in b\}$
2. $a \,/\, b = \{(x, y) \mid \forall z \, (y, z) \in b \Rightarrow (x, z) \in a\}$

One can visualise residuals in **RRA** with the following triangles:

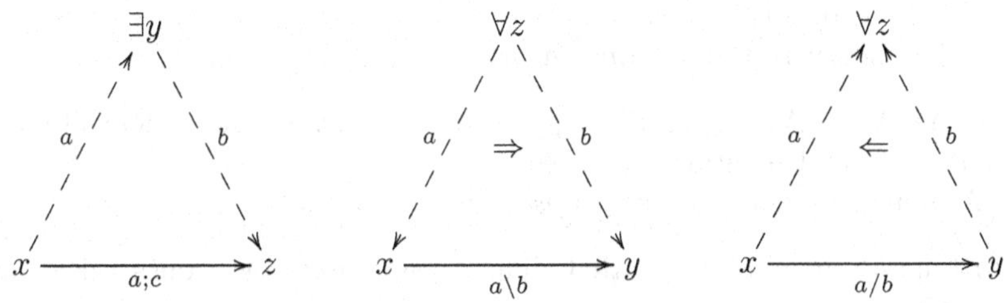

Thus, in particular, every relation algebra is a residuated lattice.

2.3 Join Semilattice-Ordered Semigroups

A *join semilattice-ordered semigroup* is an algebra $\mathcal{A} = \langle A, ;, + \rangle$ such that $\langle A, ; \rangle$ is a semigroup, $\langle A, + \rangle$ is a join-semilattice, and the following identities hold, for all $a, b, c \in A$:

1. $a; (b + c) = a; b + a; c,$
2. $(a + b); c = a; c + b; c.$

A join semilattice-ordered semigroup is also a poset and ordering is defined as $a \leq b$ iff $a + b = b$.

Definition 3. *A representation R of a join semilattice-ordered semigroup $\mathcal{A}$ is a one-to-one map $R : \mathcal{A} \to 2^{D \times D}$ (where D is a non-empty base set) such that*

1. $(a + b)^R = a^R \cup b^R,$
2. $(a; b)^R = a^R; b^R.$

A join semilattice-ordered semigroup $\mathcal{A}$ is *representable*, if there exists a representation $R : \mathcal{A} \to 2^{D \times D}$ for some non-empty base set D.

2.4 Order-theoretic Definitions

Let us also remind the reader several order-theoretic notions, see [4, Chapter 1] for more details. Let $\langle P, \leq \rangle$ be a partial order. An upper cone generated by x is the set $\uparrow x = \{a \in P \mid x \leq a\}$. Let $A \subseteq P$, then $\uparrow A = \bigcup_{x \in A} \uparrow x = \{a \in P \mid \exists x \in P \; x \leq a\}$. The set of all upper cones of a poset $\langle P, \leq \rangle$ is denoted as $\mathrm{Up}(P)$. Given $a \in P$, the *lower cone* generated by a is a subset $\downarrow a = \{x \in P \mid x \leq a\}$. The lower cone generated by a subset is defined similarly.

A *closure operator* on a poset $\langle P, \leq \rangle$ is a monotone map $j : P \to P$ such that for all $a \in P$ we have $a \leq ja = jja$.

2.5 Pseudo-elementary Classes

The following definitions are due to [13, Section 9]. Let $\mathcal{K}$ be a class of structures of a first-order signature $\mathcal{L}$. $\mathcal{K}$ is called a *pseudo-elementary* class if there are:

1. a two-sorted language $\mathcal{L}^s$ with disjoint sorts $\mathbf{a}$ and $\mathbf{r}$ that contains all symbols of $\mathcal{L}$ as $\mathbf{a}$-sorted symbols,
2. an $\mathcal{L}^s$-theory T, the defining theory.

such that $\mathcal{K} = \{\mathcal{M}^{\mathbf{a}} \restriction_{\mathcal{L}} \mid \mathcal{M} \models T\}$. More generally, a pseudo-elementary class is a reduct of an elementary class, see [5].

A pseudo-elementary class is *pseudo-universal* if

1. a function symbol in $\mathcal{L}^s$ that differs from copies of $\mathcal{L}$ ones takes values in sort $\mathbf{r}$,
2. Every sentence in T is built from atomic formulas and negated-atomic formulas using $\vee$, $\wedge$, $\forall x^{\mathbf{a}}$, $\forall x^{\mathbf{r}}$, $\exists x^{\mathbf{r}}$.

We are going to use the following fact to axiomatise the class of representable join semilattice-ordered semigroups, see [13, Corollary 9.15, Theorem 9.28]:

Theorem 1.

1. *If $\mathcal{K}$ is a pseudo-universal class, then $\mathcal{K}$ is elementary and universally axiomatisable.*
2. *Let $\mathcal{K} = \{\mathcal{M}^{\mathbf{a}} \restriction_{\mathcal{L}} \mid \mathcal{M} \models T\}$ be a pseudo-universal class of $\mathcal{L}$-structures, where T is an $\mathcal{L}^s$-theory and $\mathcal{L}$, $\mathcal{L}^s$, T are recursively enumerable. Then there exists the set of $\mathcal{L}$-sentences $\{\eta_n\}_{n<\omega}$ from T such that $\mathcal{A} \in \mathcal{K}$ iff $\mathcal{A} \models \{\eta_n\}_{n<\omega}$. That is, $\{\eta_n\}_{n<\omega}$ axiomatises $\mathcal{K}$.*

3 The Finite Representation Property for Residuated Semigroups

The problem we are interested in is the following [13, Problem 19.17]:

Does $\mathbf{R}(;, \backslash, /, \leq)$ have the finite representation property?

The class $\mathbf{R}(;, \backslash, /, \leq)$ consists of the following structures, here is the explicit definition:

Definition 4. *Let A be a set of binary relations on some base set W such that $R = \bigcup A$ is transitive and W is a domain of R. A relational residuated semigroup is an algebra $\mathcal{A} = \langle A, ;, \backslash, /, \subseteq \rangle$ where, for all $a, b \in A$:*

1. $a;b = \{(x,z) \mid \exists y \in W \; ((x,y) \in a \;\&\; (y,z) \in b)\}$,
2. $a \backslash b = \{(x,y) \mid \forall z \in W \; ((z,x) \in a \Rightarrow (z,y) \in b)\}$,
3. $a \,/\, b = \{(x,y) \mid \forall z \in W \; ((y,z) \in b \Rightarrow (x,z) \in a)\}$,
4. $a \leq b$ *iff* $a \subseteq b$.

A residuated semigroup is called *representable* if it is isomorphic to some algebra that belongs to $\mathbf{R}(;,\backslash,/,\leq)$.

Definition 5. *Let $\tau = \{;,\backslash,/,\leq\}$, let $\mathcal{A}$ be a τ-structure and X a base set. An* interpretation *R over a base X maps every $a \in \mathcal{A}$ to a binary relation $a^R \subseteq X \times X$. A* representation *of $\mathcal{A}$ is an interpretation R that interprets operations and $\leq$ as above.*

Andréka and Mikulás proved the representation theorem for residuated semigroups ([1]) in the step-by-step fashion. See this paper to learn more about step-by-step representations in general [11]. The representation theorem for residuated semigroups obviously implies that the class $\mathbf{R}(;,\backslash,/,\leq)$ is finitely axiomatisable. As it is well known, the logic of such structures is the Lambek calculus [21], so we also have the relational completeness of the Lambek calculus. With our result, we also have a version of the finite model property for the Lambek calculus since one can refute any unprovable sequent in some finite relational residuated semigroup over a finite base. This is a corollary of our result and the fact that the Lambek calculus is complete w.r.t finite residuated semigroups, see [6, Chapter 7, Sect. 7.4] to have an even stronger result.

It is sufficient to show that any finite residuated semigroup has a representation over a finite base in order to show that $\mathbf{R}(;,\backslash,/,\leq)$ has the finite representation property. For that, we will use the representation of residuated semigroups as subalgebras of quantales and the relational representation of quantales.

A quantale is a complete lattice-ordered semigroup. That is, a binary operation respects suprema in both arguments. Quantales have been introduced by Mulvey to provide a noncommutative generalisation of locales, see [25].

Definition 6. *A* quantale *is a structure $\mathcal{Q} = \langle Q, ;, \Sigma\rangle$ such that $\mathcal{Q} = \langle Q, \Sigma\rangle$ is a complete lattice, where Σ denotes an infinite join, $\langle Q, ;\rangle$ is a semigroup, and the following conditions hold for all $a \in Q$ and $A \subseteq Q$:*

1. $a \,;\, \Sigma A = \Sigma\{a;q \mid q \in A\}$,
2. $\Sigma A \,;\, a = \Sigma\{q;a \mid q \in A\}$.

Definition 7. *Given a quantale $\mathcal{Q} = \langle Q, ;, \Sigma\rangle$, a set of* generators *is a subset $G \subseteq \mathcal{Q}$, if*

1. *For all $q \in Q$ one has $q \leq \Sigma\{g \in G \mid g \leq q\}$,*
2. *For all $g \in G$ and $q_1, q_2 \in \mathcal{Q}$, $g \leq q_1;q_2$ implies $g \leq q_1;r$ for some $r \in G$ with $r \leq q_2$.*

The existence of a set of generators for an arbitrary quantale has been shown here [3, Lemma 3.12].

Note that any quantale is a residuated semigroup as well. Given a quantale $\mathcal{Q} = \langle Q, ;, \Sigma\rangle$, One may express residuals with supremum and product as follows for all $a, b \in Q$:

1. $a \setminus b = \Sigma\{c \in Q \mid a;c \leq b\}$,
2. $a \,/\, b = \Sigma\{c \in Q \mid b;c \leq a\}$.

It is readily checked that residuals are unique.

A quantic nucleus is a closure operator on a quantale. Such an operator is a noncommutative generalisation of a nucleus operator from locale theory. The following definition and the proposition below are due to [26, Definition 3.1.1, Theorem 3.1.1] respectively.

Definition 8. *A* quantic nucleus *on a quantale* $\langle A, ;, \Sigma\rangle$ *is a mapping* $j : A \to A$ *such that* j *a closure operator satisfying* $ja; jb \leq j(a; b)$.

Proposition 1. *Let* $\mathcal{A} = \langle A, ;, \Sigma\rangle$ *be a quantale and* j *a quantic nucleus, the set* $\mathcal{A}_j = \{a \in A \,|\, ja = a\}$ *forms a quantale, where* $a ;_j b = j(a; b)$ *and* $\Sigma_j A = j(\Sigma A)$ *for all* $a, b \in A_j$ *and* $A \subseteq \mathcal{A}_j$.

One can embed any residuated semigroup into some quantale with using Dedekind-MacNeille completion (see, for example, [28]) as follows. According to Goldblatt [7], residuated semigroups have the following representation based on quantic nuclei and the Galois connection. We need the construction for the solution, so we discuss it briefly. See Goldblatt's paper to have a complete argument in more detail [7].

Let $\mathcal{A} = \langle A, \leq, ;, \backslash, /\rangle$ be a residuated semigroup. Then $\langle \mathcal{P}(A), ;, \bigcup\rangle$ is a quantale with pairwise product of subsets.

Let $X \subseteq A$. We put lX and uX as the sets of lower and upper bounds of X in A. We also put $mX = luX$. Note that the lower cone of an arbitrary x is m-closed, that is, $m(\downarrow x) = \downarrow x$.

$m : \mathcal{P}(A) \to \mathcal{P}(A)$ is a closure operator and the set

$$(\mathcal{P}(A))_m = \{X \in \mathcal{P}(S) \,|\, mX = X\}$$

forms a complete lattice with $\Sigma_m \mathcal{X} = m(\bigcup \mathcal{X})$ and $\Pi_m = \bigcap \mathcal{X}$, see [4, Theorem 7.3]. The key observation is that m is a quantic nucleus on $\mathcal{P}(A)$, that is, $mA; mB \subseteq m(A; B)$. We refer here to the aforementioned paper by Goldblatt. Thus, according to Proposition 1, $\langle (\mathcal{P}(A))_m, \subseteq, ;_m\rangle$ is a quantale itself since m is a quantic nucleus.

We define a map $f_m : \mathcal{A} \to (\mathcal{P}(A))_m$ such that $f_m : a \mapsto \downarrow a$. This map is well-defined since any lower cone generated by a point is m-closed. Moreover, f_m preserves products, residuals, and existing suprema. In particular, f_m is a residuated semigroup embedding. As a result, we have the following representation theorem [7, Corollary 2].

Theorem 2. *Every residuated semigroup is isomorphic to the subalgebra of some quantale.*

In turn, quantales are representable with quantales of binary relations. The notion of a relational quantale has been introduced by Brown and Gurr to represent quantales as quantales of relations [3].

Definition 9. *Let* A *be a non-empty set. A* relational quantale *on* A *is an algebra* $\langle R, \subseteq, ;\rangle$, *where*

1. $R \subseteq \mathcal{P}(A \times A)$,

2. $\langle R, \subseteq \rangle$ *is a complete join-semilattice,*
3. ; *is a relational composition that respects all suprema in both coordinates.*

The uniqueness of residuals in any quantale implies the following fact.

Proposition 2. *Let* $\mathcal{A}$ *be a relational quantale over a base set* X*, then for all* $a, b \in \mathcal{A}$

1. $a \setminus b = \{(x, y) \in X^2 \mid \forall z \in X((z, x) \in a \Rightarrow (z, y) \in b)\}$,
2. $a \,/\, b = \{(x, y) \in X^2 \mid \forall z \in X((y, z) \in b \Rightarrow (x, z) \in b)\}$.

Now let us discuss the representation theorem for quantales. Let $\mathcal{Q}$ be a quantale, Q its carrier, and $\langle G \rangle$ a set of its generators. Given $a \in \mathcal{Q}$, define the binary relation $\hat{a} \subseteq Q \times Q$ as:

$$\hat{a} = \{(g, p) \mid g \in \langle G \rangle, p \in Q \;\; g \leq a; p\}$$

Denote $\widehat{Q}$ as $\{\hat{a} \mid a \in \mathcal{Q}\}$.

The mapping $a \mapsto \hat{a}$ satisfies the following conditions:

1. $a \leq b$ iff $\hat{a} \subseteq \hat{b}$,
2. $\widehat{\Sigma A} = \Sigma \widehat{A}$, $\hat{a}; \hat{b} = \widehat{a; b}$, and $\langle \widehat{\mathcal{Q}}, \subseteq, \Sigma \rangle$ is a complete lattice,
3. $\langle \widehat{\mathcal{Q}}, \subseteq, ; \rangle$ is a relational quantale,
4. $\mathcal{Q}$ is isomorphic to $\langle \widehat{\mathcal{Q}}, \subseteq, ; \rangle$ and $a \mapsto \hat{a}$ is a quantale isomorphism.

We summarise the construction above with the following theorem proved by Brown and Gurr, see [3, Theorem 3.11].

Theorem 3. *Every quantale* $\mathcal{Q} = \langle Q, ;, \Sigma \rangle$ *is isomorphic to a relational quantale on* Q *as a base set.*

Let $\mathcal{A}$ be a residuated semigroup and $\mathcal{Q}_{\mathcal{A}}$ a quantale of Galois closed subsets of $\mathcal{A}$. $\widehat{\mathcal{Q}_{\mathcal{A}}}$ is the corresponding relational quantale. Let us define an interpretation $R : \mathcal{A} \to \widehat{\mathcal{Q}_{\mathcal{A}}}$ such that:

$$R : a \mapsto a^R = \widehat{\downarrow a}$$

According to the lemma below, such an interpretation is a representation. As we have already said above, the function $a \mapsto \downarrow a$ is order-preserving and it commutes with products and residuals.

Lemma 1. *Let* $\mathcal{A}$ *be a residuated semigroup, then the interpretation* $R : \mathcal{A} \to \widehat{\mathcal{Q}_{\mathcal{A}}}$ *such that* $R : a \mapsto a^R = \widehat{\downarrow a}$ *is a representation.*

Proof. By Theorem 2, $\mathcal{A}$ emdeds to $\mathcal{Q}_{\mathcal{A}}$, but by Theorem 3, $\mathcal{Q}_{\mathcal{A}}$ is isomorphic to $\widehat{\mathcal{Q}_{\mathcal{A}}}$. The fact that R is an injective homomorphism follows from the construction of the embedding of a residuated semigroup to the quantale of its Galois-stable subsets, the isomorphism of $\mathcal{Q}_{\mathcal{A}}$ with $\widehat{\mathcal{Q}_{\mathcal{A}}}$, and Proposition 2.

The lemma above implies the following statement.

Theorem 4. *Every residuated semigroup is isomorphic to the subalgebra of some relational quantale. Moreover,* $\mathbf{R}(;,\backslash,/,\leq)$ *has the finite representation property.*

Proof. Let $\mathcal{A}$ be a finite residuated semigroup. The representation of $\mathcal{A}$ as a subalgebra of the relational quantale of $\widehat{\mathcal{Q}_{\mathcal{A}}}$ belongs to $\mathbf{R}(;,\backslash,/,\leq)$ by Lemma 1. This representation has the following form:

$$\widehat{\mathcal{A}} = \langle \{\widehat{\downarrow a}\}_{a \in \mathcal{A}}, ;, \backslash, /, \subseteq \rangle$$

.

Moreover, such a representation with the corresponding relational quantale has the finite base, if the original algebra is finite. The base set of the quantale $\widehat{\mathcal{Q}_{\mathcal{A}}}$ is the set of Galois stable subsets of $\mathcal{A}$, which is finite.

4 Join Semilattice-Ordered Semigroups: The Explicit Axiomatisation

We note that a similar construction does not work for finite representable upper semilattice-ordered semigroups. From the one hand, the notions of a finite upper semilattice-ordered semigroup and finite quantale are quite close to each other. From the other hand, the relational representation of quantales does not have to represent joins as set-theoretic unions generally. Moreover, there is a countable sequence of non-representable upper semilattice-ordered semigroups with a non-representable ultraproduct, see [2, Theorem 3.1]. Thus, $\mathbf{R}(;,+)$ is not finitely axiomatisable. Although, as we will see below, this class has a universal recursively enumerable axiomatisation. For that, we characterise representability using representability games on networks. The construction is somewhat similar to the proof of [10, Proposition 5].

Definition 10. *Let* $\mathcal{A}$ *be a join-semilattice ordered semigroup. A* prenetwork *over* $\mathcal{A}$ *is a tuple* (V, E, l)*, where* V *is a set of vertices,* E *is a set of edges such that* $\langle V, E\rangle$ *is a directed graph, and* l *is a labelling function* $l : E \to \mathrm{Up}(\mathcal{A})$*.*

A prenetwork over $\mathcal{A} = (V, E, l)$ *is a* network *if the following hold:*

1. **(Saturation condition)** *For all* $u, v \in V$ *and for all* $x, y, z \in \mathcal{A}$*,* $z \in l(u, v)$ *and* $z \leq x \,;\, y$ *implies* $x \in l(u, w)$ *and* $y \in l(w, v)$ *for some* $w \in V$*.*
2. **(Coherence condition)** *For all* $u, v, w \in V$*, one has* $l(u, v); l(v, w) \subseteq l(u, w)$*.*
3. **(Join-primeness)** *For all* $u, v \in V$*,* $l(u, v)$ *is join-prime. That is, for all* $a, b \in \mathcal{A}$ *if* $a + b \in l(u, v)$*, then either* $a \in l(u, v)$ *or* $b \in l(u, v)$*.*

If $\mathcal{N}$ is a prenetwork, then we will denote its sets of nodes as $\mathrm{Nodes}(\mathcal{N})$ occasionally.

Let I be a non-empty index set and let $\{\mathcal{N}_i\}_{i \in I}$ be an indexed set of prenetworks (where each $\mathcal{N}_i = (V_i, E_i, l_i)$), then $\mathcal{N} = \bigcup\limits_{i \in I} \mathcal{N}_i$ defined as (V, E, l), where

1. $V = \bigcup_{i \in I} V_i$ and $E = \bigcup_{i \in I} E_i$.
2. $l(x, y) = \bigcup_{i \in I} l_i(x, y)$ for all $x, y \in V$.

Definition 11. *Let $n \leq \omega$ and $\mathcal{A}$ a join semilattice-ordered semigroup. A play of the game $\mathcal{G}_n(\mathcal{A})$ has n rounds and consists of n prenetworks. As usual, we have two players, $\forall$ (Abelard, he/his) and $\exists$ (Héloïse, she/her).*

1. *Round 0: $\forall$ picks $a, b \in \mathcal{A}$ such that $a \not\leq b$. $\exists$ responds with a prenetwork $\mathcal{N}_0 = (V_0 = \{x_0, x_1\}, E_0 = \{(x_0, x_1)\}, l_0)$ such that $l_0(x_0, x_1) = \uparrow a$.*
2. *Round $n+1$. Suppose, the prenetwork $\mathcal{N}_n = (V_n, E_n, l_n)$ has been played. $\forall$ has the following three options:*
   (a) **(Composition move)**: *$\forall$ picks $x, y, z \in V_n$ with $b \in l_n(x, y)$ and $c \in l_n(y, z)$. We denote such a move as $N(x, y, z, b, c)$. Then $\exists$ responds with $\mathcal{N}_{n+1} = (V_{n+1}, E_{n+1}, l_{n+1})$ such that $\mathcal{N}_{n+1}$ is the same as $\mathcal{N}_n$, but $l_{n+1}(x, z) = \uparrow (l_n(x, z) \cup \{b \,;\, c\})$.*
   (b) **(Witness move)**:
   $\forall$ picks an edge $(x, y) \in E_n$ and $d, e \in \mathcal{A}$ such that $c \leq d; e$ for $c \in l_n(x, y)$. $\exists$ has to find a witness. She has to find a z which is either a fresh node or an old one. If z is fresh, then she defines the prenetwork T, the edges of which are x, y, z with labelling:
      i. *$l_T(x, z) = \uparrow d$*
      ii. *$l_T(z, y) = \uparrow e$*
   If z is already an element of $\mathcal{A}$, then her response is similar. For her response, $\exists$ plays $\mathcal{N}_{n+1} = \mathcal{N}_n \cup T$.
   (c) **(Join move)**:
   $\forall$ picks an edge $(x, y) \in E_n$ and $c + d$ for $c, d \in \mathcal{A}$. $\exists$ has the following two alternatives for her response:
      i. *$\exists$ chooses c and responds with the prenetwork $\mathcal{N}_{n+1} = \langle V_{n+1}, E_{n+1}, l_{n+1} \rangle$, where $l_{n+1}(x, y) = \uparrow (l_n(x, y) \cup \{c\})$.*
      ii. *$\exists$ chooses b. The response is similar but $l_{n+1}(x, y) = \uparrow (l_n(x, y) \cup \{d\})$.*

$\forall$ wins the play if $b \notin l_{\mathcal{N}_i}(x, y)$ for some $i < n$. Otherwise, $\exists$ wins the play.

Let $a \in \mathcal{A}$ and $\mathcal{N}$ a network, define a game $\mathcal{G}(\mathcal{N}, \mathcal{A}, a)$ such that $\forall$ picks a in the initial round and $\mathcal{N}_0 = \mathcal{N}$. The rules of the game are the same as previously.

Lemma 2. *Let $\mathcal{A} = \langle A, ;, + \rangle$ be a join semilattice-ordered semigroup,*

1. *If $\mathcal{A}$ is representable then $\exists$ has a winning strategy in $\mathcal{G}_\omega(\mathcal{A})$.*
2. *If $|\mathcal{A}| \leq \omega$ and $\exists$ has a winning strategy in $\mathcal{G}_\omega(\mathcal{A})$ then $\mathcal{A}$ is representable.*

Proof.

1. Let $h : \mathcal{A} \to 2^{D \times D}$ be a representation of some base set $D \neq \emptyset$. $\exists$ maintains a map $'$: Nodes($\mathcal{N}$) $\to D$, where $\mathcal{N}$ is a network being played, such that $a \in l_{\mathcal{N}}(x, y)$ implies $(x', y') \in h(a)$.

2. Given $a \in \mathcal{A}$, we consider a play of the game where $\forall$ picks a and b with $a \nleq b$ in the initial round and plays $(\mathcal{N}, x, y, z, c, d)$ in the further rounds for all $x, y, z \in \text{Nodes}(\mathcal{N})$ and $c, d \in \mathcal{A}$. Here, $c \in l_N(x, y)$ and $d \in l_N(y, z)$.
$\forall$ also plays all rounds $(\mathcal{N}, x, y, c, d)$ for all $x, y \in \text{Nodes}(\mathcal{N})$ and $c, d \in \mathcal{A}$ such that there is $e \in \mathcal{A}$ such that $e \leq c; d$ and $e \in l_N(x, y)$.
$\forall$ picks also $c + d$ and vertices $x, y \in \text{Nodes}(\mathcal{N})$ for $c, d \in \mathcal{A}$.
Note that $\mathcal{A}$ is at most countable, so we can schedule all these moves. We have the following play of a game where Héloïse uses a winning strategy:

$$\mathcal{N}_0 \subseteq \mathcal{N}_1 \subseteq \mathcal{N}_2 \ldots$$

Let us put $\mathcal{N}^*(a, b) = \bigcup\limits_{i<\omega} \mathcal{N}_i$. $\mathcal{N}^*(a, b)$ is clearly a network. Let us put the following network assuming that $\mathcal{N}^*(a_1, a_2)$ and $\mathcal{N}^*(b_1, b_2)$ are disjoint for $a_1 \neq a_2$ and $b_1 \neq b_2$:

$$\mathcal{N} = \bigcup_{a,b \in \mathcal{A}, a \nleq b} \mathcal{N}^*(a, b)$$

Note that $\mathcal{N} = \langle V, E, l \rangle$ is a well-defined network since it is the disjoint union of networks.
Define $\text{rep} : \mathcal{A} \to E$ as:

$$\text{rep}(a) = \{(x, y) \mid \exists b \leq a \;\; b \in l_{\mathcal{N}}(x, y)\}$$

Let us check that rep is a representation. Let us show that $\text{rep}(a + b) = \text{rep}(a) \cup \text{rep}(b)$ Suppose $(x, y) \in \text{rep}(a + b)$. That is, there exists $c \leq a + b$ with $c \in l_{\mathcal{N}}(x, y)$, so does $a + b$ since $l_{\mathcal{N}}$ is an upper cone. $a + b \in l_{\mathcal{N}}(x, y)$, that is,

$$a + b \in \bigcup_{\substack{c_1, c_2 \in \mathcal{A} \\ c_1 \nleq c_2}} l_{\mathcal{N}^*(c_1, c_2)}(x, y)$$

That is, there is $c \in \mathcal{A}$ with such that $a + b \in l_{\mathcal{N}^*(c_1,c_2)}(x, y)$, but $l_{\mathcal{N}^*(c_1,c_2)}(x, y)$ is join-prime, so we have either $a \in l_{\mathcal{N}^*(c_1,c_2)}(x, y)$ or $b \in l_{\mathcal{N}^*(c_1,c_2)}(x, y)$. Thus, $\text{rep}(a + b) \subseteq \text{rep}(a) \cup \text{rep}(b)$.
Suppose for the converse, $(x, y) \in \text{rep}(a)$. We need $(x, y) \in \text{rep}(a + b)$. In other words, we have some $c \in \mathcal{A}$ with $c \leq a$ and $c \in l_{\mathcal{N}}(x, y)$. We have $c \leq a \leq a + b$, so $(x, y) \in \text{rep}(a + b)$.
Let us show that $\text{rep}(a; b) = \text{rep}(a); \text{rep}(b)$.
Suppose $(x, y) \in \text{rep}(a; b)$. We need some z with $(x, z) \in \text{rep}(a)$ and $(z, y) \in \text{rep}(b)$. There is $c \leq a; b$ with $c \in l_{\mathcal{N}}(x, y)$. That is, there are $a_1, a_0 \in \mathcal{A}$ and $\mathcal{N}_i$ such that $c \in l_{\mathcal{N}_i}(x, y)$ where $\forall$ plays (a_1, a_0) for the initial round. By the condition, $\forall$ makes the witness moves and $\exists$ responds with a witness. Her response is a node z such that $l_{\mathcal{N}_{i+1}}(x, z) = \uparrow (l_{\mathcal{N}_i}(x, z) \cup \{a\})$ and $l_{\mathcal{N}_{i+1}}(z, y) = \uparrow (l_{\mathcal{N}_i}(z, y) \cup \{b\})$. The inclusion $\text{rep}(a; b) \subseteq \text{rep}(a); \text{rep}(b)$ holds since all witness moves have been played.

Suppose $(x, y) \in \text{rep}(a); \text{rep}(b)$. We need $(x, y) \in \text{rep}(a; b)$. There exists $z \in \text{Nodes}(\mathcal{N})$ with $(x, z) \in \text{rep}(a)$ and $(z, y) \in \text{rep}(b)$. So, there are c, d such that $c \leq a$ with $c \in l_{\mathcal{N}}(x, z)$ and $d \leq b$ with $d \in l_{\mathcal{N}}(z, y)$. We also know that $l_{\mathcal{N}}(x, z); l_{\mathcal{N}}(z, y) \subseteq l_{\mathcal{N}}(x, y)$ because all composition moves have been played. So $c; d \in l_{\mathcal{N}}(x, y)$. That makes $(x, y) \in \text{rep}(a; b)$ since $c; d \leq a; b$.
For injectivity, suppose $a \leq b$ and $(x, y) \in \text{rep}(a)$, that is, there is $c \leq a$ such that $c \in l_{\mathcal{N}}(x, y)$, but $c \leq a \leq b$, so $(x, y) \in \text{rep}(b)$.
Suppose $a \not\leq b$, then there are $x, y \in \text{Nodes}(\mathcal{N}(a, b))$ such that $a \in l_{\mathcal{N}}(x, y)$ and $b \notin l_{\mathcal{N}}(x, y)$. These elements are x_0, x_1 that $\exists$ picks as her response in the zero round. $\exists$ has a winning strategy, so $b \notin l(x_0, x_1)$, but $(x, y) \in \text{rep}(a)$, but $(x, y) \notin \text{rep}(b)$.

The following proposition is a version of [13, Proposition 7.24] and the right-to-left part is proved using König's lemma [17, Exercise 5.6.5].

Proposition 3. *Let $\mathcal{A}$ be a join semilattice-ordered semigroup and $\mathcal{N}$ a network, iff $\exists$ has a winning strategy in $\mathcal{G}_n(\mathcal{A}, \mathcal{N})$ for all $n < \omega$ iff she has a winning strategy in $\mathcal{G}_\omega(\mathcal{A}, \mathcal{N})$.*

Our purpose is to axiomatise of $\mathbf{R}(;, +)$ with a recursively enumerable set of universal formulas. See [13, Chapter 9] for the discussion in detail to have a more general methodology.

Definition 12. *Let* Var $= \{v_0, v_1, \dots\}$ *be a set of variables. The set of terms is generated by the following grammar:*

$$t_1, t_2 ::= v \mid (t_1 + t_2) \mid (t_1; t_2)$$

Definition 13. *A* term network *is a finite network $\langle V, E, l \rangle$, where $\langle V, E \rangle$ is a directed graph and $l : E \to 2^{Term}$ is a labelling function such that every $l(x, y)$ is finite for all $(x, y) \in E$.*

Let $\mathcal{A}$ be a join semilattice-ordered semigroup and $\vartheta : \text{Var} \to \mathcal{A}$ a valuation. The value of complex terms is defined inductively for $a, b \in T$:

1. $(a; b)^\vartheta = a^\vartheta; b^\vartheta$
2. $(a + b)^\vartheta = a^\vartheta + b^\vartheta$

Let $\mathcal{N} = \langle V, E, l \rangle$ be a term network, $\mathcal{A}$ be a join-semilattice ordered semigroup and $\vartheta : \text{Var} \to \mathcal{A}$ a valuation. Let us define the prenetwork $\mathcal{N}^\vartheta$ with the same edges and vertices with labelling $l^\vartheta(x, y) = \uparrow \vartheta[l_{\mathcal{N}}(x, y)]$. We define the following three extensions of $\mathcal{N}$ reflecting the composition, witness, and join moves respectively:

1. Let $x, y \in \text{Nodes}(\mathcal{N})$ and let t be a term. $\mathcal{N}_c$ is the extension of $\mathcal{N}$, where $\text{Nodes}(\mathcal{N}_c) = \text{Nodes}(\mathcal{N})$ and $l_{\mathcal{N}_c}(x, y) = l_{\mathcal{N}}(x, y) \cup \{t\}$ and $l_{\mathcal{N}_c}(u, v) = l_{\mathcal{N}}(u, v)$ for all $u \neq x$ and $v \neq y$. We denote this network as $\mathcal{N}_c(\mathcal{N}, x, y, t)$.

2. Let $x, y \in \text{Nodes}(\mathcal{N})$, let z be a node (regardless of whether z is fresh or not), and t_1, t_2 any terms. Let us define a network T such that $\text{Nodes}(T) = \{x, y, z\}$. We define labelling as $l_T(x, y) = \{t_1\}$ and $l_T(y, z) = \{t_2\}$. So we put $\mathcal{N}_w = \mathcal{N} \cup T$. We denote this network as $\mathcal{N}_w(\mathcal{N}, x, y, z, t_1, t_2)$.
3. Let $x, y \in \text{Nodes}(\mathcal{N})$ and let t_1, t_2 be terms. We define $T_i = \langle \{x, y\}, \{(x, y)\}, l_{T_i} \rangle$, where $l_{T_i}(x, y) = l_{\mathcal{N}}(x, y) \cup \{t_i\}$ for $i = 1, 2$. So $\mathcal{N}_{j_1} = \mathcal{N} \cup T_1$ and $\mathcal{N}_{j_2} = \mathcal{N} \cup T_2$.

Lemma 3. *For all $n < \omega$ there exists a first-order sentence ρ_n such that $\exists$ has a winning strategy in $\mathcal{G}_n(\mathcal{A})$ iff $\mathcal{A} \models \rho_n$.*

Proof. As usual, for each $n < \omega$ we construct a formula σ_n claiming that $\exists$ has a winning strategy in the game of lenght n. To be more precise, our purpose is to have

$$\exists \text{ has a winning strategy in } \mathcal{G}_n(\mathcal{N}^{\vartheta}, \mathcal{A}, \vartheta(v)) \text{ if and only if } \mathcal{A} \models \sigma_n(\mathcal{N}, v)$$

where $\mathcal{A}$ is a join semilattice-ordered semigroup, $\vartheta : \text{Var} \to \mathcal{A}$ is a variable assignment, and $\mathcal{N}$ is a term network.

We define the following sequence of formulas $\{\sigma_n\}_{n<\omega}$ inductively:

1. $\sigma_0(\mathcal{N}, v) = \bigwedge\limits_{a \in l_{\mathcal{N}}(x,y)} \neg(a \leq v)$
   $\sigma_0(\mathcal{N}, v)$ merely claims that $\exists$ has a winning strategy in the zero length game.
2. Suppose $\sigma_n(\mathcal{N}, v)$ are already constructed for some $n < \omega$. Let us define a formula σ_{n+1} claiming that $\exists$ always has a proper response for a network $\mathcal{N}$ being played.
   $\sigma_{n+1}(\mathcal{N}, v)$ is defined as follows:

$$\sigma_{n+1}(\mathcal{N}, v) = \sigma_{n+1_c}(\mathcal{N}, v) \wedge \sigma_{n+1_w}(\mathcal{N}, v) \wedge \sigma_{n+1_j}(\mathcal{N}, v)$$

   where
   - $\sigma_{n+1_c}(\mathcal{N}, v) = \bigwedge\limits_{\substack{x,y,z \in \text{Nodes}(\mathcal{N}) \\ t_1 \in l_{\mathcal{N}}(x,y) \\ t_2 \in l_{\mathcal{N}}(y,z)}} \sigma_n(\mathcal{N}_c(x, z, t_1, t_2), v)$
   - $\sigma_{n+1_w}(\mathcal{N}, v) = \bigwedge\limits_{\substack{x,y \in \text{Nodes}(\mathcal{N}) \\ t \in l_{\mathcal{N}}(x,y)}} \forall u_1, u_2 (t \leq u_1; u_2 \to \bigvee\limits_{w \in \text{Nodes}(\mathcal{N}) \cup \{z\}} \mathcal{N}_c(x, y, w, u_1, u_2))$, where $z \notin \text{Nodes}(\mathcal{N})$.
   - $\sigma_{n+1_j}(\mathcal{N}, v) = \forall a \, \forall b (v = a + b \to \bigwedge\limits_{x,y \in \text{Nodes}(\mathcal{N})} \sigma_n(\mathcal{N}_{j_1}(\mathcal{N}, x, y, a), v) \vee \sigma_n(\mathcal{N}_{j_2}(\mathcal{N}, x, y, b), v))$

So, $\exists$ has a winning strategy iff these formulas are true under the valuation ϑ since the formulas $\{\sigma_n\}_{n<\omega}$ encode the presence of a winning strategy for $\exists$ in every finite round.

Let v_0 be any variable, $\mathcal{N}_{v_0}$ denotes the term network having the form $\langle \{\{x_0, x_1\}, \{(x_0, x_1)\}, l\} \rangle$, where $l(x, y) = \{v_0\}$. We define the following sequence of formulas $(\rho_n)_{n<\omega}$:

$$\rho_n = \forall v_0 \forall v_1 (\neg(v_0 \leq v_1) \to \sigma(\mathcal{N}_{v_0}, v_0))$$

This inductive sequence of formulas provides us the explicit axiomatisation of the class of representable join semilattice-ordered semigroups.

Theorem 5. *A join semilattice-ordered semigroup $\mathcal{A}$ is representable iff $\mathcal{A} \models \{\rho_n\}_{n<\omega}$. Moreover, $\mathbf{R}(;,+)$ has a recursively enumerable universal axiomatisation.*

Proof. Let us define a two sorted language with sorts **a** (algebra) and **r** (representation). $\mathbf{R}(;,+)$ clearly forms a pseudo-elementary class, see [14, Introduction] for more details. Moreover, this class is pseudo-universal and it satisfies the condition of the second item of Theorem 1.

By Proposition 3, Lemma 2, and Lemma 3, a countable join semilattice-ordered semigroup $\mathcal{A}$ is representable iff $A \models \{\rho_n\}_{n<\omega}$. Suppose $\mathcal{A}$ is uncountable. The class is pseudo-elementary, so it is closed under elementary equivalence, so, by the downward Löwenheim-Skolem theorem [17, Corollary 3.1.5], we can take $\mathcal{A}_0 \preceq \mathcal{A}$, a countable elementary substructure of $\mathcal{A}$. Then $\mathcal{A}_0 \models \{\rho_n\}_{n<\omega}$ iff $\mathcal{A} \models \{\rho_n\}_{n<\omega}$. Therefore, if $\mathcal{A}_0$ is representable, so is $\mathcal{A}$.

As we have already discussed, the finite representation property for $(;,+)$-structures remains an open question. If the solution is positive, then the problem of representability for finite join semilattice-ordered semigroups is decidable since finite representability and recursive axiomatisability imply decidability.

Acknowledgements. The author would like to thank Robin Hirsch, Ian Hodkinson, Stepan Kuznetsov, Jaš Šemrl, Valentin Shehtman, and his supervisor Ilya Shapirovsky for valuable comments. The author is also grateful to the reviewers whose comments improved the original version of the paper.

References

1. Andréka, H., Mikulás, S.: Lambek calculus and its relational semantics: completeness and incompleteness. J. Logic Lang. Inform. **3**(1), 1–37 (1994)
2. Andréka, H., Mikulás, S.: Axiomatizability of positive algebras of binary relations. Algebra Univers. **66**(1–2), 7 (2011)
3. Brown, C., Gurr, D.: A representation theorem for quantales. J. Pure Appl. Algebra **85**(1), 27–42 (1993)
4. Davey, B.A., Priestley, H.A.: Introduction to Lattices and Order. Cambridge University Press, Cambridge (2002)
5. Eklof, P.C.: Ultraproducts for algebraists. In: Studies in Logic and the Foundations of Mathematics, vol. 90, pp. 105–137. Elsevier (1977)
6. Galatos, N., Jipsen, P., Kowalski, T., Ono, H.: Residuated Lattices: an Algebraic Glimpse at Substructural Logics. Elsevier, Amsterdam (2007)
7. Goldblatt, R.: A Kripke-Joyal semantics for noncommutative logic in quantales. Adv. Modal Logic **6**, 209–225 (2006)
8. Hirsch, R., Semrl, J.: Demonic lattices and semilattices in relational semigroups with ordinary composition. In: 2021 36th Annual ACM/IEEE Symposium on Logic in Computer Science (LICS), pp. 1–10. IEEE Computer Society, Los Alamitos, CA, USA (July 2021). https://doi.org/10.1109/LICS52264.2021.9470509, https://doi.ieeecomputersociety.org/10.1109/LICS52264.2021.9470509

9. Hirsch, R.: The finite representation property for reducts of relation algebra. Manuscript, September (2004)
10. Hirsch, R.: The class of representable ordered monoids has a recursively enumerable, universal axiomatisation but it is not finitely axiomatisable. Logic J. IGPL **13**(2), 159–171 (2005)
11. Hirsch, R., Hodkinson, I.: Step by step-building representations in algebraic logic. J. Symbolic Logic **62**, 225–279 (1997)
12. Hirsch, R., Hodkinson, I.: Representability is not decidable for finite relation algebras. Trans. Am. Math. Soc. **353**(4), 1403–1425 (2001)
13. Hirsch, R., Hodkinson, I.: Relation Algebras by Games. Elsevier, Amsterdam (2002)
14. Hirsch, R., Mikulás, S.: Representable semilattice-ordered monoids. Algebra Univers. **57**(3), 333–370 (2007)
15. Hirsch, R., Mikulás, S.: Ordered domain algebras. J. Appl. Logic **11**(3), 266–271 (2013)
16. Hirsch, R., Šemrl, J.: Finite representability of semigroups with demonic refinement. Algebra Univers. **82**(2), 1–14 (2021). https://doi.org/10.1007/s00012-021-00718-5
17. Hodges, W.: Model Theory. Cambridge University Press, Cambridge (1993)
18. Hodkinson, I., Mikulás, S.: Axiomatizability of reducts of algebras of relations. Algebra Univers. **43**(2–3), 127–156 (2000)
19. Hodkinson, I., Venema, Y.: Canonical varieties with no canonical axiomatisation. Trans. Am. Math. Soc. **357**(11), 4579–4605 (2005)
20. Jönsson, B., Tarski, A.: Boolean algebras with operators, i, ii. Am. J. Math. **73**, 891–939 (1951)
21. Lambek, J.: The mathematics of sentence structure. Am. Math. Mon. **65**(3), 154–170 (1958)
22. Lyndon, R.C.: The representation of relational algebras. Ann. Math. **51**, 707–729 (1950)
23. McLean, B., Mikulás, S.: The finite representation property for composition, intersection, domain and range. Int. J. Algebra Comput. **26**(06), 1199–1215 (2016)
24. Monk, D.: On representable relation algebras. Mich. Math. J. **11**(3), 207–210 (1964)
25. Mulvey, C.J.: & suppl. Rend. Circ. Mat. Palermo II **12**, 99–104 (1986)
26. Rosenthal, K.I.: Quantales and their applications, vol. 234, Longman Scientific and Technical (1990)
27. Šemrl, J.: Domain range semigroups and finite representations. In: Fahrenberg, U., Gehrke, M., Santocanale, L., Winter, M. (eds.) RAMiCS 2021. LNCS, vol. 13027, pp. 483–498. Springer, Cham (2021). https://doi.org/10.1007/978-3-030-88701-8_29
28. Theunissen, M., Venema, Y.: MacNeille completions of lattice expansions. Algebra Univers. **57**(2), 143–193 (2007)
29. Venema, Y.: Atom structures and Sahlqvist equations. Algebra Univers. **38**(2), 185–199 (1997)
30. Zaretskii, K.: The representation of ordered semigroups by binary relations. Izvestiya Vysshikh Uchebnykh Zavedenii. Matematika **6**, 48–50 (1959)

Between Turing and Kleene

Sam Sanders(✉)

Department of Philosophy II, RUB Bochum,
Universitätsstrasse 150, 44780 Bochum, Germany
sam.sanders@rub.de

Abstract. Turing's famous 'machine' model constitutes the first intuitively convincing framework for *computing with real numbers*. Kleene's computation schemes S1–S9 extend Turing's approach to *computing with objects of any finite type*. Both frameworks have their pros and cons and it is a natural question if there is an approach that marries the best of both the Turing and Kleene worlds. In answer to this question, we propose a considerable extension of the scope of Turing's approach. Central is a fragment of the Axiom of Choice involving *continuous* choice functions, going back to Kreisel-Troelstra and intuitionistic analysis. Put another way, we formulate a relation 'is computationally stronger than' involving **third-order** objects that overcomes (many of) the pitfalls of the Turing and Kleene frameworks.

Keywords: Computability theory · Kleene S1–S9 · Turing machines

1 Between Turing and Kleene Computability

1.1 Short Summary

In a nutshell, we propose a sizable extension of the scope of Turing's 'machine' model of computation [50], motivated by a fragment of the Axiom of Choice involving *continuous* choice functions, going back to Kreisel-Troelstra and intuitionistic analysis [21]. In particular, we formulate a relation 'is computationally stronger than' involving **third-order** objects *but* still based on Turing computability by and large.

The interested reader will find the aforementioned extension discussed in more detail in Sect. 1.2, along with a critical discussion of the scope of our extension. The critical reader will learn about the pressing need for the aforementioned extension in Sect. 1.3. In particular, the latter section seeks to alleviate worries that existing frameworks are somehow sufficient for our (foundational) needs. The (problems involving the) representation of third-order objects via second-order ones is a particularly important 'case in point'.

Next, some elegant results in our proposed extension are listed in Sect. 2 pertaining to the following topics:

- convergence theorems for nets in the unit interval (Sect. 2.1.2),

S. Artemov and A. Nerode (Eds.): LFCS 2022, LNCS 13137, pp. 281–300, 2022.
https://doi.org/10.1007/978-3-030-93100-1_18

- covering theorems for the unit interval $\mathbb{R}$ (Sect. 2.1.3),
- the uncountability of the real numbers $\mathbb{R}$ (Sect. 2.2),
- discontinuous functions on the real numbers $\mathbb{R}$ (Sect. 2.3).

We note that all our results are part of classical mathematics, while we have found constructive mathematics highly inspiring on our journey towards this paper. We will assume familiarity with Turing-style computability theory [44] and higher-order primitive recursion like in Gödel's system T [23, p. 74]; knowledge of Kleene's higher-order computability theory, in particular the computation schemes S1–S9 (see [17,23]), is useful but not essential.

Finally, we will discuss a number of theorems of real analysis and the following remark discusses how the representations of real numbers can be done in a straightforward and non-intrusive way.

Remark 1 (Representation of real numbers). Kohlenbach's 'hat function' from [19, p. 289] guarantees that every element of $\mathbb{N}^\mathbb{N}$ defines a real number via the well-known representation of reals as fast-converging Cauchy sequences. Despite the definition of the latter being Π_1^0, a quantifier '$(\forall x \in \mathbb{R})$' amounts to a quantifier over $\mathbb{N}^\mathbb{N}$.

Moreover, Kohlenbach's 'tilde' function from [20, Def. 4.24] guarantees that '$(\forall x \in [0,1])$' also just amounts to a quantifier over $\mathbb{N}^\mathbb{N}$, despite $0 \leq_\mathbb{R} x \leq_\mathbb{R} 1$ being Π_1^0 (in addition). These functions ensure a smooth treatment of $\mathbb{R}$, $[0,1]$, and $2^\mathbb{N}$ and functions between such spaces. We will **always** assume that real numbers and $\mathbb{R} \to \mathbb{R}$-functions are given in this way, i.e. as in the aforementioned references [19,20], so as to ensure a smooth treatment.

1.2 Extending the Scope of Turing Computability

In this section, we discuss the extension of Turing computability mentioned in Sect. 1.1. In particular, we introduce this new concept in Sect. 1.2.1 and discuss its scope in Sect. 1.2.2. The reader will have a basic understanding of Turing computability theory [44] and higher-order primitive recursion like Gödel's system T [23, p. 74].

1.2.1 A New Notion of Reduction

In this section, we formulate (4), which is a relation formalising 'is computationally stronger than' involving **third-order** objects *but* still based on Turing computability. We first need some preliminaries, starting with (1).

First of all, many theorems in e.g. analysis can be given the form

$$(\forall Y : \mathbb{N}^\mathbb{N} \to \mathbb{N})(\exists x \in \mathbb{N}^\mathbb{N})A(Y, x), \tag{1}$$

where $\mathbb{N}^\mathbb{N}$ is the Baire space and $\mathbb{N}$ is the set of natural numbers. Indeed, as discussed in Remark 1, some basic primitive recursive operations relegate the coding of real numbers (via elements of $\mathbb{N}^\mathbb{N}$) to the background. Moreover, a list of theorems that can be brought in the form (1) can be found in Example 2 below, while we discuss the scope of theorems that can be brought in this form at the end of this section and in Sect. 1.2.2.

Secondly, to improve readability, one often uses type theoretic notation in (1), i.e. n^0 for type 0 objects $n \in \mathbb{N}$, x^1 for type 1 objects $x \in \mathbb{N}^{\mathbb{N}}$, and Y^2 for type 2 objects $Y : \mathbb{N}^{\mathbb{N}} \to \mathbb{N}$. We will only occasionally need type 3 objects, which map type 2 objects to natural numbers. We generally use Greek capitals $\Theta^3, \Lambda^3, \ldots$ for such objects.

Thirdly, to compare the logical strength of theorems of the form (1), one establishes results of the following form over weak systems:

$$(\forall Y^2)(\exists x^1)A(Y, x) \to (\forall Z^2)(\exists y^1)B(Z, x), \tag{2}$$

as part of Kohlenbach's *higher-order Reverse Mathematics* (see [19] for an introduction). The computational properties of (1) and (2) following S1–S9 can then be studied as follows: let Θ^3 and Λ^3 be *realisers* for the antecedent and consequent of (2) i.e. $(\forall Y^2)A(Y, \Theta(Y))$ and $(\forall Z^2)B(Z, \Lambda(Z))$.

A central computability theoretic question concerning (2) is whether a realiser Θ^3 for the antecedent of (2) computes, in the sense of S1–S9, a realiser Λ^3 for the consequent of (2), i.e. whether there is a Kleene algorithm with index $e \in \mathbb{N}$ satisfying the following:

$$(\forall \Theta^3)\big[(\forall Y^2)A(Y, \Theta(Y)) \to (\forall Z^2)B(Z, \{e\}(\Theta, Z))\big]. \tag{3}$$

Next, we list some theorems that have been studied via the above paradigm based on (3) and S1–S9.

Example 2 (Some representative theorems)

- The Lindelöf, Heine-Borel, and Vitali covering theorems involving uncountable coverings [28,31,34].
- The Lebesgue number lemma [33,34].
- The Baire category theorem [33].
- Convergence theorems for nets [38,39,41].
- Local-global principles like Pincherle's theorem [34].
- The uncountability of $\mathbb{R}$ and the Bolzano-Weierstrass theorem for countable sets in Cantor space [32,37].
- Weak fragments of the Axiom of (countable) Choice [35].
- Basic properties of functions of bounded variation, like the Jordan decomposition theorem [36].

Many more theorems are equivalent -in the sense of higher-order RM as in [19] - to the theorems in the above list, as can be found in the associated references.

Fourth, for all the reasons discussed in Sect. 1.3, we formulate a version of (3) based on Turing computability as follows:

$$(\forall Z^2, x^1)\big[A(t(Z), x) \to [\{e\}^{s(Z,x)}\downarrow \wedge B(Z, \{e\}^{s(Z,x)})]\big], \tag{4}$$

where $s^{2\to 1}, t^{2\to 2}$ are terms of Gödel's T and '$\{e\}^X$' is the e-th Turing machine with oracle $X \subset \mathbb{N}$. We note that (4) readily[1] implies (3); we discuss the generality of (4) at the end of this section.

[1] For $e \in \mathbb{N}$ and $s^{2\to 1}, t^{2\to 2}$ as in (4), define $e_0 \in \mathbb{N}$ as the Kleene algorithm such that $\{e_0\}(\Theta, Z) := \{e\}^{s(Z,\Theta(t(Z)))}$, which is total by assumption.

In line with the nomenclature of computability theory, we call the antecedent and consequent of (2) 'problems' and say that

solving the problem $(\forall Z^2)(\exists y^1)B(Z,x)$ ***N*-reduces** *to solving the problem* $(\forall Y^2)(\exists x^1)A(Y,x)$

in case (4) holds for the parameters mentioned. We view the N-reduction relation as 'neutral' between the Turing and Kleene framework and the reader readily verifies that N-reduction is transitive. In case the term $s(Z,x)$ can be replaced by a term $u(x)$, i.e. the latter has no access to Z, we refer to (4) as **strong** N-reduction.

Finally, the critical reader may wonder about the generality of (4). The latter is quite general, for the following two reasons.

- It is an empirical observation based on [28–37] that *positive* results in S1–S9 computability theory can be witnessed by terms of Gödel's T of low complexity. In this light, there is no real loss of generality if we use terms of Gödel's T as in (4).
- A theorem of (third-order) ordinary mathematics generally has the form (1), *unless* the former implies the existence of a discontinuous function on $\mathbb{R}$. In the latter case, an 'indirect' treatment is still possible via the so-called *Grilliot's trick*, which we sketch in Sect. 2.3.1.

Like the reader, we feel that the second item deserves a more detailed explanation, which is in Sect. 1.2.2. Regarding the first item, intellectual honesty compels us to admit that many of our S1–S9 results are witnessed by terms of Gödel's T *additionally involving* Feferman's search operator (already found in Hilbert-Bernays [15]) defined for any f^1 as:

$$\mu(f) := \begin{cases} \text{the least } n^0 \text{ such that } f(n) = 0 & (\exists m^0)(f(m) = 0) \\ 0 & \text{otherwise} \end{cases}. \tag{5}$$

While not strictly necessary always, it is convenient to have access to μ^2 as we then do not have to worry how spaces like $[0,1]$ or $2^{\mathbb{N}}$ are represented. Based on this observation, we introduce the following:

solving the problem $(\forall Z^2)(\exists y^1)B(Z,x)$ **μ*N*-reduces** *to solving the problem* $(\forall Y^2)(\exists x^1)A(Y,x)$

in case (4) holds for the parameters mentioned except that $t(Z)$ is replaced by $t(Z,\mu^2)$. Then 'strong' μN-reduction is defined similarly.

Finally, one could study (4) for other extensions of Gödel's T, e.g. involving 'minimization' (see [23, §5.1.5]), but (4) seems more salient.

1.2.2 Continuous and Discontinuous Functionals

We discuss the motivation behind our notion of N-reduction and establish its scope. To this end, we have to make the following classical case distinction.

- If a given third-order theorem is **consistent** with Brouwer's *continuity theorem* that all functions on $\mathbb{R}$ are continuous [7], then we can **directly** analyse it via N-reduction.
- If a given third-order theorem implies the existence of a discontinuous function on $\mathbb{R}$, we can **indirectly** analyse it via N-reduction based on Grilliot's trick, where the latter is sketched in Sect. 2.3.1.

To make sense of the above, we first sketch the 'standard' higher-order generalisation of (second-order) comprehension, exemplified by Kleene's $\exists^2$ as in (6). We then discuss another (less famous) formulation of comprehension, called the *neighbourhood function principle* as in Definition 4, a fragment of the Axiom of Choice involving *continuous* choice functions, going back to intuitionistic analysis [21,47].

First of all, the commonplace *one cannot fit a round peg in a square hole* has an obvious counterpart in computability theory: a type 2 functional cannot be the oracle of a Turing machine. Nonetheless, a *continuous* type 2 functional can be *represented* by a type 1 *Kleene associate* as in Definition 3, where we employ the same[2] notations as in [18]. Associates *do* 'fit' as oracles of Turing machines.

Definition 3 (Kleene associate from [18])

- *A function* α^1 *is a* neighbourhood function *if*
  - $(\forall \beta^1)(\exists n^0)(\alpha(\overline{\beta}n) > 0)$ *and*
  - $(\forall \sigma^{0^*}, \tau^{0^*})(\alpha(\sigma) > 0 \to \alpha(\sigma * \tau) = \alpha(\sigma))$.
- *A function* α^1 *is a Kleene* associate *for* Y^2 *if*
  - $(\forall \beta^1)(\exists n^0)(\alpha(\overline{\beta}n) > 0)$ *and*
  - $(\forall \beta^1, n^0)($ *n is least s.t.* $\alpha(\overline{\beta}n) > 0 \to \alpha(\overline{\beta}n) = Y(\beta) + 1)$.

As in [18, §4], we additionally assume that an associate is a neighbourhood function, as the former can readily be converted to the latter.

Hence, we should specify that a **discontinuous** type two functional cannot be the oracle of a Turing machine. Now, the archetypal example of a discontinuous function is Kleene's quantifier $\exists^2$ defined as:

$$(\forall f^1)\big[\, \underline{(\exists n^0)(f(n) = 0)} \;\leftrightarrow \exists^2(f) = 0\big]. \tag{6}$$

Clearly, (6) is the higher-order version of *arithmetical comprehension* (see e.g. [43, III]) stating that $\{n \in \mathbb{N} : A(n)\}$ exists for arithmetical formulas A, which includes the underlined formula in (6). We point out *Grilliot's trick*, a method for (effectively) obtaining $\exists^2$ from discontinuous functionals on e.g. $\mathbb{R}$ or $2^{\mathbb{N}}$, as also discussed in Sect. 2.3.1. To our own surprise, this kind of effective result is essentially the prototype of (4), as discussed in Sect. 2.3.2.

Now, as noted above, Kleene's $\exists^2$ can decide the truth of arithmetical formulas. In general, for a formula class Γ, one can study higher-order functionals

[2] In particular, σ^{0^*} is a finite sequence in $\mathbb{N}$ with length $|\sigma|$ and we assume the well-known coding of such finite sequences by natural numbers. Moreover, $\overline{f}n$ is the finite sequence $(f(0), \dots, f(n-1))$ for any f^1 and n^0, and any f^{0^*} in case $|f| \leq n$.

that decide the truth of formulas $\gamma \in \Gamma$. Examples are Kleene's quantifiers $\exists^n$ [23, Def. 5.4.3] and the Feferman-Sieg functionals ν_n from [8, p. 129], which we shall however not need.

Secondly, we consider the *neighbourhood function principle* NFP from [48], studied in [21,47] under a different name.

Definition 4 [NFP]. *For any formula $A(n^0)$, we have*

$$(\forall f^1)(\exists n^0)A(\overline{f}n) \to (\exists \gamma \in K_0)(\forall f^1)A(\overline{f}\gamma(f)), \tag{7}$$

where '$\gamma \in K_0$' means that γ^1 is a (total) Kleene associate.

Clearly, (7) is a fragment of the Axiom of Choice involving continuous choice functions. Not as obvious is that NFP is a 'more constructive' formulation of the comprehension axiom (see Remark 5 below). We also note that NFP involving third-order parameters has the form (1), namely for a formula $A(n^0, Y^2)$ with all parameters shown, (7) yields

$$(\forall Y^2)(\exists \gamma^1)\big[(\forall f^1)(\exists n^0)A(\overline{f}n, Y) \to [\gamma \in K_0 \wedge (\forall f^1)A(\overline{f}\gamma(f))]\big]. \tag{8}$$

Now, NFP proves the Lindelöf lemma [48], inspiring Theorem 12.

Finally, we can combine the above as follows: the case distinction from the beginning of this section distinguishes between whether a given theorem $\mathfrak{T}$ in the language of third-order arithmetic implies the existence of a discontinuous function (on $\mathbb{R}$ or $2^{\mathbb{N}}$), or not. It is then an empirical observation based on [28–37] that for this theorem $\mathfrak{T}$, **either** the theorem $\mathfrak{T}$ implies the existence of $\exists^2$ via the aforementioned Grilliot's trick, **or** $\mathfrak{T}$ is provable from a fragment of NFP where A may include third-order parameters.

In the former case, the theorem $\mathfrak{T}$ can be analysed 'indirectly' using N-reduction, namely via Grilliot's trick, as discussed in Sect. 2.3. In case the theorem $\mathfrak{T}$ is provable from a fragment NFP (with third-order parameters), we can generally bring $\mathfrak{T}$ in the form (1) and hence analyse it *directly* via N-reduction. In light of (8), fragments of NFP can always be analysed via N-reduction.

In conclusion, second-order comprehension has been generalised to higher types in two (more-or-less-known ways) ways, namely as follows.

- Formulate 'characteristic functionals' like $\exists^2$ from (6) that decide the truth of certain formulas.
- Formulate NFP as in Definition 4 for formulas involving higher-order parameters and variables.

If a given theorem implies the existence of $\exists^2$, we can analyse it 'indirectly' via N-reduction, namely via Grilliot's trick. If a given theorem is provable from NFP involving third-order fragments, we can (readily) analyse it via N-reduction. In other words, if a third-order theorem is **consistent** with Brouwer's *continuity theorem* that all functions on $\mathbb{R}$ are continuous [7], then we can analyse it directly via N-reduction.

Finally, we show that NFP classically follows from comprehension and vice versa, assuming a fragment of the induction axiom.

Remark 5 (NFP and comprehension). To obtain NFP from comprehension modulo coding of finite sequences, let X be such that $\sigma \in X \leftrightarrow A(\sigma^{0^*})$ for any finite sequence σ^{0^*} in $\mathbb{N}$. Then define $\gamma(\sigma) := |\sigma| + 1$ in case $\sigma \in X$, and 0 otherwise. Assuming the antecedent of (7), this yields a (total) Kleene associate. By definition, γ also satisfies the consequent of (7).

To obtain comprehension from NFP, suppose towards a contradiction that comprehension is false, i.e. there is some formula $A(n)$ such that

$$(\forall X \subset \mathbb{N})(\exists n \in \mathbb{N})\big[[n \in X \wedge \neg A(n)] \vee [A(n) \wedge n \notin X]\big]. \tag{9}$$

Now apply NFP to (9) (coding $X \subset \mathbb{N}$ as elements of $2^{\mathbb{N}}$) to obtain $\gamma \in K_0$. The latter has an upper bound $k_0 \in \mathbb{N}$ on $2^{\mathbb{N}}$, i.e. $n \in \mathbb{N}$ in (9) is bounded by k_0. However, the induction axiom readily proves 'finite comprehension' as follows:

$$(\forall k \in \mathbb{N})(\exists X \subset \mathbb{N})(\forall n \leq k)\big[n \in X \leftrightarrow A(n)\big]. \tag{10}$$

Hence, for $k = k_0 + 1$, (10) yields a contradiction.

1.3 The Need for an Extension of Turing Computation

We argue why the extension of Turing computation sketched in Sect. 1.2 is necessary and even most welcome, as follows.

- Higher-order objects are 'coded' as reals so as to accommodate their study via Turing machines. It has recently been established that this 'coding practise' yields very different results compared to Kleene's approach, even for basic objects like *functions of bounded variation* (Sect. 1.3.1)
- The conceptual complexity of Kleene's extension of Turing computability is considerable, while the extension to 'infinite time' Turing machines is too general for our purposes (Sect. 1.3.2).

Put another way, N-reduction is an attempt at formulating a relation 'is computationally stronger than' for third-order statements that overcomes the above pitfalls, namely the conceptual complexity of Kleene's S1–S9 and the problems associated with second-order representations.

1.3.1 Computing with Second-Order Representations

We show that there are huge differences between 'computing with higher-order objects' and 'computing with *representations of higher-order objects*', even for basic objects like functions of bounded variation on $[0, 1]$.

Now, various[3] research programs have been proposed in which higher-order objects are *represented/coded* as real numbers or similar representations, so as

[3] Examples of such frameworks are: reverse mathematics [43,46], constructive analysis [2, I.13], [4], predicative analysis [12], and computable analysis [52]. Bishop's constructive analysis is not based on Turing computability *directly*, but one of its 'intended models' is (constructive) recursive mathematics, as discussed in [6]. One aim of Feferman's predicative analysis is to capture Bishop's approach.

to make them amenable to the Turing framework. It is then a natural question whether there is any significant difference[4] between the Kleene S1–S9 approach or the Turing-approach-via-codes.

Continuous functions being well-studied (see Footnote 4) in this context, Dag Normann and the author have investigated *functions of bounded variation*, which have **at most** countably many points of discontinuity [36]. A central result is the *Jordan decomposition theorem* which implies that $f : [0,1] \to \mathbb{R}$ of bounded variation on $[0,1]$ satisfies $f = g - h$ on $[0,1]$ for monotone $g, h : [0,1] \to \mathbb{R}$. We have the following results.

- In case $f : [0,1] \to \mathbb{R}$ of bounded variation is given via a *second-order* representation, then the monotone $g, h : [0,1] \to \mathbb{R}$ such that $f = g - h$, can be computed from finite iterations of the Turing jump with f as a parameter by [22, Cor. 10].
- A *Jordan realiser* $\mathcal{J}$ takes as input $f : [0,1] \to \mathbb{R}$ of bounded variation and outputs $\mathcal{J}(f) = (g, h)$, i.e. monotone $g, h : [0,1] \to \mathbb{R}$ with $f = g - h$ on $[0,1]$. No Jordan realiser is computable (S1–S9) in any type 2 functional by [37, Theorem 3.9].

Regarding the second item, a Jordan realiser is therefore not computable from (finite iterations of) $\exists^2$, the higher-order counterpart of the Turing jump. The same holds for S^2_k, which is a type two functional that can decide Π^1_k-formulas (involving first- and second-order parameters). The usual proof of the Jordan decomposition theorem implies that Kleene's $\exists^3$ computes a Jordan realiser. But $\exists^3$ implies full second-order arithmetic, and the same holds for the combination of all S^2_k.

In conclusion, there is a **huge** difference in the computational hardness of the Jordan decomposition theorem depending on whether we use representations or not. However, this theorem deals with functions of bounded variation, a class 'very close' to the class of continuous functions. Hence, (Turing) computing with representations, interesting as it may be, is completely different from (Kleene) computing with actual higher-order objects. In this light, there is a clear need for a notion like N-reduction that allows us to compute with actual higher-order objects while staying close to Turing computability.

1.3.2 On Higher-Order Computation

We argue that the conceptual complexity of Kleene's S1–S9 is considerable, while the extension to 'infinite time' Turing machines is too general for our purposes (Sect. 1.3.2).

First of all, as noted above, Turing's famous 'machine' model constitutes the first intuitively convincing framework for *computing with real numbers* [50] while Kleene's S1–S9 extend Turing's approach to *computing with objects of any finite type* [17,23].

[4] The *fan functional* constitutes an early *natural* example of this difference: it has a computable code but is not S1–S9 computable (but S1–S9 computable in Kleene's $\exists^2$ from Sect. 1.2.2). The fan functional computes a modulus of uniform continuity for continuous functions on Cantor space; details may be found in [23].

We have studied or made extensive use of Kleene's S1–S9 computability theory in [28–37]. In our opinion, while vastly more general in scope, Kleene's S1–S9 has the following conceptual drawbacks.

- Turing computability boasts the elementary 'Kleene T-predicate' (see e.g. [44, p. 15]) where $T(e, x, y)$ intuitively expresses that y codes the computation steps of the e-th Turing machine program with input x. There is no such construct for S1–S9.
- Kleene's *recursion theorem* is one of the most elegant and important results in Turing computability [44, p. 36] and is derived from first principles. By contrast, Kleene's schemes S1–S8 formalise higher-order primitive recursion (only), while S9 essentially hard-codes the recursion theorem for S1–S9.
- Natural space and time constraints can be formulated for Turing machines, yielding a canonical complexity theory [42]; to the best of knowledge, no such canonical theory exists for higher-order computation in general or S1–S9 in particular.
- Even basic questions concerning S1–S9 computability theory can be challenging. We have formulated a most basic example in Sect. 2.2 concerning the *uncountability of* $\mathbb{R}$, arguably one of the most basic properties of the real numbers, which nonetheless yields very hard problems regarding S1–S9 computability.

In conclusion, the previous items suggest that the much greater scope of S1–S9 comes at the cost of conceptual clarity and causes technical difficulties. It is then a natural question whether we can find a 'sweet spot' between the conceptual clarity of Turing computability on one hand, and the generality of S1–S9, leading us to N-reduction.

Secondly, an *infinite time Turing machine* (ITTM) [14] is a generalisation of Turing computability involving infinite time or space. Welsh provides an overview in [53] and Dag Normann studies non-montone inductive definitions and the connection to ITTMs in [27].

In particular, Normann shows that ITTMs can outright compute many of the functionals introduced in [28,29,34], including realisers for the covering lemmas due to Vitali, Heine-Borel, and Lindelöf. However, all these functionals are not S1–S9 computable in any type two functional, i.e. the former are 'hard to compute' (see [28,29,34]). As a result, ITTMs yield 'too strong' a baseline framework for our purposes.

2 Some Results

We establish some results based on our freshly minted notion of N-reduction from Sect. 1.2, namely concerning the following topics.

- Convergence theorems for nets (Sects. 2.1.2 and 2.1.3).
- Covering theorems (Sects. 2.1.3).
- The uncountability of $\mathbb{R}$ (Sect. 2.2).
- Discontinuous functions on $\mathbb{R}$ and Grilliot's trick (Sect. 2.3).

The below just constitutes an illustrative first collection of examples: we do not claim our results to be particularly deep or ground-breaking. We do point out that the above items yield functionals that are, like the Jordan realisers from Sect. 1.3.1, hard to compute in that no type 2 functional can (S1–S9) compute them, while $\exists^3$ can.

Finally, the curious reader of course wonders what the counterpart of the Turing jump is for N-reduction. We believe this to be the 'J' operation discussed in Sect. 2.1.2.

2.1 Nets and Computability Theory

We study basic properties of *nets* via N-reduction. Nets are a generalisation of sequences, and the latter hark back to the early days of computability theory [45]. Filters provide an alternative to nets, but will not be discussed here for reasons discussed in Remark 9.

2.1.1 Nets, a Very Short Introduction

Nets are the generalisation of the concept of *sequence* to possibly uncountable index sets, nowadays called *nets* or *Moore-Smith sequences.* These were first described in [24] and then formally introduced by Moore and Smith in [25] and by Vietoris in [51]. These authors also established the generalisation to nets of various basic theorems due to Bolzano-Weierstrass, Dini, and Arzelà [25, §8–9] and [51, §4].

One well-know application is the formulation of fundamental topological notions like compactness in terms of nets, as pioneered in [3], while Kelley's textbook [16] is standard. Tukey's monograph [49] builds a similar framework, based on very specific nets, called *phalanxes*, where the index sets consist of finite subsets ordered by inclusion. We now list some basic definitions.

Definition 6. *A set $D \neq \emptyset$ with a binary relation '$\preceq$' is* directed *if*

a. $\preceq$ is transitive, i.e. $(\forall x, y, z \in D)([x \preceq y \wedge y \preceq z] \to x \preceq z)$,
b. for $x, y \in D$, there is $z \in D$ such that $x \preceq z \wedge y \preceq z$,
c. $\preceq$ is reflexive, i.e. $(\forall x \in D)(x \preceq x)$.

For a directed set $(D, \preceq)$ and a topological space X, any mapping $x : D \to X$ is a net *in X. We denote $\lambda d.x(d)$ as '$(x_d)_{d\in D}$' or '$x_d : D \to X$' to suggest the connection to sequences. The directed set $(D, \preceq)$ is not always explicitly mentioned together with a net $x_d : D \to X$.*

The following definitions readily generalise from the sequence notion.

Definition 7 [Convergence of nets]**.** *If $x_d : D \to X$ is a net, we say that it* converges *to the limit* $\lim_d x_d = y \in X$ *if for every neighbourhood U of y, there is $d_0 \in D$ such that for all $e \succeq d_0$, $x_e \in U$.*

Definition 8 [Increasing nets]**.** *A net $x_d : D \to \mathbb{R}$ is* increasing *if $a \preceq b$ implies $x_a \leq_{\mathbb{R}} x_b$ for all $a, b \in D$.*

Now, we shall mostly use nets where the index set consists of finite sets of real numbers ordered by inclusion, i.e. Tukey's 'phalanxes' from [49]. As noted in Remark 1, real numbers can readily be represented via elements of Baire space using primitive recursive operations. Thus, such phalanxes are essentially nets indexed by $\mathbb{N}^{\mathbb{N}}$. The notion of 'sub-sequence' of course generalises to 'sub-net' (see e.g. [41]), but we do not need this (slightly technical) notion here.

Finally, we discuss an alternative to nets and why it is not suitable here.

Remark 9 (Nets and filters). For completeness, we discuss the intimate connection between *filters* and nets. Now, a topological space X is compact if and only if every *filter base* has a *refinement* that converges to some point of X, which follows by [1, Prop. 3.4].

Whatever the meaning of the previous italicised notions, the similarity to the Bolzano-Weierstrass theorem for nets is obvious, and not a coincidence: for every net $\mathfrak{r}$, there is an associated filter base $\mathfrak{B}(\mathfrak{r})$ such that if the erstwhile converges, so does the latter to the same point; one similarly associates a net $\mathfrak{r}(\mathfrak{B})$ to a given filter base $\mathfrak{B}$ with the same convergence properties (see [1, §2]).

Hence, filters provide an alternative to nets, but we have chosen to work with nets for the following reasons, where the second one is the most pressing.

- Nets have a greater intuitive clarity compared to filters, in our opinion, due to the similarity between nets and sequences.
- Nets are 'more economical' in terms of ontology: consider the aforementioned filter base $\mathfrak{B}(\mathfrak{r})$ associated to the net $\mathfrak{r}$. By [1, Prop. 2.1], the base has strictly higher type than the net. The same holds for $\mathfrak{r}(\mathfrak{B})$ versus $\mathfrak{B}$.
- The notion of *refinement* mirrors the notion of sub-net [1, §2]. The former is studied in [40] in the context of paracompactness; the associated results suggest that the notion of sub-net works better in weak systems.

On a conceptual note, the well-known notion of *ultrafilter* corresponds to the equivalent notion of *universal net* [1, §3]. On a historical note, Vietoris introduces the notion of *oriented set* in [51, p. 184], which is exactly the notion of 'directed set'. He proceeds to prove (among others) a version of the Bolzano-Weierstrass theorem for nets. Vietoris also explains that these results are part of his dissertation, written in the period 1913–1919, i.e. during his army service for the Great War.

2.1.2 Nets and Convergence

We obtain a first result concerning N-reduction and convergence theorems for nets. In particular, as promised above, we connect the latter to the following operation, which is central and seems to play the role of the Turing jump: for given Y^2, define

$$J(Y) := \{n \in \mathbb{N} : (\exists f^1)(Y(f, n) = 0)\}.$$

We now have Theorem 10 where C is Cantor space ordered via the lexicographic ordering $\leq_{\mathsf{lex}}$, i.e. the notion of 'increasing net in C' is obvious following Definition 8. We note that subsets of $\mathbb{N}^{\mathbb{N}}$ or $\mathbb{R}$ are given by characteristic functions, well-known from measure and probability theory and going back one hundred plus of years [11].

Theorem 10. *The following strongly N-reduce to one and other:*

- *for all Y^2, there is $X \subset \mathbb{N}$ such that $X = J(Y)$,*
- *a monotone net in C indexed by Baire space, has a limit.*

Proof. To show that the second item strongly N-reduces to the first one, let $f_d : D \to C$ be an increasing net in C indexed by Baire space and consider the formula $(\exists d \in D)(f_d \geq_{\mathsf{lex}} \sigma * 00\dots)$, where σ^{0^*} is a finite binary sequence. The latter formula is equivalent to a formula of the form $(\exists g^1)(Y(g,n) = 0)$ where Y has the form $t(\lambda d.f_d, n)$ for a term t of Gödel's T. Now use $J(Y)$ to define the limit $f = \lim_d f_d$, as follows: $f(0)$ is 1 if $(\exists d \in D)(f_d \geq_{\mathsf{lex}} 100\dots)$ and zero otherwise. One then defines $f(n+1)$ in terms of $\overline{f}n$ in the same way. Note that we only used $J(Y)$ to define f, i.e. we have a strong N-reduction.

For the remaining case, fix some Y^2 and let w^{1^*} be a sequence of elements in $\mathbb{N}^{\mathbb{N}}$. Define $f_w : D \to C$ as $f_w := \lambda k.F(w,k)$ where $F(w,k)$ is 1 if $(\exists i < |w|)(Y(w(i),k) = 0)$, and zero otherwise. Then $\lambda w^{1^*}.f_w$ is a monotone net (phalanx) in C indexed by Baire space (modulo coding). In case $\lim_w f_w = f$, then it is readily verified that:

$$(\forall n^0)\big[(\exists g^1)(Y(g,n) = 0) \leftrightarrow f(n) = 1\big]. \tag{11}$$

In the notation of (4), the net $\lambda w^{1^*} f_w$ has the form $t(Y)(w)$ while s does not depend on Y, i.e. we have a strong N-reduction. □

The reader is warned that not all N-reduction results are as elegant.

2.1.3 Nets and Compactness

We connect the Heine-Borel theorem and convergence theorems for nets via N-reduction.

First of all, the Heine-Borel theorem, aka *Cousin's lemma*, [5,10] pertains to open-cover compactness, which we study for the unit interval. Clearly, each $\Psi : [0,1] \to \mathbb{R}^+$ yields a 'canonical' covering $\cup_{x \in [0,1]} B(x, \Psi(x))$, which must have a finite sub-covering. This yields the principle HBU, which has the form (1).

$$(\forall \Psi : [0,1] \to \mathbb{R}^+)(\exists x_0, \dots, x_k \in [0,1])\big([0,1] \subset \cup_{i \leq k} B(x_i, \Psi(x_i))\big). \tag{HBU}$$

The reals in HBU are hard to compute (S1–S9) in terms of Ψ, as shown in [28,31], as no type two functional can perform this task. Computing a *Lebesgue number*[5] is similarly hard as shown in [34]. Nonetheless, HBU seems stronger than the *Lebesgue number lemma* expressing that a Lebesgue number exists for any $\Psi : [0,1] \to \mathbb{R}^+$. We believe that Theorem 11 expresses this fundamental difference.

[5] The notion of *Lebesgue number* is familiar from topology (see e.g. [26, p. 175]) and amounts to the following: for a metric space (X,d) and an open covering O of X, the real number $\delta > 0$ is a Lebesgue number for O if every subset Y of X with $\mathsf{diam}(Y) := \sup_{x,y \in Y} d(x,y) < \delta$ is contained in some member of the covering.

Theorem 11

- HBU *μN-reduces to: for a monotone convergent net in* $[0,1]$ *indexed by Baire space, there is a modulus*[6] *of convergence.*
- *The* Lebesgue number lemma *strongly μN-reduces to: a monotone net in* $[0,1]$ *indexed by Baire space, has a limit.*

Proof. For the first part, fix $\Psi : [0,1] \to \mathbb{R}^+$ and define the following where w^{1^*} is a finite sequence of reals:

$$x_w := \begin{cases} 1 & (\forall q \in \mathbb{Q} \cap [0,1])(q \in \cup_{i<|w|} B(w(i), \Psi(w(i)))) \\ B(w)/2 & \text{otherwise} \end{cases}.$$

Here, $B(w)$ is the left-most end-point in $[0,1]$ of the intervals of the form $B(w(i), \Psi(B(w(i))))$ for $i \leq k$ that is not covered by the union. Note that $B(w)$ and x_w are readily defined using μ^2. Modulo coding of reals, $\lambda w^{1^*}.x_w$ can be viewed as a monotone net (phalanx) indexed by Baire space and we must have $\lim_w x_w = 1$. If $(w_k)_{k\in\mathbb{N}}$ is a modulus of convergence, then $|x_{w_2} - 1| < \frac{1}{4}$ by definition, implying $x_{w_2} = 1$. Hence, $\cup_{i<|w_2|} B(w(i), \Psi(w(i)))$ covers $[0,1] \cap \mathbb{Q}$. Now adjoin to w_2 all the points $w_2(i) \pm \Psi(w_2(i))$ for $i < |w_2|$, to obtain a covering of $[0,1]$. This 'adjoining' takes the form of $s(\Psi, w_2)$ while x_w takes the form $t(\Psi, \mu^2)(w)$ for terms s, t of Gödel's T, using the notation from (4).

For the second part, replace the output 1 by $\frac{3}{4} + \frac{1}{2^{N+3}}$ in the first case of x_w, where N is as follows: adjoin to w all the points $w(i) \pm \Psi(w(i))$ for $i < |w|$, to obtain a covering of $[0,1]$. Now use μ^2 to find $N \in \mathbb{N}$ such that $\frac{1}{2^N}$ is a Lebesgue number for the latter covering. Note that the modified net is still monotone as extending w can only increase the associated Lebesgue number. Clearly, any cluster point of the modified net is found in $(\frac{3}{4}, 1)$. A straightforward unbounded search can now recover a Lebesgue number from the cluster point of the net *without* access to Ψ, i.e. we have a strong μN-reduction. □

In light of the first part of the previous proof, the 'post-processing' term s in (4) seems necessary as a Turing machine cannot evaluate a third-order functional at a given point due to type restrictions.

As shown in [41], the existence of a modulus of convergence as in the first item of the theorem requires a fragment of the Axiom of Choice (AC) beyond ZF. In fact, one readily shows that the former existence statement N-reduces (and vice versa) to the following fragment of AC:

$$(\forall Y^2)\big[(\forall n^0)(\exists f^1)(Y(f,n)=0) \to (\exists Z^{0\to 1})(\forall n^0)(Y(Z(n),n)=0)\big],$$

where we exclude the trivial case $(\exists f^1)(\forall n^0)(Y(f,n)=0)$.

Finally, we connect the Lebesgue number lemma and NFP as follows.

Theorem 12. *The* Lebesgue number lemma *strongly N-reduces to* NFP *for* $A(n) \equiv (\exists f^1)(Y(f,n)=0)$ *for any* Y^2.

[6] A modulus of convergence for a net $x_d : D \to \mathbb{R}$ with $\lim_d x_d = x$ is a sequence $(d_k)_{k\in\mathbb{N}}$ with $(\forall k \in \mathbb{N})(\forall d \succeq d_n)(|x_d - x| < \frac{1}{2^k})$.

Proof. By Remark 1, quantifying over $2^{\mathbb{N}}$ or $[0,1]$ amounts to nothing more than quantifying over Baire space. To see this, define $\mathfrak{b} : \mathbb{N}^{\mathbb{N}} \to 2^{\mathbb{N}}$ as follows: $\mathfrak{b}(f)(n) := 0$ if $f(n) = 0$, and 1 otherwise. Also, define $\mathfrak{r}(f) := \sum_{n=0}^{\infty} \frac{\mathfrak{b}(f)(n)}{2^n}$ as the real in $[0,1]$ coded by $f \in \mathbb{N}^{\mathbb{N}}$. For $\Psi : \mathbb{R} \to \mathbb{R}^+$, the following formula is trivial (take $g = f$ and large n):

$$(\forall f \in \mathbb{N}^{\mathbb{N}})(\exists n \in \mathbb{N})\big[(\exists g \in \mathbb{N}^{\mathbb{N}})[B(\mathfrak{r}(f), \tfrac{1}{2^n}) \subset B(\mathfrak{r}(g), \Psi(\mathfrak{r}(g)))]\big],$$

which merely expresses that for every $x \in [0,1]$, there is $n \in \mathbb{N}$ and $y \in [0,1]$ such that $B(x, \frac{1}{2^n}) \subset B(y, \Psi(y))$. Applying NFP with parameter Ψ, we obtain $\gamma \in K_0$ such that

$$(\forall f \in \mathbb{N}^{\mathbb{N}})\big[(\exists g \in \mathbb{N}^{\mathbb{N}})[B(\mathfrak{r}(f), \tfrac{1}{2^{\gamma(f)}}) \subset B(\mathfrak{r}(g), \Psi(\mathfrak{r}(g)))]\big].$$

Now compute an upper bound for γ on $2^{\mathbb{N}}$, using the Kleene associate for the fan functional [23, §8.3.2]. This upper bound yields the required Lebesgue number, which only depends on γ^1, not on Ψ, i.e. we have obtained a *strong* N-reduction. □

We conjecture that HBU does not strongly μN-reduce to the fragment of NFP from Theorem 12.

2.2 On the Uncountability of $\mathbb{R}$

We study one of the most (in)famous properties of $\mathbb{R}$, namely its uncountability, established by Cantor in 1874 as part of his/the first set theory paper [9]. The following two principles were first studied in [32,37].

- NIN: there is no injection from $[0,1]$ to $\mathbb{N}$.
- **Cantor's theorem**: for a set $A \subset [0,1]$ and $Y : [0,1] \to \mathbb{N}$ injective on A, there is $x \in ([0,1] \setminus A)$.

A trivial manipulation of definitions shows that NIN and Cantor's theorem are logically equivalent. We however have the following theorem and associated Conjecture 14.

Theorem 13

- *The problem* NIN *N-reduces to the Heine-Borel theorem* HBU.
- *Cantor's theorem N-reduces to the Heine-Borel theorem* HBU *restricted to Baire class* 2 *functions.*

Proof. For the first part, fix $Z : [0,1] \to \mathbb{N}$ and define $t(Z)(x) := \frac{1}{2^{Z(x)+1}}$ motivated by the notation in (4). In case $x_0, \ldots, x_k \in [0,1]$ is a finite sub-covering of $\cup_{x\in[0,1]} B(x, t(Z)(x))$, there are $i, j \leq k$ with

$$Z(x_i) = Z(x_j) \wedge x_i \neq x_j. \tag{12}$$

Indeed, in case there are no $i, j \leq k$ as in (12), then the measure of $\cup_{i\leq k} B(x_i, t(Z)(x_i))$ is at most $\sum_{n=0}^{k} \frac{1}{2^{i+1}} < 1$, contradicting the fact that

$\cup_{i\leq k}B(x_i, t(Z)(x_i))$ covers $[0,1]$. In light of (12), given the finite sequence $s(Z, x_0, \ldots, x_k)$ defined as $x_0, t(Z)(x_0), \ldots, x_k, t(Z)(x_k)$, we can perform an unbounded search (on a Turing machine) to find $i, j \leq k$ and $l \in \mathbb{N}$ such that $t(Z)(x_i) =_{\mathbb{Q}} t(Z)(x_j)$ and $[|x_i - x_j|](l) >_{\mathbb{Q}} \frac{1}{2^l}$, where $[z](m)$ is the approximation of $z \in \mathbb{R}$ up to $\frac{1}{2^{m+1}}$. Hence, we also obtain the consequent of (4) for the case at hand.

For the second part, fix $A \subset [0,1]$ and $Y : [0,1] \to \mathbb{N}$ such that Y is injective on A. Now consider the following:

$$t(Y, A)(x) := \begin{cases} \frac{1}{2^{Y(x)+5}} & x \in A \\ \frac{1}{8} & x \notin A \end{cases}.$$

One readily shows that $t(Y, A) : \mathbb{R} \to \mathbb{R}$ is Baire class 2, as it only has countably many points of discontinuity by definition. For a finite sub-covering $x_0, \ldots, x_k \in [0,1]$ of $\cup_{x\in[0,1]}B(x, t(Y, A)(x))$, there must be $j \leq k$, with $x_j \notin A$. Indeed, as in the previous paragraph, the measure of $\cup_{i\leq k}B(x_i, t(Y, A)(x_i))$ is otherwise at most $\sum_{n=0}^{k} \frac{1}{2^{i+5}} < 1$, a contradiction. One can effectively decide whether $t(Y, A)(x_i) < \frac{1}{8}$ or $t(Y, A)(x_i) > \frac{1}{16}$ for $i \leq k$, i.e. one readily finds a $j \leq k$ with $x_j \notin A$. □

In light of the previous proof, the 'post-processing' term s in (4) again seems necessary as a Turing machine cannot evaluate a third-order functional at a point due to type restrictions.

Based on the previous proof, we conjecture the following.

Conjecture 14. *The problem* NIN *does **not** N-reduce to the Heine-Borel theorem* HBU *restricted to Baire class* 2 *functions, nor to the full Lebesgue number lemma.*

2.3 Discontinuous Functions

We show that a representative equivalence from the Reverse Mathematics literature involving $(\exists^2)$ gives rise to N-reductions between the members of the equivalence. That N-reduction applies here was surprising to us, as the existence of a discontinuous function like $\exists^2$ does not have the syntactic form (1).

A central role is played by *Grilliot's trick*, a method for (effectively) obtaining $\exists^2$ from a discontinuous function [13]. We discuss this trick in some detail in Sect. 2.3.1, while the connection between this trick and N-reduction is discussed in Sect. 2.3.2.

2.3.1 Grilliot's Trick

In a nutshell, *Grilliot's trick* is a method for effectively obtaining $\exists^2$ from a discontinuous function, say on $\mathbb{N}^{\mathbb{N}}$ or $\mathbb{R}$. Clearly, $\exists^2$ is discontinuous at $11\ldots$, making the former functional a kind of 'canonical' discontinuous function.

First of all, Grilliot's paper [13] pioneers the aforementioned method, nowadays called Grilliot's trick; we refer to [23, Remark 5.3.9] for a discussion of the general background and history. We note that Kohlenbach formalises Grilliot's trick in a weak logical system (namely his 'base theory' RCA_0^ω) in [19, §3].

Secondly, Kohlenbach's rendition of Grilliot's trick [19, §3] is quite easy to understand conceptually. Indeed, assume we have a function $F : \mathbb{R} \to \mathbb{R}$ and a sequence $(x_n)_{n\in\mathbb{N}}$ with $\lim_{n\to\infty} x_n = x$ such that $\lim_{n\to\infty} F(x_n) \neq F(x)$, i.e. F is not sequentially continuous at x. Then there is a term t^3 of Gödel's T of low complexity such that $E(f) := \lambda f^1.t(F, \lambda n.x_n, x, f)$ is Kleene's $\exists^2$ as in (6). All technical details, including the exact definition of t, are found in [19, §3].

Thirdly, Kohlenbach uses Grilliot's trick in [19, §3] to show that e.g. the following sentence implies the existence of $\exists^2$:

$$\underline{(\exists\varepsilon)(\forall g \in L([0,1]))}[\varepsilon(g) \in [0,1] \wedge (\forall y \in [0,1])(g(y) \leq g(\varepsilon(g)))]. \tag{13}$$

Here, $\varepsilon(g)$ is a real in $[0,1]$ where the Lipschitz-continuous[7] function $g : [0,1] \to \mathbb{R}$ with constant 1 attains its maximum. We note that the underlined quantifiers can be brought in the form[7] '$(\exists Y^2)(\forall \alpha^1)$', which can also be obtained by representing continuous functions via second-order codes.

2.3.2 Discontinuous Functions and N-reduction

In this section, we discuss the connection between Grilliot's trick from Sect. 2.3.1 and N-reduction. In particular, we show that the proof of [19, Prop. 3.14], establishing the equivalence (13) $\leftrightarrow$ ($\exists^2$), gives rise to N-reductions involving (13) and ($\exists^2$).

First of all, consider (13) from Sect. 2.3.1. The functional ε from (13) yields a discontinuous function on $\mathbb{R}$, which yields $\exists^2$ in turn, following the proof of [19, Prop. 3.14]. If we make all steps in the latter proof explicit[8], we obtain a term t of Gödel's T of low complexity such that

$$(\forall\varepsilon)\big[(\forall g)A(g,\varepsilon(g)) \to (\forall f^1)B(t(\varepsilon), f)\big], \tag{14}$$

where $A(g,x)$ expresses that $x \in [0,1]$ is a real where the Lipschitz-continuous function $g : [0,1] \to \mathbb{R}$ with Lipschitz constant 1 attains its maximum; the formula $(\forall f^1)B(\exists^2, f)$ is (6), i.e. the specification of $\exists^2$. Clearly, (14) implies by contraposition that:

$$(\forall\varepsilon, f^1)\big[\neg B(t(\varepsilon), f) \to (\exists g)\neg A(g,\varepsilon(g))\big], \tag{15}$$

which is 'almost' the definition of N-reduction as in (4). Indeed, '$(\exists g)$' in (15) is essentially a quantifier over $\mathbb{R}$ by Footnote 7, whence $(\forall\varepsilon)$ can be viewed as

[7] A function $g : [0,1] \to \mathbb{R}$ is *Lifschitz-continuous* with constant 1 on $[0,1]$ if $(\forall x, y \in [0,1])(|g(x) - g(y)| < |x - y|)$. Hence, to (effectively) recover the graph of g, it suffices to have access to the sequence $(g(q))_{q\in\mathbb{Q}\cap[0,1]}$.

[8] The construction of a discontinuous function on $\mathbb{R}$ in the proof of [19, Prop. 3.14] depends on whether $\varepsilon(g_0) \in [0, \frac{1}{2}]$ or $\varepsilon(g_0) \in [\frac{1}{2}, 1]$, where g_0 is the constant 0 function and ε as in (13). This non-effective case distinction can be replaced by an effective case distinction whether $\varepsilon(g_0) < \frac{3}{4}$ or $\varepsilon(g_0) > \frac{1}{4}$. The proof in the first case goes through unmodified, while one replaces yx in the second case by $-yx$.

a quantifier $(\forall Z^2)$. Furthermore, a detailed inspection of the proof that (13) implies the existence of $\exists^2$ in [19, Prop. 3.14], reveals the following: this proof still goes through if we restrict (13) to a sentence of the form:

$$(\exists \varepsilon)(\forall n^0))[\varepsilon(g_n) \in [0,1] \wedge (\forall q \in [0,1] \cap \mathbb{Q})(g_n(q) \leq g_n(\varepsilon(g_n)))], \tag{16}$$

for some effective[9] sequence of functions $(g_n)_{n\in\mathbb{N}}$ all in $L([0,1])$. In case the formula in square brackets in (16) is false for some $n \in \mathbb{N}$, an unbounded search will yield this number. Hence, we can replace '$(\exists g)\neg A(g, \varepsilon(g)$' in (15) by

$$\{e\}^{s(\varepsilon,f)} \downarrow \wedge \neg A(\{e\}^{s(\varepsilon,f)}, \varepsilon(\{e\}^{s(\varepsilon,f)}))$$

for some index $e \in \mathbb{N}$ and term s of Gödel's T, which is exactly (4). The details are somewhat tedious, but we nonetheless can say that the negation of (13) N-reduces to the negation of $(\exists^2)$.

Finally, the usual 'interval-halving' proof of the existence of a maximum of a continuous function on $[0,1]$, can be done using $\exists^2$, yielding a term t of Gödel's T such that:

$$(\forall E^2)\big[(\forall f^1)B(E,f) \to (\forall g)A(g, t(E)(g))\big]. \tag{17}$$

The contraposition of (17) then has the same form as (15). One readily obtains an index $e \in \mathbb{N}$ and term s of Gödel's T with

$$(\forall E^2, g)\big[\neg A(g, t(E)(g)) \to [\{e\}^{s(E,g)} \downarrow \wedge \neg B(E, \{e\}^{s(E,g)})]\big], \tag{18}$$

as one only needs to decide $g(r) \geq g(q)$ for $r, q \in [0,1] \cap \mathbb{Q}$ to find a maximum of a (Lipschitz) continuous function $g : [0,1] \to \mathbb{R}$. Hence, an unbounded search on a Turing machine will find f^1 with $\neg B(E,f)$. We note that (18) is a case of N-reduction of the negation of $(\exists^2)$ to the negation of (13).

In conclusion, we observe that the negation of $(\exists^2)$ will N-reduce to the negation of (13), and vice versa. Thus, it perhaps makes sense to drop the 'negation of' here and distinguish between (1) and its negation in the definition of N-reduction.

Acknowledgements. I thank Anil Nerode for his most helpful advise. My research was kindly supported by the *Deutsche Forschungsgemeinschaft* via the DFG grant SA3418/1-1. I thank the anonymous referees for their suggestions, which have greatly improved this paper.

References

1. Bartle, R.G.: Nets and filters in topology. Am. Math. Mon. **62**, 551–557 (1955)
2. Beeson, M.J.: Foundations of Constructive Mathematics: Metamathematical Studies. Ergebnisse der Mathematik und ihrer Grenzgebiete, vol. 6. Springer, Heidelberg (1985). https://doi.org/10.1007/978-3-642-68952-9

[9] The join of the sequences $(qx - q)_{q\in\mathbb{Q}\cap[0,1]}$ and $(-qx)_{q\in\mathbb{Q}\cap[0,1]}$ suffices.

3. Birkhoff, G.: Moore-smith convergence in general topology. Ann. Math. **38**(1), 39–56 (1937)
4. Bishop, E.: Foundations of Constructive Analysis. McGraw-Hill, New York (1967)
5. Borel, E.: Leçons sur la théorie des fonctions. Gauthier-Villars, Paris (1898)
6. Bridges, D., Richman, F.: Varieties of Constructive Mathematics. LMS Lecture Note Series 97. Cambridge University Press (1987)
7. Brouwer, L.E.J.: Collected Works. Philosophy and Foundations of Mathematics, vol. 1. North-Holland Publishing Co., Amsterdam (1975)
8. Buchholz, W., Feferman, S., Pohlers, W., Sieg, W.: Iterated Inductive Definitions and Subsystems of Analysis: Recent Proof-Theoretical Studies. LNM, vol. 897. Springer, Heidelberg (1981). https://doi.org/10.1007/BFb0091894
9. Cantor, G.: Ueber eine eigenschaft des inbegriffs aller reellen algebraischen zahlen. J. Reine Angew. Math. **77**, 258–262 (1874)
10. Cousin, P.: Sur les fonctions de n variables complexes. Acta Math. **19**, 1–61 (1895)
11. Dirichlet, L.P.G.: Über die Darstellung ganz willkürlicher Funktionen durch Sinus- und Cosinusreihen. Repertorium der physik, von H.W. Dove und L. Moser, bd. 1 (1837)
12. Feferman, S.: How a little bit goes a long way: predicative foundations of analysis. In: 2013 Unpublished Notes from 1977–1981 with Updated Introduction. https://math.stanford.edu/~feferman/papers/pfa(1).pdf
13. Grilliot, T.J.: On effectively discontinuous type-2 objects. J. Symb. Log. **36**, 245–248 (1971)
14. Hamkins, J.D., Lewis, A.: Infinite time Turing machines. J. Symb. Log. **65**, 567–604 (1998)
15. Hilbert, D., Bernays, P.: Grundlagen der Mathematik. II. Die Grundlehren der mathematischen Wissenschaften, vol. 50. Springer, Heidelberg (1970). https://doi.org/10.1007/978-3-642-86896-2
16. Kelley, J.L.: General Topology. Graduate Texts in Mathematics, vol. 27. Springer, New York (1975). Reprint of the 1955th edn
17. Kleene, S.C.: Recursive functionals and quantifiers of finite types. i. Trans. Am. Math. Soc. **91**, 1–52 (1959)
18. Kohlenbach, U.: Foundational and mathematical uses of higher types. In: Lecture Notes in Logic, vol. 15, 92–116. ASL (2002)
19. Kohlenbach, U.: Higher order reverse mathematics. In: Lecture Notes in Logic, vol. 21, pp. 281–295. ASL (2005)
20. Kohlenbach, U.: Applied Proof Theory: Proof Interpretations and their Use in Mathematics. Springer Monographs in Mathematics. Springer, Heidelberg (2008). https://doi.org/10.1007/978-3-540-77533-1
21. Kreisel, G., Troelstra, A.S.: Formal systems for some branches of intuitionistic analysis. Ann. Math. Logic **1**, 229–387 (1970)
22. Kreuzer, A.P.: Bounded variation and the strength of Helly's selection theorem. Log. Meth. Comput. Sci. **10**(4), 1–23 (2014)
23. Longley, J., Normann, D.: Higher-Order Computability. Theory and Applications of Computability. Springer, Heidelberg (2015). https://doi.org/10.1007/978-3-662-47992-6
24. Moore, E.H.: Definition of limit in general integral analysis. PNAS **1**(12), 628–632 (1915)
25. Moore, E.H., Smith, H.: A general theory of limits. Am. J. Math. **44**, 102–121 (1922)
26. Vietoris, L.: Topology. Monatshefte für Mathematik und Physik **39**(1), A17–A19 (1932). https://doi.org/10.1007/BF01699114

27. Normann, D.: Computability and non-monotone induction, p. 41. arXiv arXiv:2006.03389 (2020, submitted)
28. Normann, D., Sanders, S.: Nonstandard analysis, computability theory, and their connections. J. Symb. Log. **84**(4), 1422–1465 (2019)
29. Normann, D., Sanders, S.: On the mathematical and foundational significance of the uncountable. J. Math. Log. (2019). https://doi.org/10.1142/S0219061319500016
30. Normann, D., Sanders, S.: Representations in measure theory. arXiv arXiv:1902.02756 (2019, submitted)
31. Normann, D., Sanders, S.: The strength of compactness in computability theory and nonstandard analysis. Ann. Pure Appl. Log. **170**(11), 102710 (2019)
32. Normann, D., Sanders, S.: On the uncountability of R, p. 37. arXiv arXiv:2007.07560 (2020, submitted)
33. Normann, D., Sanders, S.: Open sets in reverse mathematics and computability theory. J. Log. Comput. **30**(8), 40 (2020)
34. Normann, D., Sanders, S.: Pincherle's theorem in reverse mathematics and computability theory. Ann. Pure Appl. Log. **171**(5), 102788 (2020)
35. Normann, D., Sanders, S.: The axiom of choice in computability theory and reverse mathematics. J. Log. Comput. **31**(1), 297–325 (2021)
36. Normann, D., Sanders, S.: Betwixt turing and Kleene. In: Artemov, S., Nerode, A. (eds.) LFCS 2022. LNCS, vol. 13137, pp. 236–252. Springer, Cham (2022)
37. Normann, D., Sanders, S.: On robust theorems due to Bolzano, Weierstrass, and Cantor in reverse mathematics, p. 30. arXiv arXiv:2102.04787 (2021, submitted)
38. Sanders, S.: Nets and reverse mathematics. In: Manea, F., Martin, B., Paulusma, D., Primiero, G. (eds.) CiE 2019. LNCS, vol. 11558, pp. 253–264. Springer, Cham (2019). https://doi.org/10.1007/978-3-030-22996-2_22
39. Sanders, S.: Reverse mathematics and computability theory of domain theory. In: Iemhoff, R., Moortgat, M., de Queiroz, R. (eds.) WoLLIC 2019. LNCS, vol. 11541, pp. 550–568. Springer, Heidelberg (2019). https://doi.org/10.1007/978-3-662-59533-6_33
40. Sanders, S.: Reverse mathematics of topology: dimension, paracompactness, and splittings. Notre Dame J. Form. Log. **61**(4), 537–559 (2020)
41. Sanders, S.: Nets and reverse mathematics: a pilot study. Computability **10**(1), 31–62 (2021)
42. Arora, S., Barak, B.: Computational Complexity: A Modern Approach. Cambridge University Press (2009)
43. Simpson, S.G.: Subsystems of Second Order Arithmetic, 2nd edn. Perspectives in Logic. Cambridge University Press (2009)
44. Soare, R.I.: Recursively Enumerable Sets and Degrees. Perspectives in Mathematical Logic. Springer, Heidelberg (1987)
45. Specker, E.: Nicht konstruktiv beweisbare sätze der analysis. J. Symb. Log. **14**, 145–158 (1949)
46. Stillwell, J.: Reverse Mathematics, Proofs from the Inside Out. Princeton University Press (2018)
47. Troelstra, A.S.: Choice Sequences: A Chapter of Intuitionistic Mathematics. Oxford Logic Guides. Clarendon Press, Oxford (1977)
48. Troelstra, A.S., van Dalen, D.: Constructivism in Mathematics. Vol. I. Studies in Logic and the Foundations of Mathematics, vol. 121. North-Holland (1988)
49. Tukey, J.W.: Convergence and Uniformity in Topology. Annals of Mathematics Studies, vol. 2. Princeton University Press, Princeton, N.J. (1940)

50. Turing, A.: On computable numbers, with an application to the Entscheidungsproblem. Proc. Lond. Math. Soc. **42**, 230–265 (1936)
51. Vietoris, L.: Stetige mengen. Monatsh. Math. Phys. **31**(1), 173–204 (1921)
52. Weihrauch, K.: Computable Analysis: An Introduction. TTCSAES, Springer, Heidelberg (2000). https://doi.org/10.1007/978-3-642-56999-9
53. Welch, P.D.: Transfinite Machine Models, pp. 493–529 (2014)

Propositional Dynamic Logic with Quantification over Regular Computation Sequences

Igor Sedlár(✉)

Institute of Computer Science, The Czech Academy of Sciences, Prague, Czech Republic
sedlar@cs.cas.cz

Abstract. We extend test-free regular propositional dynamic logic with operators expressing combinations of existential and universal quantifiers quantifying over computation sequences represented by a given regular expression and states accessible via these computation sequences. This extended language is able to express that there is a computation sequence represented by a given regular expression that leads only to states where a given formula is satisfied, or that for all computation sequences represented by a given regular expression there is a state accessible via the computation sequence where a given formula is satisfied. Such quantifier combinations are essential in expressing, for instance, that a given non-deterministic finite automaton accepts all words of a given regular language or that there is a specific sequence of actions instantiating a plan expressed by a regular expression that is guaranteed to accomplish a certain goal. The existential-universal quantifier combination is modelled by neighborhood functions. We prove that a rich fragment of our logic is decidable and *EXPTIME*-complete by embedding the fragment into deterministic propositional dynamic logic.

Keywords: Finite automata · Planning · Propositional dynamic logic · Quantification

1 Introduction

Propositional dynamic logic, PDL, is a well-known modal logic originally introduced to formalize propositional-level reasoning about correctness properties of imperative programs [4,6]. PDL has been used since in a number of other settings such as reasoning about action [8,9] or planning [13,14]. Multi-agent epistemic logic with common knowledge [3] is a fragment of PDL.

In PDL, modal operators are indexed by *programs*, which are expressions built up from a countable set of atomic programs using the operations of *choice* $\cup$, *composition* ;, *Kleene star* * and, for each formula φ, *test* $\varphi?$. Programs are interpreted semantically as binary relations on a set of states, where $\cup$ corresponds to union, ; to relational composition, * to reflexive-transitive closure and

S. Artemov and A. Nerode (Eds.): LFCS 2022, LNCS 13137, pp. 301–315, 2022.
https://doi.org/10.1007/978-3-030-93100-1_19

φ? is interpreted as the identity relation on the set of states satisfying φ. In *test-free* PDL the test operator is omitted and *deterministic* PDL requires atomic programs to be interpreted as partial functions. *Regular* PDL over a finite alphabet Σ is a version of test-free PDL where Σ replaces the set of atomic programs and two zeroary program operators $\mathbf{0}$ and $\mathbf{1}$ are added, corresponding semantically to the empty relation and the identity relation, respectively. In effect, programs in regular PDL over Σ are regular expressions over Σ. Regular PDL embeds into PDL, where $\mathbf{1}$ is represented by $\top$? and $\mathbf{0}$ by $\bot$?. Satisfiability (and validity) problems for all these variants of PDL are decidable and *EXPTIME*-complete.

It is easily observed that pairs (M, x), where M is a finite Kripke model for regular PDL over Σ and x is a state in the model, correspond to non-deterministic finite automata (NFA) over Σ where a fixed propositional variable represents the set of accepting states. It can be shown that acceptance of words by NFAs can be represented by formulas: for each word w over Σ, there is a formula φ_w such that any (M, x) satisfies φ_w iff the NFA represented by (M, x) accepts w. However, this does not hold for acceptance of *regular languages:* there is no formula φ_α such that any (M, x) satisfies φ_α iff the automaton represented by (M, x) accepts all strings in the language determined by the regular expression α. A similar deficiency is encountered when planning-related interpretations of PDL are considered: if actions are non-deterministic, then there is no formula of PDL saying that there is *some* course of action instantiating a given structured plan (represented by a regular expression) such that *all* states reachable by that course of action satisfy a certain goal formula.

To represent these notions, one needs to extend PDL with modal operators that express a combination of existential and universal quantification over words belonging to the language determined by a regular expression on the one hand and sets of states reachable by relations corresponding to such words on the other hand. In this paper we introduce **QPDL**, an extension of regular PDL with modal formulas of the form $\langle\, \alpha\,]\varphi$ saying that there is a word w in the language determined by α such that all states reachable via the relation corresponding to w satisfy φ. Dual formulas $[\, \alpha\, \rangle\varphi$, defined as $\neg\langle\, \alpha\,]\neg\varphi$, say that for all words w in the language determined by α there is a state reachable via the relation corresponding to w that satisfies φ. The $\langle\, \alpha\,]$-operators are represented semantically in terms of α-indexed neighborhood functions (of the kind used also in concurrent PDL [12] and Game logic [11]) that are specifically related to the relations corresponding to α.

Our main technical result is a decidability result for **qPDL**, the fragment of **QPDL** where $\mathbf{1}$ is omitted and the reflexive-transitive closure operator * is replaced by a transitive closure operator $^+$; it is also shown that **qPDL** is *EXPTIME*-complete. The result is established by embedding **qPDL** into deterministic PDL.

The paper is structured as follows. After discussing some background material on regular expressions in Sect. 2, we introduce **QPDL** and **qPDL** in Sect. 3. In Sect. 4 we briefly discuss two notions that the language of **QPDL** is able to express, namely, that a given non-deterministic finite automaton accepts all strings in the language determined by a given regular expression, and that there is a way to execute an abstract plan, represented by a regular expression, in a way that guarantees that a specific goal, expressed by a formula, is achieved. In Sect. 5 we establish our decidability and complexity result on **qPDL**. In Sect. 6 we briefly discuss the problems that arise when our proof technique is applied to **QPDL** itself. The concluding section summarizes the paper and outlines some interesting problems we leave for future work.

2 Regular Computation Sequences

Definition 1. *Fix a finite alphabet Σ. The set of* expressions *over Σ, denoted $Ex_1(\Sigma)$, is defined by the following grammar:*

$$\alpha := a \mid \mathbf{0} \mid \mathbf{1} \mid \alpha \cup \alpha \mid \alpha ; \alpha \mid \alpha^+$$

where $a \in \Sigma$. We define $\alpha^* := \mathbf{1} \cup \alpha^+$. The set of $\mathbf{1}$-free expressions *over Σ, denoted $Ex(\Sigma)$, is the set of $\alpha \in Ex_1(\Sigma)$ such that $\mathbf{1}$ does not occur in α. For all non-empty $\Gamma = \{a_1, \ldots, a_n\} \subseteq \Sigma$ we define $\bigcup \Gamma := a_1 \cup \ldots \cup a_n$ (assuming some fixed bracketing).*

The set $Ex_1(\Sigma)$ is virtually identical to the set of regular expressions over Σ; our mode of presentation is determined by the fact that $Ex(\Sigma)$ will be used in our main technical result.

In what follows, we use the standard notation Σ^* to denote the set of all finite sequences of elements of Σ (words over Σ); the empty sequence is denoted as ϵ. If $w, u \in \Sigma^*$, then wu denotes the concatenation of w and u. Σ^+ is the set of all finite non-empty sequences of elements of Σ (non-empty words over Σ).

Definition 2. *For each $\alpha \in Ex_1(\Sigma)$, we define $L(\alpha) \subseteq \Sigma^*$, the* language determined by α, *as follows:*

$$L(a) = \{a\} \qquad L(\mathbf{0}) = \emptyset$$
$$L(\mathbf{1}) = \{\epsilon\} \qquad L(\alpha \cup \beta) = L(\alpha) \cup L(\beta)$$
$$L(\alpha ; \beta) = L(\alpha) \cdot L(\beta) \qquad L(\alpha^+) = \bigcup_{n>0} L^n(\alpha)$$

where $L_1 \cdot L_2 = \{wu \mid w \in L_1 \;\&\; u \in L_2\}$, $L^1(\alpha) = L(\alpha)$ and $L^{n+1}(\alpha) = L^n(\alpha) \cdot L(\alpha)$.

For example $L(a^+; (\mathbf{1} \cup (b \cup c)))$ is the set of all non-empty words over the alphabet $\{a, b, c\}$ where all symbols but possibly the last one are a's; words a, ab, aac belong to this language but ca does not.

In what follows we usually do not distinguish between α and w if $L(\alpha) = \{w\}$ (e.g. between the regular expression $(a; b); c$ and the string abc). We will often write $\alpha\beta$ instead of $\alpha; \beta$.

If elements of Σ are seen as representing (executions of) atomic *programs* (or, more generally, *instructions* or *actions*), then $L(\alpha)$ can be seen as a set of *computation sequences*—finite sequences of (executions of) atomic programs (instructions, actions)—determined by the regular expression α. Such computation sequences may be called *regular computation sequences.*

3 QPDL and Its 1-Free Fragment qPDL

Definition 3. *A* Σ-frame *is a pair* (S, R)*, where* S *is a non-empty set and* $R : \Sigma \to (S \to 2^S)$ *such that* $R(a)(x)$ *is at most countable for each* $a \in \Sigma$ *and* $x \in S$*. A* neighborhood Σ-frame *is a tuple* $\mathfrak{F} = (S, R, N)$ *where* (S, R) *is a* Σ*-frame and* $N : \Sigma \to (S \to 2^{2^S})$ *such that*

$$N(a)(x) = \{R(a)(x)\}.$$

In each neighborhood frame $\mathfrak{F}$*, the functions* R *and* N *are extended to domain* $Ex_{\mathbf{1}}(\Sigma) \cup \Sigma^*$ *as indicated in Fig. 1. (We define* $R(\alpha) := \{(x, y) \mid y \in R(\alpha)(x)\}$ *and* $N(\alpha) = \{(x, Y) \mid Y \in N(\alpha)(x)\}$*. For* $P \in \{R, N\}$*,* $\xi \in Ex_{\mathbf{1}}(\Sigma) \cup \Sigma^*$ *and* $X \subseteq S$ *we define* $P(\xi)(X) := \bigcup_{x \in X} P(\xi)(x)$ *and* $P(\xi)(\{X_i \mid i \in I\}) = \bigcup\{P(\xi)(X_i) \mid i \in I\}$*. We also define* $P(\alpha)^n(x)$ *by fixing* $P(\alpha)^1(x) := P(\alpha)(x)$ *and* $P(\alpha)^{n+1}(x) := P(\alpha)(P(\alpha)^n(x))$*. We will often write* P_ξ *instead of* $P(\xi)$*).*

$$\begin{aligned} R(\mathbf{1}) &= \mathrm{id}_S = R(\epsilon) & N(\mathbf{1})(x) &= \{\{x\}\} = N(\epsilon)(x) \\ R(\mathbf{0}) &= \emptyset & N(\mathbf{0})(x) &= \emptyset \\ R(\alpha \cup \beta) &= R(\alpha) \cup R(\beta) & N(\alpha \cup \beta)(x) &= N(\alpha)(x) \cup N(\beta)(x) \\ R(\alpha\beta) &= R(\alpha) \circ R(\beta) & N(\alpha\beta)(x) &= N(\beta)((N(\alpha)(x)) \\ R(\alpha^+) &= \bigcup_{n>0} R(\alpha)^n & N(\alpha^+)(x) &= \bigcup_{n>0} N(\alpha)^n(x) \\ R(wa) &= R(w) \circ R(a) & N(wa)(x) &= N(a)(N(w)(x)) \end{aligned}$$

Fig. 1. Accessibility relations and neighborhood functions for regular expressions and words over Σ.

It is easily seen that $R(\alpha)(x)$ is at most countable for each $\alpha \in Ex_{\mathbf{1}}(\Sigma)$ and $x \in S$. This "image-countability" property will be essential in Sec. 5.2. Note that, for all $\alpha \in Ex_{\mathbf{1}}(\Sigma)$,

$$R_\alpha = \{(x, y) \mid \exists w \in L(\alpha) R_w xy\} \tag{1}$$

It follows that if $L(\alpha) = L(\beta)$, then $R_\alpha = R_\beta$. We will use this fact later when reasoning about R "using properties of regular expressions".

Proposition 1. *For each neighborhood Σ-frame, each x in the frame and each $\alpha \in Ex_1(\Sigma)$:*

$$N_\alpha(x) = \{R_w(x) \mid w \in L(\alpha)\} \tag{2}$$

$$R_\alpha(x) = \bigcup N_\alpha(x) \tag{3}$$

Proof. We prove (2) by induction on the complexity of α. The base holds by definition and the cases of the induction step corresponding to $\mathbf{1}, \mathbf{0}$ and $\cup$ are trivial. The case for ; is established as follows: $N_{\alpha\beta}(x) = N_\beta(N_\alpha(x)) = \bigcup\{N_\beta(X) \mid X \in N_\alpha(x)\} = \bigcup\{\bigcup\{N_\beta(y) \mid R_w xy\} \mid w \in L(\alpha)\} = \{R_{wu}(x) \mid w \in L(\alpha) \ \& \ u \in L(\beta)\}$. The case for $^+$ follows from the fact that $N_{\alpha^n}(x) = \{R_w(x) \mid w \in L(\alpha^n)\}$, which is established by induction on n (the induction step is established similarly as the case for ; above). Claim (3) follows from (1) and (2). □

Proposition 1 entails that we could have equivalently used (2) as a definition of N in terms of R. Instead, we opted for a more "constructive" definition of N.

Example 1. An example of a Σ-frame for $\Sigma = \{a, b\}$, based on the $\mathbb{N} \times \mathbb{N}$ matrix, is shown in Fig. 2. The frame is extended to a neighborhood frame by defining $N(a)((n, m)) = \{\{(n + 1, m)\}\}$ and $N(b)((n, m)) = \{\{(n, m + 1)\}\}$. Hence, for example,

$$N((a \cup b)^+)((n, m)) \\ = \Big\{\{(n', m') \mid n \leq n' \ \& \ m \leq m' \ \& \ n' + m' = k + (n + m)\} \Big| \ k > 0\Big\}$$

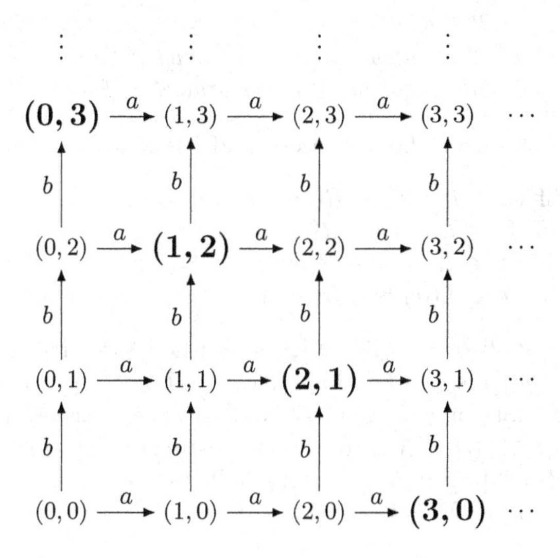

Fig. 2. An example of a Σ-frame.

That is, $X \in N((a \cup b)^+)((n,m))$ iff X is the set of pairs (n', m') that can be reached from (n,m) by k steps along the a, b arrows (for some $k > 0$). The highlighted states are reachable from $(0,0)$ in 3 steps.

Fix a countable set Pr of propositional variables. The set of $Ex_{\mathbf{1}}(\Sigma)$-*formulas*, denoted $Fm_{\mathbf{1}}(\Sigma)$, is defined by the following grammar:

$$\varphi := p \mid \neg\varphi \mid \varphi \wedge \varphi \mid [\,\alpha\,]\varphi \mid \langle\,\alpha\,]\varphi$$

where $p \in Pr$ and $\alpha \in Ex_{\mathbf{1}}(\Sigma)$. Boolean operators $\top, \bot, \vee, \to$ and $\leftrightarrow$ are defined as usual; moreover, we define $\langle\,\alpha\,\rangle\varphi := \neg[\,\alpha\,]\neg\varphi$ and $[\,\alpha\,\rangle\varphi := \neg\langle\,\alpha\,]\neg\varphi$. A formula $\varphi \in Ex_{\mathbf{1}}(\Sigma)$ is a $Ex(\Sigma)$-formula iff $\mathbf{1}$ does not occur in φ; the set of $Ex(\Sigma)$-formulas will be denoted as $Fm(\Sigma)$. $Ex_{\mathbf{1}}(\Sigma)$-formulas and $Ex(\Sigma)$ formulas will be often referred to as *formulas* and $\mathbf{1}$-*free formulas*, respectively.

Definition 4. *A* neighbourhood Σ-model *based on* $\mathfrak{F}$ *is* $\mathfrak{M} = (\mathfrak{F}, V)$ *where* V *is a function from* Pr *to* 2^S.

For each $\mathfrak{M}$, *the* satisfaction relation $\models_{\mathfrak{M}}$ *between states of* $\mathfrak{M}$ *and formulas is defined as follows:*

- $x \models_{\mathfrak{M}} p$ *iff* $x \in V(p)$;
- $x \models_{\mathfrak{M}} \neg\varphi$ *iff* $x \not\models_{\mathfrak{M}} \varphi$;
- $x \models_{\mathfrak{M}} \varphi \wedge \psi$ *iff* $x \models_{\mathfrak{M}} \varphi$ *and* $x \models_{\mathfrak{M}} \psi$;
- $x \models_{\mathfrak{M}} [\,\alpha\,]\varphi$ *iff* $y \models_{\mathfrak{M}} \varphi$ *for all* $y \in R_\alpha(x)$;
- $x \models_{\mathfrak{M}} \langle\,\alpha\,]\varphi$ *iff* $X \models_{\mathfrak{M}} \varphi$ *for some* $X \in N_\alpha(x)$.

(We write $X \models_{\mathfrak{M}} \varphi$ *iff* $z \models_{\mathfrak{M}} \varphi$ *for all* $z \in X$.*) A formula* φ *is* valid in $\mathfrak{M}$ *iff* $x \models_{\mathfrak{M}} \varphi$ *for all* x *in* $\mathfrak{M}$; *formula* φ *is valid in* $\mathfrak{F}$ *iff it is valid in all* $\mathfrak{M}$ *based on* $\mathfrak{F}$. **QPDL** *is the set of formulas valid in all neighborhood* Σ-*frames;* **qPDL** *is the set of* $\mathbf{1}$-*free formulas valid in all neighborhood* Σ-*frames.*

Remark 1. It follows from the definitions and Proposition 1 that

- $x \models_{\mathfrak{M}} [\,\alpha\,]\varphi$ iff $\forall w \in L(\alpha) \forall y \in R(w)(x) : y \models_{\mathfrak{M}} \varphi$;
- $x \models_{\mathfrak{M}} \langle\,\alpha\,]\varphi$ iff $\exists w \in L(\alpha) \forall y \in R(w)(x) : y \models_{\mathfrak{M}} \varphi$;
- $x \models_{\mathfrak{M}} \langle\,\alpha\,\rangle\varphi$ iff $\exists w \in L(\alpha) \exists y \in R(w)(x) : y \models_{\mathfrak{M}} \varphi$;
- $x \models_{\mathfrak{M}} [\,\alpha\,\rangle\varphi$ iff $\forall w \in L(\alpha) \exists y \in R(w)(x) : y \models_{\mathfrak{M}} \varphi$.

Hence, $[\,\alpha\,]\varphi$ is satisfied in x iff $[\,w\,]\varphi$ is satisfied for all $w \in L(\alpha)$; $\langle\,\alpha\,]\varphi$ is satisfied in x iff $[\,w\,]\varphi$ is satisfied for some $w \in L(\alpha)$; $\langle\,\alpha\,\rangle\varphi$ is satisfied in x iff $\langle\,w\,\rangle\varphi$ is satisfied for some $w \in L(\alpha)$; and $[\,\alpha\,\rangle\varphi$ is satisfied in x iff $\langle\,w\,\rangle\varphi$ is satisfied for all $w \in L(\alpha)$. (As a result, for instance, $\langle\,a^+\,]p$ is equivalent to an infinite "disjunction" $[\,a\,]p \vee [\,a^2\,]p \vee \ldots$.) It follows from (3) that $[\,\alpha\,]\varphi$ is satisfied in x iff $X \models_{\mathfrak{M}} \varphi$ for all $X \in N_\alpha(x)$.

Example 2. Let us return to the frame given in Example 1. Formula $[\,(a \cup b)^+\,]p$ is satisfied in, say, $(0,0)$ iff p holds in all (n,m) such that $n + m > 0$, that is, in all elements of the matrix but possibly $(0,0)$. Formula $\langle\,(a \cup b)^+\,\rangle p$ is satisfied in $(0,0)$ iff p is satisfied in some (n,m) such that $n + m > 0$. On the other hand,

$\langle (a \cup b)^+]p$ is satisfied in $(0,0)$ iff there is a set of (n',m'), where $n'+m'=k$ for some $k>0$, such that p is satisfied in *all* elements of the set. That is, there is a number $k>0$ of steps such that p holds in all elements we can get to from $(0,0)$ by k steps. (Informally, you are "guaranteed" to end up in a p-state after k rounds of "choosing" between "going right" and "going up".) For instance, if p holds precisely in the highlighted states in Fig. 2, then $\langle (a \cup b)^+]p$ holds in $(0,0)$, but $[(a \cup b)^+]p$ does not; $\langle (a \cup b)^+ \rangle p$ would hold also if p was satisfied only in some highlighted states. Finally, $[(a \cup b)^+ \rangle p$ is satisfied in $(0,0)$ iff, for all $k>0$, p is satisfied in *some* element of the matrix reachable from $(0,0)$ in k steps. (For each k, there is a sequence of k choices that brings you to a p-state.)

Remark 2. It can be argued that the $\mathbf{1}$-free fragment of the language is still quite expressive. First, it is known that each regular language L over Σ such that $\epsilon \notin L$ is determined by some $\alpha \in Ex(\Sigma)$. Second, $[\alpha^*]\varphi$ is equivalent to $\varphi \wedge [\alpha^+]\varphi$. Hence, for instance, $Fm(\Sigma)$ is still able to represent the fact that φ is common knowledge in a group of agents $\Gamma \subseteq \Sigma$, namely, by means of the formula $\varphi \wedge [(\bigcup \Gamma)^+]\varphi$.

Proposition 2. *The following are valid in each neighborhood model:*

1. $\langle a]\varphi \leftrightarrow [a]\varphi$;
2. $\langle \mathbf{0}]\varphi \leftrightarrow \bot$;
3. $\langle \mathbf{1}]\varphi \leftrightarrow \varphi$;
4. $\langle \alpha \cup \beta]\varphi \leftrightarrow (\langle \alpha]\varphi \vee \langle \beta]\varphi)$;
5. $\langle \alpha\beta]\varphi \rightarrow \langle \alpha]\langle \beta]\varphi$;
6. $\langle \alpha^*]\varphi \rightarrow \varphi \vee \langle \alpha]\langle \alpha^*]\varphi$.

Proof. We prove the last two items. If $x \models \langle \alpha\beta]\varphi$, then there is a word $w \in L(\alpha\beta)$ such that $x \models [w]\varphi$ (see Remark 1). But we know that in this case $w = uv$ for some $u \in L(\alpha)$ and $v \in L(\beta)$. Hence, $x \models [u][v]\varphi$, and so $x \models \langle \alpha]\langle \beta]\varphi$. ($x \models [u]\langle \beta]\varphi$ since for all y that are u-accessible from x, there is a word $v_y \in L(\beta)$, namely v, such that $y \models [v_y]\varphi$.)

Assume that $x \models \langle \alpha^*]\varphi$, that is, $\langle \mathbf{1} \cup \alpha^+]\varphi$. It follows that there is $n \geq 0$ and $w \in L(\alpha^n)$ such that $x \models [w]\varphi$. Hence, either $x \models \varphi$ or there is $m>0$ and $u \in L(\alpha^m)$ such that $x \models [u]\varphi$. If follows that there is $v \in L(\alpha)$ such that $x \models [v]\langle \alpha^*]\varphi$, i.e. $x \models \langle \alpha]\langle \alpha^*]\varphi$.

Remark 3. The converse implications to the last two items of Proposition 2 do not hold. The reason why this is the case is illustrated by the counterexample to

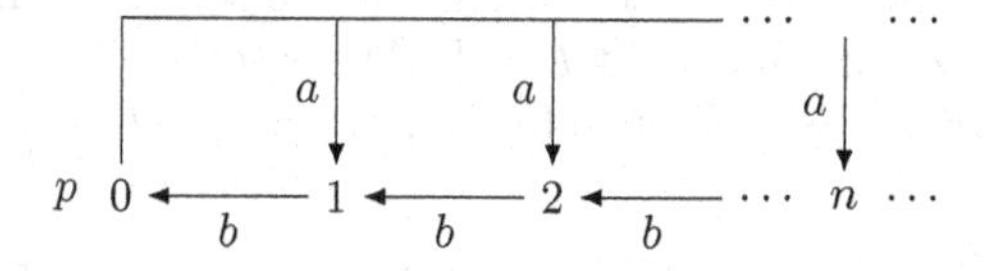

Fig. 3. A counterexample to $\langle a]\langle b^+]p \rightarrow \langle ab^+]p$.

$\langle a \rbrack \langle b^+ \rbrack p \rightarrow \langle ab^+ \rbrack p$ shown in Fig. 3. In that model, $R_a xy$ iff $x = 0$ and $y > x$, and $R_b xy$ iff $x > 0$ and $y = x - 1$. It is assumed that p is satisfied only in 0. Then, for all $n > 0$ there is $w_n \in b^+$ such that $n \models [w_n]p$, namely $w_n = b^n$. However, there is no $w \in L(b^+)$ such that $n \models [w]p$ *for all* $n > 0$. (A finite countermodel exists as well.)

Remark 4 (Similar formalisms). Leivant [7] introduced an extension of **PDL** with program quantification, which he also denoted as **QPDL**. Leivant's quantification over programs is *unrestricted:* $(\forall a)\,\varphi$ is satisfied in x within a model iff φ is satisfied in x within each model that results from the original model by fixing the interpretation of a to be an arbitrary binary relation. Leivant proved that the validity problem for his logic of unrestricted program quantification is wildly undecidable – the set of valid formulas is not in the analytical hierarchy. On the other hand, the notion of program quantification used in this paper is *restricted:* we bound the quantifier by a given regular expression, using, in effect, quantifiers $(\forall w \in L(\alpha))$ and $(\exists w \in L(\alpha))$, combined with modal operators $[w]$; see Remark 1.

Our semantics for the $\exists\forall$ sequence-state quantifier $\langle \alpha \rbrack$ is reminiscent of the semantics of a number of known logics. Neighborhood functions for programs, including the definitions of N_α for complex α employed here, are used in concurrent PDL [12], which also uses formulas equivalent to our $\langle \alpha \rbrack \varphi$. However, concurrent PDL lacks the interplay between N_α and "standard relational accessibility" via α which is crucial in our framework. A similar remark applies also to Game logic [11].

It is interesting to note that there is a relation between the framework presented here and epistemic logic with names [2,5], containing epistemic operators $E_\mathbf{n}$ and $S_\mathbf{n}$, where **n** is a "name", possibly referring to a number of agents, reading "all agents named **n** know that..." and "some agent named by **n** knows that...". Our framework generalizes this construction to the setting where "names" are regular expressions (i.e. they are structured in a particular way) and "agents" are computation sequences.

4 Expressivity

4.1 Expressing Acceptance Properties of NFA

If (S, R) is a Σ-frame, then (S, Σ, s_0, R, F), where $s_0 \in S$ and $F \subseteq S$, can be seen as a nondeterministic finite automaton (NFA), where s_0 is the starting state and F is the set of accepting states. Recall that a NFA A *accepts* a string $w \in \Sigma^*$ iff there is $x \in F$ such that $R_w s_0 x$. Let $L(A)$ be the set of strings accepted by A. Recall that $L = L(\alpha)$ for some α iff $L = L(A)$ for some A. Note that we can see A as a *pointed neighborhood* Σ*-model* $(\mathfrak{M}^A, s_0)$, where $F = V(q)$ for some fixed $q \in Pr$.

Proposition 3. *For all* α *and* A, $L(\alpha) \subseteq L(A)$ *iff* $s_0 \models_{\mathfrak{M}^A} [\alpha\rangle q$.

4.2 Ways to Execute a Plan Successfully

In a more planning-related fashion, regular expressions can be seen as *abstract structured plans* of action, e.g. ab^+c as "do a once, then b some finite non-zero times, and then c once". Computation sequences can be seen as representing sequential executions of specific actions by an agent. A computation sequence $w \in L(\alpha)$ can be seen as a specific course of action instantiating the plan α. In situations where actions are nondeterministic (for instance, the outcome of an action may depend also on "outside influences"), computation sequences do not yield a specific state as an outcome, but rather a set of states that may result from the given computation sequence. Boolean formulas may be used to represent specific desirable states of the world, or *goals.* A practically interesting question concerning abstract plans is whether there is a course of action instantiating a given plan that *guarantees* that a given goal is achieved. This is exactly what formulas of the form $\langle\, \alpha\,]\varphi$ express in our setting: there is a computation sequence represented by α (a concrete way to execute the abstract plan α) such that all states that may possibly result from executing the sequence satisfy φ.

5 Decidability and Complexity of qPDL

In this section we prove that the satisfiability (and validity) problem for **qPDL** is decidable and *EXPTIME*-complete. The result is obtained via an embedding into deterministic PDL, **DPDL**, [1,10].

5.1 DPDL

Let Δ be a countable alphabet; we may assume without loss of generality that $\Sigma \subseteq \Delta$. The sets of Δ-*programs* and Δ-*formulas* are defined by mutual induction as follows:

$$Pm(\Delta) \quad \pi := r \mid \pi \cup \pi \mid \pi ; \pi \mid \pi^* \mid \varphi?$$
$$Fm(\Delta) \quad \varphi := p \mid \neg\varphi \mid \varphi \wedge \varphi \mid [\, \pi\,]\varphi$$

where $r \in \Delta$ and $p \in Pr$. The usual definitions of the other Boolean operators and $\langle\, \pi\, \rangle$ are assumed.

Definition 5. *A* Δ-model *is a tuple* $M = (S, Q, V)$, *where* S *is a non-empty set,* Q *is a function* $\Delta \to 2^{S\times S}$ *such that* $Q(r)$ *is a partial function for all* r, *and* $V : Pr \to 2^S$. Q *is extended to a function* $Pm(\Delta) \to 2^{S\times S}$ *and* V *to a function* $Fm(\Delta) \to 2^S$ *as indicated in Fig. 4. (As usual,* $Q(\pi)^0 = \mathrm{id}_S$ *for all* π.*)* φ *is valid in* M *iff* $V(\varphi) = S$. **DPDL** *is the set of* Δ-*formulas valid in all* M.

We will often write $x \models_M \varphi$ instead of $x \in V(\varphi)$.

Theorem 1. *The satisfiability (and validity) problem for* **DPDL** *is decidable and EXPTIME-complete.*

Proof. See [1,15]. □

$$
\begin{aligned}
Q(\pi_1 \cup \pi_2) &= Q(\pi_1) \cup Q(\pi_2) \\
Q(\pi_1;\pi_2) &= Q(\pi_1) \circ Q(\pi_2) \\
Q(\pi^*) &= \bigcup_{n\geq 0} Q(\pi)^n \\
Q(\varphi?) &= \mathrm{id}_{V(\varphi)}
\end{aligned}
\qquad
\begin{aligned}
V(\neg\varphi) &= S \setminus V(\varphi) \\
V(\varphi \wedge \psi) &= V(\varphi) \cap V(\psi) \\
V([\,\pi\,]\varphi) &= \{x \mid Q(\pi)(x) \subseteq V(\varphi)\}
\end{aligned}
$$

Fig. 4. Interpretation of programs and formulas in models for deterministic PDL.

5.2 An Embedding of qPDL into DPDL

Definition 6. *Fix $d \in \Delta \setminus \Sigma$ and define $e := d^+$. Define a function $t : Ex(\Sigma) \cup Fm(\Sigma) \to Pm(\Delta) \cup Fm(\Delta)$ as follows:*

$$
\begin{aligned}
t(a) &= a \\
t(\mathbf{0}) &= \bot? \\
t(\alpha \cup \beta) &= t(\alpha) \cup t(\beta) \\
t(\alpha\beta) &= t(\alpha)e\,t(\beta) \\
t(\alpha^+) &= t(\alpha)(e\,t(\alpha))^*
\end{aligned}
\qquad
\begin{aligned}
t(p) &= p \\
t(\neg\varphi) &= \neg t(\varphi) \\
t(\varphi \wedge \psi) &= t(\varphi) \wedge t(\psi) \\
t([\,\alpha\,]\varphi) &= [\,t(\alpha)\,][\,e\,]t(\varphi) \\
t(\langle\,\alpha\,]\varphi) &= \langle\,t(\alpha)\,\rangle[\,e\,]t(\varphi)
\end{aligned}
$$

The translation reflects the following idea of "breaking down" the neighborhood function, used in the proofs below: any set $X_i \in N(\alpha)(x)$ can be thought of as the set of states reachable from an auxiliary state x_i (itself reachable from x via $t(\alpha)$) via a finite sequence of executions of an auxiliary deterministic program d (given our image-countability assumption). Reflecting this perspective, $[\,\alpha\,]$ can be "broken down" to two boxes, $[\,t(\alpha)\,][\,d^+\,]$ ("all states in all $X_i \in N(\alpha)(x)$ satisfy..."), and $\langle\,\alpha\,]$ into a diamond $\langle\,t(\alpha)\,\rangle$ and a box $[\,d^+\,]$ ("all states in some $X_i \in N(\alpha)(x)$ satisfy...").

We prove first that for each M there is a neighborhood Σ-model $\mathfrak{M}$ such that, for all $\varphi \in Fm(\Sigma)$, φ is valid in $\mathfrak{M}$ iff $t(\varphi)$ is valid in M. It follows that $\varphi \in$ **qPDL** only if $t(\varphi) \in$ **DPDL**.

Definition 7. *Let $M = (S, Q, V)$ be a Δ-model. We define $M' = (S, R, N, V)$ by*

- *$R(a) := Q(a) \circ Q(e)$;*
- *$N_a(x) := \{R_a(x)\}$.*

R and N are extended to $Ex(\Sigma) \cup \Sigma^*$ as in Fig. 1.

Lemma 1. *For all $\alpha \in Ex(\Sigma)$, $Q(t(\alpha)e) = R(\alpha)$.*

Proof. Induction on α. The base case holds by definition. The cases of the induction step corresponding to $\mathbf{0}$, $\cup$ and ; are easy (the latter two use properties of regular expressions – which can obviously be used in the context of Δ-frames as well). The case corresponding to $^+$ is established as follows:

$$\begin{aligned} Q(t(\alpha^+)\, e) &= Q(t(\alpha)(e\, t(\alpha))^*\, e) \\ &= \bigcup_{n>0} Q(t(\alpha)\, e)^n = \bigcup_{n>0} R(\alpha)^n = R(\alpha^+) \end{aligned}$$

□

Note that $t(\alpha)$ for each $\alpha \in Ex(\Sigma)$ can be regarded as a regular expression over the alphabet $\Sigma_t = \Sigma \cup \{d, \top?, \bot?\}$, so the notation $L(t(\alpha))$ makes sense.[1] Note that $\epsilon \notin L(\alpha)$ for all $\alpha \in Ex(\Sigma)$. Moreover, for $w = a_1 \dots a_n$, $t(w) = \pi_1 \dots \pi_{2n-1}$ such that the i-th odd position is occupied by a_i and each even position is occupied by e. Thus $u \in L(t(w))$ iff

$$u = a_1 d^{m_1} a_2 d^{m_2} \dots d^{m_{n-1}} a_n \tag{4}$$

This means that $u_1 \in L(t(w_1))$ and $u_2 \in L(t(w_2))$ iff $u_1 d^n u_2 \in L(t(w_1 w_2))$ for all $n > 0$. Note also that $(x, y) \in Q_{t(\alpha)}$ iff there is $u \in L(t(\alpha))$ such that $(x, y) \in Q_u$ (by appeal to (1) on $Ex_1(\Sigma_t)$).

Lemma 2. *For all* $\alpha \in Ex(\Sigma)$, $Q_{t(\alpha)} = \bigcup_{w \in L(\alpha)} Q_{t(w)}$.

Proof. Induction on α. The base case is trivial. From the induction step, we prove only the cases for ; and $^+$. The former is established as follows:

$$\begin{aligned} Q_{t(\alpha\beta)} &= \left(\bigcup_{w \in L(\alpha)} Q_{t(w)} \right) \circ Q_e \circ \left(\bigcup_{u \in L(\beta)} Q_{t(u)} \right) \\ &= \bigcup \Big\{ Q(w' d^n u') \,\Big|\, w' \in L(t(w)) \;\&\; w \in L(\alpha) \;\&\; n > 0 \\ &\qquad \& \; u' \in L(t(u)) \;\&\; u \in L(\beta) \Big\} \\ &= \bigcup \{ Q_{t(wu)} \mid w \in L(\alpha) \;\&\; u \in L(\beta) \} = \bigcup \{ Q_{t(v)} \mid v \in L(\alpha\beta) \} \end{aligned}$$

In a similar fashion, we can prove by induction on n that $Q_{t(\alpha^n)} = \bigcup\{Q_{t(v)} \mid v \in L(\alpha^n)\}$. The claim for $^+$ follows from this. □

Lemma 3. *For all* $\alpha \in Ex(\Sigma)$ *and all* $x \in S$, $N_\alpha(x) = \{Q_e(y) \mid Q_{t(\alpha)} xy\}$.

Proof.

$$\begin{aligned} N_\alpha(x) &= \{R_w(x) \mid w \in L(\alpha)\} && \text{(by Prop. 1)} \\ &= \{Q_{t(w)e}(x) \mid w \in L(\alpha)\} && \text{(by Lemma 1)} \\ &= \{Q_e(y) \mid Q_{t(w)} xy \;\&\; w \in L(\alpha)\} \\ &= \{Q_e(y) \mid Q_{t(\alpha)} xy\} && \text{(by Lemma 2)} \end{aligned}$$

□

[1] Take $Ex_1(\Sigma_t)$ and apply Definition 2.

Lemma 4. *For all $\varphi \in Fm(\Sigma)$, $t(\varphi)$ is valid in M iff φ is valid in M'.*

Proof. We prove by induction on the complexity of φ that, for all $x \in S$, $x \models_M t(\varphi)$ iff $x \models_{M'} \varphi$. The base case holds by definition and the Boolean cases of the induction step follow easily from the induction hypothesis. The claim for $[\alpha]\varphi$ follows from Lemma 1. The claim for $\langle \alpha]\varphi$ follows from Lemma 3. □

Now we prove the other direction: for each neighborhood model $\mathfrak{M}$ there is a Δ-model M such that if φ is not valid in $\mathfrak{M}$, then $t(\varphi)$ is not valid in M. It follows that $\varphi \in$ **qPDL** if $t(\varphi) \in$ **DPDL**.

Recall our assumption that in each neighborhood Σ-model, $R_a(x)$ is at most countable for each $x \in S$ and $a \in \Sigma$. This means that for each a and x, there is a bijection θ between $R_a(x)$ and either $\mathbb{Z}^+$ or $\{k \mid 0 < k < m\}$ for some $m > 0$. It the former case, it is clear what *n-th element* of $R_a(x)$ means (namely, $\theta(n)$), for all $n > 0$. In the latter case, let us stipulate that the n-th element of $R_a(x)$ does not exist, for all n, in case $R_a(x)$ is empty and, if the cardinality of $R_a(x)$ is $m > 0$, then the n-th element of $R_a(x)$ is $\theta(m)$ if m divides n and $\theta(n \bmod m)$ otherwise. (Hence, for example, if $m = 3$, then the 4th element is $\theta(1)$ and the 6th element is $\theta(m)$.)

Definition 8. *Let $\mathfrak{M} = (S, R, N, V)$ be a neighborhood Σ-model. Let $U = \{u \mid u \in L(t(w))$ for some $w \in \Sigma^+\}$, i.e. the set of words of the form (4). Let $T = \{\epsilon\} \cup (U \cdot L(e))$ and $W = (S \times U) \cup (S \times T)$. Note that $w \in T$ iff $w = \epsilon$ or $w = vad^n$ for some $v \in T$, $a \in \Sigma$ and $n > 0$. We write xw instead of (x, w) for $w \in U \cup T$.*

We define a partial function $s : (S \times T) \to S$ by induction on the length of $w \in T$ as follows:

- $s(x\epsilon) = x$;
- $s(xwad^n)$ *is the n-th element of* $R_a(s(w))$ *if it exists;*
- $s(xwad^n)$ *is undefined if* $R_a(s(w)) = \emptyset$.

We define $\mathfrak{M}' = (W, Q, V')$ where

- $Q_a = \{(xw, xwa) \mid x \in S \ \& \ w \in T\}$;
- $Q_d = \{(xwa, xwad) \mid x \in S \ \& \ w \in T \ \& \ a \in \Sigma \ \& \ R_a(s(xw)) \neq \emptyset\} \cup \{(xwad^n, xwad^{n+1}) \mid x \in S \ \& \ w \in T \ \& \ a \in \Sigma \ \& \ n > 0\}$;
- $Q_r = \emptyset$ *for all* $r \in \Delta \setminus (\Sigma \cup \{d\})$;
- $V'(p) = \{xw \mid w \in T \ \& \ s(w) \in V(p)\}$.

Lemma 5. *If $x = s(zw)$ and $y = s(zu)$ for some $z \in S$ and $w, u \in T$, then $R_\alpha xy$ iff $Q_{t(\alpha)e}(zw, zu)$.*

Proof. Induction on the complexity of α. We assume throughout that $x = s(zw)$ and $y = s(zu)$ for some $z \in S$ and $w, u \in T$. $Q_{t(a)e}(zw, zu)$ iff $Q_{d^+}(zwa, zu)$ iff $u = wad^n$ for some $n > 0$ and $s(zu) \in R_a(x)$ iff $R_a xy$. The induction step is easy and we omit it. (In the case for ; we use the obvious fact that $Q_\pi(zw, z'u)$ only if $z = z'$.) □

Lemma 6. *For all $x, y \in S$, $w \in T$ and $\alpha \in Ex(\Sigma)$, if $R_\alpha s(xw)y$, then $y = s(xu)$ for some $u \in T$.*

Lemma 7. *For all $x \in S$, $w \in T$ and $\alpha \in Ex(\Sigma)$, $N_\alpha(s(xw)) = \{Q_e(xu) \mid Q_{t(\alpha)}(xw, xu)\}$.*

Proof. This lemma is established quite similarly as Lemma 3, using Proposition 1, Lemma 5 and the claim that $Q_{t(\alpha)} = \bigcup_{w \in L(\alpha)} Q_{t(w)}$, which is established exactly as Lemma 2.

Lemma 8. *For all $\varphi \in Fm(\Sigma)$, all $\mathfrak{M}$ and all x in $\mathfrak{M}$: $x \models_{\mathfrak{M}} \varphi$ iff $zw \models_{\mathfrak{M}'} t(\varphi)$ for all $z \in S$ and $w \in T$ such that $x = s(zw)$.*

Proof. Induction on the complexity of φ. The base holds by definition and the Boolean cases of the induction step follow easily from the induction hypothesis. The case corresponding to $[\alpha]\varphi$ follows from Lemma 5 and the case corresponding to $\langle \alpha]\varphi$ follows from Lemma 7. □

Theorem 2. $\varphi \in$ **qPDL** *iff* $t(\varphi) \in$ **DPDL**.

Proof. Lemmas 4 and 8.

Theorem 3. **qPDL** *is decidable and the* **qPDL***-satisfiability problem is EXPTIME-complete.*

Proof. Theorems 1 and 2, and the fact that the translation function t is polynomial. (In particular, **qPDL**-sat is in *EXPTIME* by Theorem 2, since $\neg\varphi \notin$ **qPDL** iff $\neg t(\varphi) \notin$ **DPDL**, and **qPDL**-sat is *EXPTIME*-hard since, for $\langle \alpha]$-free formulas φ, $\varphi \in$ **qPDL** iff φ is valid in test-free PDL, and the latter is *EXPTIME*-hard [4].)

6 Discussion

A similar and somewhat simpler argument can be used to embed **QPDL** into **DPDL**$'$, the logic arising from a variant of Δ-models where the atomic program d is non-deterministic (and it plays the role of e in the embedding) and, in addition, the property that $Q(cd) = \mathrm{id}_S$ is assumed for some atomic program c. However, we are not able to use this embedding to establish decidability of **QPDL** since it is not clear at this point if the proof techniques used in [1,15] to establish decidability of **DPDL** can be applied in some form to **DPDL**$'$ as well.

7 Conclusion

We argued that the language of regular PDL is not sufficiently strong to express specific statements involving $\exists\forall$ or $\forall\exists$ combinations of quantifiers over computation sequences determined by a given regular expression and states in a model reachable via such computation sequences. Examples include statements that a particular NFA accepts all words in a given regular language or that some finite number of iterations of an action is guaranteed to accomplish a given goal. We introduced **QPDL**, an extension of regular PDL using a language containing modal operators expressing these quantifier combinations. We showed that **qPDL**, the **1**-free fragment of **QPDL**, is decidable and *EXPTIME*-complete. These results were obtained by embedding **qPDL** into deterministic PDL.

Several interesting problems are left for future research. Most importantly, we would like to settle the question of decidability of **QPDL**. A viable route to achieve this is to determine if **DPDL**′ is decidable. Another attractive problem is to find a sound and complete axiomatization of **QPDL**. The cursory observations on valid formulas in Sect. 3 may provide a starting point. Finally, one might find it interesting to study **QPDL** extended with tests.

Acknowledgements. The author is grateful to the anonymous referees for their comments. This work was supported by the long-term strategic development financing of the Institute of Computer Science (RVO:67985807).

References

1. Ben-Ari, M., Halpern, J.Y., Pnueli, A.: Deterministic propositional dynamic logic: finite models, complexity, and completeness. J. Comput. Syst. Sci. **25**(3), 402–417 (1982). https://doi.org/10.1016/0022-0000(82)90018-6
2. Bílková, M., Christoff, Z., Roy, O.: Revisiting epistemic logic with names. In: Halpern, J., Perea, A. (eds.) Proceedings of the 18th Conference on Theoretical Aspects of Rationality and Knowledge (TARK 2021), pp. 39–54 (2021). https://doi.org/10.4204/EPTCS.335.4
3. Fagin, R., Halpern, J.Y., Moses, Y., Vardi, M.Y.: Reasoning About Knowledge. MIT Press, Cambridge (1995)
4. Fischer, M.J., Ladner, R.E.: Propositional dynamic logic of regular programs. J. Comput. Syst. Sci. **18**, 194–211 (1979). https://doi.org/10.1016/0022-0000(79)90046-1
5. Grove, A.J., Halpern, J.Y.: Naming and identity in epistemic logics. Part I: the propositional case. J. Logic Comput. **3**(4), 345–378 (1993). https://doi.org/10.1093/logcom/3.4.345
6. Harel, D., Kozen, D., Tiuryn, J.: Dynamic Logic. MIT Press, Cambridge (2000)
7. Leivant, D.: Propositional dynamic logic with program quantifiers. Electron. Notes Theoret. Comput. Sci. **218**, 231–240 (2008). https://doi.org/10.1016/j.entcs.2008.10.014. Proceedings of the 24th Conference on the Mathematical Foundations of Programming Semantics (MFPS XXIV)
8. Meyer, J.J.C.: A different approach to deontic logic: deontic logic viewed as a variant of dynamic logic. Notre Dame J. Formal Logic **29**(1), 109–136 (1987). https://doi.org/10.1305/ndjfl/1093637776
9. Meyer, J.J.C.: Dynamic logic for reasoning about actions and agents. In: Minker, J. (ed.) Logic-Based Artificial Intelligence. SECS, vol. 597, pp. 281–311. Springer, Boston (2000). https://doi.org/10.1007/978-1-4615-1567-8_13
10. Parikh, R.: Propositional logics of programs: systems, models, and complexity. In: Proceedings of the 7th ACM SIGPLAN-SIGACT Symposium on Principles of Programming Languages, POPL 1980, pp. 186–192. Association for Computing Machinery, New York (1980). https://doi.org/10.1145/567446.567464

11. Parikh, R.: The logic of games and its applications. In: Karplnski, M., van Leeuwen, J. (eds.) Topics in the Theory of Computation, North-Holland Mathematics Studies, vol. 102, pp. 111–139. North-Holland (1985). https://doi.org/10.1016/S0304-0208(08)73078-0
12. Peleg, D.: Concurrent dynamic logic. J. ACM **34**(2), 450–479 (1987). https://doi.org/10.1145/23005.23008
13. Rosenschein, S.J.: Plan synthesis: a logical perspective. In: Proceedings of the Seventh International Joint Conference on Artificial Intelligence (IJCAI 1981), pp. 331–337 (1981)
14. Spalazzi, L., Traverso, P.: A dynamic logic for acting, sensing, and planning. J. Logic Comput. **10**(6), 787–821 (2000). https://doi.org/10.1093/logcom/10.6.787
15. Vardi, M.Y., Wolper, P.: Automata-theoretic techniques for modal logics of programs. J. Comput. Syst. Sci. **32**(2), 183–221 (1986). https://doi.org/10.1016/0022-0000(86)90026-7

Finite Generation and Presentation Problems for Lambda Calculus and Combinatory Logic

Rick Statman(✉)

Rick Statman, Pittsburgh, PA 15213, USA
statman@cs.cmu.edu

Abstract. We solve several finite generation and presentation problems for lambda calculus and combinatory logic.

1 Introduction

Although we prefer Dana Scott's use of "Combinatory Algebra" for "Combinatory Logic" this has come to take on a technical meaning. So, we will stick with CL. In 1937 Church [4] formulated lambda calculus as a semigroup. His ideas were pursued by Curry and Feys [5], and later by Bohm [2] (see also [1, p. 532]) and Dezani [6]. If lambda terms in some way represent functions, then such a presentation based on composition is a natural complement to the presentation based on application. So, we have problems based on composition as well as application.

In 1986 [11] the author proved that the problem of deciding if a finite set of normal lambda terms forms an applicative basis is recursively unsolvable. The problem for finite sets of normal proper lambda terms remains unresolved to this day. However, Broda and Damas [3] have proved that the linear case is decidable. There is a similar problem for the monoid, but here no finite set of proper lambda terms can generate the entire monoid. This is because no such set generates a normal closed lambda-I term with a proper closed subterm. Here we prove that

(a) the membership problem for the submonoid generated by a finite set of regular proper lambda terms is decidable, and
(b) the membership problem for the submonoid generated by a finite set of normal lambda terms is undecidable. In particular, it is undecidable if a given finite set of normal lambda terms generates the whole monoid.

In [1, 21.1, p. 532] Barendregt asks if the lambda calculus semigroup under beta-eta conversion is finitely presented. Of course, there is the corresponding question for application and the similar ones for CL. Here we answer all but the first

S. Artemov and A. Nerode (Eds.): LFCS 2022, LNCS 13137, pp. 316–326, 2022.
https://doi.org/10.1007/978-3-030-93100-1_20

(c) no extension of the linear lambda calculus by finitely many true equations between closed lambda terms can present full beta-eta conversion, and
(d) no extension of super-linear combinatory logic by finitely many true equations between applicative combinations of combinators can present full weak-beta conversion.

Here “true” means resp. beta-eta conv. and weak-beta conv. Superlinear combinatory logic extends linear CL by the addition of the axioms appropriate for the combinator monoid. As a consequence we obtain

(e) no extension of super-linear combinatory logic by finitely many true equations between applicative combinations of combinators can present the combinatory monoid.

2 Finite Generation

We begin with a simple problem. A regular proper lambda term M is a normal form

$$\lambda a.\lambda x_1 \cdots x_n.aX_1 \cdots X_m$$

where each X_i is an applicative combination of the variables $x_1, \ldots, x_n$. Here m is the degree of M. These form a submonoid of the monoid of all lambda terms under beta-eta conversion. If \$ is a finite set of regular proper lambda terms then let s be the maximum of the degrees of members of \$. Let P be any product of members of \$. Then beta-eta normal form of P can be obtained by eta postponement. It is easy to see that any eta reduction takes place on a term of degree at most s. Now we assume that every member of \$ has been eta expanded to degree exactly s. Given \$, M we wish to decide if M is in the submonoid generated by \$. For this we can assume that the degree of M is not larger than s and we consider the individual eta expansions of M up to degree s separately. For each such expansion we encode the problem into the corresponding problem for Cartesian monoids (Statman [13], Theorem 3) as follows.

To each term $P = \lambda a.\lambda y_1 \cdots y_p.aY_1 \cdots Y_q$ we associate a CQ expression $\#P$ as follows. First, for subterms;

- $\#(y_i) = R * \cdots * R * L$ for i occurrences of R,
- $\#(UV) = \langle \#U, \#V \rangle$ for applicative combinations of the y_j, and finally
- $\#P = \langle L, \langle \#Y_1, \langle \cdots \langle \#(Y_{q-1}), \#(Y_q) \rangle \cdots \rangle \rangle \rangle$.

Set $\#\$ = \{\#P \mid P \in \$\}$. Then it is easy to see that

Proposition 1. *\$ generates M iff $\#\$$ generates $\#M$.*

Thus by Statman [13] theorem 3 the problem is decidable.

We now consider the general case where \$ is a finite set of beta-eta normal forms. For each Godel number e we construct terms X, Y, Z such that

(i) X, Y, Z are beta-eta normal terms with one free variable x, and

(ii) if $\{e\}(e)$ diverges then $[I/x]X$, $[I/x]Y$, $[I/x]Z$ are all unsolvable, and
(iii) if $\{e\}(e)$ converges then
- $[I/x]X$ beta-eta conv. S
- $[I/x]Y$ beta-eta conv. K
- $[I/x]Z$ beta-eta conv. I.

Next, we define terms L, M, N, P as follows.

- $L := \lambda a \lambda u.a(u(\lambda x.X)I)$
- $M := \lambda a \lambda u.a(u(\lambda x.Y)I)$
- $N := \lambda a \lambda u.\lambda yz.a(u(\lambda x.Z)I(yz))$
- $P := \lambda a \lambda u.\lambda v \lambda b.(\lambda z.b(v(\lambda x.Z)Iz)((\lambda y.a(u(\lambda x.Z)Iy)))$.

In addition we have $Q := CII$. Finally set

$$O = \{L, M, N, P, Q\}.$$

Now assume $\{e\}(e)$ diverges. We write OMEGA for a generic unsolvable term. We use the following notation;

$$a(\text{OMEGA})+ := a(\text{OMEGA})Z_1 \cdots Z_m$$

for arbitrary $Z_1 \cdots Z_m$.

$$z_i+ := z_{i,1} \ldots z_{i,s(i)}$$

for the ith sequence of $s(i)$ variables. Set

$$J := \lambda z.\lambda v \lambda b.(b(v(\lambda x.Z)Iz);$$

we almost get a numeral system.
Facts:

1. If R is a product of L, M, N, and P, then R beta-eta conv. to the form

$$\lambda au.\lambda z_1 + .J(\cdots(\lambda z_n + .J(\lambda z + .a(uZ_1..Z_m)))\cdots)$$

where Z_1 is $\lambda x.X$, $\lambda x.Y$, or $\lambda x.Z$, and Z_2 is I. The number of occurrences of J can be 0.
2. If R is as in (1) then $Q^n * R$ beta-eta conv.
   - $\lambda a.\lambda z_1 + .J(\cdots(\lambda z_n + .J(\lambda z + .a(\text{OMEGA})+)\cdots)$
   - $\lambda a \lambda b.b(\text{OMEGA})$, or
   - $\lambda a.\text{OMEGA}$
3. If R is a product of lambda terms of the form

$$\lambda a.\lambda z_1 + .J(\cdots(\lambda z_n + .J(\lambda z + .a(\text{OMEGA})+)))\cdots)$$

and

$$\lambda au.\lambda z_1 + .J(\cdots(\lambda z_n + .J(\lambda z + .a(uZ_1..Z_m)))\cdots)),$$

where Z_1 is $\lambda x.X$, $\lambda x.Y$, or $\lambda x.Z$, and Z_2 is I, and if there is at least one of the first type then every occurrence of the first bound variable is in a context ___(OMEGA)+. Diagrammatically, $\lambda a. \cdots$ a(OMEGA)+$\cdots$.

Lemma 1. *If* $\{e\}(e)$ *diverges then* CIS *is not generated by* O.

Proof. Suppose not and R is a product on members of O which beta-eta converts to CIS. We may suppose that R has the form

$$Q^{s(1)} * T_1 * Q^{s(2)} * T_2 * \cdots * Q^{s(n)} * T_n * Q^k$$

where for $i = 2, \ldots, n$, s(i) ¿ 0 and for $j = 1, \ldots, n$, T_j does not contain Q. Now if $s(i) > 0$, $Q^{s(i)} * T_i$ falls under Fact (2). It cannot be the case that $Q^{s(i)} * T_i$ beta-eta conv. to $\lambda a \lambda b.b$(OMEGA), or to λa.OMEGA since these together form a semigroup ideal not containing CIS. Thus

$$Q^{s(1)} * T_1 * Q^{s(2)} * T_2 * \cdots * Q^{s(n)} * T_n$$

falls under Fact (3). Now, if every occurrence of the first bound variable is in a context __(OMEGA)+, then R beta-eta converts to a lambda term with the same property, since Q^k beta-eta conv. to $\lambda a.aI \cdots I$. Thus R beta-eta converts to a product of terms of the form

$$\lambda au.\lambda z_1 + .J(\cdots(\lambda z_n + .J(\lambda z + .a(uZ_1..Z_m)))\cdots)$$

followed by Q^k. Clearly no such term beta-eta converts to CIS.

Theorem 1. O *generates the monoid iff* $\{e\}(e)$ *converges.*

Proof. First suppose that $\{e\}(e)$ converges. Then O generates B, CI, CIS, and CIK. In particular, $Q * Q * P$ beta-eta conv.

$$\lambda ab.(\lambda z.bz)(\lambda y.ay)$$

eta conv. $\lambda ab.ba$. Thus by Church's theorem ([1, 21.1.2, p. 533]) O generates the entire monoid. Conversely, if $\{e\}(e)$ diverges then O does not generate CIS by lemma 1.

3 Finite Presentation

We begin with some preliminaries. The equation calculus is a formal system logically complete for proving equations $X = Y$ from sets of equations \$ such that if $U = V$ is in \$ then so is $V = U$. The axioms and rules of the equation calculus are the following

(ref)	$X = X$	
(axE)	$U = V$	provided $U = V$ is in E
(sub)	$[U/u]\,(X = Y)$	and $U = V \Rightarrow [V/u]\,(X = Y)$

This system is normally formulated only for first order terms without bound variables. So it is perfectly suited to CL. We can also use it for lambda calculus since we assume that the substitution $[U/u]$ in (sub) does not bind any variable in U. We will not quibble about alpha conversion and allow it freely. We write $\$ \vdash X = Y$ if $X = Y$ is provable from \$ in the equation calculus.

An \$ computation is a sequence of terms

$$X_0, \ldots, X_n$$

such that for each $i = 0, \ldots, n-1$ there exists X, $U = V$ in \$, and u with

$$\begin{aligned} X_i &= [U/u]X \\ X_{i+1} &= [V/u]X \end{aligned}$$

where u appears in X at most once. The following is an old observation from [10].

Fact: $\$ \vdash X = Y$ iff there is an \$ computation beginning with X and ending with Y.

Now we will be interested in \$ which extend a particular set. The linear lambda calculus consists of all lambda terms together with the congruence $\sim$ generated by the following

(bta)	$(\lambda x.X)Y \sim [Y/x]X$	provided x occurs exactly once in X
(eta)	$\lambda x.(Xx) \sim X$	provided x does not occur in X
(ref)	$X \sim X$	
(sym)	$X \sim Y$	$\Rightarrow Y \sim X$
(tra)	$X \sim Y$ and $Y \sim Z$	$\Rightarrow X \sim Z$
(app)	$X \sim Y$ and $U \sim V$	$\Rightarrow (XU) \sim (YV)$
(abs)	$X \sim Y$	$\Rightarrow \lambda x.X \sim \lambda x.Y$

Now let E be the set $\{P = Q \mid P \sim Q\}$. Now the redex-contractum relation

(bta)	$(\lambda x.X)Y \rightarrow [Y/x]X$	provided x occurs exactly once in X
(eta)	$\lambda x.(Xx) \rightarrow X$	provided x does not occur in X

generates the 1-step reduction relation $\rightarrow$, and this, in turn, generates the multi-step reduction relation $\twoheadrightarrow$, and the term congruence lin. conv. It is a well known fact that

$$E \vdash X = Y \quad \text{iff } X \text{ lin. conv. } Y.$$

Next we consider a couple of easy cases. CL first. Suppose that we have a finite set of combinators; for example, B, C, K, I, W. In addition, suppose that we have a finite set of closed equations between applicative combinations of these combinators; for example, the Eqs. A.1–A.6 [1, 7.3.15, p. 161]. Then this set of equations cannot present weak-beta conversion. This is because provability from this set, without the use of the reduction rules for the given combinators, is decidable by [9].

Let \$ be a finite set of closed equations between lambda terms. We have the following Jacopini type theorem [7, 12] for linear lambda calculus, with additional axioms \$.

Lemma 2. *For closed lambda terms P, Q we have $\$ \cup E \vdash P = Q$ iff there exist closed terms $M_1, \ldots, M_m$ and equations $P_1 = Q_1, \ldots, P_m = Q_m$ in \$*

$$\begin{aligned}
&\text{s.t.}\\
&P \text{lin.conv.} M_1 P_1\\
&M_1 Q_1 \text{lin.conv.} M_2 P_2\\
&\quad\vdots\\
&M_{m-1} Q_{m-1} \text{lin.conv.} M_m P_m\\
&M_m Q_m \text{lin.conv.} Q.
\end{aligned}$$

Proof. Given a proof of $X = Y$ from $\$ \cup E$ we construct an $\$ \cup E$ computation Y from X with the additional property that if in the transition

$$X_i = [U/u]X$$
$$X_{i+1} = [V/u]X$$

we have U lin.conv. V then U is X.

Examples

Now the following are provable in the equation calculus from E alone with the lambda terms

$$B := \lambda abc.a(bc)$$
$$C^* := \lambda ab.ba$$
$$I := \lambda a.a.$$

They are organized into three distinct groups.

Combinatory:

(1) $Ix = x$
(2) $C^* xy = yx$
(3) $Bxyz = x(yz)$
(4) $Lx_1 \cdots x_n = X$ for any other linear combinator L with redex/contractum rule $Lx_1 \cdots x_n \to X$

Monoid:

(5) $B(Bxy)z = Bx(Byz)$
(6) $B(Bx)(By) = B(Bxy)$
(7) $BI = I$
(8) $BxI = x$

Inverse Properties:

(9) $B(B(Bx))B = BB(Bx)$

(10) $B(C^*x)(By) = By(C^*x)$
(11) $B(C^*x)C^* = x$
(12) $B(C^*y)(B(C^*x)B) = C^*(xy)$.

Given a lambda term X and an occurrence of lambda λx in X we set $|\lambda x| =$ the number of occurrences of x bound by λx, and we define —X— = the maximum of all $|\lambda x|$ in X. We observe

Lemma 3. *Suppose that there is an integer n s.t. whenever $M = N$ belongs to \$ we have $|M|$ and $|N| < n$. Then if $\$ \cup E \vdash P = Q$ we have*

$$|Q| < \max\{n, |P| + 1\}.$$

Proof. Suppose $\$ \cup E \vdash P = Q$. We apply Jacopini's theorem to get closed terms $M_1, \ldots, M_m$ and $P_1 = Q_1, \ldots, P_m = Q_m$ in \$

s.t.
$P \text{lin.conv.} M_1 P_1$
$M_1 Q_1 \text{lin.conv.} M_2 P_2$
$\vdots$
$M_{m-1} Q_{m-1} \text{lin.conv.} M_m P_m$
$M_m Q_m \text{lin.conv.} Q.$

Now we note that either

$$|M_i Q_i| = |M_i| < |M_i P_i| + 1$$

or

$$|M_i Q_i| = |Q_i| < n$$

and the lemma follows by induction.

Corollary 1. *No finite \$ can prove all true P beta-eta conv. Q.*

Remark 1. Even though (1)–(12) includes the full monoid of lambda calculus the corollary does not mean that the monoid has no finite presentation. Given a finite set of generators, like C^*S, C^*K, B, C^*, I, the P and Q would be linear in the generators.

Finally, we settle finite representability for CL. First an intermezzo. The following questions seems very natural. Take a finite set of weak-beta congruence classes of S, K combinations. For each class take all the equations between terms in that class and let \$ be the union of this finite number of infinite sets of equations. Is it possible to find a finite set such that for all closed M, N

$$M \text{weak} - \text{betaconv.} N \text{iff} \$ \models M = N?$$

It turns out that the answer to this question will recur in the general case for CL. Let \$ be determined as above by a finite set of weak-beta classes $C_1, \ldots, C_n$.

Lemma 4. *Suppose that* $\$ \vdash M = N$. *Then there exists a context* $P[x_1, \ldots, x_n]$ *and pairs* P_1, Q_1 *in* $C_1, \ldots, P_n, Q_n$ *in* C_n *such that*

$$M = P[P_1, \ldots, P_n],$$

and

$$N = P[Q_1, \ldots, Q_n].$$

Proof. Suppose that $\$ \vdash M = N$. Then there is an \$ computation beginning with M and ending with N. But overlapping steps

$$X_i = [U/u]X$$
$$X_{i+1} = [[W/v]V/u]X$$
$$X_{i+2} = [[Y/v]V/u]X,$$

and

$$X_i = [[Y/v]V/u]X$$
$$X_{i+1} = [[W/v]V/u]X$$
$$X_{i+2} = [U/u]X,$$

can be replaced by

$$X_i = [U/u]X$$
$$X_{i+2} = [[Y/v]V/u]X,$$

and

$$X_i = [[Y/v]V/u]X$$
$$X_{i+2} = [U/u]X.$$

Proposition 2. *There is no finite set \$ such that*

$$M \text{weak} - \text{betaconv.} N \text{iff} \$ \vdash M = N.$$

Proof. It suffices to find an S redex $SPQR$ such that

(1) None of SP, P, Q and R belong to one of the weak-beta congruence classes $C_1, \ldots, C_n$
(2) $SPQR$ has a weak-beta head normal form.

For if $\$ \vdash SPQR = T$ then T has the form $LP'Q'R'$, where $\$ \vdash L = S$ and $P = P'$ and $Q = Q'$ and $R = R'$, which is not in head normal form. Finding such an $SPQR$ is easy.

We now consider the following problem. Suppose that we are given a finite set of true equations between applicative combinations of combinators; that is, equations where the left hand side weak-beta converts to the right hand side.

Can this set together with the reduction rules for all linear combinators prove all true (weak-beta) equations?

We will show that the answer is no. Indeed we can do somewhat better.

The set of superlinear reduction rules is defined as follows. They are simply the equations from the above examples presented as reductions.

Combinatory:

(1) $Ix \rightarrowtail x$
(2) $C^*xy \rightarrowtail yx$
(3) $Bxyz \rightarrowtail x(yz)$
(4) $Lx_1 \cdots x_n \rightarrowtail X$ reduction rules for other proper linear combinators; these can be added at will

Monoid:

(5) $B(Bxy)z \rightarrowtail Bx(Byz)$
(6) $B(Bx)(By) \rightarrowtail B(Bxy)$
(7) $BI \rightarrowtail I$
(8) $BxI \rightarrowtail x$

Inverse Properties:

(9) $B(B(Bx))B \rightarrowtail BB(Bx)$
(10) $B(C^*x)(By) \rightarrowtail By(C^*x)$
(11) $B(C^*x)C^* \rightarrowtail x$
(12) $B(C^*y)(B(C^*x)B) \rightarrowtail C^*(xy)$.

Next define height and weight of a lambda term as follows.

$$\begin{aligned}
\text{height}(x) &= 1\\
\text{height}(\lambda x.X) &= 1 + \text{height}(X)\\
\text{height}((XY)) &= \text{height}(X) + \text{height}(Y)\\
\text{weight}(x) &= 2\\
\text{weight}(\lambda x.X) &= 1 + \text{weight}(X)\\
\text{weight}((XY)) &= (2^{\text{weight}(X)}) + \text{weight}(Y).
\end{aligned}$$

Note 1. Let $t(1) = 2$ and $t(n+1) = 2^{t(n)}$ then the maximum weight of a closed term of height n is less than $t(n)$.

Lemma 5. *Every $\rightarrowtail$ reduction sequence terminates. $\rightarrowtail$ has the weak diamond property.*

Proof. Let the rank of X be (height(X), weight(X)) and these are ordered lexicographically. $\rightarrowtail$ reductions reduce rank. Only (5) needs to be checked. For the weak diamond property there are only 9 critical pairs worth noting. They are (3) and (5)–(8), (3) and (9)–(12), and (5) and (8), and these are easily verified.

Corollary 2. $\twoheadrightarrow$ *is Church-Rosser.*

Now suppose that we are given a finite set of true equations between applicative combinations of combinators. Without loss of generality we can assume that

(i) the equations are weak-beta redex/contractum pairs,
(ii) the combinators are from among B, C^*, I, W^*, K by combinatory completeness.

Now, in addition, we will assume that the redexes of the weak-beta redex/contractum pairs are $\rightarrowtail$ normal. $\rightarrowtail$ will be added back later. Thus the redex/contractum pairs have the form

(iv) W^*M weak-beta red. MM, or
(v) KMN weak-beta red. M.

Let the resulting set be \$. We write '$\mapsto$' for 'weak-beta red.'

Next we take a variant of the Knuth-Bendix [8] completion of \$. The set \$+ is obtained by applying the following closure operations to the initial \$.

$$F[G] \mapsto H[G], G \mapsto J \Rightarrow F[J] \mapsto H[J]$$
$$F[G] \mapsto H[G], G \mapsto J \Rightarrow F[J] \mapsto H[J].$$

We obtain the following

Lemma 6. $\mapsto\!\!\!\twoheadrightarrow$ *restricted to \$+ is Church-Rosser.*

Proof. The proof in [1, 3.2, p. 59] adapts well here. It is often called "the parallel reduction proof."

Lemma 7. *Let* $\rightarrowtail\!\!\!\mapsto$ *be the union of* $\rightarrowtail$ *unrestricted and* $\mapsto$ *restricted to \$+. Then* $\rightarrowtail\!\!\!\mapsto$ *is Church-Rosser.*

Proof. By design $\rightarrowtail$ commutes with $\mapsto$ restricted to \$+. So, we can apply Rosen's theorem [1, p. 64].

Now if $F \mapsto H$ is member of \$+ then there exists a member of \$ with a redex weak-beta conv. to F. There are only finitely many of these. Thus there are infinitely many MM normal w.r.t. the reductions

$$\rightarrowtail \cup \mapsto$$

s.t.

$$W^*M$$

is normal w.r.t. $\mapsto$ restricted to \$+. For example, $M := B(K(\cdots(BK)\cdots))$ Thus by the Church-Rosser theorem $W^*M = MM$ is not provable from the original finite set of true equations.

Finally we show that the combinatory monoid is not finitely presented. More precisely we do not consider full eta conv. for the combinator case which would require either the combinatory axioms A.1–A.6 in [1, p. 157] or strong reduction. Instead we add to weak-beta reduction the above Eqs. (1)–(12). These equations give us Church's finitely generated, recursively presented monoid in combinator form.

Theorem 2. *Church's combinator monoid is not finitely presentable.*

Proof. Suppose such a presentation exists. Without loss of generality we can assume it is on the generators C^*S, C^*K, I, B, and C^*. Then there exists a finite set of weak-beta combinatory reduction pairs which together with the Eqs. (1)–(12) proves the equations in the presentation. But by Eqs. (11) and (12) this contradicts the remarks of the previous paragraph.

References

1. Barendregt, H.P.: The Lambda Calculus: Its Syntax and Semantics. Studies in Logic and the Foundations of Mathematics, vol. 103. North-Holland (1985)
2. Böhm, C.: Personal Communication
3. Broda, S., Damas, L.: On combinatory complete sets of proper combinators. J. Funct. Program. **7**(6), 593–612 (1997)
4. Church, A.: Combinatory logic as a semigroup. Bull. Am. Math. Soc. **43**, 333 (1937)
5. Curry, H.B., Feys, R.: Combinatory Logic, vol. 1. North Holland (1958)
6. Dezani-Ciancaglini, M.: Characterization of normal forms possessing inverse in the lambda-beta-eta-Calculus. Theor. Comput. Sci. **2**(3), 323–337 (1976). https://doi.org/10.1016/0304-3975(76)90085-2
7. Jacopini, G.: A condition for identifying two elements of whatever model of combinatory logic. In: Böhm, C. (ed.) λ-Calculus and Computer Science Theory. LNCS, vol. 37, pp. 213–219. Springer, Heidelberg (1975). https://doi.org/10.1007/BFb0029527
8. Knuth, D.E., Bendix, P.B.: Simple Word Problems in Universal Algebras, pp. 263–297. Pergamon Press Ltd. (1970)
9. Kozen, D.: Complexity of finitely presented algebras. In: Hopcroft, J.E., Friedman, E.P., Harrison, M.A. (eds.) Proceedings of the 9th Annual ACM Symposium on Theory of Computing, Boulder, Colorado, USA, 4–6 May 1977, pp. 164–177. ACM (1977). https://doi.org/10.1145/800105.803406
10. Statman, R.: Herbrand's Theorem and Gentzen's Notion of a Direct Proof. In: Barwise, J. (eds.) Handbook of Mathematical Logic, pp. 897–912. North Holland (1977)
11. Statman, R.: On translating lambda terms into combinators; the basis problem. In: Proceedings of the Symposium on Logic in Computer Science, LICS 1986, Cambridge, Massachusetts, USA, 16–18 June 1986, pp. 378–382. IEEE Computer Society (1986)
12. Statman, R.: Consequences of Jacopini's theorem: consistent equalities and equations. In: Girard, J.-Y. (ed.) TLCA 1999. LNCS, vol. 1581, pp. 355–364. Springer, Heidelberg (1999). https://doi.org/10.1007/3-540-48959-2_25
13. Statman, R.: Products in a category with only one object. In: Spivak, D.I., Vicary, J. (eds.) Proceedings of the 3rd Annual International Applied Category Theory Conference 2020, ACT 2020. EPTCS, Cambridge, USA, 6–10 July 2020, vol. 333, pp. 347–353 (2020). https://doi.org/10.4204/EPTCS.333.24

Exact and Parameterized Algorithms for Read-Once Refutations in Horn Constraint Systems

K. Subramani(✉) and Piotr Wojciechowski

LDCSEE, West Virginia University, Morgantown, WV, USA
{k.subramani,pwjociec}@mail.wvu.edu

Abstract. In this paper, we discuss exact and parameterized algorithms for the problem of finding a read-once refutation in an unsatisfiable Horn Constraint System (HCS). Recall that a linear constraint system $\mathbf{A} \cdot \mathbf{x} \geq \mathbf{b}$ is said to be a Horn constraint system, if each entry in $\mathbf{A}$ belongs to the set $\{0, 1, -1\}$ and at most one entry in each row of $\mathbf{A}$ is positive. In this paper, we examine the importance of constraints in which more variables have negative coefficients than have positive coefficients. There exist several algorithms for checking whether a Horn constraint system is feasible. To the best of our knowledge, these algorithms are not certifying, i.e., they do not provide a certificate of infeasibility. Our work is concerned with providing a specialized class of certificates called "read-once refutations". In a read-once refutation, each constraint defining the HCS may be used at most once in the derivation of a refutation. The problem of checking if an HCS has a read-once refutation (*HCS ROR*) has been shown to be **NP-hard**. We analyze the HCS ROR problem from two different algorithmic perspectives, viz., parameterized algorithms and exact exponential algorithms.

1 Introduction

This paper is focused on the task of designing exact and parameterized algorithms for the problem of finding a read-once refutation in an unsatisfiable Horn constraint system (HCS ROR). Recall that a linear constraint system $\mathbf{A} \cdot \mathbf{x} \geq \mathbf{b}$ is said to be a Horn constraint system, if each entry in $\mathbf{A}$ belongs to the set $\{0, 1, -1\}$ and at most one entry in each row of $\mathbf{A}$ is positive. Horn constraint systems find applications in a number of practical domains, including power systems, abstract interpretation, and econometrics. On account of their wide applicability, there exist a number of algorithms to check if an HCS is feasible [3,20]. To the best of our knowledge, these algorithms are not certifying. In particular, they do not provide an easily checkable "no"-certificate (refutation)

This research was supported in part by the Air-Force Office of Scientific Research through Grant FA9550-19-1-0177 and in part by the Air-Force Research Laboratory, Rome through Contract FA8750-17-S-7007.

S. Artemov and A. Nerode (Eds.): LFCS 2022, LNCS 13137, pp. 327–345, 2022.
https://doi.org/10.1007/978-3-030-93100-1_21

when given infeasible instances. It must be noted though, that negative certificates can be obtained through a general purpose linear programming algorithm using Farkas' lemma [6]. However, in general, these certificates are not "read-once". A read-once refutation is one in which each constraint is used by at most one inference step. Read-once refutations are "visualizable", especially in the presence of a single inference rule (see Sect. 2). Such refutations are invaluable from the perspective of **explaining** the infeasibility of a constraint system.

The HCS ROR problem has been shown to be **NP-hard** [14]. Accordingly, we focus on parameterized and exact approaches for this problem. The goal is to design algorithms that are efficient when certain parameters are "small". To achieve this, we use, as a parameter, the number of constraints in the HCS which have more variables with negative coefficient than variables with positive coefficient.

The rest of this paper is organized as follows: Sect. 2 formally describes the problems under consideration in this paper. The motivation for our work and related approaches in the literature are discussed in Sect. 3. In Sect. 4, we detail an **FPT** algorithm for the read-once refutation problem in Horn constraints. Section 5 describes an exact exponential algorithm for the same problem. In Sect. 6, we relate literal-once refutations to read-once refutations in Horn constraint systems. In Sect. 7, we provide lower bounds on kernel size for the HCS ROR problem, using the number of constraints with at least 3 variables as a parameter. We conclude in Sect. 8 by summarizing our contributions.

2 Statement of Problems

In this section, we describe the problems under consideration. We are concerned with read-once refutations in Horn constraint systems.

Definition 1. *A constraint of the form* $a_i \cdot x_i - \sum_{x_j \in S} x_j \geq b$*, where* $a_i \in \{0, 1\}$*,* S *is a set of variables, and* $b \in \mathbb{Z}$*, is called a* **Horn constraint**.

Example 1. The following are Horn constraints:

$$x_1 - x_2 - x_4 \geq 3 \quad x_2 - x_4 \geq 5 \quad - x_3 - x_4 \geq 2.$$

Note that if only one coefficient in a Horn constraint is non-zero, then the constraint is called an **absolute constraint**. If the coefficient of this constraint is 1, then it is a **positive** absolute constraint. If a Horn constraint has more negative coefficients than positive coefficients, then it is called a **net-negative** constraint. Additionally, b is the **defining constant** of the constraint.

Definition 2. *A conjunction of Horn constraints is called a* **Horn Constraint System (HCS)**. *An HCS can be represented in matrix form as* $\mathbf{A} \cdot \mathbf{x} \geq \mathbf{b}$.

In HCSs, we are interested in refutations which prove that the system has no rational solutions. Such a refutation is called a **linear refutation**. In such a refutation, each new constraint is derived by the addition rule.

Definition 3. *The ADD rule is given as:*

$$\textbf{ADD}: \frac{\sum_{i=1}^{n} a_i \cdot x_i \geq b_1 \qquad \sum_{i=1}^{n} a'_i \cdot x_i \geq b_2}{\sum_{i=1}^{n} (a_i + a'_i) \cdot x_i \geq b_1 + b_2} \tag{1}$$

We refer to Rule (1) as the **addition (ADD) rule**.

Example 2. Consider the constraints $x_1 - x_2 - x_3 \geq 2$ and $x_2 - x_4 \geq -1$. Applying the ADD rule to these constraints lets us derive the constraint $x_1 - x_3 - x_4 \geq 1$.

A sequence of applications of the ADD rule that proves the infeasibility of an HCS $\mathbf{H}$, (by deriving the contradiction $0 \geq b$ where $b > 0$) is known as a **linear refutation**.

We are interested in refutations in which each constraint is used at most once. Such a refutation is called a read-once refutation.

Definition 4. *A* **read-once** *refutation is a refutation in which each constraint can be used in at most one inference.*

Example 3. Consider the HCS $\mathbf{H}$ defined by System (2).

$$l_1 : x_1 - x_2 - x_3 \geq 0 \quad l_2 : x_2 - x_3 \geq -1 \quad l_3 : x_3 - x_1 \geq 1 \quad l_4 : x_3 \geq 1 \tag{2}$$

System (2) has the following read-once refutation:

1. Apply the ADD rule to l_1 and l_2 to get $l_5 : x_1 - 2 \cdot x_3 \geq -1$.
2. Apply the ADD rule to l_5 and l_3 to get $l_6 : -x_3 \geq 0$.
3. Apply the ADD rule to l_6 and l_4 to get the contradiction $0 \geq 1$.

Note that this applies to constraints present in the original system as well as those derived as a result of previous inferences. If a constraint can be re-derived from a different set of input constraints, then the constraint can be reused.

There is a more restrictive form of read-once refutation known as literal-once refutation.

Definition 5. *A* **literal-once** *refutation is a refutation in which each constraint can be used at most once and no two input constraints can share a literal (x_i or $-x_i$).*

Example 4. Note that the read-once refutation in Example 3 is not literal-once since it reuses the literal x_3 (and $-x_3$). However, consider the HCS $\mathbf{H}$ defined by System (3).

$$l_1 : x_1 - x_2 - x_3 \geq 0 \quad l_2 : x_2 \geq -1 \quad l_3 : x_3 - x_1 \geq 2 \tag{3}$$

System (3) has the following literal-once refutation:

1. Apply the ADD rule to l_1 and l_2 to get $l_4 : x_1 - x_3 \geq -1$.
2. Apply the ADD rule to l_4 and l_3 to get the contradiction $0 \geq 1$.

Note that in a literal-once refutation, two derived constraints can share a literal. However, since no input constraints can share a literal, these constraints must be part of a single chain of derivations.

In this paper, we are interested in measuring the number of constraints with three or more variables used by a refutation. We will refer to these as "long" constraints. Accordingly, we define the length of a read-once refutation as follows:

Definition 6. *The* **length** *of a read-once refutation R of an HCS* **H** *is the number of constraints in* **H** *used by R with three or more variables.*

Example 5. The read-once refutation in Example 3 uses one constraint with three or more variables. This is the constraint $l_1 : x_1 - x_2 - x_3 \geq 0$. Thus, the read-once refutation in Example 3 has length 1.

We can now define the problems examined in this paper.

Definition 7. *HCS ROR_D: Given a Horn constraint system* **H**, *does* **H** *have a read-once refutation?*

Example 6. Let **H** be the HCS defined by System (4).

$$l_1 : x_1 - x_2 \geq 1 \quad l_2 : -x_1 - x_2 \geq 0 \quad l_3 : x_2 \geq 0 \tag{4}$$

Note that l_1 is the only constraint in **H** with a positive defining constant. Thus, l_1 must be in any refutation R of **H**. To cancel the literal x_1 from l_1, R must also use the constraint l_2. Note that both l_1 and l_2 use the literal $-x_2$. However, the only constraint in **H** with the literal x_2 is l_3. Thus, any refutation of **H** must use the constraint l_3 at least twice. This means that **H** does not have a read-once refutation.

Definition 8. *HCS LOR_D: Given an HCS* **H**, *does* **H** *have a literal-once refutation?*

Example 7. Let **H** be the HCS defined by System (5).

$$\begin{array}{c} l_1 : x_1 - x_2 \geq 1 \quad l_2 : -x_1 - x_2 \geq 0 \quad l_3 : x_2 \geq 0 \\ l_4 : x_2 - x_3 \geq 0 \quad l_5 : x_3 \geq 0 \end{array} \tag{5}$$

H has the following read-once refutation:

1. Apply the ADD rule to l_1 and l_2 to get $l_6 : -2 \cdot x_2 \geq 1$.
2. Apply the ADD rule to l_3 and l_6 to get $l_7 : -x_2 \geq 1$.
3. Apply the ADD rule to l_4 and l_7 to get $l_8 : -x_3 \geq 1$.
4. Apply the ADD rule to l_5 and l_8 to get the contradiction $0 \geq 1$.

Note that l_1 is the only constraint in **H** with a positive defining constant. Thus, l_1 must be in any refutation R of **H**. To cancel the literal x_1 from l_1, R must also use the constraint l_2. Note that both l_1 and l_2 use the literal $-x_2$. Thus, **H** does not have a literal-once refutation.

Definition 9. *HCS ROR$_{OptD}$: Given an HCS* **H** *and a positive integer k, does* **H** *have a read-once refutation of length at most k?*

Example 8. Let **H** be the HCS defined by System (6).

$$l_1 : x_1 - x_2 - x_3 \geq 0 \quad l_2 : x_2 - x_1 \geq 1 \quad l_3 : x_3 \geq 0$$
$$l_4 : -x_2 \geq 0 \quad l_5 : x_1 - x_3 \geq 0 \tag{6}$$

H has the following read-once refutation:

1. Apply the ADD rule to l_1 and l_2 to get $l_6 : -x_3 \geq 1$.
2. Apply the ADD rule to l_3 and l_6 to get the contradiction $0 \geq 1$.

Note that this refutation uses two inferences and one long constraint. Thus, this refutation has length 1. However, this is not a shortest read-once refutation of **H**.

H also has the following read-once refutation:

1. Apply the ADD rule to l_2 and l_4 to get $l_7 : -x_1 \geq 1$.
2. Apply the ADD rule to l_5 and l_7 to get $l_8 : -x_3 \geq 1$.
3. Apply the ADD rule to l_3 and l_8 to get the contradiction $0 \geq 1$.

Note that this refutation uses three inferences and no long constraints. Thus, this refutation has length 0. This is a shortest read-once refutation of **H**.

Definition 10. *HCS LOR$_{OptD}$: Given an HCS* **H** *and a positive integer k, does* **H** *have a literal-once refutation of length at most k?*

Example 9. Note that the refutations in Example 8 are also literal-once refutations.

The principal contributions of this paper are as follows:

1. An **FPT** algorithm for the HCS ROR$_D$ problem parameterized by the number of net-negative constraints. This algorithm can also find the shortest read-once refutation (Sect. 4).
2. An exact exponential algorithm for the HCS ROR$_D$ problem. This algorithm can also find the shortest read-once refutation. This algorithm extends the **FPT** algorithm by utilizing non-trivial properties of the set of net-negative constraints (Sect. 5).
3. A reduction from the HCS LOR$_D$ problem to the HCS ROR$_D$ problem (Sect. 6).
4. Establishing that the HCS ROR$_{OptD}$ problem does not have a polynomial sized kernel when parameterized by the length (number of long constraints) of the refutation (Sect. 7).

3 Motivation and Related Work

In this section, we briefly motivate the read-once refutation problem in Horn constraint systems and discuss related approaches in the literature.

This paper discusses algorithms for determining read-once refutations in a subset of polyhedral constraint systems known as Horn constraint systems. Polyhedral constraint systems find applications in a number of domains in operations research and combinatorial optimization. Refutations are negative certificates in that they provide an explanation as to why a constraint system is unsatisfiable. Any problem in the class **NP** $\cap$ **coNP** has both "positive" and "negative" certificates, which are succinct (short, polynomial in the size of the input constraint system) [11]. Problems which are **NP-complete** have short positive certificates, but negative certificates must be superpolynomial unless **NP** = **coNP** [12]. It follows that even if a negative certificate for an **NP-complete** problem is obtained, it is not possible to *verify* its correctness in time polynomial in the size of the input constraint system. One technique of overcoming this issue is to study **incomplete** (weak) but sound refutation systems for constraint systems. The idea underlying incomplete proof systems is to sacrifice completeness for efficient decidability. One such refutation system called *read-once resolution* refutation was introduced in [13]. That paper focuses on read-once refutations of Boolean formulas and refers to the procedure as read-once resolution. [13] shows that the problem of checking if a CNF formula has a read-once refutation is **NP-complete**. Read-once refutation has been studied extensively in the literature for a wide variety of constraint systems [14,18,19].

HCSs are used in both program verification [9] and as part of Satisfiability Modulo Theories (SMT) solvers. The field of program verification uses Horn constraints both in their own right and because of their use in SMT solvers [4]. Due to their use in SMT solvers, Horn constraint systems are also used for bounded model checking, infinite state systems, and test-case generation [5]. Additionally, [2,17] provide an in depth description of the use of Horn constraint systems in the field of program verification.

Both the linear ROR and integer ROR problems have been studied for HCSs. In contrast to Horn formulas, not every system of Horn constraints has a read-once refutation. In fact, the linear ROR problem for HCSs is more closely related to the problem of finding a read-once unit resolution refutation of a Horn formula [14]. In [14], both the linear ROR and integer ROR problems for HCSs were shown to be **NP-hard**. Read-once cutting plane proof systems were explored in [15].

We have examined various properties of HCSs and Horn formulas in a number of other papers. In [16], we looked at resolution refutations of Horn formulas. In this paper, we showed that the problem of finding the shortest read-once resolution refutation of a Horn formula is **NP-hard**. Additionally, we established that the problem of checking if a 2-Horn formula with m clauses has a read-once resolution refutation can be solved in $O(m^2)$ time.

4 A Parameterized Algorithm

In this section, we describe a parameterized algorithm for the HCS ROR_D problem.

The algorithms in this paper rely on the following property of read-once linear refutations of HCSs.

Lemma 1. *Let* $\mathbf{H}$ *be an HCS.* $\mathbf{H}$ *has a read-once linear refutation* R*, if and only if there exists a set* $\mathbf{H}_R \subseteq \mathbf{H}$*, such that summing the constraints in* $\mathbf{H}_R$ *results in a constraint of the form* $0 \geq b_R$ *where* $b_R > 0$*.*

Proof. Assume that $\mathbf{H}$ has a read-once refutation R. Since R is a read-once refutation, the constraint derived by R is a constraint of the form $0 \geq b_R$ where $b_R > 0$. Let $\mathbf{H}_R$ be the set of constraints in $\mathbf{H}$ used by R. Since addition is commutative and associative, summing all of the constraints in $\mathbf{H}_R$ results in the constraint $0 \geq b_R$. Since $b_R > 0$, this constraint has the desired form.

Now assume that there exists a set $\mathbf{H}_R \subseteq \mathbf{H}$, such that summing all of the constraints in $\mathbf{H}_R$ results in a constraint of the form $0 \geq b_R$ where $b_R > 0$. Note that this summation imposes an order on the constraints in $\mathbf{H}_R$. For each i, let l_i be the ith constraint in this summation. Let R be the following sequence of applications of the ADD rule:

1. Apply the ADD rule to the constraints l_1 and l_2 to derive the constraint d_1.
2. For each $i = 1, \ldots, (n-2)$, apply the ADD rule to d_i and l_{i+2} to derive the constraint d_{i+1}.

Note that each constraint in $\mathbf{H}_R$ is used exactly once in this summation. Since addition is associative, the constraint d_{n-1} is the result of summing all the constraints in $\mathbf{H}_R$ together. Thus, d_{n-1} is the constraint $0 \geq b_R$. Since $b_R > 0$, R is a read-once refutation. □

Note that Lemma 1 applies to general linear systems, and not just HCSs.

Let $\mathbf{H}$ be a system of Horn constraints. We can partition $\mathbf{H}$ into $\mathbf{H_1}$, the set of constraints with at least as many positive coefficients as negative coefficients and $\mathbf{H_2}$, the set of net-negative constraints. Note that $\mathbf{H_1}$ consists of difference constraints and positive absolute constraints. We now construct an **FPT** for the HCS ROR_D problem parameterized by $k = |\mathbf{H_2}|$.

Let S be an arbitrary subset of $\mathbf{H_2}$, and let l_S be the constraint derived from summing the constraints in S. Let b_S be the defining constant of l_S. If there is a read-once refutation of $\mathbf{H}$ using all of the constraints in S and none of the constraints in $\mathbf{H_2} \setminus S$, then we can derive a constraint which cancels all variables from the constraint l_S using the constraints in $\mathbf{H_1}$. Our algorithm uses a flow network to construct this derivation.

From $\mathbf{H}$ and S, we can create the weighted flow network $\mathbf{G}_S$ as follows:

1. Create the vertices s_0, s_1, and t.
2. For each variable x_i in $\mathbf{H}$ create the vertex x_i.

3. For each constraint of the form $x_i - x_j \geq b$ in $\mathbf{H_1}$, create an edge from x_j to x_i of cost $-b$ and capacity 1.
4. For each constraint of the form $x_i \geq b$ in $\mathbf{H_1}$ create an edge from s_1 to x_i with cost $-b$ and capacity 1.
5. Let a_i be the coefficient of x_i in the constraint l_S.
6. For each variable x_i:
   (a) If $a_i > 0$, then create an edge from s_0 to x_i with cost 0 and capacity a_i.
   (b) If $a_i < 0$, create an edge from x_i to t with cost 0 and capacity $-a_i$.
7. Create an edge from s_0 to s_1 with cost 0 and capacity $-\sum_{i=1}^{n} a_i$. Since each constraint in S is net-negative, $-\sum_{i=1}^{n} a_i \geq 0$

This graph construction is utilized by Algorithm 4.1 to find a read-once refutation of an HCS.

Algorithm 4.1. FPT algorithm for the HCS ROR$_D$ problem

Input: HCS $\mathbf{H}$ with m constraints over n variables.
Output: true if $\mathbf{H}$ has a read-once refutation, **false** otherwise.

1: **procedure** HCS-ROR-D($\mathbf{H}$)
2: Create sets $\mathbf{H_1} := \emptyset$ and $\mathbf{H_2} := \emptyset$.
3: **for** (each constraint l in $\mathbf{H}$) **do**
4: **if** (l is a net-negative constraint) **then**
5: Add l to $\mathbf{H_2}$.
6: **else**
7: Add l to $\mathbf{H_1}$.
8: **for** (each subset $S \subseteq \mathbf{H_2}$) **do**
9: Let $l_S = \sum_{i=1}^{n} a_i \cdot x_i \geq b_S$ be the constraint obtained by summing the constraints in S.
10: Construct the flow network $\mathbf{G}_S$ from l_S and $\mathbf{H_1}$.
11: **if** ($-\sum_{x_i : a_i < 0} a_i$ units of flow can be pushed from s_0 to t in $\mathbf{G}_S$ with total cost less than b_S) **then**
12: **return true.**
13: **return false.**

4.1 Correctness

We now show that Algorithm 4.1 correctly determines if an HCS $\mathbf{H}$ has a read-once refutation.

Let X_S be the set of variables, such that $a_i < 0$. We will show that $\mathbf{H}$ has a read-once refutation using the constraints in S, if and only if we can push $-\sum_{x_i \in X_S} a_i$ units of flow from s_0 to t with cost less than b_S.

Lemma 2. *Let* $\mathbf{H}$ *be an HCS and* S *be a subset of* $\mathbf{H_2}$. *Let* $l_S = \sum_{i=1}^{n} a_i \cdot x_i \geq b_S$ *be the constraint obtained by summing the constraints in* S, *and let* X_S *be the set of variables, such that* $a_i < 0$. $\mathbf{H}$ *has a read-once refutation that uses all of the constraints in* S *and no other constraints from* $\mathbf{H_2}$, *if and only if* $\sum_{x_i \in X_S} -a_i$ *units of flow can be pushed from* s_0 *to* t *in* $\mathbf{G}_S$ *with cost less than* b_S.

Proof. First, assume that $\mathbf{H}$ has a read-once refutation R that uses all of the constraints in S, and none of the constraints in $\mathbf{H_2} \setminus S$.

Let $\mathbf{H}_R$ be the set of constraints in $\mathbf{H}$ used in R. Since R is a read-once refutation of $\mathbf{H}$, from Lemma 1, summing all of the constraints in $\mathbf{H}_R$ results in a constraint of the form $0 \geq b_R$ where $b_R > 0$. Note that $S \subseteq \mathbf{H}_R$, thus the constraints in $\mathbf{H}_R \setminus S$ sum to produce the constraint $-\sum_{i=1}^{n} a_i \cdot x_i \geq b_R - b_S$. Additionally note that $\mathbf{H}_R \setminus S \subseteq \mathbf{H_1}$.

We will use $\mathbf{H}_R$ to construct the desired flow in $\mathbf{G}_S$. This is done as follows:

1. For each constraint $l_r \in \mathbf{H}_R \setminus S$:
   (a) $l_r \in \mathbf{H_1}$, thus l_r is of the form $x_i - x_j \geq b$ or l_r is of the form $x_i \geq b$.
   (b) If l_r is of the form $x_i - x_j \geq b$, then push 1 unit of flow along the edge from x_j to x_i.
   (c) If l_r is of the form $x_i \geq b$, then push 1 unit of flow along the edge from s_1 to x_i.
2. For each variable x_i:
   (a) If $a_i > 0$, then push a_i units of flow along the edge from s_0 to x_i.
   (b) If $a_i < 0$, then push $-a_i$ units of flow along the edge from x_i to t.
3. Push $-\sum_{i=1}^{n} a_i$ units of flow along the edge from s_0 to s_1.

Observe the following:

1. For each x_i, such that $a_i > 0$, there are a_i units of flow leaving s_0 along the edge to x_i. There are an additional $-\sum_{i=1}^{n} a_i$ units of flow leaving s_0 along the edge to s_1. Thus, there are a total of

$$\sum_{x_i \notin X_S} a_i - \sum_{i=1}^{n} a_i = - \sum_{x_i \in X_S} a_i$$

   units of flow leaving s_0 as desired.
2. For each x_i, such that $a_i < 0$, there are $-a_i$ units of flow entering t along the edge from x_i. Thus, there are a total of $-\sum_{x_i \in X_S} a_i$ units of flow entering t as desired.
3. For each x_i:
   (a) If $a_i = 0$, then there is no edge from s_0 to x_i and no edge from x_i to t. Thus, the only flow through x_i comes along edges corresponding to constraints in $\mathbf{H}_R \setminus S$.
   Note that x_i has a coefficient of $-a_i = 0$ in the constraint obtained by summing the constraints in $\mathbf{H}_R \setminus S$. Thus, the number of constraints in $\mathbf{H}_R \setminus S$ that use the literal x_i equals the number of constraints that use the literal $-x_i$. Thus, the amount of flow entering x_i is equal to the amount of flow leaving x_i.

(b) If $a_i > 0$, then there is no edge from x_i to t. Additionally, there are a_i units of flow entering x_i along the edge from s_0.
Note that x_i has a coefficient of $-a_i < 0$ in the constraint obtained by summing the constraints in $\mathbf{H}_R \setminus S$. Thus, there are a_i more constraints in $\mathbf{H}_R \setminus S$ that use the literal $-x_i$ than constraints that use the literal x_i. This means that there are a net a_i units of flow leaving x_i along edges corresponding to constraints in $\mathbf{H}_R \setminus S$. Recall that a_i units of flow enter x_i along the edge from s_0. Thus, the amount of flow entering x_i is equal to the amount of flow leaving x_i.

(c) If $a_i < 0$, then there is no edge from s_0 to x_i. Additionally, there are $-a_i$ units of flow leaving x_i along the edge to t.
Note that x_i has a coefficient of $-a_i > 0$ in the constraint obtained by summing the constraints in $\mathbf{H}_R \setminus S$. Thus, there are $-a_i$ more constraints in $\mathbf{H}_R \setminus S$ that use the literal x_i than constraints that use the literal $-x_i$. Thus, there is a net $-a_i$ units of flow entering x_i along edges corresponding to constraints in $\mathbf{H}_R \setminus S$. Recall that $-a_i$ units of flow leave x_i along the edge to t. Thus, the amount of flow entering x_i is equal to the amount of flow leaving x_i.

4. There are $-\sum_{x_i \in X_S} a_i$ units of flow leaving s_0 and entering t. Additionally, the net flow through each x_i is 0. Thus, the net flow through the only remaining vertex s_1, must also be 0. This means that the flow constructed is valid.

Observe that each edge corresponding to a constraint of the form $x_i - x_j \geq b$ or $x_i \geq b$ in $\mathbf{H}_R \setminus S$ contributes $-b$ to the total cost of the flow. Additionally, the cost of the flow along the remaining edges is 0. Thus, the total cost of the flow is $b_S - b_R < b_S$ as desired.

Now assume that $\mathbf{G}_R$ has a flow F which pushes $\sum_{x_i \in X_S} -a_i$ units of flow from s_0 to t with total cost less than b_S. Construct a subset $\mathbf{H}_R$ of $\mathbf{H}$ as follows:

1. Add each constraint $l \in S$ to $\mathbf{H}_R$.
2. For each constraint $l \in \mathbf{H}_1$, if flow is being pushed along the edge corresponding to l in $\mathbf{G}_S$, then add the constraint l to $\mathbf{H}_R$. Note that these constraints form $\mathbf{H}_R \setminus S$.

We now show that summing the constraints in $\mathbf{H}_R$ results in a constraint of the form $0 \geq b_R$ where $b_R > 0$. This is equivalent to showing that summing the constraints in $\mathbf{H}_R \setminus S$ results in the constraint $-\sum_{i=1}^{n} a_i \cdot x_i \geq b_R - b_S$ where $b_R > 0$.

Observe the following:

1. For each x_i, such that $a_i > 0$, there is an edge of capacity a_i from s_0 to x_i. There is an additional edge of capacity $\sum_{i=1}^{n} -a_i$ from s_0 to s_1. By construction, there are no additional edges leaving s_0. Thus, there is a total capacity of
$$\sum_{x_i \notin X_S} a_i - \sum_{i=1}^{n} a_i = - \sum_{x_i \in X_S} a_i$$
leaving s_0. Since this is the amount of flow that needs to leave s_0, each of these edges needs to be filled to capacity.

2. For each x_i, such that $a_i < 0$, there is an edge of capacity $-a_i$ from x_i to t. By construction, there are no additional edges entering t. Thus, there is a total capacity of $-\sum_{x_i \in X_S} a_i$ entering t. Since this is the amount of flow that needs to enter t, each of these edges needs to be filled to capacity.
3. For each x_i:
   (a) If $a_i = 0$, then there is no edge from s_0 to x_i and no edge from x_i to t. Thus, the only flow through x_i comes along edges corresponding to constraints in $\mathbf{H}_R \setminus S$.
   Note that the amount of flow entering x_i is equal to the amount of flow leaving x_i. Thus, the number of constraints in $\mathbf{H}_R \setminus S$ that use the literal x_i equals the number of constraints that use the literal $-x_i$. This means that x_i has a coefficient of $-a_i = 0$ in the constraint obtained by summing the constraints in $\mathbf{H}_R \setminus S$ as desired.
   (b) If $a_i > 0$, then there is no edge from x_i to t. Additionally, there are a_i units of flow entering x_i along the edge from s_0.
   Note that the amount of flow entering x_i is equal to the amount of flow leaving x_i. Recall that, a_i units of flow enter x_i along the edge from s_0. Thus, there is a net a_i units of flow leaving x_i along edges corresponding to constraints in $\mathbf{H_1}$. By construction, all of these edges are in $\mathbf{H}_R \setminus S$. This means that there are a_i more constraints in $\mathbf{H}_R \setminus S$ that use the literal $-x_i$ than constraints that use the literal x_i. Thus, x_i has a coefficient of $-a_i$ in the constraint obtained by summing the constraints in $\mathbf{H}_R \setminus S$ as desired.
   (c) If $a_i < 0$, then there is no edge from s_0 to x_i. Additionally, there are $-a_i$ units of flow leaving x_i along the edge to t.
   Note that the amount of flow entering x_i is equal to the amount of flow leaving x_i. Recall that, $-a_i$ units of flow leave x_i along the edge to t. Thus, there is a net $-a_i$ units of flow entering x_i along edges corresponding to constraints in $\mathbf{H_1}$. By construction, all of these edges are in $\mathbf{H}_R \setminus S$. This means that there are $-a_i$ more constraints in $\mathbf{H}_R \setminus S$ that use the literal x_i than constraints that use the literal $-x_i$. Thus, x_i has a coefficient of $-a_i$ in the constraint obtained by summing the constraints in $\mathbf{H}_R \setminus S$ as desired.

Thus, summing the constraints in $\mathbf{H}_R \setminus S$ results in a constraint of the form

$$-\sum_{i=1}^{n} a_i \cdot x_i \geq b_R - b_S.$$

All that remains is to show that $b_R > 0$.

Observe that each edge with non-zero flow and non-zero cost b corresponds to a constraint of the form $x_i - x_j \geq -b$ or $x_i \geq -b$ in $\mathbf{H}_R \setminus S$. Since summing these constraints results in a constraint of the form $-\sum_{i=1}^{n} a_i \cdot x_i \geq b_R - b_S$, the total cost of the flow is $(b_S - b_R)$. Since the total cost of the flow is less than b_S, we must have that $b_R > 0$ as desired. □

Theorem 1. *Algorithm 4.1 returns* **true**, *if and only if the HCS* $\mathbf{H}$ *has a read-once refutation.*

Proof. First, assume that $\mathbf{H}$ has a read-once refutation R. Let S_R be the subset of constraints in $\mathbf{H_2}$ used by R. Additionally, let $l_{S_R} = \sum_{i=1}^{n} a_i \cdot x_i \geq b_{S_R}$ be the constraint obtained by summing the constraints in S_R.

From Lemma 2, $\sum_{x_i \in X_{S_R}} -a_i$ units of flow can be pushed from s_0 to t in $\mathbf{G}_{S_R}$ with cost less than b_{S_R}. Thus, the **if** statement on line 11 of Algorithm 4.1 will be satisfied when $S = S_R$. Consequently Algorithm 4.1 will return **true**.

If $\mathbf{H}$ has no read-once refutation, then there is no $S \subseteq \mathbf{H_2}$, such that $\mathbf{H}$ has a read-once refutation using all of the constraints in S and no other constraints in $\mathbf{H_2}$. Thus, from Lemma 2, there is no subset $S \subseteq \mathbf{H_2}$, such that $\sum_{x_i \in X_S} -a_i$ units of flow can be pushed from s_0 to t in $\mathbf{G}_S$ with cost less than b_S. This means that the **if** statement on line 11 of Algorithm 4.1 is never satisfied. Consequently, Algorithm 4.1 will return **false**. □

Note that every long constraint in $\mathbf{H}$ is a net-negative constraint. Thus, every long constraint in $\mathbf{H}$ is in the set $\mathbf{H_2}$. Consequently, to find the shortest read-once refutation of $\mathbf{H}$ (see Definition 6), we simply need to order the subsets of $\mathbf{H_2}$ by the number of long constraints they contain.

4.2 Resource Analysis

We now analyze the time and space requirements of Algorithm 4.1.

Let $\mathbf{H}$ be an HCS with m constraints over n variables. Constructing the subsets $\mathbf{H_1}$ and $\mathbf{H_2}$ is performed by the **for** loop on line 3 of Algorithm 4.1. This takes $O(m)$ time.

Let $k = |\mathbf{H_2}|$. Thus, there are 2^k possible subsets of $\mathbf{H_2}$. This means that there are 2^k iterations of the **for** loop on line 8 of Algorithm 4.1.

In each iteration of the **for** loop on line 8, it takes $O(m+n)$ time to construct $\mathbf{G}_S$. Note that $\mathbf{G}_S$ has $O(n)$ vertices and $O(m)$ edges. The check on line 11 of Algorithm 4.1 can be performed by solving the minimum cost flow problem. This can be accomplished in $O(m \cdot n \cdot (m + n \cdot \log n) \cdot \log n)$ time [1].

Thus, Algorithm 4.1 runs in $O(2^k \cdot m \cdot n \cdot (m + n \cdot \log n) \cdot \log n)$ time. This is an **FPT** algorithm for the HCS ROR_D problem parameterized by the number of net-negative constraints in $\mathbf{H}$.

Algorithm 4.1 needs to store the sets $\mathbf{H_1}$ and $\mathbf{H_2}$. Note that each constraint has $O(n)$ variables and there are m total constraints. Thus, this requires $O(m \cdot n)$ space. Each iteration of the **for** loop on line 8 requires the set $S \subseteq \mathbf{H_2}$, the graph $\mathbf{G}_S$, and the constraint l_S. The set S requires $O(m)$ space since we only need to indicate which constraints in $\mathbf{H_2}$ are in S. Additionally, the graph $\mathbf{G}_S$ requires $O(m+n)$ space, and the constraint l_S requires $O(n)$ space. Note that S, $\mathbf{G}_S$, and l_S are only used by a single iteration of the **for** loop on line 8. Thus, this space can be reused by subsequent iterations of the **for** loop. Thus, we only need to simultaneously store $\mathbf{H_1}$, $\mathbf{H_2}$, S, $\mathbf{G}_S$ and l_S. Consequently, Algorithm 4.1 uses $O(m \cdot n)$ space.

5 An Exact Exponential Algorithm

In this section, we describe an exact exponential algorithm for the HCS ROR$_D$ problem.

Let **H** be an HCS with m constraints over n variables. From Lemma 1, we get a brute force $O^*(2^m)$ algorithm for finding a read-once refutation of **H**. This algorithm proceeds as follows:

1. For each subset S of **H**, if summing the constraints in S results in a contradiction, then return that **H** has a read-once refutation.
2. If no such subset exists, then return that **H** has no read-once refutation.

We will utilize some additional properties of Horn constraints to improve the running time of this algorithm.

We now show that the **FPT** algorithm in Sect. 4 can be converted to an exact exponential algorithm for HCSs.

Let **H** be an HCS. To convert the **FPT** algorithm to an exact exponential algorithm, we will establish a bound on the number of constraints in $\mathbf{H_2}$ that can appear in a read-once refutation.

Lemma 3. *Let* **H** *be an HCS and let* $\mathbf{H_2}$ *be the set of net-negative constraints in* **H**. *Any read-once refutation* R *of* **H** *cannot use more constraints from* $\mathbf{H_2}$ *than positive absolute constraints.*

Proof. Let $\mathbf{H}_R$ be the set of constraints in **H** used by R. For each constraint $l_r \in \mathbf{H}$ let c_r be the sum of the coefficients in l_r. Consider the set of constraints $\mathbf{H}_R \cap \mathbf{H_2}$. Note that $c_r \leq -1$ for each $l_r \in \mathbf{H}_R \cap \mathbf{H_2}$. Thus, $\sum_{l_r \in \mathbf{H}_R \cap \mathbf{H_2}} c_r \leq -|\mathbf{H}_R \cap \mathbf{H_2}|$. The only constraints in **H** with $c_r > 0$ are positive absolute constraints. For each such constraint $c_r = 1$. Note that $\sum_{l_r \in \mathbf{H_R}} c_r = 0$. Thus, $\mathbf{H_R}$ must contain at least $\sum_{l_r \in \mathbf{H}_R \cap \mathbf{H_2}} -c_r \geq |\mathbf{H}_R \cap \mathbf{H_2}|$ positive absolute constraints. □

From Lemma 3, we can limit Algorithm 4.1 to only check subsets S of $\mathbf{H_2}$, such that $|S| \leq m - k$. This results in Algorithm 5.1.

As with Algorithm 4.1, Algorithm 5.1 can find the shortest read-once refutation of H by ordering the subsets of $\mathbf{H_2}$ by the number of long constraints they contain (see Definition 6).

5.1 Resource Analysis

Since $k = |\mathbf{H_2}|$, **H** has at most $(m - k)$ absolute constraints. Thus, we only need to consider subsets S of $\mathbf{H_2}$, such that $|S| \leq m - k$. The total number of such subsets is at most $\sum_{j=0}^{m-k} \binom{k}{j}$.

Note that this only matters when $k \geq \frac{m}{2}$. Otherwise, we still consider all 2^k subsets of $\mathbf{H_2}$. In this case the running time is $O^*(2^{\frac{m}{2}}) \subseteq O^*(1.42^m)$.

Theorem 2. $\mathbf{H_2}$ *has at most* $2^{(m+1) \cdot \frac{5+\sqrt{5}}{10}}$ *subsets of size at most* $(m - k)$.

Algorithm 5.1. Exact exponential algorithm for the HCS ROR$_D$ problem

Input: HCS $\mathbf{H}$ with m constraints over n variables.

Output: **true** if $\mathbf{H}$ has a read-once refutation, **false** otherwise.

1: **procedure** HCS-ROR-D($\mathbf{H}$)
2: Create sets $\mathbf{H_1} := \emptyset$ and $\mathbf{H_2} := \emptyset$.
3: **for** (each constraint l in $\mathbf{H}$) **do**
4: **if** (l is a net-negative constraint) **then**
5: Add l to $\mathbf{H_2}$.
6: **else**
7: Add l to $\mathbf{H_1}$.
8: **for** (each subset $S \subseteq \mathbf{H_2}$, such that $|S| \leq m - |\mathbf{H_2}|$) **do**
9: Let $l_S = \sum_{i=1}^{n} a_i \cdot x_i \geq b_S$ be the constraint obtained by summing the constraints in S.
10: Construct the flow network $\mathbf{G}_S$ from l_S and $\mathbf{H_1}$.
11: **if** ($-\sum_{x_i \in X_S} a_i$ units of flow can be pushed from s_0 to t in $\mathbf{G}_S$ with total cost less than b_S) **then**
12: **return true.**
13: **return false.**

Proof. For a fixed value of m, we will find the value of k that maximizes $\sum_{j=0}^{m-k} \binom{k}{j}$. Assume without loss of generality that k is even. From [10], when $k \geq 2 \cdot (m-k)$:

$$\sum_{j=0}^{m-k} \binom{k}{j} \leq 2^{k-1} \cdot \frac{\binom{k}{m-k+1}}{\binom{k}{\frac{k}{2}}} = \frac{2^{k-1} \cdot (\frac{k}{2})! \cdot (\frac{k}{2})!}{(m-k+1)! \cdot (2 \cdot k - m - 1)!}.$$

This can be bounded from above and below by Stirling's Formula [7]. This gives us the following bound:

$$\frac{2^{k-1} \cdot e \cdot \sqrt{\frac{k}{2}} \cdot (\frac{k}{2 \cdot e})^{\frac{k}{2}} \cdot e \cdot \sqrt{\frac{k}{2}} \cdot (\frac{k}{2 \cdot e})^{\frac{k}{2}}}{\sqrt{2 \cdot \pi \cdot (m+1-k)} \cdot (\frac{m+1-k}{e})^{m-k+1} \cdot \sqrt{2 \cdot \pi \cdot (2 \cdot k - m - 1)} \cdot (\frac{2 \cdot k - m - 1}{e})^{2 \cdot k - m - 1}}.$$

Simplifying we get

$$\sum_{j=0}^{m-k} \binom{k}{j} \leq \frac{e^2 \cdot k^{k+1}}{8 \cdot \pi \cdot (m+1-k)^{m+1.5-k} \cdot (2 \cdot k - m - 1)^{2 \cdot k - m - 0.5}}.$$

We want to find a value of k that maximizes this function. Note that this value of k also maximizes the natural log of this function. Thus, we want to find

k which maximizes

$$(k+1) \cdot \ln k + \ln \frac{e^2}{8 \cdot \pi} - (m + 1.5 - k) \cdot \ln(m + 1 - k)$$
$$- (2 \cdot k - m - 0.5) \cdot \ln(2 \cdot k - m - 1).$$

This happens when

$$\ln k + \frac{k+1}{k} + \ln(m+1-k) + \frac{m+1.5-k}{m+1-k}$$
$$- 2 \cdot \ln(2 \cdot k - m - 1) - 2 \cdot \frac{2 \cdot k - m - 0.5}{2 \cdot k - m - 1} = 0.$$

When k and m are large we have that $k \gg 1$. Additionally since $k > 2 \cdot (m-k)$, we have that $2 \cdot k \gg m$. Note that if $(m-k)$ is constant then $\sum_{j=0}^{m-k} \binom{k}{j}$ is polynomial in both m and k. Thus, without loss of generality, we can also assume that $m - k \gg 1$. This means we can make the following simplifications:

$$\frac{k+1}{k} = 1, \quad \frac{m+1.5-k}{m+1-k} = 1, \text{ and } \frac{2 \cdot k - m - 0.5}{2 \cdot k - m - 1} = 1.$$

Thus, we can simplify the above equation to get

$$2 \cdot \ln(2 \cdot k - (m+1)) = \ln k + \ln(m + 1 - k).$$

This happens when $5 \cdot k^2 - 5 \cdot (m+1) \cdot k + (m+1)^2 = 0$. Solving for k results in $k = (m+1) \cdot \frac{5 \pm \sqrt{5}}{10}$. When we add the assumption that $k \geq \frac{m}{2}$, we get that the maximum occurs when $k = (m+1) \cdot \frac{5+\sqrt{5}}{10} \approx 0.72 \cdot (m+1)$.

Thus, we have that

$$\max_{k=\frac{m}{2} \ldots m} \Big(\sum_{j=0}^{m-k} \binom{k}{j}\Big) \leq \max_{k=\frac{m}{2} \ldots m} 2^{k-1} \cdot \frac{\binom{k}{m-k+1}}{\binom{k}{\frac{k}{2}}} \leq 2^{0.72 \cdot (m+1) - 1} \cdot \frac{\binom{0.72 \cdot (m+1)}{0.28 \cdot (m+1)}}{\binom{0.72 \cdot (m+1)}{0.36 \cdot (m+1)}}$$
$$\leq 2^{0.72 \cdot (m+1)} \leq 1.66^{m+1}.$$

Note that this bound only applies when $k > 2 \cdot (m-k)$. However, if $k \leq 2 \cdot (m-k)$ then $k \leq \frac{2}{3} \cdot m$. Thus,

$$\sum_{j=0}^{m-k} \binom{k}{j} \leq \sum_{j=0}^{k} \binom{k}{j} = 2^k \leq 2^{\frac{2}{3} \cdot m} \leq 1.59^m.$$

This is a lower bound than the 1.66^{m+1} bound. Thus, the 1.66^{m+1} bound applies even when $k \leq 2 \cdot (m-k)$. □

From Theorem 2, there are $O(1.66^m)$ subsets $S \subseteq \mathbf{H_2}$, such that $|S| \leq m-k$. Thus, this algorithm runs in time $O^*(1.66^m)$. This is an improvement over the $O^*(2^m)$ brute force approach.

The exact exponential algorithm uses $O(m \cdot n)$ space, just like Algorithm 4.1.

6 Literal-Once Refutations

In this section, we show that the HCS LOR_D problem can be transformed to the HCS ROR_D problem by a polynomial, many to one reduction. Thus, Algorithm 4.1 can be used to solve the HCS LOR_D problem.

Let $\mathbf{H}$ be an HCS with m constraints over n variables. We construct the HCS $\mathbf{H}'$ as follows:

1. For each variable x_i in $\mathbf{H}$ create the variables x_i^+ and x_i^-. Additionally create the constraint $x_i^- - x_i^+ \geq 0$.
2. Add every constraint in $\mathbf{H}$ to $\mathbf{H}'$ replacing each instance of the literal x_i with x_i^+ and each instance of the literal $-x_i$ with $-x_i^-$. Note that every constraint remains Horn.

We now show that $\mathbf{H}$ has a literal-once refutation, if and only if $\mathbf{H}'$ has a read-once refutation.

Theorem 3. *The HCS $\mathbf{H}$ has a literal-once refutation, if and only if $\mathbf{H}'$ has a read-once refutation.*

Proof. First, assume that $\mathbf{H}$ has a literal-once refutation R. We construct a read-once refutation R' of $\mathbf{H}'$ as follows:

1. For each constraint of the form $x_i - \sum_{x_j \in S} x_j \geq b$ used by R, add the constraints $x_i^+ - \sum_{x_j \in S} x_j^- \geq b$ and $x_i^- - x_i^+ \geq 0$ to R'. Note that $x_i - \sum_{x_j \in S} x_j \geq b$ is the only constraint in R which uses the literal x_i, thus the constraint $x_i^- - x_i^+ \geq 0$ is only added to R' once.
2. For each constraint of the form $-\sum_{x_j \in S} x_j \geq b$ used by R, add the constraint $-\sum_{x_j \in S} x_j^- \geq b$ to R'. This is precisely the original constraint from $\mathbf{H}$ but with each variable x_i replaced by x_i^-.
3. Summing the constraints $x_i^- - x_i^+ \geq 0$ and $x_i^+ - \sum_{x_j \in S} x_j^- \geq b$ results in the constraint $x_i^- - \sum_{x_j \in S} x_j^- \geq b$. This is precisely the original constraint from $\mathbf{H}$ but with each variable x_i replaced by x_i^-. Since R is a literal-once refutation, it is also read-once. Thus, R' is a read-once refutation of $\mathbf{H}'$.

Now assume that $\mathbf{H}'$ has a read-once refutation R'. We construct a literal-once refutation R of $\mathbf{H}$ as follows:

1. For each constraint of the form $x_i^+ - \sum_{x_j \in S} x_j^- \geq b$ used by R', add the constraint $x_i - \sum_{x_j \in S} x_j \geq b$ to R. Note that the only constraint in $\mathbf{H}'$ with the literal $-x_i^+$ is $x_i^- - x_i^+ \geq 0$. Thus, at most one constraint in R' uses the literal x_i^+. Consequently, $x_i - \sum_{x_j \in S} x_j \geq b$ is the only constraint in R which uses the literal x_i.
2. Summing the constraints $x_i^- - x_i^+ \geq 0$ and $x_i^+ - \sum_{x_j \in S} x_j^- \geq b$ results in the constraint $x_i^- - \sum_{x_j \in S} x_j^- \geq b$. This is precisely the original constraint from $\mathbf{H}$ but with each variable x_i replaced by x_i^-.

3. For each constraint of the form $-\sum_{x_j \in S} x_j^- \geq b$ used by R', add the constraint $-\sum_{x_j \in S} x_j \geq b$ to R. This is precisely the original constraint from $\mathbf{H}'$ but with each variable x_i^- replaced by x_i. Since R is a read-once refutation, so is R'. Recall that R has at most once constraint that uses the literal x_i. This means that R has at most once constraint that uses the literal $-x_i$. Thus, R is a literal-once refutation of $\mathbf{H}$. □

After the HCS LOR$_D$ problem is reduced to the HCS ROR$_D$ problem, Algorithm 4.1 or Algorithm 5.1 can be run on the resulting HCS $\mathbf{H}'$ to determine if the original HCS $\mathbf{H}$ has a literal-once refutation. Let $\mathbf{H}$ have m constraints over n variables with k long constraints. Note that, by construction, $\mathbf{H}'$ has $m' = (m + n)$ constraints over $n' = 2 \cdot n$ variables. Since all added constraints have only two non-zero coefficients, $\mathbf{H}'$ still has k long constraints. Thus, running Algorithm 4.1 on $\mathbf{H}'$ takes $O(2^k \cdot (m + n) \cdot n \cdot (m + n \cdot \log n) \cdot \log n)$ time. Additionally, running Algorithm 5.1 on $\mathbf{H}'$ takes $O^*(1.66^{m+n})$ time.

7 A Lower Bound on Kernel Size

In this section, we show that the HCS ROR$_{OptD}$ problem does not have a kernel whose size is polynomial in k, the length of the refutation (see Definition 6), unless some well-accepted complexity theoretic assumptions fail. This is done through the use of an OR-distillation [8].

Definition 11. *Let P and Q be a pair of problems and let $t : \mathbb{N} \rightarrow \mathbb{N} \setminus \{0\}$ be a polynomially bounded function. Then a* t**-bounded OR-distillation** *from P into Q is an algorithm that for every S, given as input $t(S)$ strings $x_1, \ldots, x_{t(S)}$ with $|x_j| = S$ for all j:*

1. *runs in polynomial time, and*
2. *outputs a string y of length at most $t(S) \cdot \log S$, such that y is a* **yes** *instance of Q, if and only if x_j is a* **yes** *instance of P for some $j \in \{1, \ldots, t(S)\}$.*

If any **NP-hard** problem has a t-bounded OR-distillation, then **coNP** $\subseteq$ **NP/poly** [8]. If **coNP** $\subseteq$ **NP/poly**, then $\mathbf{\Sigma_3^P} = \mathbf{\Pi_3^P}$ [21]. Thus, the polynomial hierarchy would collapse to the third level.

Theorem 4. *The HCS ROR$_{OptD}$ problem does not have a polynomial sized kernel unless* **coNP** $\subseteq$ **NP/poly**.

Proof. We will prove this by showing that if the HCS ROR$_{OptD}$ problem has a polynomial sized kernel, then there exists a t-bounded OR-distillation from the HCS ROR$_{OptD}$ problem into itself.

For each j, let $\mathbf{H}_j$ be an HCS with m constraints over n variables. Let $b_{max} > 0$ be such that the defining constants of every constraint in every $\mathbf{H}_j$ fall between b_{max} and $-b_{max}$. We have that $S = |\mathbf{H}_j| = m \cdot n + m \cdot \log b_{max}$.

Assume that for some constant c, the HCS ROR$_{OptD}$ problem has a kernel of size k^c. Let $t(S) = S^c$. Note that $t(S)$ is a polynomial.

For each $j = 1 \ldots t(S)$, let $\mathbf{H}_j$ be an HCS with m constraints over n variables, such that $|\mathbf{H}_j| = S$. From, these HCSs we can create a new HCS $\mathbf{H}$ with $t(S) \cdot m$ constraints over $t(S) \cdot n$ variables, such that: For each $j = 1 \ldots t(S)$, constraints $l_{1+m\cdot(j-1)}$ through $l_{m\cdot j}$ use variables $x_{1+n\cdot(j-1)}$ through $x_{n\cdot j}$ and correspond to the constraints in HCS $\mathbf{H}_j$.

Note that no constraint in $\mathbf{H}$ corresponding to a constraint in $\mathbf{H}_j$ shares variables with a constraint in $\mathbf{H}$ corresponding to a constraint in $\mathbf{H}_{j'}$, $j' \neq j$. Thus, any read-once refutation of $\mathbf{H}$ corresponds to a read-once refutation of $\mathbf{H}_j$ for some $j \in \{1, \ldots, t(S)\}$. Consequently, $\mathbf{H}$ has a read-once refutation of length k, if and only if $\mathbf{H}_j$ has a has a read-once refutation of length k for some $j \in \{1, \ldots, t(S)\}$.

Let $\mathbf{H}'$ be a kernel of $\mathbf{H}$, such that $|\mathbf{H}'| \leq k^c$. Since $k \leq m \leq S$, we have that $|\mathbf{H}'| \leq k^c \leq S^c = t(S)$. Additionally, $\mathbf{H}'$ has a read-once refutation of length k, if and only if $\mathbf{H}_j$ has a read-once refutation of length k for some $j \in \{1, \ldots, t(S)\}$. Thus, we have a t-bounded OR-distillation from the HCS ROR$_{OptD}$ problem to itself. This cannot happen unless **coNP** $\subseteq$ **NP/poly**. □

8 Conclusion

In this paper, we focused on the problem of checking if a Horn constraint system has a read-once refutation under the ADD inference rule. Previous research had established the **NP-hardness** of this problem [14]. We focused on the design of algorithms. In particular, we showed that the problem is **FPT** and devised an exact exponential algorithm that is asymptotically superior to the brute-force approach of checking every combination of constraints. We also showed that the HCS ROR$_D$ problem does not admit a polynomial sized kernel when parameterized by the length of the refutation. Note that we defined the length of a refutation as the number of long (three variable) constraints in the refutation. We also studied the literal-once refutation problem in HCSs and reduced the HCS LOR$_D$ problem to the HCS ROR$_D$ problem.

References

1. Armstrong, R.D., Jin, Z.: A new strongly polynomial dual network simplex algorithm. Math. Program. **78**(2), 131–148 (1997)
2. Bjørner, N., Gurfinkel, A., McMillan, K.L., Rybalchenko, A.: Horn clause solvers for program verification. In: Fields of Logic and Computation II - Essays Dedicated to Yuri Gurevich on the Occasion of His 75th Birthday, pp. 24–51 (2015)
3. Chandrasekaran, R., Subramani, K.: A combinatorial algorithm for Horn programs. Discret. Optim. **10**, 85–101 (2013)
4. de Moura, L., Owre, S., Rueß, H., Rushby, J., Shankar, N.: The ICS decision procedures for embedded deduction. In: Basin, D., Rusinowitch, M. (eds.) IJCAR 2004. LNCS (LNAI), vol. 3097, pp. 218–222. Springer, Heidelberg (2004). https://doi.org/10.1007/978-3-540-25984-8_14
5. Duterre, B., de Moura, L.: The YICES SMT solver. Technical report, SRI International (2006)

6. Farkas, G.: Über die Theorie der Einfachen Ungleichungen. Journal für die Reine und Angewandte Mathematik **124**(124), 1–27 (1902)
7. Feller, W.: An Introduction to Probability Theory and Its Applications, vol. 1 and 2. Wiley, Hoboken (1970)
8. Fomin, F.V., Lokshtanov, D., Saurabh, S., Zehavi, M.: Kernelization: Theory of Parameterized Preprocessing. Cambridge University Press, Cambridge (2019)
9. Fouilhe, A., Monniaux, D., Périn, M.: Efficient generation of correctness certificates for the abstract domain of polyhedra. In: Logozzo, F., Fähndrich, M. (eds.) SAS 2013. LNCS, vol. 7935, pp. 345–365. Springer, Heidelberg (2013). https://doi.org/10.1007/978-3-642-38856-9_19
10. Gallier, J.: Discrete Mathematics. UTX, 1st edn. Springer, New York (2011). https://doi.org/10.1007/978-1-4419-8047-2
11. Garey, M.R., Johnson, D.S.: Computers and Intractability: A Guide to the Theory of NP-Completeness. W. H. Freeman Company, San Francisco (1979)
12. Haken, A.: The intractability of resolution. Theoret. Comput. Sci. **39**(2–3), 297–308 (1985)
13. Iwama, K., Miyano, E.: Intractability of read-once resolution. In: Proceedings of the 10th Annual Conference on Structure in Complexity Theory (SCTC 1995), Los Alamitos, CA, USA, June 1995, pp. 29–36. IEEE Computer Society Press (1995)
14. Kleine Büning, H., Wojciechowski, P., Chandrasekaran, R., Subramani, K.: Restricted cutting plane proofs in horn constraint systems. In: Herzig, A., Popescu, A. (eds.) FroCoS 2019. LNCS (LNAI), vol. 11715, pp. 149–164. Springer, Cham (2019). https://doi.org/10.1007/978-3-030-29007-8_9
15. Büning, H.K., Wojciechowski, P.J., Subramani, K.: New results on cutting plane proofs for Horn constraint systems. In: 39th IARCS Annual Conference on Foundations of Software Technology and Theoretical Computer Science, FSTTCS 2019, Bombay, India, 11–13 December 2019, pp. 43:1–43:14 (2019)
16. Kleine Büning, H., Wojciechowski, P., Subramani, K.: Read-once resolutions in horn formulas. In: Chen, Y., Deng, X., Lu, M. (eds.) FAW 2019. LNCS, vol. 11458, pp. 100–110. Springer, Cham (2019). https://doi.org/10.1007/978-3-030-18126-0_9
17. Komuravelli, A., Bjørner, N., Gurfinkel, A., McMillan, K.L.: Compositional verification of procedural programs using Horn clauses over integers and arrays. In: Formal Methods in Computer-Aided Design, FMCAD 2015, Austin, Texas, USA, 27–30 September 2015, pp. 89–96 (2015)
18. Subramani, K.: Optimal length resolution refutations of difference constraint systems. J. Autom. Reason. (JAR) **43**(2), 121–137 (2009)
19. Subramani, K., Wojciechowki, P.: A polynomial time algorithm for read-once certification of linear infeasibility in UTVPI constraints. Algorithmica **81**(7), 2765–2794 (2019)
20. Subramani, K., Worthington, J.: Feasibility checking in Horn constraint systems through a reduction based approach. Theor. Comput. Sci. **576**, 1–17 (2015)
21. Yap, C.K.: Some consequences of non-uniform conditions on uniform classes. Theoret. Comput. Sci. **26**(3), 287–300 (1983)

Dialectica Logical Principles

Davide Trotta[1(✉)], Matteo Spadetto[2], and Valeria de Paiva[3]

[1] University of Pisa, Pisa, Italy
trottadavide92@gmail.com
[2] University of Leeds, Leeds, UK
matteo.spadetto.42@gmail.com
[3] Topos Institute, Berkeley, USA
valeria@topos.institute

Abstract. Gödel' s Dialectica interpretation was designed to obtain a relative consistency proof for Heyting arithmetic, to be used in conjunction with the double negation interpretation to obtain the consistency of Peano arithmetic. In recent years, proof theoretic transformations (so-called proof interpretations) that are based on Gödel's Dialectica interpretation have been used systematically to extract new content from proofs and so the interpretation has found relevant applications in several areas of mathematics and computer science. Following our previous work on 'Gödel fibrations', we present a (hyper)doctrine characterisation of the Dialectica which corresponds exactly to the logical description of the interpretation. To show that we derive the soundness of the interpretation of the implication connective, as expounded on by Spector and Troelstra, in the categorical model. This requires extra logical principles, going beyond intuitionistic logic, namely Markov's Principle (MP) and the Independence of Premise (IP) principle, as well as some choice. We show how these principles are satisfied in the categorical setting, establishing a tight (internal language) correspondence between the logical system and the categorical framework. This tight correspondence should come handy not only when discussing the applications of the Dialectica already known, like its use to extract computational content from (some) classical theorems (proof mining), its use to help to model specific abstract machines, etc. but also to help devise new applications.

Keywords: Dialectica interpretation · Markov and independence of premise principles · Categorical logic

1 Introduction

Categorical logic is the branch of mathematics in which tools and concepts from category theory are applied to the study of mathematical logic and its connections to theoretical computer science. In broad terms, categorical logic represents

Research supported by the project MIUR PRIN 2017FTXR IT-MaTTerS (Trotta), by a School of Mathematics EPSRC Doctoral Studentship (Spadetto), and by AFOSR grant FA9550-20-10348 (de Paiva).

S. Artemov and A. Nerode (Eds.): LFCS 2022, LNCS 13137, pp. 346–363, 2022.
https://doi.org/10.1007/978-3-030-93100-1_22

both syntax and semantics by a category, and an interpretation by a functor. The categorical framework provides a rich conceptual background for logical and type-theoretic constructions. In many cases, the categorical semantics of a logic provides a basis for establishing a correspondence between theories in the logic and instances of an appropriate kind of category. A classic example is the correspondence between theories of $\beta\eta$-equational logic over simply typed lambda calculus and Cartesian closed categories. Categories arising from theories via term-model constructions can usually be characterised up to equivalence by a suitable universal property. This has enabled proofs of meta-theoretical properties of logics by means of an appropriate categorical algebra. One defines a suitable internal language naming relevant constituents of a category, and then applies categorical semantics to turn assertions in a logic over the internal language into corresponding categorical statements. The goal is to obtain 'internal language theorems' that allow us to pass freely from the logic/type theory to the categorical universe, in such a way that we can solve issues in whichever framework is more appropriate.

Several kinds of categorical universe are available. Our previous joint work on Gödel's Dialectica Interpretation [4] used the fibrational framework expounded by Jacobs in [9]. The identification of syntax-free notions of quantifier-free formulae using categorical concepts is the key insight to our results in [29]. This identification, besides explaining how Gödel's Dialectica interpretation works as a double completion under products and coproducts, is itself of independent interest, as it deepens our ability to think about first-order logic, using categorical notions. Here we show that the notions introduced in our previous paper correspond to well-known (non-intuitionistic but) constructive principles underlying Gödel's Dialectica interpretation.

2 Logical Principles in the Dialectica Interpretation

Gödel's Dialectica interpretation [4,5] associates to each formula ϕ in the language of arithmetic its *Dialectica interpretation* ϕ^D, a formula of the form:

$$\phi^D = \exists u.\forall x.\phi_D$$

which tries to be *as constructive as possible.* The most complicated clause of the translation (and, in Gödel's words, "the most important one") is the definition of the translation of the implication connective $(\psi \to \phi)^D$. This involves two logical principles which are usually not acceptable from an intuitionistic point of view, namely a form of the *Principle of Independence of Premise* (IP) and a generalisation of *Markov's Principle* (MP). The interpretation is given by:

$$(\psi \to \phi)^D = \exists V, X.\forall u, y.(\psi_D(u, X(u, y)) \to \phi_D(V(u), y)).$$

The motivation provided in the collected works of Gödel for this translation is that given a witness u for the hypothesis ψ_D one should be able to obtain a witness for the conclusion ϕ_D, i.e. there exists a function V assigning a witness

$V(u)$ of ϕ_D to every witness u of ψ_D. Moreover, this assignment has to be such that from a counterexample y of the conclusion ϕ_D we should be able to find a counterexample $X(u, y)$ to the hypothesis ψ_D. This transformation of counterexamples of the conclusion into counterexamples for the hypothesis is what gives Dialectica its essential bidirectional character.

We first recall the technical details behind the translation of $(\psi \to \phi)^D$ ([4]) showing the precise points in which we have to employ the non-intuitionistic principles (MP) and (IP). First notice that $\psi^D \to \phi^D$, that is:

$$\exists u.\forall x.\psi_D(u, x) \to \exists v.\forall y.\phi_D(v, y) \tag{1}$$

is *classically* equivalent to:

$$\forall u.(\forall x.\psi_D(u, x) \to \exists v.\forall y.\phi_D(v, y)). \tag{2}$$

If we apply a special case of the **Principle of Independence of Premise**, namely:

$$(\forall x.\theta(x) \to \exists v.\forall y.\eta(v, y)) \to \exists v.(\forall x.\theta(x) \to \forall y.\eta(v, y)) \tag{IP*}$$

we obtain that (2) is equivalent to:

$$\forall u.\exists v.(\forall x.\psi_D(u, x) \to \forall y.\phi_D(v, y)). \tag{3}$$

Moreover, we can see that this is equivalent to:

$$\forall u.\exists v.\forall y.(\forall x.\psi_D(u, x) \to \phi_D(v, y)). \tag{4}$$

The next equivalence is motivated by a generalisation of **Markov's Principle**, namely:

$$\neg\forall x.\theta(u, x) \to \exists x.\neg\theta(u, x). \tag{MP}$$

By applying (MP) we obtain that (4) is equivalent to:

$$\forall u.\exists v.\forall y.\exists x.(\psi_D(u, x) \to \phi_D(v, y)). \tag{5}$$

To conclude that $\psi^D \to \phi^D = (\psi \to \phi)^D$ we have to apply the **Axiom of Choice** (or **Skolemisation**), i.e.:

$$\forall y.\exists x.\theta(y, x) \to \exists V.\forall y.\theta(y, V(y)) \tag{AC}$$

twice, obtaining that (5) is equivalent to:

$$\exists V, X.\forall u, y.(\psi_D(u, X(u, y)) \to \phi_D(V(u), y)).$$

This analysis (from Gödel's Collected Works, page 231) highlights the key role the principles (IP), (MP) and (AC) play in the Dialectica interpretation of implicational formulae. The role of the axiom of choice (AC) has been discussed from a categorical perspective both by Hofstra [7] and in our previous work [29]. We examine the two principles (IP) and (MP) in the next subsections.

2.1 Independence of Premise

In logic and proof theory, the **Principle of Independence of Premise** states that:

$$(\theta \to \exists u.\eta(u)) \to \exists u.(\theta \to \eta(u))$$

where u is not a free variable of θ. While this principle is valid in classical logic (it follows from the law of the excluded middle), it does not hold in intuitionistic logic, and it is not generally accepted constructively [2]. The reason why the principle (IP) is not generally accepted constructively is that, from a constructive perspective, turning any proof of the premise ϕ into a proof of $\exists u.\eta(u)$ means turning a proof of θ into a proof of $\eta(t)$ where t is a witness for the existential quantifier depending on the proof of θ. In particular, the choice of the witness *depends* on the proof of the premise θ, while the (IP) principle tell us, constructively, that the witness can be chosen independently of any proof of the premise θ.

In the Dialectica translation we only need a particular version of the (IP) principle:

$$(\forall y.\theta(y) \to \exists u.\forall v.\eta(u,v)) \to \exists u.(\forall y.\theta(y) \to \forall v.\eta(u,v)) \qquad \text{(IP*)}$$

which means that we are asking (IP) to hold not for every formula, but only for those formulas of the form $\forall y.\theta(y)$ with θ quantifier-free. We recall a useful generalisation of the (IP*) principle, namely:

$$(\theta \to \exists u.\eta(u)) \to \exists u.(\theta \to \eta(u)) \qquad \text{(IP)}$$

where θ is $\exists$-free, i.e. θ contains neither existential quantifiers nor disjunctions (of course, it is also assumed that u is not a free variable of θ). Therefore, the condition that IP holds for every formula of the form $\forall y.\theta(y)$ with $\theta(y)$ quantifier-free is replaced by asking that it holds for every formula *free from the existential quantifier.*

This formulation of (IP) is introduced in [19] where, starting from the observation that intuitionistic finite-type arithmetic is closed under the independence of premise rule for $\exists$-free formula (IPR), it is proved that a similar result holds for many set theories including Constructive Zermelo-Fraenkel Set Theory (CZF) and Intuitionistic Zermelo-Fraenkel Set Theory (IZF). The **Independence of Premise Rule** for $\exists$-free formula (IPR) that we use in this paper, which is the same as the one in [19], states that:

$$\text{if } \vdash \theta \to \exists u.\eta(u)) \text{ then } \vdash \exists u.(\theta \to \eta(u)) \qquad \text{(IPR)}$$

where θ is $\exists$-free.

2.2 Markov's Principle

Markov's Principle is a statement that originated in the Russian school of constructive mathematics. Formally, Markov's principle is usually presented as the statement:

$$\neg\neg\exists x.\phi(x) \to \exists x.\phi(x)$$

where ϕ is a quantifier-free formula. Thus, MP in the Dialectica interpretation, namely:

$$\neg\forall x.\phi(x) \rightarrow \exists x.\neg\phi(x) \tag{MP}$$

with $\phi(x)$ a quantifier-free formula, can be thought of as a generalisation of the Markov Principle above. As remarked in [2], the reason why MP is not generally accepted in constructive mathematics is that in general there is no reasonable way to choose constructively a witness x for $\neg\phi(x)$ from a proof that $\forall x.\phi(x)$ leads to a contradiction. However, in the context of Heyting Arithmetic, i.e. when x ranges over the natural numbers, one can prove that these two formulations of Markov's Principle are equivalent. More details about the computational interpretation of Markov's Principle can be found in [17]. We recall the version of **Markov's Rule** (MR) corresponding to Markov's Principle:

$$\text{if } \vdash \neg\forall x.\phi(x) \text{ then } \vdash \exists x.\neg\phi(x) \tag{MR}$$

where $\phi(x)$ is a quantifier-free formula.

3 Logical Doctrines

One of the most relevant notions of categorical logic which enabled the study of logic from a pure algebraic perspective is that of a *hyperdoctrine*, introduced in a series of seminal papers by F.W. Lawvere to synthesise the structural properties of logical systems [10–12]. Lawvere's crucial intuition was to consider logical languages and theories as fibrations to study their 2-categorical properties, e.g. connectives, quantifiers and equality are determined by structural adjunctions. Recall from [10,23] that a *hyperdoctrine* is a functor:

$$P\colon \mathcal{C}^{\mathrm{op}} \longrightarrow \mathbf{Hey}$$

from the opposite of a Cartesian closed category $\mathcal{C}$ to the category of Heyting algebras **Hey** satisfying some further conditions: for every arrow $A \xrightarrow{f} B$ in $\mathcal{C}$, the homomorphism $P_f\colon P(B) \longrightarrow P(A)$ of Heyting algebras, where P_f denotes the action of the functor P on the arrow f, has a left adjoint $\exists_f$ and a right adjoint $\forall_f$ satisfying the Beck-Chevalley conditions. The intuition is that a hyperdoctrine determines an appropriate categorical structure to abstract both notions of first order theory and of interpretation.

Semantically, a hyperdoctrine is essentially a generalisation of the contravariant *powerset functor* on the category of sets:

$$\mathcal{P}\colon \mathbf{Set}^{\mathrm{op}} \longrightarrow \mathbf{Hey}$$

sending any set-theoretic arrow $A \xrightarrow{f} B$ to the inverse image functor $\mathcal{P}B \xrightarrow{\mathcal{P}f=f^{-1}} \mathcal{P}A$. However, from the syntactic point of view, a hyperdoctrine can be seen as the generalisation of the so-called *Lindenbaum-Tarski algebra* of well-formed formulae of a first order theory. In particular, given a first order theory $\mathcal{T}$ in a first order language $\mathcal{L}$, one can consider the functor:

$$\mathcal{LT}\colon \mathcal{V}^{\mathrm{op}} \longrightarrow \mathbf{Hey}$$

whose base category $\mathcal{V}$ is the *syntactic* category of $\mathcal{L}$, i.e. the objects of $\mathcal{V}$ are finite lists $\overrightarrow{x} := (x_1, \ldots, x_n)$ of variables and morphisms are lists of substitutions, while the elements of $\mathcal{LT}(\overrightarrow{x})$ are given by equivalence classes (with respect to provable reciprocal consequence $\dashv\vdash$) of well-formed formulae in the context $\overrightarrow{x}$, and order is given by the provable consequences with respect to the fixed theory $\mathcal{T}$. Notice that in this case an existential left adjoint to the weakening functor $\mathcal{LT}_\pi$ is computed by quantifying existentially the variables that are not involved in the substitution given by the projection (by duality the right adjoint is computed by quantifying universally).

Recently, several generalisations of the notion of a Lawvere hyperdoctrine were considered, and we refer for example to [13,15,16] or to [8,24] for higher-order versions. In this work we consider a natural generalisation of the notion of hyperdoctrine, and we call it simply a *doctrine*. A **doctrine** is just a functor:

$$P\colon \mathcal{C}^{\mathrm{op}} \longrightarrow \mathbf{Pos}$$

where the category $\mathcal{C}$ has finite products and **Pos** is the category of posets.

Depending on the categorical properties enjoyed by P, we get P to model the corresponding fragments of first order logic formally in a way identical to the one for $\mathcal{P}$, which we call a *generalised Tarski semantics* and which continues to be complete. Again, the syntactic intuition behind the notion of doctrine $P\colon \mathcal{C}^{\mathrm{op}} \longrightarrow \mathbf{Pos}$ remains the same, one should think of $\mathcal{C}$ as the category of contexts associated to a given type theory. Given such a context A, the elements of the posets $P(A)$ represent the predicates in context A and the order relation of $P(A)$ represents the relation of syntactic provability (with respect to the fragment of first order logic modelled by P). Arrows $B \xrightarrow{f} A$ of $\mathcal{C}$ represent (finite lists of) terms-in-context:

$$b : B \mid f(b) : A$$

in such a way that the functor P_f models the substitution by the (finite list of) term(s) f. For instance, if $\alpha \in PA$ represents a formula in context $a : A \mid \alpha(a)$, then $P_f(\alpha) \in P(B)$ represents the formula $b : B \mid \alpha(f(b))$ in context B obtained by substituting f into α.

Now we recall from [13,15,26] the notions of existential and universal doctrines, and we refer to [23] for a detailed introduction to the theory of doctrines and hyperdoctrines. For further insights and applications to higher-order logic or realizability, we refer to [8,20,24].

Definition 1. *A doctrine* $P\colon \mathcal{C}^{op} \longrightarrow \mathbf{Pos}$ *is* **existential** *(resp.* **universal***) if, for every* A_1 *and* A_2 *in* $\mathcal{C}$ *and every projection* $A_1 \times A_2 \xrightarrow{\pi_i} A_i$, $i = 1, 2$, *the functor:*

$$PA_i \xrightarrow{P_{\pi_i}} P(A_1 \times A_2)$$

has a left adjoint $\exists_{\pi_i}$ *(resp. a right adjoint* $\forall_{\pi_i}$*), and these satisfy the* **Beck-Chevalley condition***: for any pullback diagram*

$$\begin{array}{ccc} X' & \xrightarrow{\pi'} & A' \\ {\scriptstyle f'}\downarrow & & \downarrow{\scriptstyle f} \\ X & \xrightarrow[\pi]{} & A \end{array}$$

with π *and* π' *projections, for any* β *in* $P(X)$ *the equality:*

$$\exists_{\pi'} P_{f'}\beta = P_f \exists_\pi \beta \text{ (resp. } \forall_{\pi'} P_{f'}\beta = P_f \forall_\pi \beta \text{)}$$

holds (however, observe that the inequality $\exists_{\pi'} P_{f'}\beta \leq P_f \exists_\pi \beta$ (resp. $\forall_{\pi'} P_{f'}\beta \geq P_f \forall_\pi \beta$) *always holds).*

If a doctrine $P\colon \mathcal{C}^{\mathrm{op}} \longrightarrow \mathbf{Pos}$ is existential and $\alpha \in P(A \times B)$ is a formula-in-context $a : A, b : B \mid \alpha(a,b)$ and $A \times B \xrightarrow{\pi_A} A$ is the product projection on the component A, then $\exists_{\pi_A}\alpha \in PA$ represents the formula $a : A \mid \exists b : B.\alpha(a,b)$ in context A. Analogously, if the doctrine P is universal, then $\forall_{\pi_A}\alpha \in PA$ represents the formula $a : A \mid \forall b : B.\alpha(a,b)$ in context A. This interpretation is sound and complete for the usual reasons: this is how classic Tarski semantics can be characterised in terms of categorical properties of the powerset functor $\mathcal{P}\colon \mathbf{Set}^{\mathrm{op}} \longrightarrow \mathbf{Pos}$.

One of the most interesting aspects of this categorical approach to logic is that there is categorical equivalence between logical theories and doctrines, via the so-called *internal language* of a doctrine [14,23]. The internal language of a doctrine P essentially constitutes a syntax endowed with a semantics induced by P itself: there is a way to interpret every sequent in the fragment of first-order logic modelled by P into a categorical statement involving P. This interpretation is sound and complete; this is precisely why we can deduce properties of P through a purely syntactical procedure. We define the following notation for this syntax, taking advantage of these equivalent ways of reasoning about doctrines and logic.

Notation. From now on, we shall employ the logical language provided by the *internal language* of a doctrine and write:

$$a_1 : A_1, \ldots, a_n : A_n \mid \phi(a_1, \ldots, a_n) \vdash \psi(a_1, \ldots, a_n)$$

instead of:

$$\phi \leq \psi$$

in the fibre $P(A_1 \times \cdots \times A_n)$. Similarly, we write:

$$a : A \mid \phi(a) \vdash \exists b : B.\psi(a,b) \text{ and } a : A \mid \phi(a) \vdash \forall b : B.\psi(a,b)$$

in place of:

$$\phi \leq \exists_{\pi_A}\psi \text{ and } \phi \leq \forall_{\pi_A}\psi$$

in the fibre $P(A)$. Also, we write $a : A \mid \phi \dashv\vdash \psi$ to abbreviate $a : A \mid \phi \vdash \psi$ and $a : A \mid \psi \vdash \phi$. Substitutions via given terms (i.e. reindexings and weakenings) are modelled by pulling back along those given terms. Applications of propositional connectives are interpreted by using the corresponding operations in the fibres of the given doctrine. Finally, when the type of a quantified variable is clear from the context, we will omit the type for the sake of readability.

4 Logical Principles via Universal Properties

It is possible to characterise, in terms of weak universal properties, those predicates of a doctrine that are free from a quantifier. In the following definitions, we pursue this idea of defining those elements of an existential doctrine $P\colon \mathcal{C}^{\text{op}} \longrightarrow \mathbf{Pos}$ which are *free from the left adjoints* $\exists_\pi$. This idea was originally introduced in [27] and, independently, in [3], and then further developed and generalised in the fibrational setting in [29].

Definition 2. *Let $P\colon \mathcal{C}^{op} \longrightarrow \mathbf{Pos}$ be an existential doctrine and let A be an object of $\mathcal{C}$. A predicate α of the fibre $P(A)$ is said to be an* **existential splitting** *if it satisfies the following weak universal property: for every projection $A \times B \xrightarrow{\pi_A} A$ of $\mathcal{C}$ and every predicate $\beta \in P(A \times B)$ such that $\alpha \leq \exists_{\pi_A}(\beta)$, there exists an arrow $A \xrightarrow{g} B$ such that:*

$$\alpha \leq P_{\langle 1_A, g\rangle}(\beta).$$

Existential splittings stable under re-indexing are called *existential-free elements.* Thus we introduce the following definition:

Definition 3. *Let $P\colon \mathcal{C}^{op} \longrightarrow \mathbf{Pos}$ be an existential doctrine and let I be an object of $\mathcal{C}$. A predicate α of the fibre $P(I)$ is said to be* **existential-free** *if $P_f(\alpha)$ is an existential splitting for every morphism $A \xrightarrow{f} I$.*

Employing the presentation of doctrines via internal language, we require that for the formula $i : I \mid \alpha(i)$ to be free from the existential quantifier, whenever $a : A \mid \alpha(f(a)) \vdash \exists b : B.\beta(a, b)$, for some term $a : A \mid f(a) : I$, then there is a term $a : A \mid g(a) : B$ such that $a : A \mid \alpha(f(a)) \vdash \beta(a, g(a))$.

Observe that in general we always have that $a : A \mid \beta(a, g(a)) \vdash \exists b : B.\beta(a, b)$, in other words $P_{\langle 1_A, g\rangle}\beta \leq \exists_{\pi_A}\beta$. In fact, it is the case that $\beta \leq P_{\pi_A}\exists_{\pi_A}\beta$ (as this arrow of $P(A \times B)$ is nothing but the unit of the adjunction $\exists_{\pi_A} \dashv P_{\pi_A}$), hence a re-indexing by the term $\langle 1_A, g\rangle$ yields the desired inequality. Therefore, the property that we are requiring for $i : I \mid \alpha(i)$ turns out to be the following: whenever there are proofs of $\exists b : B.\beta(a, b)$ from $\alpha(f(a))$, at least one of them factors through the canonical proof of $\exists b : B.\beta(a, b)$ from $\beta(a, g(a))$ for some term $a : A \mid g(a) : B$. This fact implies that, while freely adding the existential quantifiers to a doctrine, we do not add a new sequent $\alpha \vdash \exists b.\beta(b)$ (where α and $\beta(b)$ are predicates in the doctrine we started from) as long as we do not allow a sequent $\alpha \vdash \beta(g)$ as well, for some term g (see [28] for more details). For the proof-relevant versions of this definition we refer to [29].

We dualise the previous Definitions 2 and Definition 3 to get the corresponding ones for the universal quantifier.

Definition 4. *Let* $P\colon \mathcal{C}^{\mathrm{op}} \longrightarrow \mathbf{Pos}$ *be a universal doctrine and let* A *be an object of* $\mathcal{C}$. *A predicate* α *of the fibre* $P(A)$ *is said to be a* **universal splitting** *if it satisfies the following weak universal property: for every projection* $A \times B \xrightarrow{\pi_A} A$ *of* $\mathcal{C}$ *and every predicate* $\beta \in P(A \times B)$ *such that* $\forall_{\pi_A}(\beta) \leq \alpha$, *there exists an arrow* $A \xrightarrow{g} B$ *such that:*

$$P_{\langle 1_A, g\rangle}(\beta) \leq \alpha.$$

Definition 5. *Let* $P\colon \mathcal{C}^{\mathrm{op}} \longrightarrow \mathbf{Pos}$ *be a universal doctrine and let* I *be an object of* $\mathcal{C}$. *A predicate* α *of the fibre* $P(I)$ *is said to be* **universal-free** *if* $P_f(\alpha)$ *is a universal splitting for every morphism* $A \xrightarrow{f} I$.

The property we require of the formula $i : I \mid \alpha(i)$, so that it is free from the universal quantifiers, is that whenever $a : A \mid \forall b : B.\beta(a,b) \vdash \alpha(f(a))$, for some term $a : A \mid f(a) : I$, then there is a term $a : A \mid g(a) : B$ such that $a : A \mid \beta(a, g(a)) \vdash \alpha(f(a))$.

Definition 6. *Let* $P\colon \mathcal{C}^{\mathrm{op}} \longrightarrow \mathbf{Pos}$ *be a doctrine. If* P *is existential, we say that* P *has* **enough existential-free predicates** *if, for every object* I *of* $\mathcal{C}$ *and every predicate* $\alpha \in PI$, *there exist an object* A *and an existential-free object* β *in* $P(I \times A)$ *such that* $\alpha = \exists_{\pi_I}\beta$.

Analogously, if P *is universal, we say that* P *has* **enough universal-free predicates** *if, for every object* I *of* $\mathcal{C}$ *and every predicate* $\alpha \in PI$, *there exist an object* A *and a universal-free object* β *in* $P(I \times A)$ *such that* $\alpha = \forall_{\pi_I}\beta$.

Now we can introduce a particular kind of doctrine called a *Gödel doctrine.* This definition works as a synthesis of our process of categorification of the logical notions.

Definition 7. *A doctrine* $P\colon \mathcal{C}^{\mathrm{op}} \longrightarrow \mathbf{Pos}$ *is called a* **Gödel doctrine** *if:*

1. *the category* $\mathcal{C}$ *is cartesian closed;*
2. *the doctrine* P *is existential and universal;*
3. *the doctrine* P *has enough existential-free predicates;*
4. *the existential-free objects of* P *are stable under universal quantification, i.e. if* $\alpha \in P(A)$ *is existential-free, then* $\forall_\pi(\alpha)$ *is existential-free for every projection* π *from* A;
5. *the sub-doctrine* $P'\colon \mathcal{C}^{\mathrm{op}} \longrightarrow \mathbf{Pos}$ *of the existential-free predicates of* P *has enough universal-free predicates.*

The fourth point of the Definition 7 above implies that, given a Gödel doctrine $P\colon \mathcal{C}^{\mathrm{op}} \longrightarrow \mathbf{Pos}$, the sub-doctrine $P'\colon \mathcal{C}^{\mathrm{op}} \longrightarrow \mathbf{Pos}$, such that $P'(A)$ is the poset of existential-free predicates contained in $P(A)$ for any object A of $\mathcal{C}$, is a universal doctrine. From a purely logical perspective, requiring existential-free elements to be stable under universal quantification is quite natural since this can be also read as *if* $\alpha(x)$ *is an existential-free predicate, then* $\forall x : X.\alpha(x)$ *is again an existential-free predicate.*

An element α of a fibre $P(A)$ of a Gödel doctrine P that is both an existential-free predicate and a universal-free predicate in the sub-doctrine P' of existential-free elements of P is called a **quantifier-free predicate** of P. In order to simplify the notation, but also to make clear the connection with the logical presentation in the Dialectica interpretation, we will use the notation α_D to indicate an element α which is a quantifier-free predicate. Applying the definition of a Gödel doctrine we obtain the following result.

Theorem 8. *Let $P\colon \mathcal{C}^{\mathrm{op}} \longrightarrow \mathbf{Pos}$ be a Gödel doctrine, and let α be an element of $P(A)$. Then there exists a quantifier-free predicate α_D of $P(I \times U \times X)$ such that:*

$$i : I \mid \alpha(i) \dashv\vdash \exists u : U.\forall x : X.\alpha_D(i, u, x).$$

This theorem shows that in a Gödel doctrine every formula admits a presentation of the precise form used in the Dialectica translation.

Now we show that employing the properties of a Gödel doctrine we can provide a complete categorical description and presentation of the chain of equivalences involved in the Dialectica interpretation of the implicational formulae. In particular, we show that the crucial steps where (IP) and (MP) are applied are represented categorically via the notions of existential-free element and universal-free element.

Let us consider a Gödel fibration $P\colon \mathcal{C}^{\mathrm{op}} \longrightarrow \mathbf{Pos}$ and two quantifier-free predicates $\psi_D \in P(U \times X)$ and $\phi_D \in P(V \times Y)$. First notice that the following equivalence follows by definition of left adjoint functor (for sake of readability we omit the types of quantified variables as we anticipated in the previous section):

$$- \mid \exists u.\forall x.\psi_D(u, x) \vdash \exists v.\forall y.\phi_D(v, y) \iff u : U \mid \forall x.\psi_D(u, x) \vdash \exists v.\forall y.\phi_D(v, y) \tag{6}$$

Now we employ the fact that the predicate $\forall x.\psi_D(u, x)$ is existential-free in the Gödel doctrine, obtaining that there exists an arrow $U \xrightarrow{f_0} V$, such that:

$$u : U \mid \forall x.\psi_D(u, x) \vdash \exists v.\forall y.\phi_D(v, y) \iff u : U \mid \forall x.\psi_D(u, x) \vdash \forall y.\phi_D(f_0(u), y)$$

Then, since the universal quantifier is right adjoint to the weakening functor, we have that:

$$u : U \mid \forall x.\psi_D(u, x) \vdash \forall y.\phi_D(f_0(u), y) \iff u : U, y : Y \mid \forall x.\psi_D(u, x) \vdash \phi_D(f_0(u), y).$$

Now we employ the fact that $\phi_D(f_0(u), y)$ is universal-free in the subdoctrine of existential-free elements of P. Notice that since $\psi_D(u, x)$ is a quantifier-free element of the Gödel doctrine, we have that $\forall x.\psi_D(u, x)$ is existential free. Recall that this follows from the fact that in every Gödel doctrine, existential-free elements are stable under universal quantification (this is the last point Definition 7). Therefore we can conclude that there exists an arrow $U \times Y \xrightarrow{f_1} X$ of $\mathcal{C}$ such that:

$$u : U, y : Y \mid \forall x.\psi_D(u, x) \vdash \phi_D(f_0(u), y) \iff u : U, y : Y \mid \psi_D(u, f_1(u, y)) \vdash \phi_D(f_0(u), y) \tag{7}$$

Then, combining the equivalence (6) and (7), we obtain the following equivalence:

$$- \mid \exists u.\forall x.\psi_D(u,x) \vdash \exists v.\forall y.\phi_D(v,y) \iff$$
$$\text{there exist } (f_0, f_1) \text{ s.t. } u:U, y:Y \mid \psi_D(u, f_1(u,y)) \vdash \phi_D(f_0(u), y).$$

The arrow $U \xrightarrow{f_0} V$ represents the *witness function*, i.e. it assigns to every witness u of the hypothesis a witness $f_0(u)$ of the thesis, while the arrow $U \times Y \xrightarrow{f_1} X$ represents the *counterexample function.* Notice that while the witness function $f_0(u)$ depends only of the witness u the counterexample function $f_1(u,y)$ depends on a witness of the hypothesis and a counterexample of the thesis. This is a quite natural fact because, considering the constructive point of view, the counterxample has to be relative to a witness validating the thesis.

This provides a proof of the following theorem which establishes the connection between Gödel doctrines and the Dialectica interpretation. Notice that for the sake of clarity, but also to keep the presentation closer to the original one, in the previous paragraph we have considered formulae $\exists u.\forall x.\psi_D(u,x)$ with no free-variables. However, the previous arguments can be easily generalised also for the case of formulae of the form $\exists u.\forall x.\psi_D(u,x,i)$, i.e. with free-variables i. In this case one needs to change just the domains of the functions f_0 and f_1, since they are allowed to depend also on the free-variables.

Theorem 9. *Let* $P\colon \mathcal{C}^{op} \longrightarrow \mathbf{Pos}$ *be a Gödel doctrine. Then for every* $\psi_D \in P(I \times U \times X)$ *and* $\phi_D \in P(I \times V \times Y)$ *quantifier-free predicates of* P *we have that:*

$$i:I \mid \exists u.\forall x.\psi_D(i,u,x) \vdash \exists v.\forall y.\phi_D(i,v,y)$$

if and only if there exists $I \times U \xrightarrow{f_0} V$ *and* $I \times U \times Y \xrightarrow{f_1} X$ *such that:*

$$u:U, y:Y, i:I \mid \psi_D(i,u,f_1(i,u,y)) \vdash \phi_D(i, f_0(i,u), y).$$

This theorem shows that the notion of Gödel doctrine encapsulates in a pure form the basic mathematical feature of the Dialectica interpretation, namely its interpretation of implication, which corresponds to the existence of functionals of types $f_0 : U \to V$ and $f_1 : U \times Y \to X$ as described. One should think of this as saying that a proof of $\exists u.\forall x.\psi_D(i,u,x) \to \exists v.\forall y.\phi_D(i,v,y)$ is obtained by transforming to $\forall u.\exists v.\forall y.\exists x.(\psi_D(i,u,x) \to \phi_D(i,v,y))$, and then Skolemizing along the lines explained in the Sect. 2 and by Troelstra [4]. So, combining Theorems 8 and 9 we have strong evidence that the notion of Gödel doctrine really provides a categorical abstraction of the main concepts involved in the Dialectica translation. Now we show that this kind of doctrine embodies also the *logical principles* involved in the translation. The first principle we consider it the axiom of choice (AC) also sometimes called the principle of Skolemisation. Since the following theorem is the proof-irrelevant version of the proof we refer to [29, Prop. 2.8] for the detailed proof.

Theorem 10. *Every Gödel doctrine* $P\colon \mathcal{C}^{op} \longrightarrow \mathbf{Pos}$ *validates the* **Skolemisation principle**, *that is:*

$$a_1 : A_1 \mid \forall a_2.\exists b.\alpha(a_1,a_2,b) \dashv\vdash \exists f.\forall a_2.\alpha(a_1,a_2,fa_2)$$

where $f : B^{A_2}$ *and* fa_2 *denote the evaluation of* f *on* a_2, *whenever* $\alpha(a_1,a_2,b)$ *is a predicate in the context* $A_1 \times A_2 \times B$.

Remark 11. In the proof of Theorem 10 we do not need the property *5.* of Definition 7. That is why, according to [29], one calls a Skolem doctrine a doctrine satisfying all of the properties satisfied by a Gödel doctrine, except for the *5.* one.

Recall that the notion of Dialectica category introduced in [21] has been generalised to the fibrational setting in [7], and then, in particular, we can consider the proof-irrelevant construction associating a doctrine $\mathfrak{Dial}(P)$ to a given doctrine P:

Dialectica construction. Let $P\colon \mathcal{C}^{\mathrm{op}} \longrightarrow \mathbf{Pos}$ be a doctrine whose base category $\mathcal{C}$ is cartesian closed. The **dialectica doctrine** $\mathfrak{Dial}(P)\colon \mathcal{C}^{\mathrm{op}} \longrightarrow \mathbf{Pos}$ is defined as the functor sending an object I into the poset $\mathfrak{Dial}(P)(I)$ defined as follows:

- **objects** are quadruples (I,X,U,α) where I,X and U are objects of the base category $\mathcal{C}$ and $\alpha \in P(I \times X \times U)$;
- **partial order:** we stipulate that $(I,U,X,\alpha) \leq (I,V,Y,\beta)$ if there exists a pair (f_0,f_1), where $I \times U \xrightarrow{f_0} V$ and $I \times U \times Y \xrightarrow{f_1} X$ are morphisms of $\mathcal{C}$ such that:
$$\alpha(i,u,f_1(i,u,y)) \leq \beta(i,f_0(i,u),y).$$

In [29] we proved that a fibration is an instance of the Dialectica construction if and only if it is a Gödel fibration, and to prove this result we employ the decomposition of the Dialectica monad as a free-simple-product completion followed by the free-simple-coproduct completion of fibrations. So we can deduce the same result for the proof-irrelevant version here simply as a particular case.

However, notice that employing Theorems 8 and 9 we have another simpler and more direct way for proving such correspondence, because Theorem 9 states that the order defined in the fibres of a Gödel doctrine is exactly the same order defined in a dialectica doctrine. The idea is that if P is a Gödel doctrine and P' is the subdoctrine of quantifier-free elements of P it is easy to check that the assignment $P(I) \xrightarrow{(-)^D} \mathfrak{Dial}(P')(I)$ sending $\alpha \mapsto (I,X,U,\alpha_D)$ where α_D is the quantifier-free element such that $\alpha(i) \dashv\vdash \exists u \forall x \alpha_D(i,u,x)$ (which exists by Theorem 8), provides an isomorphism of posets by Theorem 9, and it can be extended to an isomorphism of existential and universal doctrines.

Theorem 12. *Every Gödel doctrine* P *is equivalent to the Dialectica completion* $\mathfrak{Dial}(P')$ *of the full subdoctrine* P' *of* P *consisting of the quantifier-free predicates of* P.

Therefore, we have that Theorem 12 provides another way of thinking about Dialectica doctrines (or Dialectica categories) since it underlines the logical properties that a doctrine has to satisfy in order to be an instance of the Dialectica construction.

5 Logical Principles in Gödel Hyperdoctrines

Gödel doctrines provide a categorical framework that generalises the principal concepts underlying the Dialectica translation, such as the existence of witness and counterexample functions whenever we have an implication $i : I \mid \exists u.\forall x.\psi_D(u,x,i) \vdash \exists v.\forall y.\phi_D(v,y,i)$. The key idea is that, intuitively, the notion of *existential-quantifier-free* objects can be seen as a reformulation of the *independence of premises rule*, while *product-quantifier-free* objects can be seen as a reformulation of *Markov's rule*. Notice that in the proof of Theorem 9 existential and universal free elements play the same role that (IP) and (MP) have in the Dialectica interpretation of implicational formulae.

The main goal of this section is to formalise this intuition showing the exact connection between the rules (IPR) and (MR) and Gödel doctrines. So, first of all we have to equip Gödel doctrines with the appropriate Heyting structure in the fibres in order to be able to formally express these principles. Therefore, we have to consider Gödel hyperdoctrines.

Definition 13. *A hyperdoctrine* $P\colon \mathcal{C}^{op} \longrightarrow \mathbf{Hey}$ *is said a* **Gödel hyperdoctrine** *when* P *is a Gödel doctrine.*

From a logical perspective, one might want the quantifier-free predicates to be closed with respect to all of the propositional connectives (or equivalently that P is the dialectica completion of a hyperdoctrine itself - see [28]), since this is what happens in logic. However, we do not need such a strong condition here. We only require in the next statements that $\bot$ is quantifier-free and/or that $\top$ is existential free.

Theorem 14. *Every Gödel hyperdoctrine* $P\colon \mathcal{C}^{op} \longrightarrow \mathbf{Hey}$ *satisfies the* **Rule of Independence of Premise**, *i.e. whenever* $\beta \in P(A \times B)$ *and* $\alpha \in P(A)$ *is a existential-free predicate, it is the case that:*

$$a : A \mid \top \vdash \alpha(a) \to \exists b.\beta(a,b) \text{ implies that } a : A \mid \top \vdash \exists b.(\alpha(a) \to \beta(a,b)).$$

Proof. Let us assume that $a : A \mid \top \vdash \alpha(a) \to \exists b.\beta(a,b)$. Then it is the case that $a : A \mid \alpha(a) \vdash \exists b.\beta(a,b)$. Since $\alpha(a)$ is free from the existential quantifier, it is the case that there is a term in context $a : A \mid t(a) : B$ such that:

$$a : A \mid \top \vdash \alpha(a) \to \beta(a,t(a)).$$

Therefore, since:

$$a : A \mid \alpha(a) \to \beta(a,t(a)) \vdash \exists b.(\alpha(a) \to \beta(a,b))$$

(as this holds for any predicate $\gamma(a,-)$ in place of the predicate $\alpha_D(a) \to \beta(a,-)$) we conclude that:

$$a : A \mid \top \vdash \exists b.(\alpha(a) \to \beta(a,b)).$$

Notice that Theorem 14 formalises precisely the intuition that the notion of existential-free element can be seen as a reformulation of the independence of premises rule: in a Gödel hyperdoctrine we have that existential-free elements are *exactly* elements satisfying the independence of premises rule.

Theorem 15. *Every Gödel hyperdoctrine* $P\colon \mathcal{C}^{op} \longrightarrow \mathbf{Hey}$ *satisfies the following* **Modified Markov's Rule**, *i.e. whenever* $\beta_D \in P(A)$ *is a quantifier-free predicate and* $\alpha \in P(A \times B)$ *is an existential-free predicate, it is the case that:*

$$a : A \mid \top \vdash (\forall b.\alpha(a,b)) \to \beta_D(a) \text{ implies that } a : A \mid \top \vdash \exists b.(\alpha(a,b) \to \beta_D(a)).$$

Proof. Let us assume that $a : A \mid \top \vdash (\forall b.\alpha(a,b)) \to \beta_D(a)$. Then it is the case that $a : A \mid (\forall b.\alpha(a,b)) \vdash \beta_D(a)$. Hence, since β_D is quantifier-free and α is existential-free, there exists a term in context $a : A \mid t(a) : B$ such that:

$$a : A \mid \top \vdash \alpha(a,t(a)) \to \beta_D(a)$$

therefore, since:

$$a : A \mid \alpha(a,t(a)) \to \beta(a) \vdash \exists b.(\alpha(a,b) \to \beta_D(a))$$

we can conclude that:

$$a : A \mid \top \vdash \exists b.(\alpha(a,b) \to \beta_D(a)).$$

While for the case of (IPR) we have that existential-free elements of a Gödel hyperdoctrine correspond to formulae satisfying (IPR), we have that the elements of a Gödel doctrine that are quantifier-free, i.e. universal-free in the subdoctrine of existential-free elements, are exactly those satisfying a modified Markov's Rule by Theorem 15. Moreover, notice that this Modified Markov's Rule is exactly the one we need in the equivalence between (4) and (5) in the interpretation of the implication in Sect. 2. Alternatively, in order to get this equivalence one requires β_D to satisfy the law of excluded middle and the usual Markov's Rule (see Corollary 16), as these two assumptions yield the Modified Markov's Rule. In particular, any boolean doctrine (a hyperdoctrine modelling the law of excluded middle) satisfies the Modified Markov's Rule (see Remark 17).

To obtain the usual Markov Rule as corollary of Theorem 15, we simply have to require the bottom element $\bot$ of a Gödel hyperdoctrine to be *quantifier-free.*

Corollary 16. *Every Gödel hyperdoctrine* $P\colon \mathcal{C}^{op} \longrightarrow \mathbf{Hey}$ *such that* $\bot$ *is a quantifier-free predicate satisfies* **Markov's Rule**, *i.e. for every quantifier-free element* $\alpha_D \in P(A \times B)$ *it is the case that:*

$$b : B \mid \top \vdash \neg\forall a.\alpha_D(a,b) \text{ implies that } b : B \mid \top \vdash \exists a.\neg\alpha_D(a,b).$$

Proof. It follows by Theorem 15 just by replacing β_D with $\bot$, that is quantifier-free by hypothesis.

Remark 17. Any boolean doctrine satisfies the Rule of Independence of Premises and the (Modified) Markov Rule. In general these are not satisfied by a usual hyperdoctrine, because they are not satisfied by intuitionistic first-order logic. It turns out that *the logic modelled by a Gödel hyperdoctrine is right in-between intuitionistic first-order and classical first-order logic*: it is powerful enough to guarantee the equivalences in Sect. 2 that justify the Dialectica interpretation of the implication.

Remark 18. We observe that Theorem 15 and Theorem 14 deal with the validity of the *rule versions* of (IP) and (MP), and not the usual presentation in form of axioms or *principles.* As pointed out in [19], even if HA validates these rules, in general, these are not valid in an arbitrary intuitionistic theory, so it becomes interesting to find out which are the intuitionist theories that validate these rules. The validity of these rules in arbitrary Gödel hyperdoctrines have two main consequences: first, since HA validates these rules and Gödel's Dialectica interpretation was originally introduced to provide proofs of the relative consistency of HA, the fact that Gödel doctrines validate these rules too *underscores how faithful the modelling is.* If Gödel doctrines or Dialectica categories didn't validate these rules, it would be hard to say that these categorical constructions abstract the main features of the logical translation, since they could not be employed for giving proofs of relative consistency of HA.

Secondly, we have that validating these rules suggests that the internal logic of Gödel hyperdoctrines could represent an interesting family of theories being intuitionistic, but at the same time they validate the rule version of (IP) and (MP) as HA.

We conclude by presenting two other results about the Rule of Choice and the Counterexample Property previously defined in [28], which follow directly from the definitions of existential-free and universal-free elements.

Corollary 19. *Every Gödel hyperdoctrine* $P\colon \mathcal{C}^{op} \longrightarrow \mathbf{Hey}$ *such that* $\bot$ *is a quantifier-free object satisfies the* **Counterexample Property**, *that is, whenever:*

$$a : A \mid \forall b.\alpha(a,b) \vdash \bot$$

for some predicate $\alpha(a,b) \in P(A \times B)$, *then it is the case that:*

$$a : A \mid \alpha(a, g(a)) \vdash \bot$$

for some term in context $a : A \mid g(a) : B$.

Corollary 20. *Every Gödel hyperdoctrine* $P\colon \mathcal{C}^{op} \longrightarrow \mathbf{Hey}$ *such that* $\top$ *is existential-free satisfies the* **Rule of Choice**, *that is, whenever:*

$$a : A \mid \top \vdash \exists b.\alpha(a,b)$$

for some existential-free predicate $\alpha \in P(A \times B)$*, then it is the case that:*

$$a : A \mid \top \vdash \alpha(a, g(a))$$

for some term in context $a : A \mid g(a) : B$.

The rule appearing in Corollary 20 is called *Rule of Choice* in [13], while it appears as *explicit definability* in [19].

6 Conclusion

We have recast our previous fibrational based modelling of Gödel's interpretation [29] in terms of categorical (hyper)doctrines. We show that the notions we considered in our previous work (existential-free and universal-free objects) really provide a categorical explanation of the traditional syntactic notions as described in [4]. This means that we are able to mimic completely the purely logical explanation of the interpretation, given by Spector and expounded on by Troelstra [4], using categorical notions. We show how to interpret logical implications using the Dialectica transformation. Through this process we explain how we go beyond intuitionistic principles, adopting both the Independence of Premise (IP) principle and Markov's Principle (MP) as well as the axiom of choice in the logic.

Our main results show the perfect correspondence between the logical and the categorical tools, in the cases of Markov's principle (MP) and the independence of premise (IP) principle. This is very interesting by itself, as it shows that the categorical modelling really captures all the essential features of the interpretation. But it also opens new possibilities for modelling of constructive set theories (in the style of Nemoto and Rathjan [19]) and of categorical modelling of intermediate logics (intuitionistic propositional logic plus (IP) or (MK), see [1,6]). This leads into applications both into the investigation of functional abstract machines [18,22], of reverse mathematics [19] and of quantified modal logic [25].

Acknowledgements. We would like to thank Milly Maietti for ideas and discussions that inspired this work, and the anonymous referees for extremely useful comments.

References

1. Aschieri, F., Manighetti, M.: On Natural Deduction for Herbrand Constructive logics II: Curry-Howard correspondence for Markov's Principle in First-Order Logic and Arithmetic. CoRR abs/1612.05457 (2016)
2. Avigad, J., Feferman, S.: Gödel's functional (Dialectica) interpretation. In: Handbook of Proof Theory, vol. 137 (February 1999)
3. Frey, J.: Categories of partial equivalence relations as localizations. preprint (2020)
4. Gödel, K., Feferman, S., et al.: Kurt Gödel: Collected Works: Volume II: Publications 1938–1974, vol. 2. Oxford University Press, Oxford (1986)

5. Gödel, K.: Über eine bisher noch nicht benützte erweiterung des finiten standpunktes. Dialectica **12**(3–4), 280–287 (1958)
6. Herbelin, H.: An intuitionistic logic that proves Markov's principle. In: 2010 25th Annual IEEE Symposium on Logic in Computer Science, pp. 50–56 (2010)
7. Hofstra, P.: The Dialectica monad and its cousins. Models, Logics Higherdimensional Categories Tribute Work Mihály Makkai **53**, 107–139 (2011)
8. Hyland, J., Johnstone, P., Pitts, A.: Tripos theory. Math. Proc. Camb. Phil. Soc. **88**, 205–232 (1980)
9. Jacobs, B.: Categorical Logic and Type Theory, Studies in Logic and the Foundations of Mathematics, vol. 141. North Holland Publishing Company, Amsterdam (1999)
10. Lawvere, F.: Adjointness in foundations. Dialectica **23**, 281–296 (1969)
11. Lawvere, F.W.: Diagonal arguments and cartesian closed categories. In: Category Theory, Homology Theory and their Applications II. LNM, vol. 92, pp. 134–145. Springer, Heidelberg (1969). https://doi.org/10.1007/BFb0080769
12. Lawvere, F.: Equality in hyperdoctrines and comprehension schema as an adjoint functor. In: Heller, A. (ed.) New York Symposium on Application of Categorical Algebra, vol. 2, pp. 1–14. American Mathematical Society, Rhode Island (1970)
13. Maietti, M., Pasquali, F., Rosolini, G.: Triposes, exact completions, and Hilbert's ε-operator. Tbilisi Math. J. **10** (2017). https://doi.org/10.1515/tmj-2017-0106
14. Maietti, M.E.: Modular correspondence between dependent type theories and categories including pretopoi and topoi. Math. Struct. Comput. Sci. **15**(6), 1089–1149 (2005)
15. Maietti, M., Rosolini, G.: Quotient completion for the foundation of constructive mathematics. Log. Univers. **7**(3), 371–402 (2013)
16. Maietti, M., Rosolini, G.: Unifying exact completions. Appl. Categ. Structures **23**, 43–52 (2013)
17. Manighetti, M.: Computational interpretations of Markov's principle (2016)
18. Moss, S., von Glehn, T.: Dialectica models of type theory. In: 33rd Annual ACM/IEEE Symposium on Logic in Computer Science, pp. 739–748. Association for Computing Machinery, New York (2018)
19. Nemoto, T., Rathjen, M.: The independence of premise rule in intuitionistic set theories (November 2019)
20. van Oosten, J.: Realizability: An Introduction to its Categorical Side. ISSN, Elsevier Science (2008). https://books.google.it/books?id=0Fvvurmr7FsC
21. de Paiva, V.: The Dialectica categories. Categories Comput. Sci. Logic **92**, 47–62 (1989)
22. Pédrot, P.: A functional functional interpretation. In: CSL-LICS 2014 Science Logic and the Twenty-Ninth Annual ACM/IEEE Symposium on Logic in Computer Science (2014)
23. Pitts, A.M.: Categorical logic. In: Abramsky, S., Gabbay, D.M., Maibaum, T.S.E. (eds.) Handbook of Logic in Computer Science, vol. 6, pp. 39-.129. Oxford University Press, Oxford (1995)
24. Pitts, A.M.: Tripos theory in retrospect. Math. Struct. in Comp. Science **12**, 265–279 (2002)
25. Shimura, T., Kashima, R.: Cut-elimination theorem for the logic of constant domains. Math. Log. Q. **40**, 153–172 (1994)
26. Trotta, D.: The existential completion. Theor. Appl. Categories **35**, 1576–1607 (2020)
27. Trotta, D., Maietti, M.: Generalized existential completions and their regular and exact completions. Preprint (2021)

28. Trotta, D., Spadetto, M.: Quantifier completions, choice principles and applications. preprint (2020). https://arxiv.org/abs/2010.09111
29. Trotta, D., Spadetto, M., de Paiva, V.: The Gödel fibration. In: 46th International Symposium on Mathematical Foundations of Computer Science (2021), LIPIcs, vol. 202, pp. 87:1–87:16 (2021). https://doi.org/10.4230/LIPIcs.MFCS.2021.87

Small Model Property Reflects in Games and Automata

Maciej Zielenkiewicz(✉)

Institute of Informatics, University of Warsaw, Warsaw, Poland
maciekz@mimuw.edu.pl

Abstract. Small model property is an important property that implies decidability. We show that the small model size is directly related to some important resources in games and automata for checking provability.

1 Introduction

Dependent types is one of the popular logic-based approaches developed in the field of functional programming. With the help of such types it is possible to more precisely capture the behaviour of programs. Intuitionistic first order logic is the primary form of dependent types and the Curry-Howard isomorphism strictly relates functional program synthesis and construction of proofs in intuitionistic first order logic.

One of the well established ways towards understanding different aspects of logic, proofs and proof search is through correspondence with different representations, e.g. ones that are more abstract like games and tableaux or ones that are more detailed like linear logic. One of the most fruitful ideas fulfilling the pattern is the game based approach, in the spirit of Ehrenfeucht-Fraïssé games [3,4]. Another game-based technique was introduced for intuitionistic first order logic [6]. The duality between proof-search and countermodel search [1] has been interpreted there in terms of games and was used to make one unified game that yields either a proof or a Kripke countermodel.

We extend the game based approach [6] to classes that have the *finite model property* which, for algorithmically well-behaved classes, implies decidability [2, p. 240]. A stronger property, the *small model property*, that also gives an upper bound on the complexity of the satisfiability problem, is also studied. As it turns out these two properties are equivalent for many interesting classes.

We show in the current work a correspondence between the limit of the model size given by the small model property and some resources in automata and games used for the description of logic. Section 2 contains preliminaries and definitions. Section 3 discusses the automata and Theorem 1 bounds the size of the set of eigenvariables with a number dependent on the number of subformulas in the formula, the limit on the model size and the number of variables in the initial formula. Section 4 covers games and Theorem 2 shows that a strategy can be constructed that uses a number of maximal variables at most equal to the

S. Artemov and A. Nerode (Eds.): LFCS 2022, LNCS 13137, pp. 364–375, 2022.
https://doi.org/10.1007/978-3-030-93100-1_23

limit on the model size; the maximal variable would be understood as the one havina a maximal, by inclusion, set of known facts.

The paper is structured as follows. Section 2 contains preliminaries and definitions necessary to understand the following sections and discusses basic facts about the small model property. Section 3 defines a quasiorder on variables capturing the notion of variable with more facts. Using this order we show that the size of the small countermodel defined in the small model property is also a limit on the number of *maximal* variables in Afrodite strategy. Section 4 shows a limit on the size of the set of eigenvariables of an Arcadian automaton, which depends on the size of the small model, number of subformulas in the original formula and the number of its variables.

2 Preliminaries

We work in intuitionistic first-order logic with no function symbols or constants. The logic is the same as in previous works on games [6] and automata [7]. There is a set of predicates $\mathcal{P}$ and every predicate $P \in \mathcal{P}$ has a defined *arity*. First order variables are noted as X, Y, ... (with possible annotations) and form an infinite set $\mathcal{X}_1$. The formulas are understood as abstract syntax trees and the possible formulas are generated with the grammar

$$\tau, \sigma ::= P(X, \ldots, X) \mid \tau \wedge \sigma \mid \tau \vee \sigma \mid \tau \to \sigma \mid \forall X.\tau \mid \exists X.\tau \mid \bot.$$

We define the set $\mathrm{FV}(\tau)$ of free variables of a formula τ as

- $\mathrm{FV}(P(X_1, \ldots, X_n)) = \{X_1, \ldots, X_n\}$,
- $\mathrm{FV}(\tau_1 * \tau_2) = \mathrm{FV}(\tau_1) \cup \mathrm{FV}(\tau_2)$ where $* \in \{\wedge, \vee, \to\}$,
- $\mathrm{FV}(\triangledown X.\tau) = \mathrm{FV}(\tau) \backslash \{X\}$ where $\triangledown \in \{\exists, \forall\}$,
- $\mathrm{FV}(\bot) = \emptyset$.

We assume that there is an infinite set $\mathcal{X}_\mathrm{p}$ of *proof term variables* usually noted as $x, y,$... that can be used to form the following terms.

$$\begin{aligned} M, N, P ::= {} & x \mid \langle M, N\rangle \mid \pi_1 M \mid \pi_2 M \mid \lambda x : \varphi.M \mid MN \mid \lambda X M \\ & MX \mid \mathsf{in}_{1,\varphi\vee\psi} M \mid \mathsf{in}_{2,\varphi\vee\psi} M \mid \mathsf{case}\ M\ \mathsf{of}\ [x : \varphi] N,\ [y : \psi] P \mid \\ & \mathsf{pack}\ M, Y\ \mathsf{to}\ \exists X.\varphi \mid \mathsf{let}\ x : \varphi\ \mathsf{be}\ M : \exists X.\varphi\ \mathsf{in}\ N \mid \bot_\varphi M \end{aligned}$$

The free variables in terms are defined by structural recursion on the terms, i.e. $\mathrm{FV}(\lambda x : \varphi.M) = \mathrm{FV}(\varphi) \cup \mathrm{FV}(M)$. The inference rules for the logic are shown in Fig. 1.

2.1 Models

We follow the definition of Kripke model from the work of Sørensen and Urzyczyn [5]: A Kripke model is a triple $\langle C, \leq, \{\mathcal{A}_c : c \in C\}\rangle$ where $C \neq \emptyset$ is a set of states, $\leq$ is a partial order on C and $\mathcal{A}_c = \langle A_c, P_1^{\mathcal{A}_c}, \ldots, P_n^{\mathcal{A}_c}\rangle$ are *structures* such that if $c \leq c'$ then $A_c \subseteq A_{c'}$ and for all i the relation $P_i^{\mathcal{A}_c} \subseteq P_i^{\mathcal{A}_{c'}}$ holds. A *valuation* ρ maps variables to elements of A_c. The *satisfaction relation* $c, \rho \models \varphi$ is defined in the usual way:

$$\begin{array}{ll}
c,\rho \models P(t_1,\dots,t_n) & \text{iff } \mathcal{A}_c,\rho \models P(t_1,\dots,t_n) \text{ classicaly,}\\
c,\rho \models \tau\vee\sigma & \text{iff } \mathcal{A}_c,\rho \models \tau \text{ or } \mathcal{A}_c,\rho \models \sigma,\\
c,\rho \models \tau\wedge\sigma & \text{iff } \mathcal{A}_c,\rho \models \tau \text{ and } \mathcal{A}_c,\rho \models \sigma,\\
c,\rho \models \tau\to\sigma & \text{iff for all } c'\geq c \text{ if } c',\rho\models\tau \text{ then } c',\rho\models\sigma,\\
c,\rho \models \forall a\tau & \text{iff for all } c'\geq c \text{ if } \hat{a}\in\mathcal{A}_{c'} \text{ then } c',\rho[\hat{a}/a]\models\tau,\\
c,\rho \models \exists a\tau & \text{iff for some } \hat{a}\in\mathcal{A}_c,\ c,\rho[\hat{a}/a]\models\tau.
\end{array}$$

Proposition 1 (completeness, Theorem 8.6.7 of of [5]). *The Kripke models as defined above are complete for the intuitionistic predicate logic, i.e.* $\Gamma\models\tau$ *iff* $\Gamma\vdash\tau$.

2.2 The Finite Model Property and the Small Model Property

We focus on classes of formulas that have *finite model property*. Our definitions closely follow that of Börger et al. [2]:

$$\frac{}{\Gamma,x{:}\tau\vdash x:\tau}\ (var)$$

$$\frac{\Gamma\vdash M_1:\tau_1\quad \Gamma\vdash M_2:\tau_2}{\Gamma\vdash\langle M_1,M_2\rangle:\tau_1\wedge\tau_2}\ (\wedge I)$$

$$\frac{\Gamma\vdash M:\tau_1\wedge\tau_2}{\Gamma\vdash\pi_1 M:\tau_1}\ (\wedge E1)\qquad \frac{\Gamma\vdash M:\tau_1\wedge\tau_2}{\Gamma\vdash\pi_2 M:\tau_2}\ (\wedge E2)$$

$$\frac{\Gamma\vdash M:\tau_1}{\Gamma\vdash \mathsf{in}_{1,\tau_1\vee\tau_2} M:\tau_1\vee\tau_2}\ (\vee I1)\qquad \frac{\Gamma\vdash M:\tau_2}{\Gamma\vdash \mathsf{in}_{2,\tau_1\vee\tau_2} M:\tau_1\vee\tau_2}\ (\vee I2)$$

$$\frac{\Gamma\vdash M:\tau_1\vee\tau_2\quad \Gamma,x_1:\tau_1\vdash N_1:\tau\quad \Gamma,x_2:\tau_2\vdash N_2:\tau}{\Gamma\vdash \mathsf{case}\ M\ \mathsf{of}\ [x_1:\tau_1]\,N_1,\ [x_2:\tau_2]\,N_2:\tau}\ (\vee E)$$

$$\frac{\Gamma,x{:}\tau_1\vdash M:\tau_2}{\Gamma\vdash\lambda x:\tau_1.M:\tau_1\to\tau_2}\ (\to I)$$

$$\frac{\Gamma\vdash M_1:\tau_1\to\tau_2\quad \Gamma\vdash M_2:\tau_1}{\Gamma\vdash M_1M_2:\tau_2}\ (\to E)$$

$$\frac{\Gamma\vdash M:\tau}{\Gamma\vdash\lambda XM:\forall X.\tau}\ (\forall I)^*\qquad \frac{\Gamma\vdash M:\forall X.\tau}{\Gamma\vdash MY:\tau[X:=Y]}\ (\forall E)$$

$$\frac{\Gamma\vdash M:\tau[X:=Y]}{\Gamma\vdash \mathsf{pack}\,M,Y\ \mathsf{to}\ \exists X.\tau:\exists X.\tau}\ (\exists I)$$

$$\frac{\Gamma\vdash M_1:\exists X.\tau\quad \Gamma,x{:}\tau\vdash M_2:\sigma}{\Gamma\vdash \mathsf{let}\ x{:}\tau\ \mathsf{be}\ M_1{:}\exists X.\tau\ \mathsf{in}\ M_2:\sigma}\ (\exists E)^*$$

$$\frac{\Gamma\vdash M:\bot}{\Gamma\vdash \bot\!\!\!\bot_\tau M:\tau}\ (\bot E)$$

* Under the eigenvariable condition $X\notin\mathrm{FV}(\Gamma,\sigma)$.

Fig. 1. The rules of the intuitionistic first-order logic ([7])

Definition 1 (finite model property). *A class of formulas X has the* finite model property *when, for all formulas $\tau \in X$, if τ is satisfiable, there exists a finite model $\mathcal{M}$ such that $\mathcal{M} \models \tau$.*

Since all classical theories can be easily expressed as intuitionistic theories by explicitly including the law of excluded middle, so there are many interesting classes that have finite model property.

Although the finite model property in the book by Börger et al. [2] is strongly attached to decidability of a particular fragment of logic, this is not a property that implies this computational feature. The following property is what actually takes place in the fragment considered in the book.

Definition 2 (small model property). *A class X has the* small model property *when there exists a computable function s_X such that for all formulas $\tau \in X$, if τ is satisfiable, there exists a finite model $\mathcal{M}$ of size $s_X(\tau)$ such that $\mathcal{M} \models \tau$.*

This definition was used in the book [2] in the context of classical logic. It can also be used for intuitionistic first order logic, but the finiteness concerns a different but relevant notion of size. We say that a model $\mathcal{M} = \langle C, \leq, \{\mathcal{A}_c : c \in C\}\rangle$ is finite when C is finite and $\mathcal{A}_c$ is finite for all $c \in C$. The number $u = |C| + |\bigcup_{c \in C} \mathcal{A}_c|$ is the size of the model $\mathcal{M}$.

Lemma 1. *For all formulas τ from a class X that has the finite model property either $\vdash \neg\tau$ or there exists a finite model $\mathcal{M}$ and a state s such that $s, \mathcal{M} \models \tau$.*

Proof. If $\neg\tau$ or τ is true the proof follows by the completeness theorem and by definition. Otherwise there exists a model $\mathcal{M}$ of class X such that $s, \mathcal{M} \not\models \tau$, and a state $s' > s$ such that $s', \mathcal{M} \models \tau$ (if M, s and s' do not exist either τ or $\neg\tau$ would be valid in the model). Note that it does not necessarily mean that a proof for $\neg\tau$ exists. But, since Kripke models are monotonous, $(s', \mathcal{M}) \in X$ and we have a model of τ in the class X: the part of the original model starting in s'. □

Definition 3 (effective class). *We say that a model $\mathcal{M}$ is a model of a class of formulas X when for each $\varphi \in X$ it holds that $\mathcal{M} \models \varphi$.*

A class of formulas X is effective *iff, it is decidable that given a final model $\mathcal{M}$ whether $\mathcal{M}$ is a model of the class X.*

For example every class that has a finite number of axioms or axiom schemes, as well as prefix classes from the book of Börger et al. [2] is effective.

Proposition 2. *An effective class X has* finite model property *iff it has* small model property.

Proof. The implication from right to left is trivial. For the other one: we show how to compute $s_X(\tau)$. Given τ we run two processess in parallel: one generates finite models and checks whether they are models of X and then if τ is satisfied in them, and another one generates proof and checks if one of them proves $\neg\tau$. If the first process succeeds, we return the size of the model found, and if the second suceeds we return 1. Lemma 1 shows that one of the processess succeeds. This function is correct as it returns the size of a finite model if it exists, and otherwise the formula is not satisfiable, so the return value does not matter. □

In the proof above we can also make the function return the smallest finite model by enumerating the models ordered by size, but it is not needed in this paper.

3 Small Model Size and Small Afrodite Strategies

In this section we show that from a finite countermodel of a given size we can construct a small Afrodite strategy. First we introduce games, strategies and introduce tools to replace some variables in a strategy. Then we use these tools to show a limit on the set of eigenvariables in the game depending on the size of the small model, proving the Theorem 1.

3.1 Better Variables

Notation. We define *substitutions* applied to a formula: $\varphi[x/y]$ is the formula φ with all free occurrences of x replaced by y in a capture-avoiding fashion. The *disjuncts* of a formula $\alpha \vee \beta$ are α and β, and for a formula that is not a disjunction the whole formula is called a disjunct. We understand the formula $\alpha \vee \beta \vee \gamma$ to mean $\alpha \vee (\beta \vee \gamma)$, but understanding it as a disjunction of three disjuncts would also be possible with minor technical changes.

Games. We show how the small model property can be expressed in terms associated with the notion of intuitionistic games for first-order logic as defined in Sect. 5 of the work by Urzyczyn [6]. The game describes a search for a proof and has two players: Eros, tryig to prove the judgement and Afrodite, showing that it can't be proven. We write $\Gamma \vdash \tau \rightsquigarrow_{\text{move}} \Gamma' \vdash \tau'$ to state that the positions $\Gamma \vdash \tau$ and $\Gamma' \vdash \tau'$ are connected with a turn. A *game* is a sequence of *positions* connected by *turns*, i.e. a sequence $\mathcal{P}_1, \dots, \mathcal{P}_n, \dots$ such that $\mathcal{P}_i \rightsquigarrow_{\text{move}} \mathcal{P}_{i+1}$ for each $i \in \mathbb{N}$. Possible moves are shown in Fig. 2. We omit the subscript "move" when it is not needed or clear from the context. A game starts in a position $\Gamma \vdash \tau$ and begins with Eros' move, followed by Afrodite's move which determines the next turn. If Eros reaches a *final position* he wins, otherwise the game is infinite and Afrodite wins. We call $\Gamma \vdash \tau$ the *precedent* and $\Gamma' \vdash \tau'$ the *antecedent* of the move. Some turns have players associated with them: if Afrodite makes a choice in a move we call the precedent an *Afrodite's position*, and if Eros makes a choice we call it an *Eros position.* A disjunct of τ is an *aim*. In order to avoid confusion with classical provability we write $\Gamma \vdash_{\text{IFOL}} \tau$ to denote that τ is provable from Γ in first-order intuitionistic logic. If the exact proof p is important we use the notation $\Gamma \vdash_{\text{IFOL},p} \tau$.

Strategies. A strategy is a tree of nodes labeled by positions linked by edges labeled by turns, which we call *moves*. In each position either one or none of the players makes a choice. In a position with no choice the next position is determined by game rules (see Fig. 2) and the corresponding turn must appear in the strategy. For Afrodite strategy the tree consists of non-final positions and all paths are infinite as well as the tree has at least one move in each Afrodite's position and all the possible moves (up to renaming of fresh and bound variables) in Eros positions. A final position is a position in which $\tau \in \Gamma$ or $\perp \in \Gamma$. For Eros strategy all paths end at a final position and the tree has at least one move in each Eros' position and all the possible moves (up to renaming of fresh and bound variables) in Afrodite positions. It should be obvious that if Eros cannot make a move that introduces something new to the game, he is forced to replay one of the previous moves and Afrodite wins.

Ordering of Variables. Intuitively speaking we would like to capture the fact that one variable is "better" than the other if all the information that was known about the "worse" variable is kept and possibly extended with new facts. More formally we say $x_1 \preceq_\Gamma x_2$ when for every formula τ

if $\Gamma \vdash_{\mathrm{IFOL}} \tau$ then $\Gamma \vdash_{\mathrm{IFOL}} (\tau[x_1/x_2])$.

The relations $\prec_\Gamma$ and $\sim_\Gamma$ are defined in the following way:

$$x_1 \sim_\Gamma x_2 \quad \text{when} \quad x_1 \preceq_\Gamma x_2 \wedge x_2 \preceq_\Gamma x_1,$$

$$x_1 \prec_\Gamma x_2 \quad \text{when} \quad x_1 \preceq_\Gamma x_2 \wedge x_2 \not\preceq_\Gamma x_1.$$

In cases when Γ is clear we omit it for brevity.

Proposition 3. *For any Γ, the relation $\preceq_\Gamma$ is a quasiorder, but not a partial order.*

Proof. The relation $\preceq_\Gamma$ is trivially reflexive and transitivity follows immediately from definition with help of an observation that $\tau[x1/x2][x2/x3] = \tau[x1/x3]$, so it is a quasiorder.

If we choose two distinct fresh variables x_α and x_β, i.e. not in $\mathrm{FV}(\Gamma)$, we have $x_\alpha \preceq x_\beta$ and $x_\beta \preceq x_\alpha$, but $x_\alpha \neq x_\beta$, so $\preceq$ is not a partial order. □

This leads to the conclusion that the only important variables are those that are *maximal in the $\preceq$ relation*, as we can replace all the other variables with their maximal counterparts.

Proposition 4. *Let τ be a formula such that $\mathrm{FV}(\tau) = x_1, \ldots, x_n$. If $\Gamma \vdash_{\mathrm{IFOL}} \tau$, then $\Gamma \vdash_{\mathrm{IFOL}} \tau\,[x_1/x_1', \ldots, x_n/x_n']$, where, for all i, $x_i \preceq x_i'$.*

Proof. We apply the definition of $\preceq$ for each x_i in turn. □

Proposition 5. *If Eros or Afrodite has a strategy in position $\mathcal{P} = \Gamma \vdash \tau$ and if at some position $\mathcal{P}_{i_0}$ and all subsequent positions in that strategy we have $x_1' \preceq x_1, \ldots, x_n' \preceq x_n$ then we can replace all occurrences of variables x_i' with x_i at $\mathcal{P}_{i_0}$ and the same player still has strategy in position $\mathcal{P}_{i_0}$.*

Proof. The replacement is done while looking at the whole game tree and with maximum knowledge (i.e. trueness of predicates through the whole game tree) about variables, which is not a problem since our aim is to construct the strategy for Afrodite, which implies knowledge of all the possible turns.

Suppose Afrodite has a strategy in $\mathcal{P}$. Let us focus on a path in the tree of the strategy

$$\Gamma_1 \vdash \tau_1 \rightsquigarrow \Gamma_2 \vdash \tau_2 \rightsquigarrow \ldots \rightsquigarrow \Gamma_n \vdash \tau_n \rightsquigarrow \ldots$$

Each of the moves may add something to Γ, but nothing is removed and we can separate the newly added facts:

$$\Gamma_2 = \Gamma_1, \psi_1 \quad \ldots \quad \Gamma_{n+1} = \Gamma_n, \psi_n$$

Another view of the new facts would be to separately keep track of those referring to the variables x_i:

$$\Gamma_n = \Gamma_1, \Delta_n, \hat{\Delta}_n$$

where Δ_n has all the facts that reference the variables x_i and $\hat{\Delta}_n$ the others. To make the notation concise we write $\Gamma, \Delta_1, \Delta_2 \vdash \tau$ as a shorthand for $\Gamma, (\Delta_1 \cup \Delta_2) \vdash \tau$.

Instead of taking the original path we can take the following one

$$\Gamma_1 \vdash \tau_1 \rightsquigarrow \Gamma_2' \vdash \tau_2' \rightsquigarrow \ldots \rightsquigarrow \Gamma_n' \vdash \tau_n' \rightsquigarrow \ldots$$

where $\Gamma_n' = \Gamma_{n-1}, \psi_{n-1}', \psi_n' = \psi[x_i'/x_i]$ and $\tau_n' = \tau_n[x_i'/x_i]$. Or, viewed in the terms of Δs,

$$\Gamma_n' = \Gamma_1, \Delta_n', \hat{\Delta}_n$$

where $\Delta_n' = \Delta_n[x_i'/x_i]$.

To make this construction sound we need to show that $\Gamma_1, \Delta_n', \Delta_n \vdash x_i' \preceq x_i$ and that the move $(\Gamma_1, \Delta_n', \hat{\Delta}_n \vdash \tau_n') \rightsquigarrow (\Gamma_1, \Delta_{n+1}', \hat{\Delta}_{n+1} \vdash \tau_{n+1}')$ is possible. The first part follows directly from Proposition 4. For the second part we show how to adapt the original move $\Gamma_n \vdash \tau_n$. The possible moves are listed in Fig. 2. Only two of them have direct interaction with non-fresh variables: in (a4) and (b5) Eros is free to choose any variable and the replacement variables $x_1, \ldots, x_n$ are already available, so would not lead him to winning the game, otherwise he could have played this move in the original strategy.

The other case is when Eros has a strategy in $\mathcal{P}$. The substitution is almost the same as in the previous case except some nonfinal positions might become final, as the set of facts known about x_i' is bigger or equal to those that were known about x_i, as $x_i' \preceq x_i$. Final positions remain final by Proposition 4. Nonfinal positions might become final, but it only makes Eros win faster. □

3.2 Construction of the Strategy

Small Strategies of Afrodite. With the aim of relating the size of the Afrodite strategy and the size of the small model we define a notion of a small strategy. Proposition 5 suggest the following definition. Since we know that using only the *maximal* variables is sufficient in the game, we define *small strategy of Afrodite* for a formula τ from class X as a strategy that has at most $s_X(\tau)$ $\simeq$-classes of abstraction of maximal variables. Given a small countermodel $\mathcal{M}$ of a formula we aim to construct a small winning Afrodite strategy S, i.e. one that gives at least one possible response for each possible Eros' move. For a given turn $\mathsf{t} = \Gamma \vdash \tau$ we need to choose a response to Eros moves. We associate a state $s \in \mathcal{M}$ with each turn t. Our strategy has the following invariant that holds at each turn:

$$\exists_{\rho:\mathrm{FV}(\Gamma)\to\mathcal{A}_s} (\rho, s \models \Gamma) \quad \wedge \quad \forall_{\rho:\mathrm{FV}(\Gamma)\to\mathcal{A}_s} (\rho, s \models \Gamma \to \rho, s \not\models \tau), \tag{1}$$

and the sets of maximal variables corresponds to states of the small counter-model. The part of the invariant quantified with $\exists$ is called the existential part and the part quantified with $\forall$ is called the universal part.

Figure 2 lists possible moves and the choices players make. We define a strategy for Afrodite and she makes a choice in cases marked with * in the figure. Afrodite should choose in the indicated moves:

(a1) We choose $\Gamma, \gamma \vdash \tau$ when $\rho, s \models \Gamma, \gamma$.
(a2) We choose β when $\rho, s \models \Gamma, \beta$.
(b2) We choose β when $\rho, s \not\models \beta$.

In case of (b1) and (b4) the current model state s might needs to be advanced to some subsequent state to keep the invariant.

We still need to show that each move preserves the invariant.

Proposition 6. *At each position $\mathcal{P} : \Gamma \vdash \tau$ the invariant (1) holds.*

Proof. We assume the notation of Fig. 2 and show that each move preserves the invariant (1).

Moves manipulating assumptions:

*a1) If α is an assumption $\beta \to \gamma$ then Afrodite chooses between positions $\Gamma, \gamma \vdash \tau$ and $\Gamma \vdash \beta$.
*a2) If α is an assumption $\beta \vee \gamma$ then Afrodite chooses between positions $\Gamma, \beta \vdash \tau$ and $\Gamma, \gamma \vdash \tau$.
a3) If α is an assumption $\beta \wedge \gamma$ then the next position is $\Gamma, \beta, \gamma \vdash \tau$.
a4) If α is an assumption $\forall x\varphi$ then Eros chooses a variable y and the next position is $\Gamma, \varphi[y/x] \vdash \tau$.

Moves manipulating the proof goal:

a5) If α is an assumption $\exists x\varphi$ then the next position is $\Gamma, \varphi[y/x] \vdash \tau$ where y is a fresh variable.
b1) If α is an aim of the form $\beta \to \gamma$ the next position is $\Gamma, \beta \vdash \gamma$.
*b2) If α is an aim of the form $\beta \wedge \gamma$ then Afrodite chooses between positions $\Gamma \vdash \beta$ and $\Gamma \vdash \gamma$.
b3) If the aim α is an atom or a disjunction the next position is $\Gamma \vdash \alpha$.
b4) If α is an aim of the form $\forall x\varphi$ the next position is $\Gamma \vdash \varphi[y/x]$ where y is fresh.
b5) If α is an aim of the form $\exists x\varphi$ then Eros chooses a variable y and the next position is $\Gamma \vdash \varphi[y/x]$.

Fig. 2. Table of moves in position $\Gamma \vdash \tau$ for the intuitionistic game [6, Fig. 11, p. 32]. In each move Eros chooses a formula α - either an assumption or an aim, and the move is selected from this table according to the α chosen.

(a1) We have two possibilities:
- In case $\rho, s \models \Gamma, \gamma$: we choose $\Gamma, \gamma \vdash \tau$. The existential part of the invariant follows directly from the invariant of the previous step. For the universal part suppose the opposite, i.e. $\rho, s \models \tau$, so for given ρ we either have contradiction with $\rho, s \not\models \tau$ from invariant of the previous step or $\rho, s \not\models \gamma$, but then we would not choose this move for the strategy.
- Otherwise $\rho, s \not\models \Gamma, \gamma$ and we choose $\Gamma \vdash \beta$. The existential part of the invariant remains true as Γ does not change. For universal part suppose $\rho, s \models \beta$, but then $\rho, s \models \gamma \rightarrow \beta$, which is in contradiction with the invariant from the previous step.

(a2) Once again we have two possibilities:
- In case $\rho, s \models \Gamma, \beta$: we choose $\Gamma, \beta \models \tau$. The existential part of the invariant follows directly from the invariant of the previous step. The universal part is the same as in the corresponding point of the move (a1).
- Otherwise $\rho, s \not\models \Gamma, \gamma$ and the proof is the same as in the corresponding point of (a1).

(a3) The existential part is true because $\rho, s \models \beta, \gamma$ follows from $\rho, s \models \beta \wedge \gamma$. The universal part is proven by simply applying the definition of $\models$.

(a4) We can choose any value for $\rho(y)$. Existential part: from the invariant in the previous move we have $\rho, s \models \forall x \varphi$, so we apply definition of $\models$ to get $\rho, s \models \varphi[y/x]$. Universal part: suppose that $\rho, s \models \Gamma, \varphi[y/x]$ and $\rho, s \models \tau$. But this means $\rho, s \models \Gamma$ which implies have a contradiction with $\rho, s \not\models \tau$ from the previous move.

(a5) Since $\rho, s \models \exists x \varphi$ we know that there exists $\hat{x}$ such that $\rho[\hat{x}/x], s \models \varphi$. In the existential part we just need to take $\rho(y) = \rho(\hat{x})$. Universal part: identical with the universal part of (a4).

(b1) Using the assumption we have a state $s' \geq s$ such that $\rho, s' \models \Gamma, \beta$ but $\rho, s' \not\models \gamma$. We advance s to s'. The existential part is trivially true. For the universal part: suppose $\rho, s' \models \Gamma, \beta$ and $\rho, s' \models \gamma$. This is in contradiction with $\rho, s \not\models \beta \rightarrow \gamma$.

(b2) We have the following cases:
- $\rho, s \not\models \beta$: the set of assumptions does not change so the existential part is proven by applying the existential part from the previous move. For the universal part, $\rho, s \not\models \beta$ is exactly the assumption of the case under investigation.
- otherwise $\rho, s \models \beta$. We choose the position $\Gamma \vdash \gamma$; the existential part is the same as in the previous step. For the universal part suppose $\rho, s \models \gamma$: then $\rho, s \models \beta \wedge \gamma$ contradicts the invariant $\rho, s \not\models \beta \wedge \gamma$ from the previous move.

(b3) The existential part is the same as in the previous move. The universal part is the same as in the second bullet of (b2).

(b4) We can choose any value for $\rho'(y)$. The existential part is true since Γ is the same as previously and the valuation of y does not affect it. The universal part: suppose that $\rho', s \models \varphi[y/x]$. Then by definition $\rho, s \models \varphi$, which is in contradiction with the invariant from the previous move.

(b5) The existential part is the same as in the previous move. For the universal part suppose $\rho, s \models \varphi[y/x]$. Then by definition of $\models$ we have $\rho, s \models \exists x\varphi$, which is a contradiction with the invariant of the previuos step. □

The constructed strategy is *small*: elements of $\mathcal{A}_s$ correspond to $\simeq$-classes and the valuation ρ proves that all the variables fit in $s_X(\varphi)$ classes as the size of the model is $s_X(\varphi)$. This proves the following:

Theorem 1. *For all classes X that have the small model property and all formulas $\tau \in X$, if a strategy of Afrodite exists for τ then a small strategy of Afrodite for τ also exists.*

4 Small Model Size and the Arcadian Automata

Here we show a limit on resources of Arcadian automata [7] checking derivability of a formula φ from an effectively axiomatized class X that has the finite model property. Theorem 2 shows a limit on the size of the set of eigenvariables of the automaton in terms of the numbers of variables and subformulas in φ and the size of the small model. We reason only about automata that are translated from a formula as defined in Sect. 4 of [7]. In Sect. 4.1 we introduce Arcadian automata, show how to replace variables in their runs in Sect. 4.2, and limit the size of the set of eigenvariables in 4.3.

4.1 Arcadian Automata

Notation. We already know that $\simeq$-maximal variables play a crucial role. Given a set of facts Γ we denote by $\check{\Gamma}$ the set obtained from Γ by selecting only those facts γ that do have only maximal variables in $\mathrm{FV}(\Gamma)$. An *Arcadian automaton* is a tuple $\langle \mathcal{A}, Q, q^0, \varphi^0, \mathcal{I}, i, \mathrm{fv}\rangle$, where $\mathcal{A}$ is a finite tree, Q and I are sets of states and instructions with i mapping states to instructions, $\mathrm{fv} : A \to P(A)$ describes the binding of variables and q^0 and φ^0 are the inital state and node. The function fv satisfies the condition that for all v either v is a leaf or $\mathrm{fv}(v) = \bigcup_{w \in B(v)} \mathrm{fv}(w)$ where $B(v) = \{w \mid v \text{ succ } w\}$ and $\succ$ is the usual predicate of being a successor. An *instantaneous description* is $\langle q, \kappa, V, w, w', S\rangle$ where $q \in Q$ and $\kappa \in A$ are the current state and node, V is a set of eigenvariables, w and w' are interpretations of bindings and S is the store. For more details see [7].

4.2 Better Variables in Arcadian Automata

Equivalent Positions. We say that the position $\Gamma \vdash \tau$ and $\Gamma' \vdash \tau'$ are *equivalent* when $\check{\Gamma} = \check{\Gamma}'$ and $\tau = \tau'$.

Proposition 7. *Suppose $\Gamma, \hat{\Gamma} \vdash_{\mathrm{IFOL}} M : \tau$ where for some α and $\alpha' \succeq \alpha$ such that $\alpha \in \mathrm{FV}(\Gamma)$ and $\alpha \notin \mathrm{FV}(\hat{\Gamma})$. If $\Gamma = x_1 : \tau_1, \ldots, \Gamma \vdash_{\mathrm{IFOL}} x_n : \tau_n$ then $\hat{\Gamma}, \Gamma, \Gamma' \vdash_{\mathrm{IFOL}} M[x_1/x_1'] \ldots [x_n/x_n'] : \tau[\alpha/\alpha']$ where $\Gamma' = x_1' : \tau_1[\alpha/\alpha'], \ldots, x_n' : \tau_n[\alpha/\alpha']$ and $x_1', \ldots, x_n'$ are fresh variables, i.e. $x_i' \notin \mathrm{FV}(M)$.*

Proof. Proof is by induction over the length of the proof of τ. We look at the last rule in the proof. In most of the cases the conclusion follows by simple application of the inductive hypothesis, but there are three rules that change the environment, namely $(\vee E)$, $(\rightarrow I)$ and $(\exists E)$ and the proof is more subtle for them. Let us focus on the $(\rightarrow I)$ rule. If $(x_i : \tau_i) \in \hat{\Gamma}$ we do not need to change anything, in the other case we know that $(x_i : \tau_i) \in \Gamma$ and we apply the induction hypothesis and use the assumption $(x'_i : \tau_i[\alpha/\alpha']) \in \Gamma'$ for the λ-abstraction. We can now remove the variable $x_i : \tau_i$ as it is not referenced in $M[x_1/x'_1] \dots [x_n/x'_n] : \tau[\alpha/\alpha']$.

The induction base is the *var* rule, since the proof must begin with this rule, and the correctness of replacing α with α' follows immediately from the definition of Γ'. □

Proposition 8. *If* $\Gamma \vdash_{\mathrm{IFOL}} M : \tau$, α *and* α' *are variables in* M *such that* $\alpha \preceq \alpha'$ *then* $\Gamma \vdash_{\mathrm{IFOL}} M[\alpha/\alpha'] : \tau[\alpha/\alpha']$ *and* $\alpha \notin FV(\tau)$.

Proposition 9. *If* $\Gamma \vdash_{\mathrm{IFOL},p} M : \tau$ *then there exists* M' *and* p' *such that* $\Gamma \vdash_{\mathrm{IFOL},p'} M' : \tau$ *and in each step* $\Gamma' \vdash \tau'$ *of* p' *only maximal variables are mentioned in* τ'.

Proof. Apply Proposition 8 sequentially to each nonmaximal variable. □

4.3 Loquacious Runs

Note that our logic has the subformula property ([7]). This means that there is only a limited number of possible targets τ. Let us review fragments of an automaton run $p_0 \rightarrow_\alpha \Gamma \vdash \tau \rightarrow_\beta \Gamma' \vdash \tau \rightarrow_\gamma p_1$. If $\Gamma \vdash \tau$ and $\Gamma' \vdash \tau$ are equivalent, then a part of the run $\rightarrow_\beta$ is removable, i.e. there exists a run of the same automaton $p_p \rightarrow_\alpha \Gamma \vdash \tau \rightarrow_{\bar{\gamma}} \bar{p}_1$ where $\bar{\gamma}$ and $\bar{p}_1$ are obtained from γ and p_1 by replacing some variables by their maximal counterparts. Otherwise that fragment is not removable, as new fact about the maximal variables are discovered, but the number of such non-removable runs is limited: there are at most $s_X(\varphi)$ maximal variables. Suppose $v(\varphi)$ is the number of variables in φ and $f(\varphi)$ is the number of subformulas in φ. The maximum size of an environment Γ is $\mu = f(\varphi) \cdot s_X(\varphi)^{v(\varphi)}$. Each non-trivial step has to either add something to the environment or change the target τ as otherwise the previous state is repeated. We have at most μ possible targets, so after at most μ steps the target repeats and in the worst case each step introduces a new variable, so the maximum size of V in the automaton is μ^2.

This proves the following:

Theorem 2. *Let* τ *be a formula from an effective class* X *that has the small model property. For a given accepting run of an Arcadian automaton for that formula there exists an accepting run of the same automaton with the same result that has the property* $|V| \leq \mu^2$*, where* V *is the working domain of the automaton and* μ *is the maximum size of the environment defined in the previous paragraph.*

5 Conclusion

The small model size is directly related to important resources in games and automata for checking provability. In terms of games, the elements of models directly correspond to abstraction classes of maximal elements of a quasiorder on eigenvariables that captures the relation of having more information available about a variable.

For automata the number of such maximal elements can be directly related to the size of set of eigenvariables V; the dependency is exponential, caused by the necessity of representing the eigenvariables that correspond to non-maximal elements of the quasiorder.

These observations lead to an idea for implementing proof theory bases proves in a manner that would not be substantially less powerful than those based on model theory. More specifically we suggest that V should not be represented syntactically but rather as an abstraction class of the quasiorder.

References

1. van Benthem, J.: Logic in Games. Perspectives in Mathematical Logic. MIT Press, Cambridge (2014)
2. Börger, E., Grädel, E., Gurevich, Y.: The Classical Decision Problem. Perspectives in Mathematical Logic. Springer (1997)
3. Ehrenfeucht, A.: An application of games to the completeness problem for formalized theories. Fundamenta Mathematicae **49**, 129–141 (1961). https://doi.org/10.4064/fm-49-2-129-141
4. Fraïssé, R.: Sur une nouvelle classification des systèmes de relation. Comptes rendus hebdomadaires des séances de l'Académie des sciences **230**, 1022–1024 (1950)
5. Sørensen, M.H., Urzyczyn, P.: Lectures on the Curry-Howard Isomorphism, Studies in Logic and the Foundations of Mathematics, vol. 149. Elsevier (2006). https://doi.org/10.1016/S0049-237X(06)80499-4
6. Urzyczyn, P.: Intuitionistic games: determinacy, completeness, and normalization. Studia Logica **104**(5), 957–1001 (2016). https://doi.org/10.1007/s11225-016-9661-4
7. Zielenkiewicz, M., Schubert, A.: Automata theory approach to predicate intuitionistic logic. In: Hermenegildo, M.V., López-García, P. (eds.) Logic-Based Program Synthesis and Transformation - 26th International Symposium, LOPSTR 2016, 6–8 September 2016, Edinburgh, UK. Revised Selected Papers. Lecture Notes in Computer Science, vol. 10184, pp. 345–360. Springer (2016). https://doi.org/10.1007/978-3-319-63139-4_20

Author Index

Printed in the United States
by Baker & Taylor Publisher Services